Occupancy Estimation an

Inferring Patterns and Dynamics of S

Occupancy Estimation and Modeling

Inferring Patterns and Dynamics of Species Occurrence

Second Edition

Darryl I. MacKenzie
Proteus, *Dunedin, New Zealand*

James D. Nichols
Crofton MD, USA

J. Andrew Royle
USGS Patuxent Wildlife Research Center, *Laurel MD, USA*

Kenneth H. Pollock
North Carolina State University, *Raleigh NC, USA*

Larissa L. Bailey
Colorado State University, *Fort Collins CO, USA*

James E. Hines
USGS Patuxent Wildlife Research Center, *Laurel MD, USA*

ACADEMIC PRESS
An imprint of Elsevier

Academic Press is an imprint of Elsevier
125 London Wall, London EC2Y 5AS, United Kingdom
525 B Street, Suite 1800, San Diego, CA 92101-4495, United States
50 Hampshire Street, 5th Floor, Cambridge, MA 02139, United States
The Boulevard, Langford Lane, Kidlington, Oxford OX5 1GB, United Kingdom

Notices

Knowledge and best practice in this field are constantly changing. As new research and experience
broaden our understanding, changes in research methods, professional practices, or medical treatment
may become necessary.

Practitioners and researchers must always rely on their own experience and knowledge in evaluating and
using any information, methods, compounds, or experiments described herein. In using such information
or methods they should be mindful of their own safety and the safety of others, including parties for
whom they have a professional responsibility.

To the fullest extent of the law, neither the Publisher nor the authors, contributors, or editors, assume any
liability for any injury and/or damage to persons or property as a matter of products liability, negligence
or otherwise, or from any use or operation of any methods, products, instructions, or ideas contained in
the material herein.

Library of Congress Cataloging-in-Publication Data
A catalog record for this book is available from the Library of Congress

British Library Cataloguing-in-Publication Data
A catalogue record for this book is available from the British Library

ISBN: 978-0-12-814691-0

For information on all Academic Press publications
visit our website at https://www.elsevier.com/books-and-journals

Working together
to grow libraries in
developing countries

www.elsevier.com • www.bookaid.org

Publisher: John Fedor
Acquisition Editor: Anneka Hess
Editorial Project Manager: Emily Thomson
Production Project Manager: Mohanapriyan Rajendran
Designer: Victoria Pearson

Typeset by VTeX

Without data, all you are is just another person with an opinion.
Unknown

An approximate answer to the right question is worth a great deal more than a precise answer to the wrong question.
John Tukey

If a tree falls in the forest, but there is no one there to see it, does it make any noise?
Zen koan

If a bird sings in the forest, but the investigator fails to detect it, is the forest occupied?
Evan Cooch

Well George, we knocked the bastard off.
Sir Edmund Hillary, on scaling Mt Everest

Contents

3.   Fundamental Principals of Statistical Inference

Part II
Single-Species, Single-Season Occupancy Models

4.   Basic Presence/Absence Situation

7. Modeling Heterogeneous Detection Probabilities

Part III
Single-Species, Multiple-Season Occupancy Models

8. Basic Presence/Absence Situation

Part IV
Study Design

11. Design of Single-Season Occupancy Studies

Part V
Advanced Topics

13. Integrated Modeling of Habitat and Occupancy Dynamics

14. Species Co-Occurrence

15. Occupancy in Community-Level Studies

16. Final Comments

Preface

The presence or absence of a species from a collection of sampling units is a basic concept that is widely used in wildlife and ecological studies. When we first became involved in this topic in the late 1990's–early 2000's, we were more familiar with problems of estimating abundance, birth rates, survival probabilities and related demographic parameters, primarily from capture–recapture data, and did not appreciate the generality of the concept of 'occupancy'. At that time, some of us were involved with the newly formed U.S. Geological Survey's Amphibian Research and Monitoring Initiative (ARMI), advising on general study design and the types of data that should be collected as part of broad-scale monitoring programs. It quickly became apparent that it was logistically impossible to estimate changes in absolute abundance across large areas over time, so one suggestion was to simply measure the presence or absence of the species at a number of ponds or wetlands, which was soon known within ARMI as the 'proportion of area occupied'. With our background in capture–recapture methods, where estimation of capture probability is the key concept for making robust inference about abundance, survival, and other parameters, questions of whether it would be reasonable to always detect the target species within an occupied sampling unit were quick to surface. The response from the biologists was a resounding "No!" Our experience made us realize that inferences about 'occupancy' would be much improved if the collected data, and analytic methods used, accounted for the imperfect detection of the species. To us, the use of repeated 'presence/absence' surveys at each sampling unit would obviously supply the type of information required to estimate detection probabilities, but at the time of our initial collaborations with ARMI, we did not have a clear idea on the best approach to efficiently analyze the resulting data, particularly if unequal numbers of surveys were conducted at the different sampling units. Our thoughts on the topic crystallized in 2000–2001, sometimes aided by a fine pint or two of Guinness, which resulted in our first paper on the general topic in 2002 (MacKenzie et al., 2002).

Soon after, we realized that the general concept of occupancy is used much more widely than simply in monitoring as a surrogate for abundance. Despite widespread acknowledgment that in many applications a species may be detected only imperfectly, we were surprised at the lack of development of appropriate analytic techniques that explicitly accounted for detection probabilities, finding only a handful of previously published methods, with all of them being developed in a monitoring context. This led us to quickly extend our original method in an attempt to develop practical approaches for analyzing occupancy data in different situations (e.g., MacKenzie et al., 2003, 2004a; Royle and Nichols, 2003; Dorazio and Royle, 2005; MacKenzie and Royle, 2005). We recognized a substantial gap in the literature that was relevant to a wide range of applications, wherever reliable estimates of occupancy or changes in occupancy were required about species that are imperfectly detected. The simultaneous development of similar methodologies by independent groups (e.g., Tyre et al., 2003; Stauffer et al., 2004), and research into the effect of imperfectly detecting the species in different contexts, e.g., habitat modeling (Tyre et al., 2003; Gu and Swihart, 2004) and metapopulation incidence functions (Moilanen, 2002), suggested that perhaps the time was right for a more comprehensive treatment of the issues associated with estimating occupancy and related parameters in wildlife and ecological studies. The first edition of this book was our first attempt at doing so.

In the preface of the first edition, we noted that part of our motivation for developing the original book was to provide an early synthesis of the topic, in the hope that it would promote use of these methods and rapid methodological development in the area of occupancy estimation and modeling. This was based on what we had observed in other areas of ecological statistics, where a synthetic treatment of a topic had been a catalyst for both rapid assimilation of methods into practice, and rapid development of extensions and new methods, often many years after the methods were originally published during which time their usage tended to languish. Such was the case for band recovery models (Brownie et al., 1978, 1985), closed capture–recapture models (Otis et al., 1978; White et al., 1982), distance sampling (Burnham et al., 1980; Buckland et al., 1993, 2001), and open capture–recapture models (Burnham et al., 1987; Pollock et al., 1990; Lebreton et al., 1992). Since the publication of the book's first edition in 2006, occupancy models have certainly become very widely used, on a diverse set of taxa, around the world. There have been novel applications of the initial methods and numerous extensions, both large and small. While correlation is not causation, we would like to think the first edition played some role in the widespread uptake of the methods.

In 2012 we began consideration of a second edition, which seemed warranted due to the amount of new material that had been developed in a few short years, both in terms of new methods and increased understanding around the application of the methods. And thus began a journey that is finally coming to completion five years later. It has been a rewarding, and tiring, project that has allowed us to learn, and relearn, about many facets of modeling patterns and dynamics of species occurrence. While we have endeavored to provide details of many of the interesting extensions and notable applications of 'occupancy modeling' of which we are aware, there are undoubtedly some works that are deserving of publication in this book that we have overlooked, not been aware of, or simply forgot about. The methods have become so widespread that it would be a Herculean effort to stay abreast of all the occupancy modeling applications and extensions. At some point we had to draw the line on including further material to finish the book. We apologize to the reader if they have been party to relevant research that we have not included.

In the second edition of this book you will find a synthesis of most of the current literature on estimating occupancy-type metrics while explicitly accounting for detection probability. We declare that one cannot make reliable inferences about such metrics when issues related to the detectability of the species are effectively ignored, and we are also of the view that detection probabilities are likely to change over times, locations and species. Therefore, appropriate field and analytic methods should be employed to account for imperfect detection at every opportunity. We have divided this book into five parts. Part I provides readers with important introductory material containing our views of modeling and the conduct of management and/or science, and of the role models play in making inference about populations; historical and potential applications of species occurrence concepts in ecology; and an introduction to some of the fundamental statistical methods used throughout this book. We strongly advise readers not to skip these introductory chapters as they provide the necessary background, upon which we build in the latter chapters. In Part II we focus on methods for examining patterns of species occurrence using, what we refer to as, single-season, or static, occupancy models. This could be in the more common case where occupancy is defined as a binary variable (e.g., species presence/absence), or where occupancy is defined with more than two possible states. We detail potential modeling approaches that can incorporate a wide-range of issues such as imperfect detection, covariates, unequal sampling effort, spatial correlation and heterogeneity. Example of the methods are provided throughout. In Part III we switch focus to investigation of species occurrence dynamics, using extensions of the models discussed in Part II to the multi-season case, where data has been collected over longer time periods. Part IV is devoted to the topic of study design, for both the single- and multi-season cases. We have an overriding adherence to the GIGO principle – Garbage In, Garbage Out. Irrespective of model complexity, if the underlying data is of insufficient quality then any inferences based on those models may be unreliable. We view the topics covered in Part IV as the most important ones covered in this book for the practical implementation of the modeling methods discussed. All inferences are conditional upon the collected data, therefore the data is at the foundation of everything we do, and sound study design is of critical importance. Finally, in Part V we detail more complex methods that build upon the modeling techniques discussed in Parts II and III, turning our attention to integrated modeling of species and habitat dynamics, multi-species and community-level applications. We end the book with some concluding comments.

Our intended audience for this book is those at the 'coal face' of wildlife and ecological research, who may not necessarily have a strong background in statistics. We have thus taken great pains to use as little statistical jargon as possible, and provide detailed descriptions of how the models are developed, applied, and interpreted throughout, together with practical examples to illustrate the methods. This is, however, a technical topic, and in places some readers may struggle to understand the techniques exactly. Naturally, material gets more complex as the book progresses, and at some stage, many readers will benefit from reading the section on vectors and matrices in the Appendix. However, we would hope that readers will at least come away with an understanding of the general concepts and what the modeling is attempting to achieve. We believe this book should be useful to biologists internationally working in government agencies, private research institutes, and universities. It may also be a useful supplemental text for graduate courses on sampling wildlife populations or wildlife biometry. Further it would be an excellent choice, and has been used, for the basis of short courses on this specific topic.

We recognize that user-friendly software is necessary if the methods discussed in this book are to be widely adopted. As such, we have developed Windows-based software specifically for their application: Program PRESENCE, which is public domain software freely available for download from https://www.mbr-pwrc.usgs.gov/software/presence.shtml. Many techniques have also been incorporated into Program MARK (http://www.cnr.colostate.edu/~gwhite/mark/mark.htm) which was originally developed for the application of capture–recapture models to data collected on marked individuals. Since the publication of the first edition we have noticed that the statistical programming environment R, has become much more mainstream, particularly at the graduate level. There are now a number of R packages available that implemented many of the methods described in this book. Copies of the files associated with many examples we provide in the book, are available for free download from DIM's company's website http://www.proteus.co.nz.

Finally, as we stressed in the preface to the first edition, we still do not view the contents of this book as the final word on occupancy estimation and modeling, but an overview of the available methods as of 2017. Without doubt there will be further extensions and novel applications of the methods from both ourselves, and many others. Such extensions will only add to the available toolbox of appropriate analytic methods. We hope that readers will find the methods detailed in the following pages as thought-provoking and useful as we have found them to be during their development over the last 15 years.

USGS Disclaimer: Any use of trade, firm, or product names is for descriptive purposes only and does not imply endorsement by the U.S. Government. USGS employees authored Chapters 7 and 15.

Acknowledgments

As with the writing of any book, there is always a long list of groups and individuals whose support and contributions are invaluable to the authors, without whom the book would be much the poorer. We would like to thank the US Geological Survey's Inventory and Monitoring Program and Amphibian Research and Monitoring Initiative, and the Royal Society of New Zealand's Marsden Fund for the funding that supported much of the initial development of the methods in this book; Paul Dresler, Sam Droege, Paul Geissler, Sue Haseltine, Dan James, Rick Kearney, Melinda Knutson, and Catherine Langtimm for various forms of administrative support and encouragement; Paul Geissler for his pioneering work on occupancy estimation; Jason Fisher, Alan Franklin, Rocky Gutiérrez, Erin Johnson Hyde, Ullas Karanth, Bryan Manly, John Sauer, Ted Simons, Brad Stith, Nicole Sutton, and their respective field crews for supplying data and figures used throughout this book; Arthur Guinness and Ninkasi for providing the stimulus for many interesting discussions on this topic and many others; everybody who has asked us those 'simple' or 'stupid' questions that have stimulated and challenged our thoughts on this topic; and the team at Elsevier, particularly Nancy Maragiolio and Kelly Sonnack (1st edition), and Emily Thomson and Mohanapriyan Rajendran (2nd edition) for their encouragement, advice, and patience.

DIM would like to thank: David Fletcher, for introducing me to wildlife and ecological statistics; Bryan Manly, for setting my feet firmly on this path; James Speight, for a continuing source of inspiration; and my Ph.D. supervisor, Richard Barker, for all your support and encouragement, and for introducing me to the Patuxent 'mafia'. My life would never have taken this path if you had not suggested that I take a 1-year 'Pre-Doc' at Patuxent in 2000–2001. Thanks to Evan Cooch for help with LaTeX formating, and proving to me there are people in North America that appreciate the finer points of rugby; Jim Nichols, Bill Kendall, and Jim Hines for your advice and guidance in the early stages of my career; and all the friends and colleagues I have made since that have sent me data sets, challenged my thinking, and extended our original work in directions

I never would have dreamt of in 2001 when I first got involved with occupancy modeling. To my co-authors, thanks for bearing with me and my desire to use LATEX for the 2nd edition, being responsive to my requests for various things, but mostly for your fantastic contributions to this field and this book. The 2nd edition has been a long time in the making, but we finally got there! Last, but not least, to Kerry, Josh, Connor, Ollie, and Bella; thanks for your continuing support throughout your husband's and father's many absences (both real, and present but undetected) from family life for work commitments and during the writing of this book, twice.

JDN thanks Alan Franklin, Barry Noon, and Erran Seaman for first stimulating my thinking on occupancy estimation with respect to spotted owl surveys in Olympic National Park. Thanks to Russ Hall, Catherine Langtimm, and Franklin Percival, and to Ullas Karanth for resurrecting my interest in this topic in the contexts of amphibian monitoring and tiger surveys, respectively. Thanks to Judd Howell and Graham Smith for administrative support during preparation of the first edition of this book. Thanks to J. Ahumada, R. Altwegg, F. Bled, T. Bogich, C. Brehme, D. Breinenger, T. Chambert, R. Chandler, M. Clement, J. Collazo, R. Davis, M. Delampady, S. Dey, K. Dugger, M. Eaton, P. Fackler, A. Fernandez-Chacon, C. Estevo, G. Ferraz, E. Forsman, A. Gopalaswamy, V. Goswami, E. Grant, A. Green, R. Gutierrez, V. Herrmann, P. Hughes, F. Johnson, K.K. Karanth, W. Kendall, S. Kendrot, M. Kéry, S. Kumar, J. Martin, M. Mazerrole, B. McClintock, C. McIntyre, W. McShea, D. Miller, M. Mitchell, J. Moore, M. Nagy-Balda, Tim O'Brien, A. O'Connell, K. Pacifici, R. Parameshwaran, M. Pepper, J. Reid, L. Rich, V. Ruiz-Gutiérrez, M. Seamans, E. Stolen, S. Veran, C. Yackulic, and E. Zipkin for useful discussions, and to my family for their general support.

JAR thanks his many great friends and colleagues over the years who have motivated several careers worth of interesting research, especially Marc Kéry, the coauthors of this book, Bob Dorazio, Emmanuelle Cam, Sarah Converse, Angela Fuller, Ullas Karanth, Arjun Gopalaswamy, Michael Schaub, Benedikt Schmidt, Graham Smith, Mark Koneff, Bill Link, Jim Lyons, John Sauer, Scott Sillett, Paul Doherty, Evan Cooch, Linda Weir, Yuichi Yamaura, Florent Bled, Richard Chandler, Keiichi Fukaya, Beth Gardner, Dan Linden, Rahel Sollman, Chris Sutherland, Elise Zipkin, Tabitha Graves, Robin Russell, Matt Clement, Allan O'Connell, and many others!

KHP would like to thank Drs. Beth Gardner, Murray Efford, and Lyndon Brooks for extensive discussions that deepened my understanding of occupancy methods.

LLB thanks Ted Simons and Ken Pollock, my Ph.D. and M.S. advisors, who simultaneously got me to think hard about ecology, species distributions and detection, and 'the big picture'. A mixture of work and good timing led me to

Patuxent; I thank all the wonderful researchers there and the researchers that passed through there, many of which have stimulated thought and development of the methods covered in this book. I thank all people involved in the establishment and continued success US Geological Survey's Amphibian Research and Monitoring Initiative, especially Cathy Langtimm, an unsung hero who invited me to my first ARMI meeting. I thank the co-authors of this book and my colleagues, the graduate students, and post-docs at Colorado State University who continue to ask good questions and aid in occupancy model development and application. Finally, I thank Velda, Bud, and Tiffany for their encouragement and unwavering support, and Kevin for always giving me confidence in myself.

JEH would like to thank all of the folks who provided sample data and feedback on early (and future) versions of PRESENCE, and Bill Link and Gary White for help with statistical and programming problems. The development of new methods and models for occupancy analysis has made the development of the software an ongoing process. As each modification to the code introduces the possibility of new 'bugs', I would like to express my appreciation to users who have helped in 'de-bugging' the program over the years, and to those who provided data for new models. I would also like to thank Darryl MacKenzie, who developed the first version of the software and continues to help develop the R package RPresence. Finally, I thank the many folks who work/have worked with me at Patuxent over the years, making it a place where you never think about leaving, and in particular, my co-author, Jim Nichols, who saved me from a much less interesting career.

Part I

Background and Concepts

Chapter 1

Introduction

Ecology is frequently defined as the study of the distribution and abundance of plants and animals (e.g., Andrewartha and Birch, 1954; Krebs, 1972). Consequently, the practice of counting animals in order to draw inferences about their numbers and distribution has a long tradition in animal ecology and management. In his classic book, *Animal Ecology*, Charles Elton (1927, p. 173) wrote: "The study of numbers is a very new subject, and perfect methods of recording the numbers and changes in the numbers of animals have yet to be evolved." Elton then devoted six pages to the topic of animal "census" methods. In his equally influential classic, *Game Management*, Aldo Leopold (1933, p. 139) listed "Census" as the first of four steps required to initiate game management on any piece of land. He then devoted a 30-page chapter to "game census" and another 25 pages to "measurement and diagnosis of productivity", a chapter that focused on assessing vital rates and population change. Methods for counting animals have indeed evolved over the last 80 years, and animal ecologists and managers now have an impressive methodological toolbox for estimating parameters associated with animal abundance and with the vital rates that produce changes in abundance (e.g., Seber, 1973, 1982; Williams et al., 2002; Borchers et al., 2002; Newman et al., 2014; Kéry and Royle, 2016).

Today, biologists interested in understanding and managing animal populations and communities include some individuals who make full use of the methods available for drawing inferences about variation in animal numbers over time and space, and many others who do not appear to recognize the importance of appropriate inferential procedures. Because of those scientists and managers who do not take advantage of available estimation methods, the fields of animal population and community ecology, wildlife management, and conservation biology include numerous examples of substantial field efforts that do not produce reliable conclusions. These disciplines suffer not only from the failure of animal ecologists and managers to utilize the range of available methods for drawing inferences about animal abundance and associated vital rates, but also from the lack of rigorous methods for estimating other quantities that may be biologically relevant. For example, other variables that could be used to quantify the current status of a community or population (we refer to these as *state variables*) include species richness (number of species) and occupancy (proportion of an area occupied by a species or fraction of landscape units where the

Occupancy Estimation and Modeling. http://dx.doi.org/10.1016/B978-0-12-407197-1.00002-8

species is present). Prior to the early 2000's, scant attention had been devoted to estimation of these latter state variables, but since that time there has been rapid methodological development.

In this book, we emphasize the need for estimation methods that permit inference about occupancy based on detection–nondetection data and describe methods that have been developed to do so. We begin this chapter by providing brief operational definitions for some important terms, then move on to an outline of general principles for sampling animal populations, focusing on the why, what and how of such sampling. This outline is followed by a more detailed look at the critical step of using field data to discriminate among competing hypotheses about system response to environmental variation and management actions. We note different field designs that are used to generate system dynamics for such discrimination and comment on the different strengths of inference resulting from these designs. The chapter concludes with a more detailed statement of book objectives and contents.

1.1 OPERATIONAL DEFINITIONS

The methods presented in this book should be useful to biologists involved in either science or management of biological populations. Both endeavors use the following three constructs: *hypothesis*, *theory*, and *model*. These terms are not always used consistently in the literature, and therefore we provide our own operational definitions for use in this book (also see Nichols, 2001). We view a *hypothesis* simply as a plausible explanation (i.e., a 'story') about how the world, or part of it, works. For example, we would deem density-dependent recruitment for mid-continent, North American mallards (*Anas platyrhynchos*) as a hypothesis, with density-independent recruitment as an alternative, competing hypothesis (Johnson et al., 1997). Once a hypothesis has withstood repeated efforts to falsify it, to the extent that we have some faith in predictions deduced from it, the hypothesis may become a *theory* (e.g., Einstein's theory of relativity). A theory can still be disproved in the future given new data or the expansion of the part of the world to which the theory is thought to be applicable (e.g., Newtonian physics).

Very generally, we view a *model* as an abstraction of a real-world system, which can be used to describe observed system behavior and predict how the system may respond to changes or perturbations. Within this broad definition we recognize many different kinds of models (Nichols, 2001), three of which are especially useful within the context of this book. A *conceptual model* is a set of ideas about how the system of interest works, and may include one or more hypotheses or theories about the system. A *verbal model* is created by translating these ideas into words. Finally, a *mathematical model* results from translating a conceptual or verbal model into a set of mathematical equations, using defined parameters to symbolize the key processes of the system. In this book we derive mathematical expressions from our conceptual ideas about the processes that occur when collecting occupancy field data, placing particular attention on using the collected data to estimate the parameters of these models.

By following the logical progression above, note that a mathematical model is ultimately a representation of one or more hypotheses or theories about the system. Therefore, competing hypotheses can be formulated into competing mathematical models. Applying each model to the same set of available data, it may be possible to formally determine which model (and therefore which hypothesis) has a greater degree of support given the data at hand. Essentially this is an exercise in model selection. We advocate and use such an approach throughout this book.

1.2 SAMPLING ANIMAL POPULATIONS AND COMMUNITIES: GENERAL PRINCIPLES

It is our belief that many existing programs for sampling animal populations and communities are not as useful as they might be because investigators have not devoted adequate thought to fundamental questions associated with establishment of such programs. These failures have greatly reduced the value of efforts ranging from individual scientific investigations to large-scale monitoring programs. These latter programs are especially troubling, because they can require nontrivial fractions of the total funding and effort available for the conduct of science and management of animal populations and communities. Here we present some opinions about the sort of thinking that should precede and underlie good animal sampling programs. These opinions are structured around three basic questions to be addressed during the design of an animal sampling program (see Yoccoz et al., 2001): *Why?*, *What?*, and *How?*

1.2.1   Why?

Efforts to sample animal populations are generally associated with one of two main classes of endeavor, science or conservation and management (or possibly both). Science can be viewed as a process used to discriminate among competing hypotheses about system behavior, that is, discriminating among different ideas about how the world, or a part of it, works (e.g., whether recruitment to a population is density-dependent). This process typically involves mathematical models. For example, a mathematical model that could be used to represent the number of recruits to a population (r) that assumes no density dependence would be $r = N_F b$, where N_F is the number of breeding females in the population and b is the average number of female births per adult female and is viewed as a constant (with respect to current breeding female population size). A different model that conceptualizes the effect of density-dependent recruitment would be $r = N_F b(N_F)$, where $b(N_F)$ specifies a functional relationship such that number of recruits per female is a function of total female abundance. The primary use of models is to project the consequences of hypotheses, that is, to deduce predictions about system behavior (e.g., Nichols, 2001). In the case of our example, the model is used to predict the number of recruits at different levels of population density.

The key step in science then involves the confrontation of these model-based predictions with the relevant components of the real-world system (Hilborn and Mangel, 1997; Williams et al., 2002). Confidence increases for those models (and hence those underlying hypotheses) whose predictions match observed system behavior well and decreases for models that do a poor job of predicting. However, for most practical situations involving animal populations and communities, true system behavior cannot be directly observed, but must be estimated from data collected from sampling programs. Thus, sampling programs constitute a key component of scientific research.

In the conduct of management and conservation, estimates of state variables for animal populations and communities serve three distinct roles (Kendall, 2001). First, estimates of system state are needed in order to make state-dependent management decisions (e.g., Kendall, 2001; Williams et al., 2002). For example, the decision of which management action to take frequently depends upon the current population size. Second, system state is frequently contained in the objective functions (precise, usually mathematical, statements of management objectives) for managing animal populations and communities. Evaluation of the objective function is an important part of management, addressing the question "to what extent are management objectives being met?" Finally, good management requires either a single model thought to be predictive of system response to management actions or a set of models with associated weights reflecting relative degrees of confidence in their validity. The process of developing confidence in a single model or weights for members of a model set involves the confrontation of model predictions with estimates of true system response. This confrontation is the scientific component of informed management and requires animal sampling programs that provide reliable estimates of state variables and associated vital rates.

Despite the importance of being explicit about why a program for sampling animal populations or communities is needed, we believe that many studies suffer from a failure to clearly articulate specific study objectives. This is especially evident in many large-scale monitoring programs (Yoccoz et al., 2001). For example, the following objectives statements from a report on ecological monitoring programs in the United States (LaRoe et al., 1995, pp. 3–4) are fairly typical: "The goal of inventory and monitoring is to determine the status and trends of selected species or ecosystems." and "Inventory and monitoring programs can provide measures of status and trends to determine levels of ecological success or stress." The second statement implies an interest in management and conservation, but without specification of available management actions and hypotheses about system response to those actions, the statement provides little basis for monitoring program design. Thus, we advocate clear specification of monitoring program objectives.

Objective specification is facilitated by the recognition that monitoring of animal populations and communities is not a stand-alone activity of great inherent utility, but is more usefully viewed as a component of the processes of science and/or management. This recognition leads naturally to detailed consideration of exactly how the monitoring program results are to be used in the conduct of science or management or both. Such considerations lead directly to decisions about monitoring program design, whereas vague objectives that fail to specify use of program data and estimates provide little guidance for program design and can lead to endless debate about design issues.

Clearly stated objectives are not very common in wildlife management or conservation biology, but we provide some examples. The objective of a longstanding management program for mid-continent mallard ducks (*Anas platyrhynchos*) in North America is to maximize undiscounted duck harvest by sport hunters over an infinite time horizon, subject to a penalty when expected population size drops below a specified threshold (Nichols et al., 1995, 2007b; Johnson et al., 1997, 2016). In addition to the penalty for low expected population size, the infinite time horizon and the absence of discounting impose a conservation component on the objective, as harvested ducks in the distant future are valued as highly as current harvest. A program of adaptive management is used to regulate hiking near potential nesting sites of golden eagles (*Aquila chrysaetos*) in Denali National Park, Alaska (Martin et al., 2011; Fackler et al., 2014). The objective is to maximize hiking opportunity (minimize number of sites closed to hiking) subject to a constraint that the value of hiking is decreased (number of restricted sites is increased) as a function of the difference between a specified threshold and the expected number of nesting sites with successful reproduction the next season. The monitoring program informing this program is based on occupancy modeling to draw inferences about the dynamics of nest site use and success (Martin et al., 2009a). Examples of other objective functions in wildlife management and conservation include those developed for the commercial fishery of horseshoe crabs (*Limulus polyphemus*) in Delaware Bay, USA (McGowan et al., 2015a, 2015b) and habitat management for Florida scrub jays (*Aphelocoma coerulescens*) in Florida, USA (Johnson et al., 2011).

We believe that objectives specified in terms of detecting 'significant' trends in a population are particularly unhelpful from a management perspective. While identification of a population trend is interesting in some regards, it says nothing about the desired state of the population that managers should be working toward. In our experiences, when a statistically significant population trend is found, this usually leads to a series of additional questions such as "Is the recent trend part of a longer-term cyclical fluctuation in the population?" or "What has caused the trend?" Such questions often lead to calls for additional analyses or extra data before management actions should be implemented. Conversely, when a population trend is not deemed statistically significant, that is used as justification for not requiring management action at the moment. Focusing on trend detection leads to a reactionary, or retrospective, style of management where the emphasis is on what has happened and less attention on what managers want to happen. We would strongly argue that it is much more useful to frame the management objective in terms of "What do we want the population to be like in the future?" as this gives a much clearer goal for managers to work toward, enabling a proactive management style. There is still a place for retrospective analyses that may provide useful insights about the population to aid decisions about appropriate management actions, but the presence of a population trend, particularly a statistically significant trend, is largely irrelevant; if the current state of the population is different from the stated management objective, then the actions that are believed most likely to push the population in the desired direction need to be taken. Uncertainty about how the population may respond to various management actions can be formally accounted for using methods such as adaptive management (e.g., Section 1.3; Williams et al., 2002). That is, rather than using uncertainty as an excuse to do nothing, we advocate the use of methods that explicitly incorporate uncertainty into the decision making process.

1.2.2 What?

The selection of what state variable(s) and associated vital rates to estimate will depend largely on the answer to the initial question of *Why?* The selection of state variables for scientific programs will depend on the nature of the competing hypotheses and specifically on the quantities most likely to lead to discrimination among the hypotheses (i.e., for what quantities are predictions of competing hypotheses most different?). The selection of state variables for management programs will depend on the most relevant characterization of system state, on management objectives, and on the ability to discriminate among competing hypotheses about system response to management actions. Practicality must also be considered in both cases as, most likely, logistical resources will be limited.

When dealing with single species, the most commonly used state variable is abundance or population size. Estimation of abundance frequently requires substantial effort, but it is a natural choice for a state variable in studies of population dynamics and management of single-species populations. Some studies of animal abundance focus directly on changes in abundance, frequently expressed as the ratio of abundances in two sampling periods (e.g., two successive years) and termed the finite rate of population increase or population growth rate, λ. In scientific studies, mechanistic hypotheses frequently concern the vital rates responsible for changes in abundance, rates of birth (reproductive recruitment), death and movement in and out of the population. In management programs, effects of management actions on animal abundance must also occur through effects on one or more of these vital rates. Thus, many animal sampling programs involve efforts to estimate abundance and rates of birth, death and movement, for animals inhabiting some area(s) of interest.

We believe that another useful state variable in single-species population studies is occupancy, defined as the proportion of area, patches or sample units that is occupied (i.e., species presence). Sampling programs designed to estimate occupancy tend to require less effort than programs designed to estimate abundance (e.g., Tyre et al., 2001; MacKenzie et al., 2002; Manley et al., 2004). In the case of rare species, it is sometimes practically impossible to estimate abundance, whereas estimation of occupancy is still possible (MacKenzie et al., 2004b, 2005). Thus, for reasons that include expense and necessity, occupancy is sometimes viewed as a surrogate for abundance. However, there are also a number of kinds of questions for which occupancy would be the state variable of choice regardless of the effort involved in sampling. For example, metapopulation dynamics (e.g., Hanski and Gilpin, 1997; Hanski, 1999) are frequently described by patch occupancy models. So-called incidence functions (e.g., Diamond, 1975; Hanski, 1994a) relate patch occupancy to patch characteristics such as size, distance to mainland or some source of immigrants, habitat, etc. Occupancy is the natural state variable for use in studies of distribution and range (e.g., Brown, 1995; Scott et al., 2002) and is useful in the study of animal invasions (Bled et al., 2011; Yackulic et al., 2012) and even disease dynamics (McClintock et al., 2010c; Bailey et al., 2014). Patch occupancy dynamics may be described using the rate of change in occupancy over time, and the vital rates responsible for such change are patch-level probabilities of extinction and colonization. Historical, current, and proposed uses of patch occupancy as a state variable for science and management will be discussed in more detail in Chapter 2.

When scientific or conservation attention shifts to the community level of organization, many possible state variables exist. The basic multivariate state variable of community ecology is the species abundance distribution, specifying the number of individuals in each species in the community. Many derived state variables are obtained by attributing different values or weights to individuals of different species (Yoccoz et al., 2001). Several common diversity indices are computed by providing a weight of one to every individual of each species (e.g., Pielou, 1975; Patil and Taillie, 1979), but it is also possible to give additional weight to individuals of species thought to be of special importance (e.g., endemic species or species of economic value; Yoccoz et al., 2001). A state variable that is used commonly in community studies is simply species richness, the number of species within the taxonomic group of interest that is present in the community at any point in time or space. This state variable is used in scientific investigations (e.g., Boulinier et al., 1998a, 2001; Cam et al., 2002) and programs for management and conservation (e.g., Scott et al., 1993; Keddy and Drummond, 1996; Wiens et al., 1996). The vital rates responsible for changes in species richness over time are rates of local species extinction and colonization. In this book we focus largely on the state variable of occupancy, but note how these methods can also be applied where species richness-type metrics may be of interest (Chapter 15).

1.2.3 How?

Proper estimation of state variables and inferences about their variation over time and space require attention to two critical aspects of sampling animal populations, spatial variation and detectability (Fig. 1.1; Lancia et al., 1994, 2005; Thompson et al., 1998; Williams et al., 2002). Spatial variation in animal abundance is important because in large studies and most monitoring programs investigators cannot directly survey the entire area of interest. Instead, investigators must select a sample of locations to which survey methods are applied, and this selection must be done in such a way as to accomplish two things. First, selection of study locations should be based on study objectives. In the case of scientific objectives, study locations should be selected to provide the best opportunity to discriminate among the competing hypotheses of interest (see Section 1.3 for further discussion). For example, in the case of an observational study involving hypotheses about habitat variables, selected study locations might be extremes with respect to the variable(s) of interest or else might be locations at which changes in the variable(s) are anticipated. In the case of a management program, study locations should of course include the areas to which management actions are applied. Second, within larger areas selected based on study objectives, sample locations should be selected in a manner that permits inferences about the locations that are not surveyed, and hence about the entire area(s) of interest. Approaches to sampling that accom-

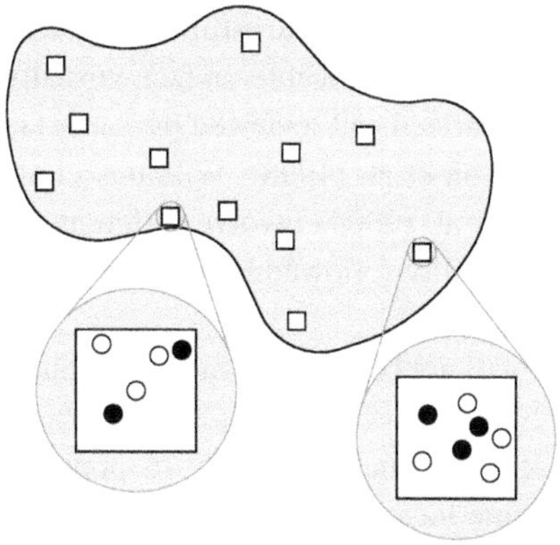

FIGURE 1.1   Illustration of the two critical aspects of sampling animal populations, spatial variation, and detectability. The shaded region indicates the area or population of interest, with the small
squares representing the locations selected for sampling. Within each sampling location, animals
will be detected (filled circles) or undetected (hollow circles) during a survey or count.

plish this inferential goal include simple random sampling, unequal probability
sampling, stratified random sampling, systematic sampling, cluster sampling,
double sampling, and various kinds of adaptive sampling (e.g., Cochran, 1977;
Thompson, 1992; Thompson and Seber, 1996).

Detectability refers to the reality that even in locations that are surveyed by
investigators, it is very common for animals and even entire species to be missed
and go undetected. Most animal survey methods yield some sort of count statistic. For example, when abundance is the quantity of interest, the count statistic
might be the number of animals caught, seen, heard, or harvested. Let N_{it} be
the true number of animals associated with an area or sample unit of interest, i,
at time t, and denote as C_{it} the associated count statistic. This statistic can be
viewed as a random variable whose expectation (i.e., the average value of the
count if we could somehow conduct the count under the exact same conditions
many times; see Chapter 3) is the product of the quantity of interest, abundance
at the surveyed location, and the detection probability associated with the count
statistic:

$$E(C_{it}) = N_{it} p_{it} \tag{1.1}$$

where p_{it} is the detection probability (probability that a member of N_{it} appears
in the count statistic, C_{it}). Estimation of N_{it} thus requires estimation of p_{it}:

$$\hat{N}_{it} = \frac{C_{it}}{\hat{p}_{it}} \tag{1.2}$$

where the 'hats' in this expression denote estimators (see Chapter 3). Expression (1.2) is very general and widely applicable. In fact, virtually all of the abundance estimation methods summarized and reviewed by Seber (1973, 1982), Lancia et al. (1994, 2005), Thompson et al. (1998), Williams et al. (2002), Borchers et al. (2002), and Nichols et al. (2009) involve different approaches to the estimation of detection probability followed by (or integrated with) application of expression (1.2).

Frequently, interest will not be in abundance itself but in relative abundance, the ratio of abundances at two locations ($\lambda_{ijt} = N_{it}/N_{jt}$, where i and j denote locations and t still denotes time), or in rate of population change, the ratio of abundances in the same location at two times ($\lambda_{it} = N_{it+1}/N_{it}$). Sometimes count statistics are treated as indices, and their ratio is used to estimate the true ratio of abundances. For example, consider the estimator $\hat{\lambda}_{it} = C_{it+1}/C_{it}$. The expectation of this estimator can be approximated using expression (1.1) as:

$$E\left(\hat{\lambda}_{it}\right) \approx \frac{N_{it+1}\,p_{it+1}}{N_{it}\,p_{it}} = \lambda_{it}\left(\frac{p_{it+1}}{p_{it}}\right). \tag{1.3}$$

As can be seen from Eq. (1.3), the ratio of counts estimates the product of the quantity of interest, λ_{it}, and the ratio of detection probabilities. If the detection probabilities are very similar for the two sample times, then the estimator will not be badly biased, but when detection probabilities differ, then the index-based estimator will be biased, with the direction of the bias depending on which detection probability is the larger. That is, an estimator of λ based only on the ratio of observed counts may be either too big or too small, and it is impossible to correct for the bias without knowledge about the values of p_{it} and p_{it+1}. If detection probability itself is viewed as a random variable, then we still require $E(p_{it}) = E(p_{it+1})$ in order for a ratio of counts to be a reasonable estimator.

Proponents of the use of count statistics as indices for estimating relative abundance typically recommend standardization of survey methods as one means of trying to insure similar detection probabilities. Standardization involves factors that are under the control of the investigator (e.g., effort, trap type, bait, season, and time of day of survey). While standardization of survey methods is usually a good idea irrespective of whether the analytic methods to be used explicitly account for detection probability or not, we believe that this approach is unlikely to produce equal detection probabilities because there are always likely to be unidentified and uncontrollable factors that influence detection probabilities (Conroy and Nichols, 1996). Sometimes it is possible to identify uncontrollable factors that could influence detection probability and incorporate them as covariates into analyses of count statistics. This approach is reasonable when dealing with factors that could only affect detection probability and not animal abundance itself (similarly for other state variables, e.g., occupancy or species richness). For example, differences in detection probabilities among observers are often incorporated into analyses of avian point count data (Link and Sauer, 1997, 2002). However, it would not be wise to use a similar approach with habitat data, as habitat would be expected to influence not only detection probability, but also animal abundance. Thus, controlling for habitat effects by incorporating them into analyses as covariates affecting detection would not be appropriate. Of course factors that we do not identify but that still affect detection probability cannot be treated as covariates either.

Another common claim supporting the use of indices is that they are relatively assumption free, unlike the methods used to actually estimate abundance (e.g., Seber, 1982; Williams et al., 2002). However, there are a large number of implicit assumptions to be made if the index is to be related to animal abundance. In fact, interpretation of an index as some indicator of true population size typically requires all the assumptions used to estimate abundance, plus the assumption that a constant fraction of the population is counted each survey. Some uses of indices require the assumption that all animals are counted during each survey. As these assumptions are unlikely to be true, we believe that indices have a very limited use in good monitoring programs. We conclude that estimation of both absolute and relative abundance requires information about detection probability (also see Lancia et al., 1994; MacKenzie and Kendall, 2002; Williams et al., 2002).

The importance of obtaining information about detection probability extends to other state variables as well. Investigations of species richness usually involve counts of the number of different species. Under some designs the counts are conducted at multiple locations within some large area to which inference is to apply, whereas other designs use counts conducted at multiple times (e.g., days) on a single area of interest (e.g., Nichols and Conroy, 1996; Williams et al., 2002). In both designs, it is recognized that some species may go undetected, and the replication (geographic or temporal) is used to estimate a species level detection probability, the probability that at least one individual of a species will be detected given that the species inhabits the area of interest. Efforts to estimate species richness from samples of animal communities are not new (Fisher et al., 1943; Preston, 1948; Burnham and Overton, 1979). Nevertheless, community ecologists have tended to ignore the issue of detection probabilities less than 1, although increasing attention has been devoted to this estimation problem since the 1990's (e.g., Chao and Lee, 1992; Bunge and Fitzpatrick, 1993; Colwell and Coddington, 1994; Walther et al., 1995; Chao et al., 1996; Nichols and Conroy, 1996; Boulinier et al., 1998b; Cam et al., 2000a; Williams et al., 2002; Dorazio and Royle, 2005; Dorazio et al., 2006; Kéry and Royle, 2008; Zipkin et al., 2009; Yamaura et al., 2016b).

Detection probability is also very relevant to the estimation of occupancy. Define occupancy, ψ, as the probability that a randomly selected site or sampling unit in an area of interest is occupied by a species (i.e., the unit contains at least one individual of the species). If x and s represent the number of occupied and surveyed units, respectively, then we can estimate occupancy as $\hat{\psi} = x/s$. However, x is not typically known. Instead, we will have a count of units where the species has been detected, but this count will likely be smaller than x, because species will not always be detected in occupied units (i.e., due to 'false absences'). Thus, we must use methods that incorporate detection probability in order to estimate x. For example, we can use an analog of expression (1.2), where the count is the number of units at which the species is detected, and the detection probability is the probability that the species is detected during sampling of an occupied unit. Occupancy can then be estimated as:

$$\hat{\psi} = \frac{\hat{x}}{s}. \tag{1.4}$$

More direct ways to estimate occupancy from appropriately collected data have been developed (e.g., MacKenzie et al., 2002; Royle and Nichols, 2003; Chapters 4 and 7), but the basic conceptual rationale underlying these approaches is the same as outlined here.

Inferences about occupancy may be misleading when detection probability is not incorporated into the methods of data analysis. Not only will naïve approaches underestimate occupancy (as above), indices intended to reflect relative occupancy could be biased (MacKenzie et al., 2005) and the effect of casual factors or variables may be underestimated (Tyre et al., 2003) or misidentified, particularly if detection probability covaries with the factors or variables thought to affect occupancy (Gu and Swihart, 2004; MacKenzie et al., 2005). Inferences about the dynamic processes that drive changes in occupancy may also be inaccurate (Moilanen, 2002; MacKenzie et al., 2003). Indeed, an important theme of this book is that robust inference about occupancy and related dynamics can only be made by explicitly accounting for detection probability.

1.3   INFERENCE ABOUT DYNAMICS AND CAUSATION

In Chapter 2, we will focus on the *What* of animal sampling programs and discuss the use of occupancy as a state variable. Much of the remainder of the book will then focus on the *How* question of sampling animal populations. That is, given interest in occupancy, how do we estimate this state variable and the vital rates responsible for its change in reasonable ways? Although we believe that this emphasis is justified by the absence of previous work and good guidance on drawing inferences about occupancy, we regret the need to abandon issues about

Why we sample animal populations. In our introductory discussion about why we might want to sample animal populations and communities, we emphasized that sampling programs are usefully viewed as components of the larger processes of science or management. In this section, we briefly discuss the manner in which results of animal sampling programs are used to draw the inferences needed for science or management. This discussion touches aspects of design that extend well beyond efforts to obtain reasonable estimates of state variables of interest.

The key step in the scientific process involves a comparison of estimates of state variables with model-based predictions associated with competing hypotheses. Such comparisons also constitute an important management use of estimates from animal sampling programs, as the ability to predict consequences of different management actions is critically important to informed management. Scientific programs include interest in responses of animal populations and communities to a variety of factors (e.g., changes in predators, competitors, weather, habitat, disease, toxins/pesticides). Management programs focus not only on responses to management actions, but also on other factors that might improve predictive abilities in order to manage a system more effectively. We would like to discriminate among competing hypotheses about the relevance of different causal factors to system dynamics with the ultimate goal of being able to predict the magnitude of the state variable(s) at time $t + 1$, given the magnitude of the state variable at time t and knowledge of the causal factors operating between times t and $t + 1$ (Williams, 1997; Williams et al., 2002).

1.3.1 Generation of System Dynamics

The scientific process usually includes some means of generating system dynamics so that estimated changes in state variables can be compared with the predictions of competing models. Multiple approaches are used to generate system dynamics in population and community ecology, and we classify these approaches broadly as true manipulative experiments, constrained designs or quasi-experiments, and observational studies (Romesburg, 1981; Skalski and Robson, 1992; Manly, 1992; Williams et al., 2002). These approaches merit brief discussion here, as they provide different strengths of inference (Fig. 1.2), i.e., how reliable our conclusions are from such studies.

Inferences are strongest when system dynamics are generated via the conduct of true manipulative experiments (see Fisher, 1947; Hurlbert, 1984; Skalski and Robson, 1992; Manly, 1992). Such experiments are characterized by replication, randomization in the assignment of different treatments (application of different hypothesized causal factors) to experimental units, and the use of a control or standard treatment group. In the context of population and commu-

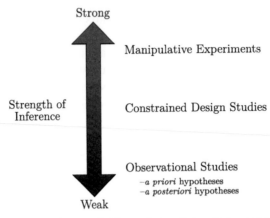

Strong

Manipulative Experiments

Strength of
Inference

Constrained Design Studies

Observational Studies
—*a priori* hypotheses
—*a posteriori* hypotheses

Weak

FIGURE 1.2 Strength of inference of different designs that could be used to generate system dynamics.

nity ecology, experimental units may be populations or communities occurring naturally or created as part of the experimental design. *Replication* refers to the application of treatments to multiple experimental units as a means of estimating the experimental error or error variance. The error variance reflects the variance among experimental units to be expected in the absence of treatment differences (i.e., the variance associated with all factors except the different treatments). *Randomization* refers to random assignment of treatments to experimental units. Randomization protects against systematic differences among experimental units receiving different treatments and represents an effort to insure that any systematic post-treatment differences among experimental units treated differently can be attributed to the treatments themselves. One treatment type is typically designated as a *control* and is used to provide a baseline against which other treatments can be compared. The use of a control group is especially useful in attributing causation to different treatments and permitting estimation of treatment effects on response variables. Manipulative experiments thus seek to reduce potential sources of ambiguity to the extent possible, yielding strong inferences about causation

True manipulative experiments are frequently difficult to perform on free-ranging animal populations and communities due to cost and practical field constraints. In many instances, we may be able to manipulate systems but may be required to do so using study designs that lack replication, randomization, or both of these features (see Green, 1979; Skalski and Robson, 1992; Williams et al., 2002). Inferences resulting from such constrained, or quasi-experimental, designs will typically not be as strong as those based on manipulative experimentation (see examples in Nichols and Johnson, 1989).

Finally, the investigator may be unable to manipulate the system at all and may be forced to rely on natural variation to generate system dynamics. For example, large-scale animal monitoring programs may provide time series of estimated state variables, and retrospective analyses can be used to try to distinguish among competing hypotheses about system dynamics (Nichols, 1991). Two general approaches to observational studies are used, and they are distinguished by the existence of *a priori* hypotheses. The observational studies that tend to be most useful to science are those for which conditional *a priori* hypotheses are specified and used to guide monitoring program design (Nichols, 2001; Williams et al., 2002). The hypotheses are conditional in the sense that changes in purported causal factors are not known *a priori* as they are when the investigator imposes a manipulation. Instead, the different hypotheses predict different relationships between suspected causal factors and system state variables, and specific predictions then emerge as changes in the causal factors occur naturally and are observed. The initial specification of the hypotheses facilitates monitoring program design, as efforts can be devoted to monitoring changes in hypothesized causal factors as well.

The other approach to observational studies involves the development of *a posteriori* hypotheses to explain observed system dynamics. Monitoring programs may yield annual estimates of quantities such as population size over relatively long time periods (e.g., 20 years), and it is commonly thought that such trajectories lead directly to an understanding of underlying population dynamics. It is a common practice to use such data with correlation and regression analyses to investigate possible relationships between population size and various environmental and management variables. The problem with this approach is that it is unlikely to yield 'reliable knowledge' (Romesburg, 1981), because there will typically be multiple *a posteriori* hypotheses that provide reasonable explanations for any observed time series (Nichols, 1991). Indeed, we tend to agree with Pirsig's (1974, p. 107) assertion that "The number of rational hypotheses that can explain any given phenomenon is infinite."

The potential for being misled by retrospective analysis of data exists for all kinds of observations (Platt, 1964; Romesburg, 1981) but is especially large for time series of estimates of state variables (e.g., population size) and related variables. One reason for this is that the state variables are not observed but are estimated, often with large sampling variances and sometimes with bias. Temporal variation in point estimates of the state variable is thus not equivalent to temporal variation in the underlying population (Link and Nichols, 1994). Another difficulty in drawing inferences from retrospective analyses of population trajectory data involves the stochastic nature of population processes. Death, for example, is typically viewed as a simple stochastic process. If a population has

100 animals at time t and if each of these animals has a probability of 0.2 of dying during the interval $(t, t + 1)$, then we do not expect exactly 80 animals to be alive at time $t + 1$. Instead, the number of survivors will be a binomial random variable with expected value 80, but with likely realized values of 78, 83, 75, etc. Reproductive processes and movement are also stochastic in nature, leading to the view of a population trajectory as a single realization of a (likely complicated) stochastic process. There is little reason for us to expect to be able to infer much about the nature of an unknown stochastic process based on a single realization of that process (Nichols, 1991). This is analogous to being handed a loaded coin, being permitted to flip it once, and then being asked to specify the probability of obtaining heads.

Another difficulty associated with inferences from retrospective studies of population monitoring data involves problems with using correlation analysis to draw inferences about the functional relationship between variables represented by time series. A clear example of such problems involves the existence of trends and monotonicity in many environmental covariates that potentially influence animal populations. Metrics of human-related environmental variables such as habitat fragmentation, habitat degradation, and pollutant levels will frequently tend to show an increasing trend over time. Correlation analyses involving two variables, each of which shows a time trend, will tend to indicate association, although this may have nothing to do with any functional relationship between the variables. In fact, the problem of conducting association analyses of two time series extends well beyond the case of monotonic trends, and such analyses frequently lead to inappropriate inferences (Yule, 1926; Barker and Sauer, 1992).

These various considerations lead us to conclude that development of *a posteriori* hypotheses based on retrospective analyses of monitoring data is an approach that necessarily results in weak inferences. Certainly we do not claim that such retrospective analyses are without value, as they can sometimes provide useful insights and ideas about system behavior. Instead our recommendation is that such analyses be viewed primarily as an approach to hypothesis generation rather than as an inferential assessment of the hypothesis as an explanation for system dynamics. We thus recommend that observational studies be guided by *a priori* hypotheses, with exploratory retrospective analyses possibly used as a means of hypothesis generation.

As noted above, distinguishing among competing hypotheses about system response to management is an important component of an informed decision process. The term *adaptive management* (e.g., Holling, 1978; Walters, 1986; Hilborn and Walters, 1992; Williams et al., 2002, 2007; Nichols et al., 2015) typically applies to management that is state-dependent and that incorporates learning about system response to management actions. It is this learning component that distinguishes adaptive management from other decision processes (Kendall, 2001; Williams et al., 2002; Johnson et al., 2015). Estimates of system state are used not only for the purpose of making state-dependent decisions, but also as a means of confronting the predictions of competing models about system response for the purpose of discriminating among their associated hypotheses. Based on objectives, potential actions, an estimate of system state, and models (with associated probabilities reflecting relative degrees of confidence), managers make the decision to take a particular action at time t. This action drives the system to a new state at time $t + 1$, and this state is identified via a monitoring program. Probabilities associated with degrees of confidence in the various system models are then updated based on the distance between estimated system state and the predictions of the competing models (Kendall, 2001; Nichols, 2001; Williams et al., 2002, 2007; Nichols et al., 2015). Although this approach to multi-model inference is used in the current applications of adaptive management with which we are most familiar (Nichols et al., 1995; Johnson et al., 1997, 2011; Williams et al., 2002, 2007; Martin et al., 2011; Fackler et al., 2014; McGowan et al., 2015a, 2015b), hypothesis-testing approaches are also possible and are also based on the distance between estimated system state and model-based predictions.

In the context of the previous discussion of approaches for generating system dynamics, the learning component of adaptive management will virtually always be manipulative, in that management actions will be imposed and system response then observed. However, attainment of management objectives is of primary importance in adaptive management, and learning is valued only to the extent that it is useful in better meeting objectives. Thus, in most applications with which we are familiar, the learning components of adaptive management exhibit the features of constrained designs. However, if management is of a spatially extended system, and if different actions are to be taken on different spatial units of the system, then a manipulative experimental approach might be taken as well.

In summary, the conduct of science requires some means of generating system dynamics for comparison with predictions of competing hypotheses. True manipulative experiments represent a study design that permits strong inferences about causation. Constrained or quasi-experimental designs involve manipulations, but the absence of either randomization or replication, or both features, does not permit the strength of inference of a true experiment. Finally, observational studies based on retrospective analyses of monitoring data involve no manipulation as part of study design and rely on natural variation in purported causal factors. These analyses tend to yield weaker inferences than manipulative studies. Within observational studies using retrospective analyses, those that test predictions of *a priori* hypotheses tend to yield stronger inferences than analyses used to generate *a posteriori* hypotheses. Adaptive management represents an informed decision process incorporating explicit efforts to learn about system responses to management actions. Because learning is not the sole objective of adaptive management, management manipulations typically follow some form of constrained design.

1.3.2   Statics and Process vs. Pattern

Inferences about causation emerge most naturally from studies of system dynamics. Scientists and managers estimate the state variable at time t, apply or observe purported causal factors operating between times t and $t + 1$, and then estimate the state variable again at time $t + 1$. However, because of the difficulties in applying manipulations to animal populations and communities and in properly estimating relevant state variables over time, animal ecologists have also tried to draw inferences about dynamics based on observations of spatial pattern at a single time, t. Brown (1995, p. 10) describes *macroecology* as a research program in ecology with "emphasis on statistical pattern analysis rather than experimental manipulation." Inferences based on such efforts have been applied to each of the state variables described above, abundance, species richness, and occupancy.

Ecologists frequently use spatial variation in abundance of animals to draw inferences about habitat quality, based on the common-sense idea that if animals are found in higher density in one habitat than others, then that habitat is likely of high quality. For such a statement to have meaning, 'quality' must be defined. In their influential work on habitat selection, Fretwell and Lucas (1969) and Fretwell (1972) defined habitat quality in terms of the fitness of organisms in that habitat. The two fundamental fitness components, survival probability and reproductive rate, are also primary determinants of population dynamics, so this definition is relevant to population ecologists and managers as well. Observations of spatial variation in animal density associated with habitat variation do not yield reliable inferences about individual fitness or dynamics of populations inhabiting such areas (e.g., Van Horne, 1983; Pulliam, 1988). Instead, such inferences require studies of system dynamics, in this case habitat-specific demography (e.g., Franklin et al., 2000), preferably in conjunction with habitat manipulations. Similar arguments hold where attempts are made to relate species presence or occurrence at locations to the habitat at those places to infer 'preferred' habitats for the species, e.g., with population-level resource selection functions (Manly et al., 2002) or so-called habitat models (Scott et al., 2002). That a species occurs more frequently in one habitat type compared to another may not be a good indicator of preference or underlying habitat quality. We suggest the underlying occupancy vital rates of local extinction/persistence and colonization would be more relevant indicators (Tyre et al., 2001).

The relationship between species richness and area is one of the oldest and most cited static relationships in ecology (e.g., Arrhenius, 1921; Preston, 1948). Hypotheses about the dynamic processes responsible for this relationship include habitat selection coupled with habitat heterogeneity (e.g., Williams, 1964) and increased probabilities of local extinction in small areas (e.g., MacArthur and Wilson, 1967). However, these two hypotheses yield similar species–area relationships, providing no basis for distinguishing between these or other mechanistic explanations (Connor and McCoy, 1979).

Occupancy appears to be used more frequently in static analyses than either of the other discussed state variables, abundance and species richness. Static analyses of occupancy data in animal ecology can be illustrated with two common applications, single-species incidence functions and multiple-species co-occurrence patterns. Incidence functions involve efforts to model dichotomous spatial occupancy pattern (presence or absence) as a function of characteristics of the sampled locations or patches. Diamond (1975) first described incidence functions in his studies of distributional ecology of birds inhabiting islands in the area of New Guinea. He grouped islands by such characteristics as land area and total avian species richness and then plotted the proportion of islands in each category (e.g., area, richness) that was occupied by a particular species. Diamond noted that some species tended to occur only on large, species-rich islands, whereas others were found only on remote, species-poor islands. Diamond (1975, p. 353) viewed the incidence function as a "'fingerprint' of the distributional strategy of a species", and used these functions to draw inferences about such processes as dispersal, habitat selection and competition (see below and Chapter 2). These inferences have been challenged based on the consistency of observed patterns with other processes (e.g., Connor and Simberloff, 1979).

Hanski (1992) adapted the incidence function for use in describing and modeling metapopulation dynamics. He noted that in an equilibrium system of many patches of similar size, the fraction of occupied patches at any point in time can be written as an explicit function of patch probabilities of extinction and colonization. He then postulated functional forms for the relationship between extinction probability and patch area and between colonization probability and patch isolation. If metapopulation dynamics can be described as a stationary Markov process, then parameters of the extinction and colonization relationships can be estimated using occupancy data from a single point in time (e.g., Hanski, 1992, 1994a, 1994b, 1998, 1999). However, the difficulties of inferring process from pattern have been noted. For example, based on analyses of year to year changes in occupancy of pikas (*Ochotona princeps*), Clinchy et al.

(2002) recommended that "simple patch occupancy surveys should not be considered as substitutes for detailed experimental tests of hypothesized population processes."

Use of occupancy data from multiple species to draw inferences about species interactions also has a long history in ecology. Some of the first statistical analyses adapted by ecologists were used to test the null hypothesis of independence of species occurrence using occupancy data for two species (Forbes, 1907; Dice, 1945; Cole, 1949). Non-independent occupancy patterns of multiple species on islands have been interpreted as evidence of competition (e.g., MacArthur, 1972; Diamond, 1975). For example, the "assembly rules" of Diamond (1975) include specification of species combinations that cannot exist for reasons of interspecific competition and are based on empirical observations of species distributions on different islands. However, such inferences about process based on observed patterns have been sharply criticized. Critics argued that rejection of predictions of neutral models developed from distributional null hypotheses should precede any attempt to develop more complicated explanatory hypotheses for static species distribution patterns (e.g., Connor and Simberloff, 1979, 1986; Simberloff and Connor, 1981). Neutral models themselves were then criticized by proponents of the original competitive hypotheses (Diamond and Gilpin, 1982; Gilpin and Diamond, 1984), neutral model proponents responded (Connor and Simberloff, 1984, 1986), and the entire issue of inference based on species distribution patterns was hotly debated (Strong et al., 1984). Such debate is not surprising, as strong disagreement is a natural consequence of weak inference, which brings us back to Pirsig's (1974) assertion about the ability to develop large numbers of plausible hypotheses to explain any given pattern.

Each of the three quantities listed as state variables of potential interest in population ecology and management (abundance, occupancy and species richness) has been investigated with respect to its distribution over space at one point in time. Identification of spatial patterns has then led to inferences about the dynamic processes that produced these patterns. However, these inferences are always very weak, as many alternative hypotheses can be invoked to explain most ecological patterns (Fig. 1.3). Our conclusions about drawing inferences about process based on snapshots of spatial pattern are simple and straightforward. First, inferences about system dynamics should be based on estimates and observations of those dynamics, and of the vital rates that produce them, whenever possible. Second, when ecologists do try to draw inferences about dynamics based on observations of static pattern, we believe that such inferences are much more likely to be useful if the specification of model-based predictions from competing or single hypotheses precedes the investigation of pattern (e.g., see Karanth et al., 2004). Brown (1995, p. 18) stated, "Macroecology seeks to

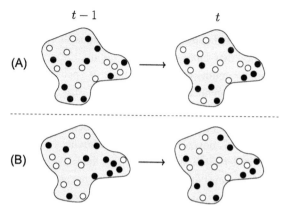

FIGURE 1.3 Illustration of how a pattern observed at time t may result from very different processes. Darkened circles represent occupied patches, and white circles represent unoccupied patches. The level of turnover between times $t - 1$ and t is much greater in scenario B.

discover, describe, and explain the patterns of variation." We recognize that such efforts can be useful, but we recommend that they be viewed as mechanisms for hypothesis generation rather than for inference and testing.

1.4   DISCUSSION

We began this chapter by asserting that many animal sampling programs, including many large-scale monitoring programs, have deficiencies resulting from failure to adequately consider three basic questions. Why do we want to sample animal populations and communities? What quantities do we want to estimate? How should we estimate the quantities of interest? In answer to the 'why' question, we suggested that animal sampling and monitoring programs should not be viewed as stand-alone activities, but as components of the larger processes of science or management. This recognition forces consideration of exactly how resulting data are to be used in these processes, and this consideration leads to program designs that maximize utility of data. The answer to the 'what' question will depend heavily on the answer to 'why', and we noted that abundance, occupancy and species richness are reasonable state variables for a variety of objectives.

The answer to the 'how' question depends on the answers to the previous two questions, but also requires attention to two basic issues. When the entire area of interest cannot be surveyed, then space must be sampled in a manner that is maximally useful to study objectives and that permits inference about the entire area of interest. Because this problem of spatial sampling characterizes a wide variety of applications in statistical inference, it has been well addressed elsewhere. Spatial sampling will be touched on throughout the other chapters, but will not be emphasized in this book. The second issue involves imperfect detection, the likelihood that surveys of animal populations and communities will not result in complete counts of all individuals or species present in surveyed locations. We present a general conceptual framework that relates the various count statistics obtained in studies of animal populations and communities to the true state variables of interest. The suite of models, methods and estimators developed in this book is developed to permit inferences about occupancy that deal adequately with detection probabilities less than 1, recognizing that failure to find evidence of a species at a location does not necessarily mean the species does not occupy the area.

Because most of the book focuses on parameter estimation, we returned to the 'why' question and the manner in which estimates of state variables are to be used in the conduct of science and management. We briefly addressed the general question of drawing inferences about system dynamics and causal factors responsible for these dynamics. Approaches to the generation of system dynamics for the purpose of conducting science include true manipulative experimentation, constrained design manipulative studies, and observational studies using retrospective analyses. Strength of inference is greatest for manipulative experiments and weakest for retrospective analyses of time series data from observational studies. Within the category of observational studies, those used to provide confrontations with predictions of *a priori* hypotheses are much more likely to be useful than those used solely to develop *a posteriori* hypotheses.

Finally, we noted that investigators sometimes try to draw inferences about system dynamics based on static looks at spatial patterns of state variables at single points in time. Such efforts to draw inferences about process based on observation of pattern have been used with all three state variables, abundance, occupancy, and species richness. However, such efforts suffer from the ability to develop many process-based hypotheses to explain the generation of any particular pattern. Previous uses of occupancy in animal ecology have relied heavily on inferences based on statics and pattern, and we note the shortcomings of this approach. In particular, we do not view the primary purpose of this book to be provision of methods for obtaining better estimates of static occupancy patterns for use in drawing inferences about dynamic processes. Instead, we also provide methods for drawing inferences about occupancy dynamics based on data covering multiple time periods.

The remainder of Part I of this book deals with other introductory and background topics. In Chapter 2, we consider both historical and proposed uses of occupancy as a state variable in studies of animal populations and communities. With each use, we emphasize the need to deal adequately with detection probabilities. Chapter 3 provides an elementary overview of the statistical concepts used throughout the book.

Part II then deals with single-species occupancy studies in which multiple locations are surveyed during a single time period or *season*. Chapter 4 presents historical approaches, a basic model and its estimators, and includes discussion of issues such as missing data, covariate modeling, goodness-of-fit tests, and consequences of violations of model assumptions. In Chapter 5 we extend the concept of occupancy from a dichotomous state variable (e.g., species presence or absence) to multiple categories of occurrence, and detail appropriate modeling approaches. In Chapter 6 a number of other important extensions are described, including inference about small areas or finite populations, false-positive detections, multi-scale occupancy, correlated surveys, staggered entry and spatial autocorrelation in occurrence. Finally in this second part, Chapter 7 focuses on the assumption violation of homogeneous detection probabilities. We present mixture models that allow for variation in detection probabilities that cannot be attributed to measured covariates. Animal abundance at a site is identified as one important source of heterogeneity in detection probability. The relationship between abundance and detection probability provides a basis for estimating abundance from occupancy survey data and for estimating occupancy itself in a manner that deals with this heterogeneity.

In Part III we focus attention on occupancy studies conducted over multiple years or seasons for the purpose of drawing inference about occupancy dynamics for a single species. Rate of change in occupancy over time is identified as a parameter of interest, and the vital rates responsible for such change, local probabilities of extinction and colonization, are also incorporated into estimation models. A range of issues is considered including estimation, covariate modeling and assumption violations. Chapter 8 details the situation where occupancy is considered a dichotomous variable, with extension to multiple occupancy categories in Chapter 9. Part III ends with a discussion of a range of further topics relevant to studies of occupancy dynamics in Chapter 10. Topics covered include false-positive detections, spatial autocorrelation in dynamic processes, investigating the nature of the dynamic processes, sensitivity of occupancy to different dynamic rate parameters and accounting for detection heterogeneity.

Part IV is dedicated to the important issue of study design, both for single-season occupancy studies, Chapter 11, and multiple seasons, Chapter 12. A good design with appropriate consideration of practical and logistical challenges is key to the success of any type of study. Poor designs tend to lead to uncertainty regarding the biological interpretation of estimates, more complex analyses, debate about results and generally weaker inferences. The importance with which we view sound study design is disproportionate to the number of pages devoted to the topic compared to the number of pages devoted to possible methods of analysis of data from such studies. High-quality data collected during a well designed study to address the objective at hand are key to making reliable statistical inferences about a system.

The final part of the book contains chapters on more advanced topics. Chapter 13 presents an integrated approach to modeling habitat and occupancy dynamics that explicitly allows for a mutualistic relationship between the two processes, i.e., habitat affects occupancy dynamics and species occupancy affects habitat dynamics. Such modeling should yield more reliable inferences about systems than modeling each as independent processes. Chapter 14 shifts emphasis to multiple species and begins with inference procedures for two species in a single year or season. Methods permit inference about dependence in probabilities of occupancy given detection probabilities that are less than 1 and that may themselves exhibit dependence on presence or detection of the other species. These methods are then extended to multiple seasons where the emphasis shifts to possible dependence of extinction and colonization probabilities of one species on the presence of the other species. Oftentimes data are being collected on multiple species, and even whole communities. In Chapter 15 we detail how occupancy modeling approaches have been applied to make community-level inferences about species richness, diversity metrics and community turnover, all while accounting for imperfect detection of species in the community. This work also provides a conceptual framework for considering species–area relationships. Finally, we end the book with brief discussion of topics that are related, but are somewhat tangential, to the main thrust of this book, and our thoughts on future research and development opportunities for species occurrence modeling (Chapter 16).

Chapter 2

Occupancy Applications

This book deals primarily with the occupancy of a sampling unit by one or more species of interest. The size and nature of the sampling unit to which the term applies may be defined either naturally or arbitrarily. For example, if interest is focused on pond-dwelling amphibians, then the pond is a likely unit of interest. If we are studying a terrestrial mammal within a national park, then the units may be defined as 1000-hectare blocks within the park. Studies of animal range over a continent might utilize the degree block as a sampling unit for occupancy. In cases where sampling units are arbitrarily, rather than naturally defined, the size of the sampling unit should depend on the nature of the question(s) being asked. Investigations of occupancy nearly always involve interest in a number of potential units, so the quantity is sometimes the number of units within a larger set of interest that is occupied. More frequently, interest will focus either on the proportion of units that is occupied or on the underlying probability that a unit within a group of units is occupied (the ψ parameter of Chapter 1).

Note there is an important distinction between 'proportion of area occupied' and 'probability of occupancy'. The *probability* can be considered as an *a priori* expectation that a particular unit will be occupied by the species as determined by some underlying process, while the *proportion* relates to a realization of that process. For example, consider a simple coin tossing experiment. The *probability* of a 'head' is a characteristic of the coin, while the *proportion* of 'heads' is determined by conducting the experiment with multiple tosses of the coin. An alternative explanation is that the probability relates to the average outcome if we were to repeat the experiment a large number of times, while the proportion relates to a single outcome of an experiment.

As the probability is generally unknown, the observed proportion can be used as an estimate of the underlying probability, and often these terms are used interchangeably (we do so ourselves in this book). In many situations the distinction is not important and, strictly speaking, most of the models we develop in this book estimate the probability of occupancy, which can be interpreted as the proportion of area occupied without penalty. However there are other situations where the distinction can be very important (i.e., when a large fraction of the population of interest is sampled), and in such situations we make sug-

Occupancy Estimation and Modeling. http://dx.doi.org/10.1016/B978-0-12-407197-1.00003-X

FIGURE 2.1 Setting up camera traps for large mammals at a sampling station in southern India. (*Eleanor Briggs*)

gestions on how the modeling can be used to make inference directly about the proportion of area occupied (Section 6.1).

The basic sampling protocol commonly used for occupancy estimation simply involves visiting units and spending time within each one looking either for individuals of the target species or for evidence that the species is present. This kind of sampling is sometimes referred to as a 'presence–absence' survey. There are no real restrictions on sampling approaches, which may include visual observations of animals, captures of animals in traps or mist nets, observations of animal tracks, detections of animal vocalizations or environmental DNA (*e*DNA), and even detections based on remote methods such as cameratraps (Fig. 2.1) and acoustic recording devices (e.g., Peterson and Dorcas, 1994). The result of such a survey is then a list of surveyed units that are 'occupied' (the species detected) and 'unoccupied' (the species not detected), respectively. In most historical work, counts of occupied units are used to compute the proportion of occupied units among all units visited, and this proportion constitutes the estimate of occupancy.

The problem with these count-based inferences is that truly occupied units may be visited, and yet no animals (or evidence of animals) are detected. Thus, units classified as 'unoccupied' based on survey efforts may in fact be occupied. There appears to be wide recognition of this problem, with such misclassified units referred to as false absences (Dunham and Rieman, 1999), false zeros

(Moilanen, 2002), false negatives (Tyre et al., 2003), pseudo-absences (Engler et al., 2004), and artefactual absences (Anderson, 2003). This problem arises from the issue of detectability, articulated in Chapter 1 with respect to estimation of both animal abundance and occupancy. This book contains a suite of methods for estimating occupancy and related parameters in a manner that deals explicitly with detectability and the likelihood that some units at which no evidence of occupancy was found, were in fact occupied by the target species.

This chapter provides an overview of different areas of investigation in ecology for which reasonable estimates of occupancy are required, hence areas of ecology where the techniques outlined in this book may be useful. Non-ecological applications where there is a similar intent to determine the presence or absence of some feature, but with the potential for false absences, are also outlined. The review provides a clear indication of the importance of occupancy to a number of important areas of inquiry in ecology, and beyond. For each class of investigation, we consider the kinds of questions being asked. We then consider possible consequences of failing to deal with detectability. That is, what sorts of inferential problems could be caused by misclassifying occupied units as unoccupied? Examples of studies that have utilized occupancy models that account for detection in these application areas are also given, and we indicate the chapters of this book that contain models used in these or similar applications and that may be useful in future work.

2.1 GEOGRAPHIC RANGE

Ecology is frequently described as the study of the distribution and abundance of organisms (Elton, 1927; Andrewartha and Birch, 1954; Krebs, 2001; also see Chapter 1). Geographic range for a single species can be viewed as the primary element describing the distributional component of ecology (Brown et al., 1996) and has been termed "the basic unit of biogeography" (MacArthur, 1972). Species ranges have long been fundamental units of analysis used to elucidate interesting ecological patterns and to address associated questions. Ranges have been the subjects of renewed interest under the macroecological research program (Brown and Maurer, 1989; Maurer, 1994; Brown, 1995, 1999; Rosenzweig, 1995; Gaston and Blackburn, 1999).

Despite the fundamental importance of *range* to questions about animal and plant distributions, explicit definitions of this term are rare, even among the scientific papers for which range is a topic of investigation (Gaston, 1991). Gaston (1991) presented two ways of defining geographic species range, but both definitions are based on records of individual locations. We will modify Gaston's (1991) two 'definitions' by focusing on the true geographic distribution of individuals of a species rather than on sample-based counts. It seems more

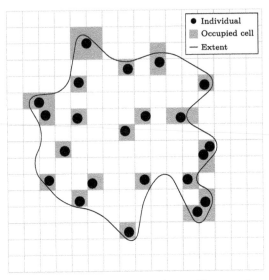

FIGURE 2.2 Illustration of two different definitions of range: (1) *extent of occurrence* defined by a line that encompasses all individuals of the species; and (2) *area of occupancy* defined by the set of occupied cells or units.

appropriate to define a true quantity of interest, and then to focus on methods of estimating this quantity, rather than mixing sampling and estimation problems with definitions. The two definitions of range are thus based on knowledge of the spatial locations of all individuals of a species at some time or interval of interest.

One approach, the *extent of occurrence*, involves drawing an imaginary line that encloses an area containing all of the individuals of the species. Stated in terms of occupancy, this area should include all of the units occupied by the species. The enclosed area should be minimal in some sense, but is very much dependent on scale and on how jagged the boundary is permitted to be. In practice, boundary lines are typically "fitted by eye" (Gaston, 1991), but we can imagine several objective ways of drawing such boundaries. The other way to define geographic range is termed *area of occupancy*. If a grid is superimposed on the area containing all individuals of a species, then area of occupancy is simply defined as the set of grid cells that contain at least one individual (Fig. 2.2). Area of occupancy differs from extent of occurrence in that the latter type of range may include cells that contain no individuals of the species (Gaston, 1991). Both definitions of geographic range are based on occupancy, and virtually all investigations of range are based on presence–absence (more properly, detection–nondetection) data of some sort.

A number of macroecological studies have used interspecific comparisons of the sizes of species ranges. Some studies report simple patterns in the distribution of range sizes. For example, most species appear to exhibit small ranges, whereas a relatively small number of species exhibit very large ranges (Willis, 1922; Rapoport, 1982; Brown, 1995; Gaston, 1994, 1998). Many investigations have noted or searched for interspecific correlates of range size. One of the oldest such correlates was pointed out by Darwin (1859), who noted that genera containing many species tended to contain "dominant species, – those which range widely over the world, are the most diffused in their own country, and are the most numerous in individuals." This relationship has been widely investigated, leading to the general inference that within related species groups, range size appears to be related to average local abundance, such that species with large ranges tend to be abundant throughout those ranges, whereas species with small ranges also tend to be less abundant (Williams, 1964; McNaughton and Wolf, 1970; Buzas et al., 1982; Hanski, 1982; Hanski et al., 1993; Bock and Ricklefs, 1983; Bock, 1984; Brown, 1984, 1995; Gotelli and Simberloff, 1987; Gaston, 1996; Gaston et al., 1997b, 1997a; Holt et al., 2002). Evidence has also been reported of a positive relationship between temporal variability of abundance and range size (Glazier, 1986). Several analyses have provided evidence of a positive relationship between range size and species characteristics such as animal body mass (Brown and Maurer, 1987; Brown, 1995; Brown et al., 1996; Gaston and Blackburn, 1996), dispersal capabilities of marine mollusks (Hansen, 1980; Brown et al., 1996), and germination patterns (specifically germination niche breadth) of weedy plant species (Brandle et al., 2003). Geographic range size has been shown to decrease with decreasing latitude (Rapoport, 1982; Stevens, 1989; Brown, 1995; Brown et al., 1996; Rohde, 1999) and decreasing elevation (Stevens, 1992; Brown, 1995; Brown et al., 1996).

Investigations of the determinants of animal distributions naturally focus on range boundaries and on the biotic and abiotic factors responsible for their locations (e.g., Bodenheimer, 1938; Connell, 1961; MacArthur, 1972; Caughley et al., 1988; Root, 1988a, 1988b; Gaston, 1990; Repasky, 1991; Brown et al., 1996). Macroecological research has also focused on the shapes of geographic ranges, with attention directed at such descriptors as perimeter/area ratios, fractal dimension and north–south versus east–west orientation (Rapoport, 1982; Brown, 1995; Brown et al., 1996). The number, size and location of holes, fragments and other discontinuities are also features of interest with respect to range shape. In particular, the number of fragments and discontinuities tends to increase near the range periphery (Rapoport, 1982; Brown, 1995; Brown et al., 1996). Within a range, abundance typically decreases from center to periphery (Whittaker, 1956; Brown, 1984; Gaston, 1994; Brown et al., 1996; Enquist et al., 1995), and local extinction and turnover are typically higher at the edges of ranges than in range interiors (Enquist et al., 1995; Curnutt et al., 1996; Mehlman, 1997; Doherty et al., 2003a).

Although many of the preceding relationships and associated references reflect a view of range as a static entity, there is substantial evidence of changes in range boundaries over both geological and ecological timescales (e.g., Udvardy, 1969; MacArthur, 1972; Hengeveld, 1990; Gaston, 1994; Enquist et al., 1995; Brown et al., 1996; Ceballos and Ehrlich, 2002; Karanth et al., 2010; Bled et al., 2011, 2013; Yackulic et al., 2012; Clement et al., 2016). Recent reductions in range are frequently associated with human activities and are thus of special conservation interest (Brown et al., 1996; Ceballos and Ehrlich, 2002; Karanth et al., 2010). A comparison of 'historic' (primarily 19th century) and more recent range data for terrestrial mammals indicated collective reductions in range size exceeding 70% for Africa, Australia, Europe and Southeast Asia (Ceballos and Ehrlich, 2002). Habitat fragmentation can be viewed as a problem of range reduction and has become an important conservation issue (e.g., Harris, 1984; Lynch and Whigham, 1984; Robbins et al., 1989).

Humans are also increasingly responsible for range extensions. In some cases, the spread of species to areas outside the original ranges has been inadvertent. For example, murid rodents (e.g., mice, *Mus*, and rats, *Rattus*) have been spread throughout the world via sailing vessels, and various pest insects have been introduced to new areas through imports of agricultural products. Some accidental introductions (e.g., brown treesnake, *Boiga irregularis*, on Guam; Fritts and Rodda, 1998) have produced dramatic extinctions and range contractions of native species. Species introductions to new areas are sometimes intentional, such as when the Acclimatization Societies of New Zealand introduced various animal species in order to provide additional hunting and fishing opportunities (e.g., Thomson, 1922; Williams, 1981). Regardless of original motivation, introductions of species into new areas by humans have led to many so-called invasions, and invasive species now present an important challenge to conservation (Mooney and Drake, 1986; DiCastri et al., 1990; Williamson, 1996). Occupancy data collected over large spatial scales for multiple years are commonly used to estimate rates of spread of exotic species (e.g., Havel et al., 2002; Wikle, 2003; Bled et al., 2011). Reintroductions and translocations are considered as a conservation tool as well and usually reflect attempts to restore species to previous ranges (Griffith et al., 1989; Chivers, 1991; Fackler et al., 2012).

Regardless of whether range is defined as *extent of occurrence* or *area of occupancy* (Gaston, 1991), the failure to detect species that are actually present in a patch or sample unit will result in biased estimates of range size and location if detection probability is ignored. Range size and extent will tend to be underestimated when species detection probabilities are < 1, with the magnitude of bias a function of sampling intensity (McArdle, 1990; Anderson, 2003). Perhaps a more serious problem involves the possibility of spurious relationships between range attributes and other quantities of interest that are induced by imperfect detection. For example, consider the interspecific relationship between range size and average abundance. Brown (1984, p. 264) put forward the hypothesis that the correlations of range size and abundance that he observed could be "simply the result of statistical sampling processes". He then rejected this hypothesis, but readers may not find his reasoning to be convincing. Buzas et al. (1982, p. 149) clearly recognized the difficulties in exploring the relationship between abundance and range size for fossil data: "Consequently, with rarely occurring species we are in a no-win situation. Because of the difficulties in detecting the presence of such a species, we cannot know whether or not its distribution is restricted to the locality where it is found or whether it is widespread, but undetected. The dilemma for the naturalist is obvious."

The problem arises because the probability of detecting a species in a location is often a direct function of the number of individuals of the species inhabiting the location. Formal expressions for this relationship are presented by Royle and Nichols (2003) and in Chapter 7, but the relationship is very intuitive. The consequence of this relationship is that occupancy will go undetected more frequently for species at low abundance than for abundant species, inducing the very relationship about which so much has been written. Of course this relationship between abundance and detection probability for occupancy does not mean that all inferences about the abundance–occupancy relationship are incorrect. It simply means that the evidence used to support this relationship is not very useful for this purpose. In addition, the relationship between abundance and detection probability will result in biased estimates of parameters specifying the relationship between occupancy and abundance. Such bias will likely have consequences for efforts to estimate abundance from raw occupancy data (Kunin, 1998; Kunin et al., 2000; He and Gaston, 2000, 2003; Warren et al., 2003; Tosh et al., 2004; Linden et al., 2017).

Other inferences about range may be influenced by detection probability as well. For example, the inference about increased numbers of fragments and discontinuities near range boundaries could easily be induced by changes in detection probabilities caused by decreased abundance near range boundaries. Local extinctions and, in some situations, turnover tend to be overestimated using raw detection data when detection probabilities are < 1 (e.g., Nichols et al., 1998a). Thus, the inferences about increased extinction probabilities and turnover near range boundaries could be induced by decreased detection probabilities produced by low abundances near boundaries (Doherty et al., 2003a; Alpizar-Jara et al., 2004). In fact, virtually any gradient in abundance can produce an apparent gradient in occupancy based on raw detection data. Of course factors other than abundance may influence detection probability as well. For example, individual animals may simply be more visible or detectable in one habitat than another, providing the potential for misleading inference about the true relationship between occupancy and habitat. Similarly, sampling effort may be higher in some areas than others (e.g., based on proximity to museums) for logistic or other reasons, again leading to the possibility of differences in occupancy in different parts of a range that do not reflect true differences in range.

The methods provided in this book have been used to address the relationships and topics reviewed above in a manner that is not nearly so vulnerable to problems produced by detection probabilities as use of raw detection–nondetection data. Occupancy models have been used to explore range limitations and dynamics for numerous taxa including birds (e.g., Yackulic et al., 2012), mammals (e.g., Bayne et al., 2008), plants (e.g., Chen et al., 2013), amphibians (e.g., Miller and Grant, 2015), and fish (e.g., Groce et al., 2012). Studies have addressed several factors discussed above including fragmentation (e.g., Collier et al., 2012), urbanization and land use change (e.g., Marcelli et al., 2012), climate (e.g., Altwegg et al., 2008; Moritz et al., 2008; Crum et al., 2017), co-occurring species (e.g., Yackulic et al., 2014; Dugger et al., 2015), and in modeling the range dynamics of native, reintroduced, or invasive species (e.g., Marcelli et al., 2012; Groce et al., 2012; Dugger et al., 2015). These studies utilize static (single-season) models (e.g., Collier et al., 2012; Bayne et al., 2008; Groce et al., 2012) or dynamic models using data that span a few years (e.g., Marcelli et al., 2012), decades (e.g., Dugger et al., 2015), or even centuries (e.g., Karanth et al., 2010; Moritz et al., 2008).

One example by Karanth et al. (2010) explored factors influencing range contractions for 25 mammal species in India over the last 200 years. Using historical records to condition on units with known species occurrence, the authors used contemporary field observations to estimate the probability that a historically occupied site was still occupied in 2006. Here the occupancy probability is equivalent to species persistence, the complement of local extinction probability. The authors explored the influence of habitat features (e.g., the presence and size of protected areas, forest cover), species biology, and social factors, including human density and cultural attitudes, on the extinction probabilities while accounting for variation in time of the historical record and unequal contemporary sampling effort. Not surprisingly, large-bodied, habitat specialists had the highest estimated extinction probabilities, ranging from 0.96 (lion, *Panthera leo*) to 0.14 (jackal, *Canis aureus*). Extinction probabilities were often lower in areas with a higher proportion of protected areas and/or forest cover, but the effects varied among species. Human population density increased extinction probabilities for approximately half of the species, but 'culturally tolerated' species did have lower extinction probabilities.

Methods utilized in these, and other, studies to estimate occupancy, and thus range size, for single species are considered in Chapters 4, 7, 10, and 11. Chapter 7 contains specific methods for dealing with the relationship between abundance and detection probability when estimating occupancy, and Chapter 16 includes discussion of the true relationship between abundance and occupancy. These methods permit not only the estimation of occupancy, but also the estimation of parameters describing relationships between occupancy and other quantities of interest. Because detection probabilities are explicitly incorporated into the models, the procedures are not vulnerable to spurious influence by variation in detection probabilities. Dynamic models discussed in Chapters 8–10 extend the analyses to multiple years or seasons, permitting inference about changes in occupancy, or species range, over time and about the probabilities of local extinction and colonization responsible for these changes.

We believe that the occupancy models described herein are ideally suited to study species ranges and range dynamics. These models were developed to use detection–nondetection data obtained from multiple units, at least some of which are surveyed multiple times (replicate visits). Here we comment briefly on alternative methods developed for (1) detection data only and (2) detection–nondetection data from single surveys.

A popular approach to inferences about species ranges and range dynamics is based on so-called presence-only data, e.g., data from museum or herbarium collections or other data sources for which only detection locations are recorded. A variety of modeling approaches has been used with presence-only data (e.g., Elith et al., 2006; Phillips et al., 2006; Elith and Leathwick, 2009; Dorazio, 2012; Royle et al., 2012). To our knowledge, all approaches to occupancy inference using presence-only data require two key assumptions (among others). We refer to the first as *random* or *representative sampling*, referencing the need to sample space randomly or in some other manner that is unbiased with respect to the true species distribution. This is a very strong assumption that is critical to reasonable inference, yet seldom met by data used with presence-only approaches (Phillips et al., 2009; Dorazio, 2012; Royle et al., 2012; Yackulic et al., 2013). The second assumption is that covariates influencing species detection probabilities do not also influence true species distribution, another strong assumption that is seldom likely to be met (Dorazio, 2012; Royle et al., 2012; Yackulic et al., 2013). Our preference for the occupancy models described in this book is based largely on their ability to yield reasonable inferences in the absence of these two assumptions. If these assumptions do happen to be met for a specific set of presence-only data, then our preference is for the approaches of Dorazio (2012) and Royle et al. (2012) (also see Lancaster and Imbens, 1996 and Lele and Keim, 2006) that provide inference about occupancy itself, rather

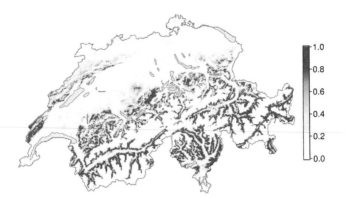

FIGURE 2.3 Example of a species distribution map created using single-season occupancy model for the willow tit (*Poecile montanus*) in Switzerland.

than about quantities proportional to occupancy. We note that the methods described in this book can be used to create species distribution maps (Chapter 4; Fig. 2.3), as is common with popular presence-only software.

Lele et al. (2012) describe an approach to inference about occupancy based on single surveys at each sample unit. Their motivation for this work was based on a reduction in survey effort and on potential difficulties meeting two assumptions underlying nearly all occupancy models described in this book, closure (the species is either present or not for all replicate surveys within a season) and independence of surveys at a site (see Chapter 4). Lele et al. (2012) (also see Moreno and Lele, 2010) show that inference about occupancy is possible with only a single visit per sample unit provided that (1) probabilities of both occupancy and detection depend on continuous-valued covariates and (2) the set of covariates that affects occupancy differs by at least one variable from the set of covariates affecting detection.

Lele et al. (2012) suggest that these requirements would be fairly common in occupancy data sets and that their approach should thus be broadly applicable. However, we recommend caution with their approach as while the above conditions may enable models to be fit to a data set and associated parameters estimated, the solution is not unique. If one was to swap the set of covariates purported for occupancy and detection probabilities, the result would be a model with an identical fit to the data, but quite different biological interpretation. This is because in the modeling approach of Lele et al. (2012), the unit-specific probabilities of occupancy (ψ_i) and detection (p_i) are always included as a product. That is, the probability of detecting the species at unit i with a single survey is $\psi_i p_i$ and the probability of not detecting the species is $1 - \psi_i p_i$. Therefore, swapping the set of covariates is equivalent to reversing the order of the multiplication, which has no effect on the product value. Furthermore, the method of Lele et al. (2012) does not work with categorical covariates which are commonly used in many areas of ecology. Hence we believe that the work of Lele et al. (2012) is an approach that could prove useful in analyzing historical data sets for which replicate sampling is not available, although we would not advocate it as an approach upon which studies should be designed. Lele et al. (2012, p. 29) concluded that "When the crucial assumptions of population closure and independence of surveys are satisfied and costs are not a major issue, then multiple survey methods will generally provide statistically more efficient estimators than a single survey-based approach." Our reasons for recommending replication are based on the simple observation that detection–nondetection data are the function of two processes, species occurrence and species detection. Use of replication provides one data source (detections and nondetections at units with at least one detection), for which variation only arises from the detection process, providing a means of inference about detection that is not confounded with occupancy. This inference about detection then enables us to decouple the separate influences of occupancy and detection on the overall detection/nondetection data set. We believe that this approach to separating occupancy and detection processes will frequently be worth potential additional costs of replicate surveys, although note that there is a variety of ways in which replicate surveys could be conducted during a single visit and without major increases in cost (Chapter 11). The approach of Lele et al. (2012) can potentially accomplish the same separation, but does so via increased reliance on model structure, likely decreasing robustness to model mis-specification.

In Section 4.4.9 we provide some discussion about the key modeling assumptions that motivated Lele et al. (2012), and the consequences of their potential violation. We also devote two chapters of this book (Chapters 11 and 12) to discussing various ways for practitioners to design studies in ways that meet the model assumptions, and also a chapter on extensions of the basic single-season occupancy model of MacKenzie et al. (2002) that can be used to accommodate some of these assumption violations (Chapter 6). Briefly, the closure assumption can be dealt with via either modeling (e.g., Rota et al., 2009; Kendall et al., 2013; Chambert et al., 2015b; Chapter 6) or design (e.g., temporal proximity of replicate surveys; Chapter 11). Similarly, the independence assumption can be dealt with using so-called removal designs (Chapter 11) or, depending on the source of dependence, the multiscale and correlated detection models described in Chapter 6 (also see Nichols et al., 2008; Hines et al., 2010, 2014; Guillera-Arroita et al., 2011). As we detail later in the book, a key realization for practitioners is that in the face of imperfect detection, more precise estimates of occupancy will often be obtained by surveying fewer units with a sufficient number of repeat surveys than trying to survey a larger number of units with fewer surveys (including only one survey).

2.2 HABITAT RELATIONSHIPS AND RESOURCE SELECTION

Studies of the determinants of range size and dynamics are necessarily very general in nature and include a variety of questions and approaches. One specific approach to the investigation of animal distribution patterns involves the concept of the ecological niche (e.g., Hutchinson, 1957) and the idea of each species having a unique set of requirements that must be provided by the habitat(s) at a location in order for the species to persist there. Some draw a distinction between the *fundamental* or *potential niche* and *realized niche* (e.g., see Vetaas, 2002). The fundamental or potential niche is a theoretical concept that refers to the set of species requirements that must be present for species persistence, in the absence of biotic interactions or other constraints. The realized niche, on the other hand, is usually viewed as the niche that actually exists after the fundamental niche has been modified by such factors as competitive interactions (e.g., resulting in exclusion of the focal species from locations that would otherwise be ideal) or perhaps dispersal constraints (inability of the focal species to colonize an area that would otherwise be ideal for species persistence). Existing distribution patterns of animals thus reflect realized niches, and occupancy modeling provides direct inference about these.

Ecologists thus seek to identify the key habitat variables to which a species responds and to develop habitat models that can be used to predict abundance, or at least occupancy, of a particular species as a function of habitat characteristics (e.g., see Verner et al., 1986; Scott et al., 2002; Royle et al., 2007b). Much of this work involves assessing species presence or absence on sample units and then asking whether presence can be modeled as a function of habitat characteristics measured on these units, i.e., what habitat variables best discriminate between locations that are and are not occupied by the species (e.g., Hirzel et al., 2002). Prior to the development of occupancy models, this approach was used to model habitat relationships of many vertebrates (e.g., amphibians, Johnson et al., 2002; mammals, Carroll et al., 1999; Reunanen et al., 2002; birds, Robbins et al., 1989; Klute et al., 2002; Tobalske, 2002; Gibson et al., 2004; fish, Dunham and Rieman, 1999; Dunham et al., 2002; reptiles, Fischer et al., 2004), invertebrates (e.g., Hanski et al., 1996; Fleishman et al., 2001; Wahlberg et al., 2002), and plants (e.g., Fertig and Reiners, 2002; Edwards et al., 2004).

Many of these habitat studies based on occupancy have been directed at conservation and management (e.g., Fuller et al., 2016) and have focused on habitat conservation and habitat change resulting from such factors as human land use and climate change. For example, a number of occupancy studies have provided evidence of the importance of area of woodland and forest habitat to occupancy of certain bird species (e.g., Moore and Hooper, 1975; Forman et al., 1976; Lynch and Whigham, 1984; Robbins et al., 1989). These and related studies have led to recommendations about both habitat conservation and the design of nature reserves (e.g., Diamond, 1975, 1976; Wilson and Willis, 1975; Robbins et al., 1989; Cabeza et al., 2004). Some studies have focused on anthropogenic determinants of occupancy such as pollutants (e.g., acid rain, Hames et al., 2002) and even human density (Chown et al., 2003).

Resource selection functions comprise a large number of specific methods that can be used in different applications (e.g., see Manly et al., 2002), with one application being to assess how a species uses resource (or sampling) units within an area of interest rather than addressing the resource use of individual animals (i.e., in the terminology of Manly et al., 2002, study design I rather than design II or III). In the resource selection context, the intent is typically to identify the relative level of use of different types of resource units. It is appropriate to consider some resource selection problems as exercises in studying the relationship between occupancy and habitat as the basic field protocols and analytic methods are usually very similar to those used above (MacKenzie, 2006).

The studies cited above are representative of the work investigating habitat relationships with so-called 'presence–absence' data. Prior to the development and application of occupancy models, virtually all this work included false absences in which units are occupied by the species of interest, but the species goes undetected. Approaches such as logistic regression are expected to yield biased results when applied to species presence–absence data in which the species is not truly absent from all units at which it goes undetected (Hirzel et al., 2002; Tyre et al., 2003; Gu and Swihart, 2004). Gu and Swihart (2004, p. 195) studied this issue via simulation and found that "logistic regression models of wildlife–habitat relationships were sensitive to even low levels of nondetection in occupancy data." Bias in estimates of parameters specifying habitat relationships are expected to be greatest when detection probabilities are themselves related to the habitat variables of interest (Gu and Swihart, 2004). For example, consider Fig. 2.4 which illustrates how detection probability can confound inferences about habitat relationships, or more generally, the effect of any occupancy covariate. Panel (A) presents the true situation where the occupancy probability at each unit increases according to the value of the habitat variable at that unit (black line), along with three different scenarios for how detection probability also varies with habitat; constant (green line: p_A); increases (red line: p_B); and decreases (blue line: p_C). Panel (B) again presents the true relationship for occupancy for reference, and also the product of occupancy and detection for each of the detection probability scenarios (i.e., ψp). While the level of bias will decrease as detection probability approaches one, the apparent relationships can be quite different to the true relationship due to the confounding of occupancy and detection probability. Therefore, estimated effect sizes from logistic regression analyses of so-called 'presence–absence' data will often be biased (i.e., the

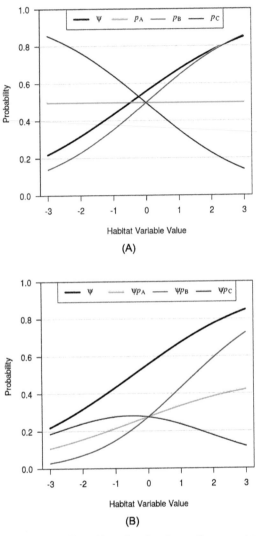

FIGURE 2.4 Illustration of the effect of imperfect detection on the apparent relationship between occupancy and a habitat variable. Panel (A) indicates the true relationship for occupancy and three scenarios for how detection probability covaries with habitat, while panel (B) presents the apparent relationship where occupancy and detection are confounded (e.g., when using logistic regression on single-survey data).

slope of the apparent relationship is not the same as the true relationship). The bias results as logistic regression simply models the relationship between habitat and where the species is *found* (a combination of occupancy and detectability), not where the species *is* (occupancy). In addition, approaches that ignore the issue of imperfect detections should yield variance estimates that are too small, as they do not incorporate all components of uncertainty (e.g., detection probabilities < 1).

Occupancy models have been used to model habitat relationships of many vertebrates (e.g., amphibians, Bailey et al., 2004; mammals, Ball et al., 2005; Karanth et al., 2010; birds, Collier et al., 2012; Henneman and Andersen, 2009; fish, Groce et al., 2012; Falke et al., 2010; reptiles, Zylstra and Steidl, 2009), invertebrates (e.g., Fernández-Chacón et al., 2014), and plants (e.g., Al-Chokhachy et al., 2013), while accounting for nondetection. MacKenzie (2006) (also see Section 4.4.12) emphasized the importance of detection probability by demonstrating how incorrect inference can arise from ignoring it. Data on the use of units by pronghorn antelope (*Antilocapra americana*) collected during two winters were analyzed using two different methods. Assuming that resource/habitat use did not change between winters, simple logistic regression (where detection in at least one winter indicated a unit was 'used', and nondetection was equated to 'unused') identified that distance from a water source appeared to the most important habitat variable, but using a method that accounted for detection probability, slope and sagebrush density were the variables considered most important for habitat use. Distance from a water source was, however, found to be the most important variable in terms of detection probability, which would explain why simple logistic regression identified it as an important variable.

The single-season models discussed in Chapters 4–7 can be viewed as logistic regression approaches to resource selection that incorporate detection probabilities. As such they provide the same sorts of inferences as historic approaches, but they do so in a manner that should be much more robust to nondetection and that properly incorporate associated uncertainty. Investigators have recognized the utility of single-season occupancy models for studying habitat relationships, and the literature now includes many examples of such uses (see citations in the previous paragraph).

The vast majority of historic work on relationships between species presence–absence and habitat has focused on static relationships based on temporal 'snapshots'. This same tendency applies to the recent use of occupancy modeling for species–habitat relationships. However, we believe that the relationship between habitat covariates and the vital rates that determine species distributions (i.e., rates of local extinction and colonization) are at least as important as static inferences and deserve increased emphasis (also see Tyre et al., 2001; Yackulic et al., 2015). Our reasons for this belief are elaborated later in this volume but include the possibility of misleading relationships based on static models when systems are not in equilibrium, and interest in projecting consequences of habitat changes for species distributions. Investigating habitat relationships in occupancy dynamics entails the modeling of probabilities of local extinction and colonization, rather than occupancy itself, as functions of habitat covariates. Systems experiencing transient dynamics can show very different static patterns of habitat and occupancy over time, even when underlying relationships between habitat and vital rates are constant over time (Yackulic et al., 2015). As a result, predictions based on static relationships at any one point in time may not be very useful. Even if a studied system happens to be in equilibrium, such that species–habitat relationships are not so dynamic, we are often interested in predicting changes in species distribution patterns expected to result from changes in habitat (e.g., associated with conservation efforts, climate change, etc.). Once again the relationship between habitat and vital rates is needed for projecting consequences of habitat change for species distributions. Indeed, an especially effective approach for projection and prediction is to model habitat dynamics and species occupancy dynamics simultaneously (e.g., Martin et al., 2010; MacKenzie et al., 2011; Falke et al., 2012; Chapter 13).

2.3 METAPOPULATION DYNAMICS

Subdivided populations connected by some degree of movement were first considered by Wright (1931, 1951), who focused on the genetic structure produced by such systems. Genetic structure in such a system will be partially determined by the nature of migration among subpopulations, with different structures expected to result from different patterns of migration (e.g., Wright's 'island model' as opposed to the 'stepping-stone' and more general 'isolation by distance' models of Kimura and Weiss, 1964; Crow and Kimura, 1970). Ecologists Andrewartha and Birch (1954) focused on local extinctions and colonizations as determinants of population dynamics in such systems. Levins (1969, 1970) formalized the concept of a metapopulation, a system of subpopulations, any of which can go extinct and later become recolonized. The metapopulation concept is now thought to provide a useful description of a large number of natural populations, and the literature on metapopulations has become substantial (e.g., see McCullough, 1996; Hanski and Gilpin, 1997; Hanski, 1999; Hanski and Gaggiotti, 2004).

Metapopulation structures involve systems of patches or units that are sometimes occupied by the species of interest, and sometimes not, depending on the dynamic processes of extinction and colonization. Indeed, the number or proportion of units that is occupied is the state variable of interest in the original models of Levins (1969, 1970) and in various subsequent models of metapopulations (e.g., Lande, 1987; Hanski, 1991, 1997, 1999). Thus, patch occupancy is of primary importance in investigations of metapopulations, as are local rates of extinction and colonization, the vital rates responsible for changes in occupancy. Until the late 20th century, occupancy estimation was viewed as a simple problem and was based on a binomial estimator of patches at which the species was detected, divided by total patches surveyed. The literature of metapopulation ecology has been more focused on estimation of local rates of extinction and colonization and on functional relationships between these vital rates and patch characteristics. Two primary approaches have been used to draw inferences about extinction and colonization, and they are distinguished by the temporal scale of the occupancy data used in estimation. An early approach was based on static detection/nondetection data for multiple units collected from a single season, whereas another is based on detection/nondetection data from multiple units collected in multiple seasons.

2.3.1 Inference Based on Single-Season Data

The single-season approach to estimation of rates of extinction and colonization precedes interest in metapopulations and has its origins in community ecology. Diamond (1975) introduced the concept of *incidence functions*, in which the probability of occurrence of bird species on islands was modeled as a function of factors (e.g., island area, distance between island and mainland) hypothesized to be relevant to species-level vital rates and, hence, to species distributions. Hanski (1991, 1992) used incidence functions to describe single species metapopulations, relating patch occupancy to patch characteristics hypothesized to influence probabilities of patch extinction and colonization. If the metapopulation system is in dynamic equilibrium, if the primary patch characteristics affecting extinction and colonization have been identified, and if the functional forms of these relationships have been adequately specified, then data on the fraction of patches occupied at a single point in time can be used to estimate the parameters defining these relationships, as well as the extinction and colonization probabilities themselves (Hanski, 1991, 1992, 1994a, 1999). Hanski (1992, p. 660) noted that direct study of the extinction and colonization processes is "expensive and time-consuming," in contrast to incidence functions "constructed from 'snap-shot' information collected during one sampling period." However, the decreased expense associated with single assessments of pattern comes at a cost, as attempts to infer process from pattern can easily lead to incorrect inferences (see Chapter 1; also Clinchy et al., 2002).

Early work on incidence function estimation of extinction and colonization assumed that colonization probability was a constant for all patches but that extinction probability decreased as a function of patch area (e.g., Hanski, 1991, 1992). Substantial interest was focused on the coefficient specifying the rate of decrease in extinction probability with increasing patch area. Later work then expressed colonization rate for specific patches as a function of distances to, and occupancy of, other patches (Hanski, 1994a, 1994b, 1998, 1999). The "rescue effect" of Brown and Kodric-Brown (1977) was considered by including the possibility of recolonization in the extinction function (Hanski, 1994a, 1999).

In addition to Hanski's direct estimation of vital rates (Hanski, 1991, 1992, 1994a, 1994b, 1999; Hanski et al., 1995; Wahlberg et al., 1996, 2002) and of parameters describing relationships between these rates and patch characteristics, several authors have drawn inferences about colonization and extinction processes indirectly by investigating the relationship between occupancy and patch characteristics. This work is thus very similar to that described above for habitat relationships, except that the patch characteristics of interest are those hypothesized to be important determinants of colonization (e.g., patch isolation) and extinction (e.g., patch area) within a metapopulation context. For example, one prediction is that, other characteristics being roughly equal, occupancy should increase as a function of patch area because of the reduced extinction probabilities of large patches. Support for this prediction comes from several studies of metapopulation systems (e.g., Lomolino et al., 1989; Peltonen and Hanski, 1991; Hanski, 1992, 1998; Hanski et al., 1995, 1996; Thomas and Hanski, 1997; Smith and Gilpin, 1997; Dunham and Rieman, 1999; Wahlberg et al., 2002; Bradford et al., 2003). Occupancy has been modeled as a function of various measures reflecting patch isolation (Hanski, 1994a, 1999) connectivity measures, distance to nearest occupied patch, etc., with the prediction that greater isolation should reduce colonization, leading to lower occupancy. Most results have provided evidence of a negative relationship between occupancy and isolation, although such evidence is not always found (Lomolino et al., 1989; Peltonen and Hanski, 1991; Hanski et al., 1995; Whitcomb et al., 1996; Smith and Gilpin, 1997; Thomas and Hanski, 1997; Hanski, 1998; Dunham and Rieman, 1999; Clinchy et al., 2002; Wahlberg et al., 2002).

Studies of metapopulation dynamics based on single-season analyses have also included information on habitat availability. Lande (1987, 1988) extended the Levins (1969, 1970) metapopulation models by designating some fraction of patches that were suitable for the species. Remaining patches were unsuitable and could not be occupied. Lande (1987) was able to compute a threshold proportion of suitable patches (termed the *extinction threshold*), such that systems with a smaller proportion of suitable patches were doomed to extinction. This type of joint occupancy and habitat modeling has received substantial attention (e.g., Ovaskainen et al., 2002; Merila and Kotze, 2003; Martin et al., 2010; MacKenzie et al., 2011) and has been applied, for example, to Northern spotted owl (*Strix occidentalis caurina*) populations in fragmented habitat (Lande, 1988; Noon and McKelvey, 1997). Hanski and Ovaskainen (2000) extended the concept and defined *metapopulation capacity*, a metric reflecting the relative capacities of different landscapes to support viable metapopulations.

This thinking has been extended to model habitat destruction as a specified fraction of suitable patches that is suddenly destroyed such that they become permanently unsuitable. Such models lead to inferences about *minimum viable metapopulations*, *minimum amount of suitable habitat*, and metapopulation capacity (Hanski et al., 1996; Hanski and Ovaskainen, 2000), as well as to related inferences about reserve designs that maximize time to metapopulation extinction (Ovaskainen et al., 2002) or minimize the probability of extinction over a finite time frame (Moilanen and Cabeza, 2002; Cabeza et al., 2004). Some work on reserve design has used probability of occurrence as a surrogate for local persistence (Araujo and Williams, 2000; Williams and Araujo, 2000; Araujo et al., 2002). A modeling effort by Ellner and Fussmann (2003) emphasized the potential importance of within-patch succession to metapopulation persistence. Such models have also been used to explore interactions among species in a community, such as competitive coexistence in two-species systems (Nee, 1994; Moilanen and Hanski, 1995) and persistence in predator–prey systems (Nee, 1994). All of these modeling efforts have not necessarily involved parameter estimation, but they are included in this section on single-season data because they employ the same models that are used for estimation based on incidence functions.

The described use of incidence function data to estimate rates of extinction and colonization assumes that incidence (probability that a patch is occupied) is known or estimated without bias. Failure to detect a species when present will result in negatively biased estimates of proportional occupancy and thus in biased estimates of rates of both extinction and colonization. Moilanen (2002, p. 524) studied three problems with data used in conjunction with incidence functions to estimate vital rates from detection–nondetection data and concluded that "all effects seen so far pale in comparison with biases caused by false zeros in the data set." He reported the possibility of substantial positive bias in estimated rates of both extinction and colonization. Functional relationships such as those between extinction probability and patch size and between colonization probability and patch isolation will be estimated with bias as well (Moilanen, 2002).

Biased estimates of proportional occupancy also influence inferences based on joint consideration of occupancy and proportion of habitat that is suitable. In addition to the problems caused by use of biased estimates of occupancy, the proportion of suitable patches may not be known in all cases, creating additional problems. For example, some insects are restricted to feeding on particular host plant species (e.g., Hanski et al., 1995; Wahlberg et al., 1996, 2002), as are various types of parasites on animals (Deredec and Courchamp, 2003). We would not expect detection probabilities for habitat patches to be equal to one in these situations. Such metrics as extinction threshold, metapopulation capacity, minimum viable metapopulation, and minimum amount of suitable habitat will all be estimated with bias in the presence of nondetections of both occupancy and unit status (suitable and not suitable).

Several authors have used single-season occupancy models to explore the relationship between occupancy and patch characteristics hypothesized to be important determinants of extinction (e.g., patch size) and colonization (e.g. patch isolation) as represented by various connectivity measurements (e.g., Collier et al., 2012). Studies have often investigated the influence of suitable (or unsuitable) habitat surrounding a focal patch (e.g., Mazerolle et al., 2005; Fairman et al., 2013) in an attempt to model scale-specific habitat availability, or identify thresholds that influence local occurrence (Groce et al., 2012). Groce et al. (2012) used static occupancy models to estimate the proportion of remaining suitable habitat and, conditionally, the occupancy of Arkansas darters (*Etheostoma cragini*) in these habitats.

Models discussed in Chapters 4–7 of this book permit unbiased estimation of occupancy in a manner that accounts for detection probabilities less than 1. While these estimates could be incorporated into the incidence function modeling of Hanski (1991, 1992, 1994a, 1998, 1999) as a means of obtaining improved estimates of rates of colonization and extinction, and of the parameters governing the relationships between these rate parameters and patch covariates, we prefer to avoid assumptions about stationarity and presumed relationships with patch characteristics and to instead estimate rates of local extinction and colonization from detection history data that cover multiple seasons or time periods (Chapters 8–10). Using such an approach, investigators can view stationarity and relationships between rate parameters and patch characteristics as hypotheses to be tested, rather than as assumptions required for estimation (see next subsection).

2.3.2 Inference Based on Multiple-Season Data

Detection–nondetection data for a set of patches over seasons or years can be used to directly estimate local extinction and colonization probabilities without requiring assumptions of system equilibrium or stationarity. In most previous work with such data, extinction rate between two seasons or years, t and $t + 1$, has been estimated by conditioning on a set of patches at which the focal species is detected at time t, and then computing the fraction of these at which the species is not detected at time $t + 1$. Similarly, colonization rate has been estimated by conditioning on the set of patches at which the focal species is not detected at time t, and then computing the fraction of these at which the species is detected at time $t + 1$. In addition to direct estimation of these rate parameters, logistic regression and related approaches have been used to relate 'observed' extinctions and colonizations to covariates such as patch area and degree of isolation. These approaches to inference (summarized by Morris and Doak, 2002) have been used in a variety of study situations including for shrews (*Sorex* spp.) on islands in Finnish lakes (Peltonen and Hanski, 1991); pikas (*Ochotona princeps*) inhabiting ore dumps created by miners in California (Smith and Gilpin, 1997; Clinchy et al., 2002); various plant species in European and North American study areas (Kéry, 2004, and references therein); European nuthatches (*Sitta europaea*) in Netherlands woodlots (Verboom et al., 1991); passerine bird species in 10 km × 10 km grid cells in Great Britain (Araujo et al., 2002); scarlet tanagers (*Piranga olivacea*) in North American forest fragments (Hames et al., 2001); various breeding bird species in woodlots of southern Finland (Haila et al., 1993); plant species found on serpentine seeps in California (Harrison et al., 2000); several amphibian species inhabiting ponds in southwestern Ontario (Hecnar and M'Closkey, 1996), southeastern Michigan (Skelly et al., 1999), and a montane park in central Spain (Martinez-Solano et al., 2003); pool frogs (*Rana lessonae*) in permanent ponds along the Baltic coast of Sweden (Sjogren-Gulve and Ray, 1996); frogs and toads at survey units throughout Wisconsin (Trenham et al., 2003), and Glanville fritillary butterflies (*Melitaea cinxia*) in dry meadow patches of southwestern Finland (Hanski, 1997; Thomas and Hanski, 1997).

Some workers have extended the conditional binomial modeling of detection–nondetection data to develop methods for estimating local extinction and colonization parameters using a sequence of such data over time, over species, or over multiple locations. Rosenzweig and Clark (1994) obtained maximum likelihood estimators for Markov process models governed by extinction and colonization rate parameters that corresponded to the period separating successive samples. Clark and Rosenzweig (1994) extended the estimation to the more difficult problem in which detection–nondetection sampling occurs at irregular intervals that do not always correspond to the time period for which the rate parameters are defined. Erwin et al. (1998) extended this modeling to include estimation for ultrastructural models of rate parameters as functions of site-specific covariates. ter Braak and Etienne (2003) applied a Bayesian approach to parameter estimation under Markov modeling of detection–nondetection data. Other approaches to estimation under this kind of patch occupancy modeling include Verboom et al. (1991) and Ferraz et al. (2003).

Estimates obtained using any of the above approaches generally will be biased by failure to detect the species at all occupied patches. Extinction rate estimates based on conditional binomial models should always be biased high, as some patches will appear to represent extinctions when this is not the case (see Kéry, 2004). Parameters expressing the relationship between extinction rate and covariates such as population size can also be biased when detection probability is not accounted for (Kéry, 2004). Binomial estimation of colonization is conditional on patches at which the species is not detected, and the species will really be present in some of these patches. Similarly, the species may be present, yet undetected, in the second sampling period as well. The result is that estimates of colonization probability will tend to be biased. Estimation based on the full Markov models will similarly tend to yield biased estimates of the rate parameters.

Most applications of occupancy models to metapopulation systems have involved multiple-season data, where investigators test the previously assumed relationships between rate parameters and patch characteristics. Authors have explored these hypotheses in systems involving amphibians (e.g., Pellet et al., 2007), mammals (e.g., Eaton et al., 2014), birds (e.g., Ferraz et al., 2007; Kennedy et al., 2011), and invertebrates (e.g., Fernández-Chacón et al., 2014). In some instances, the relationships between extinction probability and patch size and colonization probability and patch isolation are consistent with metapopulation theory (e.g., Eaton et al., 2014; Fernández-Chacón et al., 2014). In other cases, factors such as the habitat matrix between patches (Kennedy et al., 2011) or patch characteristics other than area (Pellet et al., 2007) were more influential on vital rate parameters.

Issues related to the proportion and dynamics of suitable habitat have been incorporated into occupancy applications in several ways. In highly dynamic systems, local habitat suitability and occupancy dynamics have been modeled simultaneously (Martin et al., 2010; MacKenzie et al., 2011; Falke et al., 2012; Miller et al., 2012a). Chapter 13 highlights these integrated habitat–occupancy models, where suitable habitat and/or the target species may not be detected perfectly. Markov models of habitat suitability permit estimation of transition probabilities associated with habitat changes, and the species occupancy dynamics can include constraints reflecting these habitat changes. For example, if an occupied patch at time t made the transition to an unsuitable patch at time $t + 1$, then the species extinction probability is known to be 1 and can be so constrained for that time interval. In other situations, habitat suitability is assessed directly via species occurrence, and colonization and extinction probabilities are modeled as *autologistic* functions of the current occupancy in a specified neighborhood, where an appropriate neighborhood is based on movement and dispersal distances of the focal species (Yackulic et al., 2012, 2014; Eaton et al., 2014). Finally, occupancy models described in Chapter 14 could be used in host–parasite systems, where parasitic animals or insects feed on particular host plant species, with patches of host species functioning as suitable habitat patches (e.g., Heer et al., 2013). The methods of Chapter 14 permit simultaneous inference about occupancy for both focal and host species.

2.4 LARGE-SCALE MONITORING

Large-scale monitoring programs have become popular in recent years. In some cases, monitoring is tied closely to management decisions (e.g., Nichols, 1991; Nichols et al., 1995), whereas in other cases, the reasons underlying monitoring efforts are not so clear (e.g., see comments by Krebs, 1991; Yoccoz et al., 2001; also Chapter 1). Monitoring efforts that are not so well defined are frequently viewed as surveillance tools and, as such, are focused on detection of change in animal populations. Often, density or abundance is the state variable for which estimates of change are sought. However, estimation of density and abundance often requires substantial effort (e.g., Lancia et al., 1994; Pollock et al., 2002), leading some to view occupancy as a surrogate for abundance. In territorial species, when sample units are selected to be the approximate sizes of territories, estimates of the number of occupied units are virtually equivalent to estimates of numbers of territorial animals or pairs (e.g., spotted owls, *Strix occidentalis*; Azuma et al., 1990; MacKenzie et al., 2003; MacKenzie and Nichols, 2004). Indeed, previous workers have used presence of an individual animal in a territory as a 'mark' and applied capture recapture models to detection–nondetection data from repeat samples as a means of estimating population size (e.g., Hewitt, 1967 for red-winged blackbirds, *Agelaius phoeniceus*; Thompson and Gidden, 1972 for American alligators, *Alligator mississippiensis*).

Even for species that are not territorial, occupancy is sometimes viewed as a surrogate for abundance in monitoring programs (e.g., Bart and Klosiewski, 1989). Occupancy is clearly related to abundance, as it focuses on one tail of the distribution of abundance of animals across space or patches (i.e., the portion of the distribution associated with the probability a patch contains one or more individuals: $Pr(N_i > 0)$, where N_i is abundance for patch i). Others view occupancy not as a surrogate for abundance, but as an appropriate state variable for large-scale monitoring (Hall and Langtimm, 2001; Manley et al., 2004; Noon et al., 2012). In other cases, occupancies of multiple species have been summed and species richness used as a state variable (e.g., Martinez-Solano et al., 2003; Weber et al., 2004).

Detection–nondetection surveys have been recommended or used for various large-scale monitoring programs. Such surveys are currently used throughout Switzerland to monitor multiple vertebrate and invertebrate taxa (Weber et al., 2004). They have been used and recommended for amphibians in various regions (Weir and Mossman, 2005; Muths et al., 2005; Martinez-Solano et al., 2003). Occupancy surveys of potential territory units are used to monitor spotted owls (e.g., Azuma et al., 1990; MacKenzie et al., 2003) and marbled murrelets (*Brachyramphus marmoratus*; e.g., Stauffer et al., 2002) in northwestern North America. Camera-trapping and track plates have been used by the U.S. Forest

FIGURE 2.5 Sign surveys are used to collect detection/nondetection information on tiger (*Panthera tigris*) in India. (*Ramki Sreenivasan*)

Service to monitor occupancy of fishers (*Martes pennanti*) and martins (*Martes americanus*) in the Klamath region of the western U.S. (Zielinski and Stauffer, 1996; Carroll et al., 1999). The New Zealand Department of Conservation has designed a pilot program for monitoring occupancy of an endangered insect, the Mahoenui giant weta (*Deinacrida mahoenui*), on the North Island of New Zealand (MacKenzie et al., 2004b, 2005). Zonneveld et al. (2003) discuss the design of detection–nondetection surveys for insect species such as the endangered Quino checkerspot butterfly (*Euphydryas editha quino*). Surveys based on animal sign (e.g., tracks, scat) have been recommended for large-scale monitoring of tigers (*Panthera tigris*; Fig. 2.5) and their prey species in India (Nichols and Karanth, 2002). The count-based North American breeding bird survey (Robbins et al., 1986; Sauer et al., 2013b) is sometimes analyzed from an occupancy perspective that focuses on whether or not a species is detected at a stop or route (Bart and Klosiewski, 1989).

Biased estimates of occupancy resulting from failure to detect animals that are present have the potential to lead to biased estimates of change in occupancy over time. In particular, temporal variation in detection probability will be confounded with any true temporal variation in studied populations. The problems associated with variable and unknown detection probabilities in occupancy monitoring programs generally have been recognized (e.g., Azuma et al., 1990; Zielinski and Stauffer, 1996; Hall and Langtimm, 2001; Nichols and

Karanth, 2002; Stauffer et al., 2002; Zonneveld et al., 2003; Bailey et al., 2004; Kawanishi and Sunquist, 2004; MacKenzie et al., 2003, 2004b, 2005). As with monitoring other state variables (Yoccoz et al., 2001; Pollock et al., 2002), the solution to this problem is simply to incorporate detection probabilities as parameters in models used to estimate change in occupancy over time.

The models of Part III were parameterized to directly estimate rate of change in occupancy over time while incorporating nondetection. These models have been used to analyze data from monitoring programs, such as the ones described above, to estimate occupancy trends (e.g., Kéry et al., 2010; Adams et al., 2013; Weir et al., 2009, 2014) and investigate factors associated with these trends, including invasive species (e.g., Yackulic et al., 2014; Dugger et al., 2015), global climate change (e.g., Clement et al., 2016; Grant et al., 2016), habitat management (e.g. fire and fuel reduction treatments, Sweitzer et al., 2016), and/or disease (e.g., Grant et al., 2016).

2.5 MULTI-SPECIES OCCUPANCY DATA

Most of the occupancy applications discussed thus far have focused on questions related to single species ecology. They can also be applied to studies of multiple species in which each species is considered independently. However, there are two reasons for considering analyses that permit some degree of dependence among species. The first is to exploit possible similarities among species that permit sharing of information (e.g., via common parameters or random effect distributions) and resultant stronger inference. The other reason is to directly investigate species interactions and potential effects (e.g., via competition or predation) of one species on another. As was the case for metapopulation studies, investigations of multiple species can be classified into two types of studies, those based on static patterns of occupancy and those based on occupancy dynamics.

2.5.1 Inference Based on Static Occupancy Patterns

Investigations of static occupancy patterns for multiple species may be based on data from a single season, or on data accumulated over years as species range maps. In both cases, analyses of pattern are used to draw inferences about the underlying dynamics that produced them. Use of occupancy data from multiple species to draw inferences about species interactions also has a long history in ecology (Chapter 1). Consider a two-species system and assume that survey efforts to detect occupancy have been conducted on a set of sample units. The data obtained from such a survey are typically viewed as a contingency table, with both species having been detected on some units, neither species on other

TABLE 2.1 Species incidence matrix specifying the occurrence (1) and nonoccurrence (0) of *M* different species at each of *s* specific units

Unit	Species						
	1	2	3	4	5	...	M
1	1	0	0	1	0	...	1
2	1	1	0	0	1	...	0
3	0	1	1	1	0	...	0
4	1	0	0	0	1	...	1
5	1	1	0	1	0	...	0
⋮	⋮	⋮	⋮	⋮	⋮	⋮	⋱	⋮
s	1	0	0	1	1	...	0

units, and still other units with detections of only species A and others with only species B. The null hypothesis of independence of species occurrence is then tested with a 2×2 contingency table χ^2 test, and interaction indices have been developed to reflect the strength of the alternative hypotheses that species co-occur more or less frequently than expected under the hypothesis of independence (see Forbes, 1907; Dice, 1945; Cole, 1949; Pielou, 1977; Hayek, 1994). Similar thinking has been applied to the analysis of more complex multi-species occupancy data (Table 2.1). In such multi-species systems, *null models* are typically developed to deduce occupancy patterns expected under a null hypothesis of independence, or no interspecific interactions (Connor and Simberloff, 1979, 1984, 1986; Simberloff and Connor, 1981; Diamond, 1982; Diamond and Gilpin, 1982; Gilpin and Diamond, 1982, 1984; Stone and Roberts, 1990, 1992; Kelt et al., 1995; Manly, 1995; Gotelli and Graves, 1996; Gotelli, 2000; Gotelli and McCabe, 2002).

A potential problem with attempts to draw inferences about interspecific interactions from species incidence (presence–absence) matrices involves other factors (e.g., habitat preferences and physiological tolerances) that are likely to result in nonrandom patterns of species co-occurrence, yet have nothing to do with interspecific interactions. This class of problem is inherent in all attempts to draw inferences about process based on pattern and has been recognized in previous efforts to analyze species incidence matrices (e.g., Connor and Simberloff, 1984; Gilpin and Diamond, 1984). One approach to dealing with such factors is to identify them *a priori* and incorporate them into analyses. For example, regression models have been developed to predict detections of one species as a function of both habitat variables and detections of other species (Schoener, 1974; Crowell and Pimm, 1976). Other approaches have incorporated geographic and habitat characteristics directly into null models (Kelt et al., 1995; Gotelli et al., 1997; Peres-Neto et al., 2001).

We believe that detection probabilities can have an important influence on results of observed species co-occurrence patterns based on occupancy data (MacKenzie et al., 2004a; see Chapter 14). Species frequently have different detection probabilities. In addition, we have encountered several field situations where detection probability of species A is actually thought to depend on presence, or even detection, of species B. For example, in occupancy surveys of northern spotted owls in the Pacific northwestern U.S., biologists hypothesized that presence of larger barred owls (*Strix varia*) in or near northern spotted owl (*Strix occidentalis caurina*) territories may cause northern spotted owls to vocalize less frequently. Using occupancy modeling, Bailey et al. (2009), Yackulic et al. (2014), and Dugger et al. (2015) have since provided strong evidence of lower detection probabilities of northern spotted owls in the presence of barred owls. With respect to past investigations using multi-species occupancy data, we find it surprising that such a large amount of attention has been devoted to the statistical issues associated with inferences about patterns indicative of interactions, whereas virtually no attention has been devoted to the well-known problem that species are not always detected when present.

Static occupancy data collected for multiple species have also been used to draw inferences about relationships among multi-species extinction and colonization processes that do not involve interspecific interactions. For example, occupancy data for insular locations can be tested for the existence of a *nested subset* structure, such that smaller biotas contain a nonrandom subset of larger ones (Patterson and Atmar, 1986; Patterson, 1987, 1990; Bolger et al., 1991; Wright and Reeves, 1992; Andren, 1994; Cook and Quinn, 1995, 1998; Lomolino, 1996; Wright et al., 1998; Fleishman and Murphy, 1999). Common ecological explanations for the existence of such patterns involve differences among species in susceptibility to extinction and/or ability to colonize vacant habitat. The subset of species found in the locations of lowest richness (e.g., the island at the distal end of an archipelago, or the island most distant from the mainland) may be those species with the lowest extinction probabilities or the highest rates of colonization. As in the case of testing for evidence of interspecific interactions using species incidence data, substantial effort has been devoted to statistical methods providing inferences about existence of a nested pattern in a set of insular occupancy data for multiple species (Patterson and Atmar, 1986; Wright and Reeves, 1992; Cook and Quinn, 1998; Wright et al., 1998; Cam et al., 2000a). However, only Cam et al. (2000a) used an approach for inference about nestedness that explicitly dealt with species detection probabilities. This approach is based on estimating the fraction of species in one sample unit that is present at another, despite not detecting all species at either site (Nichols et al., 1998b; Cam et al., 2000a).

It is clear that nested subset analyses that do not account for detection probability can lead to poor inferences. For example, the detection (or nondetection) of a species on an island may be as much a function of detection probability (high or low) as of the ecological processes of local extinction and colonization. The statistical approaches developed to investigate nestedness rely heavily on the appearance of 0's or 'gaps' in species incidence matrices. Such gaps can simply represent failures to detect species that are present, a possibility that has been recognized (Grayson and Livingston, 1993; Kodric-Brown and Brown, 1993) but not dealt with until the development of occupancy models (except by Cam et al., 2000a). It is not clear whether an approach based on occupancy estimates for each species will necessarily perform better than the approach of Cam et al. (2000a) based on occupancy data aggregated over species. However, it is clear that an occupancy modeling approach will permit more flexibility in modeling.

The original single-season co-occurrence model was used to test for interactions among salamander (MacKenzie et al., 2004a) and raptor species (Bailey et al., 2009), while accounting for potential dependence in the detection process (i.e., allowing the detection of each species to vary depending on the presence or detection of the other species). So-called conditional parameterizations of the co-occurrence model (Richmond et al., 2010; Waddle et al., 2010) have also been used to explore how co-occurrence of rails (family *Rallidae*) varies among wetlands of different sizes (Richmond et al., 2010), or carnivores vary along an urbanization gradient (Lewis et al., 2015a). Waddle et al. (2010) explored habitat preferences and species interactions among native and invasive treefrogs in Florida. Peoples and Frimpong (2016) studied potential facilitative relationships involving nest associations among stream fishes in North America. Specifically, they found a positive relationship between a host species (*Nocomis leptocephalus*) and nest associate species (*Chrosomus oreas* and *Clinostomus funduloides*), while accounting for abiotic factors (e.g. substrate and hydrology).

Species richness is often a state variable of interest in biodiversity studies involving few or many units. Occupancy models for multiple non-interacting species permit sharing of information for efficient estimation of species richness. Single-season occupancy models, applied to detection–nondetection data simultaneously collected for multiple species, have been used to estimate species richness and community metrics for diverse avian communities including seabirds (Flanders et al., 2015), waterbirds (Lewis et al., 2015b), and songbirds in Central America (e.g., Ruiz-Gutiérrez et al., 2010; Chandler et al., 2013), North America (e.g., Mattsson et al., 2013; Zipkin et al., 2009; Russell et al., 2009), and Europe (e.g., Kéry and Royle, 2008; Russell et al., 2015). Applications also exist for other vertebrates, including amphibians (Zipkin et al., 2012), reptiles (Hunt et al., 2013), mammals (Tobler et al., 2015), marine fish (Holt et al., 2013), invertebrates (Dorazio and Royle, 2005; Govindan and Swihart, 2015), and plants (Chen et al., 2013). Investigators in these studies have explored the influence of habitat (e.g., Mattsson et al., 2013), food resources (Lewis et al., 2015b), disturbance and land use change (Zipkin et al., 2009; Russell et al., 2009; Hunt et al., 2013), and environmental gradients (Mihaljevic et al., 2015) on local and regional metacommunity structure.

2.5.2 Inference Based on Occupancy Dynamics

Attempts to deduce process from observation of pattern are inherently difficult (Chapter 1 and above). The use of multi-species occupancy data from several points in time (e.g., years) may be more useful for drawing inferences about processes associated with interspecific interactions, especially when such studies include an event hypothesized to alter community organization. For example, Sanders et al. (2003) collected annual occupancy data for members of a native ant community on sample plots in northern California for a period of seven years. Some of the sampled areas were invaded by a non-native species, the Argentine ant (*Linepithema humile*), during the course of the study. On areas not invaded by the Argentine ant, occupancy data for native ant species showed evidence of species segregation consistent with a hypothesis of competitive interactions. However, on areas that were invaded, native ants showed evidence of segregation before invasion, yet appeared to co-occur randomly or even with aggregation following invasion (Sanders et al., 2003). This study thus provided relatively strong inferences about effects of the invasive species on community organization. Although relatively rare, experimental manipulations have been used to directly investigate the roles of interspecific interactions and habitat on species incidence patterns using occupancy data (Syms and Jones, 2000).

Multi-species occupancy data collected over multiple seasons or years can also be used to draw inferences about rates of extinction and colonization in the absence of any hypothesis about interspecific interactions. For example, a number of the studies cited in Section 2.3 deal with inferences about rates of extinction and colonization for one or more species at one or more locations over a sequence of seasons or years. In addition to the use of inference methods that ignore the issue of detection probabilities, some workers have developed estimators for rates of extinction and turnover that do incorporate detection probabilities and that are based on aggregations of species (Nichols et al., 1998a; Williams et al., 2002). For these aggregate methods, each species within a group of interest is viewed as a replicate (similar to an individual in a single-population study of survival) and assumed to exhibit a common rate of extinction or turnover. Such approaches have been used with multiple-season occupancy data for multiple species to draw inferences about effects of fragmentation on avian extinction and turnover (Boulinier et al., 1998a, 2001), the influence of sexual dichromatism on local rates of species extinction of birds (Doherty et al., 2003b), and the influence of location of species within their geographic range (edge vs. interior) on local rates of extinction and turnover (Doherty et al., 2003a; Karanth et al., 2006). This general approach has also been used with fossil data to estimate rates of global extinction and origination for various taxa (Nichols and Pollock, 1983; Conroy and Nichols, 1984; Nichols et al., 1986; Connolly and Miller, 2001a, 2001b, 2002).

When occupancy data are collected over multiple species for multiple seasons or years, failure to deal with detection probabilities leads to biased estimates of rates of both extinction and colonization (see Section 2.3). The multi-season models of Chapter 15 can be used to obtain estimates of these rate parameters for each species in a community. Inferences about rates of extinction or colonization for groups of species can then be made using an approach that deals better with interspecific variation within species groups (Chapter 15) than does the aggregated approach of Nichols et al. (1998a). Some analyses that can be conducted using the methods of Chapter 15 are not possible with the aggregation-based approach to estimation. For example, in the work of Doherty et al. (2003b) on extinction probabilities of bird species that do and do not exhibit sexual dichromatism, the authors would have preferred an analysis that dealt with phylogenetic relationships among the species while addressing the question of primary interest. Phylogeny could not be dealt with using the estimation approach based on aggregation, but it can be dealt with using occupancy-based estimates of extinction rate for each species.

Other multi-species questions not amenable to an approach based on aggregating species may be addressable using multi-species occupancy data collected over time. For example, the idea of species *guilds* is used widely in ecological work. One way of testing the guild concept, or perhaps even defining 'guild', might be to ask whether local rates of extinction and/or colonization seem to exhibit parallel temporal variation for species in a group. For example, it would be possible to fit a model to multi-species occurrence data including only the main effects for the 'species' and 'time' factors (i.e., an additive 'species + time' model) for species within an *a priori*-defined group. Good fit of such a model might provide support for the hypothesis that the species were members of a guild, whereas rejection of this model in favor of a model that includes the interaction of 'species' and 'time' (i.e., a full 'species × time' interaction model where the temporal effects are different for each species) would lead to rejection of the guild hypothesis and to the conclusion that the species were not responding to environmental variation in the same manner.

Similar to studies on species habitat relationships, many studies of species co-occurrence are based on static relationships using a temporal 'snapshot' and assuming the system is near equilibrium. Relationships between species occurrence and habitat covariates may be misleading if the system is experiencing transient dynamics (Yackulic et al., 2015). Utilizing dynamic occupancy models, where local extinction, colonization and detection probabilities are modeled as a function of the presence of another species (predator, competitor, or facilitative species) and/or habitat covariates, yields a more robust understanding of factors influencing species distributions. Additionally, such models can be used to predict changes in species distributions under future habitat, climate, or management conditions. Several authors have used dynamic co-occurrence models to explore the influence of invasive species (Yackulic et al., 2014; Dugger et al., 2015), predators (Miller et al., 2012a), or competitors (Robinson et al., 2014; Haynes et al., 2014) on species distributions, while accounting for local habitat variables. Experimental species removal studies analyzed using two-species co-occurrence models can provide strong inferences about species interactions (Diller et al., 2016). Several authors have integrated these models into management and conservation decision making (Zipkin et al., 2012; Sauer et al., 2013a; Chandler et al., 2013).

Long-term studies of communities over time have utilized dynamic multi-species models (Dorazio et al., 2010, Chapter 15) to explore how communities change due to land use and land cover changes (e.g., Goijman et al., 2015), climate (Tingley and Beissinger, 2013), and catastrophic events (Russell et al., 2015). Ruiz-Gutiérrez and Zipkin (2011) found that failing to account for detection probability in these studies can severely bias occupancy dynamics, producing misleading patterns of species persistence and colonization and misclassification of guilds. An advantage of these multi-species occupancy models is the ability to borrow information from data-rich species to provide better occupancy estimates for rare species and thus investigate factors influencing species assemblages, but this requires grouping species in a meaningful way. Pacifici et al. (2014) found that community- and species-level inferences were heavily influenced by the choice of grouping criteria and recommended that model selection be used to compare different classification approaches.

The models of Chapters 14 and 15 were developed specifically for multi-species occupancy surveys, where interest is on species-specific occupancies and/or evidence of spatial segregation or aggregation among species. The models permit inferences about these issues in the presence of detection probabilities that are < 1 and that may vary by species and depend on presence and even detection of other species. These chapters also consider multi-species occupancy data collected over seasons or years. For such data, attention may shift from single-season co-occurrence patterns to whether the vital rates of occupancy dynamics, local extinction and colonization, of one species depend on the presence of another (Chapter 14). Detection probabilities less than one will produce bias in estimates of local rates of extinction and colonization and community metrics based on presence–absence data. We thus recommend the multiple season models of Chapters 14 and 15 as a means of drawing inferences about the colonization and extinction processes for multiple species. These models should be useful for the more complicated cases in which multiple species are believed to interact (Chapter 14), as well as for the simpler case of no interspecific interactions (Chapter 15).

2.6   PALEOBIOLOGY

Paleobiologists are interested in many of the same questions that are addressed by animal ecologists, with the primary distinction being the focal time periods; the present and recent history for ecologists and the more distant past for paleobiologists. Number of extant taxa throughout geologic time and associated rates of extinction and colonization are of interest to paleobiologists, just as species richness and current rates of extinction and colonization interest community ecologists. Similarly, distribution patterns of taxa extant at different periods of time are of interest to paleobiologists, especially locations of taxonomic origin and subsequent spatial spread from these locations (e.g., Antón and Snodgrass, 2012). Paleobiologists are also interested in ecological correlates (e.g., habitat variables, competitor taxa) of distribution patterns, as are ecologists (Gray and Elliott, 2009). Thus, the parameters and relationships that are the focus of this book are relevant in a paleobiological setting, just as in ecology.

Paleobiologists face detection problems that are similar in many respects to those of field ecologists surveying extant taxa. Just as the field ecologist may 'miss' species that are present during survey efforts, the paleobiologist may not detect identifiable fossil representatives of all taxa present at the time stratum represented by the sampled material (e.g., Sepkoski Jr., 1975; Foote and Raup, 1996; Foote, 2001; Peters and Ausich, 2008; Alroy, 2010). Fossilization potential, and hence detection probabilities, are thought to vary across time and space (Brett, 1998), lending importance to dealing with detection issues. This problem of nondetection can result in biased estimates of quantities such as number of taxa extant at any time, and global probabilities of extinction and colonization. This recognition led some to recommend use of capture–recapture and occupancy approaches to inference when using data from the fossil record to estimate these quantities (e.g., see Nichols and Pollock, 1983; Connolly and Miller, 2001b; Liow and Nichols, 2010). Despite the natural application of capture–recapture and occupancy thinking to paleobiological inferences, uses of these approaches have been rare.

Perhaps a major reason for this failure to use capture–recapture and occupancy thinking when analyzing fossil data lies in the unfamiliarity of these approaches to paleobiologists and resulting misunderstandings. For example, consider the following recommendation against the use of capture–recapture models: "…capture–recapture methods address the sampling problem by trying to estimate turnover rates and raw sampling probabilities all at once using the same basic information. Doing so means that error in the sampling estimates biases the turnover rate estimates and vice versa. By contrast, my advice is to clean up the data with sampling standardization first and then compute diversity and turnover rate estimates using different methods" (Alroy, 2010, p. 76). Of course any error in dealing with sampling probabilities can produce bias whether these errors arise in independent or joint analyses. More importantly, these statements are antithetical to the current trend in ecological statistics to develop hierarchical models that deal simultaneously with both observation and ecological processes (e.g., Royle and Dorazio, 2008; Kéry and Royle, 2016; Chapter 3). Regardless of whether the data on observation probabilities are independent of those used to estimate biological process parameters, there are still substantial advantages (e.g., efficiency, proper incorporation of variance components) to combining the data sources into a single likelihood or model. This kind of thinking underlies the increasing popularity of integrated population models (e.g., Besbeas et al., 2002; Schaub and Abadi, 2011), as well as recent recommendations for dealing with false positives in occupancy modeling (Chambert et al., 2015a; Chapter 6).

The potential utility of occupancy modeling for paleobiological questions was explicitly noted by Liow and Nichols (2010) and then described and advocated by Liow (2013). Liow (2013) provided example single- and multiple-season occupancy analyses using brachiopod taxa from the Paleozoic. She relied on spatial replication with multiple samples from specific sites, as will be common for paleobiological applications. In addition to providing estimates of occupancy and local colonization, Liow (2013) emphasized the tendency of naive inference methods commonly used in paleobiology to substantially underestimate the variances and uncertainty associated with parameter estimates. We expect occupancy modeling to become common in paleobiological analyses, and the only question is how long it will take for this to occur.

2.7 DISEASE DYNAMICS

The distribution of disease and disease dynamics can be viewed as special cases of geographic range and species co-occurrence (host–pathogen interactions). Vaccinations recommended for humans traveling to various parts of the world are based on range maps of disease occurrence. Disease dynamics are frequently of great interest to epidemiologists, especially in the case of fast-spreading diseases such as West Nile Virus (Marra et al., 2004). Epidemiological models for disease dynamics have been developed to predict the spread of disease organisms across host organisms and, more generally, across space (e.g., Bailey, 1975; Anderson and May, 1991; Elliott et al., 2001). Despite the importance of such models, their use in the analysis and management of disease in wildlife populations has been limited, due, in part, to imperfect observation of the disease state for a given individual, population, or spatial sub-unit (McClintock et al., 2010c; Cooch et al., 2012), a problem also faced by human epidemiologists.

Wildlife disease ecology has gained considerable attention due to the emergence, or re-emergence, of pathogens capable of transitioning among wild, domestic, and/or human populations (Dobson and Foufopoulos, 2001; Hudson et al., 2002). During the last century, about 60% of all infectious diseases and 72% of recent emerging infectious diseases resulted from pathogens of wildlife origin (Jones et al., 2008), and these pathogens pose a substantial threat to human health and global biodiversity (Daszak et al., 2000). Wildlife disease studies aimed at understanding the effects of disease on wild populations, or pathogen prevalence and dynamics, often categorize sample units (individuals, populations, or patches of habitat) into states (e.g., susceptible, infected, post-infected). For example, Senar and Conroy (2004) and Jennelle et al. (2007) considered a simple two-state disease system where individuals were classified as either 'infected', or 'not infected'. Estimates of prevalence, defined as the proportion of individuals in a target population that are infected, and inference about disease dynamics can be severely biased if state-specific detection probabilities are ignored (Senar and Conroy, 2004; Jennelle et al., 2007; Cooch et al., 2012). Capture–recapture approaches have been utilized to assess the impact of pathogens on host demographics and estimate disease transmission in scenarios involving one or a few populations, where individuals can be marked or recognized (e.g., Senar and Conroy, 2004; Pilliod et al., 2010; Conn et al., 2012; Cooch et al., 2012). Likewise, Kendall (2009) and McClintock et al. (2010c) described the use of occupancy modeling to infer disease dynamics over space and time. If each observed individual is considered an independent site, and occupancy is defined as the presence of disease (or pathogen), then estimates of occupancy are analogous to disease/pathogen prevalence and estimates of colonization and extinction probabilities are similar to infection, or reinfection, and 'recovery' probabilities.

Additionally, occupancy approaches can be applied to different, hierarchical scales to explore factors related to the pathogen presence in multiple host populations or within defined spatial units (Kendall, 2009; McClintock et al., 2010c). Understanding factors that influence a pathogen's distribution and determining under what conditions it is transmitted among seemingly isolated host populations is a major theme in disease ecology and geographic epidemiology. In these cases, imperfect detection is possible for both the host and pathogen species, inhibiting the ability to directly observe processes of interest. This can result in misclassifying a host population or sample unit as unoccupied by the pathogen, when the pathogen is present. Part of what makes reliable inference about disease dynamics so difficult is the variety of ways that uncertainty can occur in the system, from the selecting of sample units, to the sampling of host and pathogen species, to the handling and diagnosis of the collected samples (Fig. 2.6, from McClintock et al., 2010c). Using the example of highly pathogenic Asian strain H5N1 avian influenza virus (HPAIV) in waterfowl populations that inhabit wetlands, Fig. 2.6 depicts numerous sampling processes, each with a potential for nondetection, and some sampling processes may yield false-positive detections. There is a wide array of methods used to determine pathogen or disease presence or exposure, including direct observation (e.g., Jennelle et al., 2007) and laboratory assays that utilize various sampled material (e.g., water, serum, blood, etc.). These methods can result in false-negative and false-positive detections. Quantification of false positive and false negative error rates (specificity and sensitivity) is not uncommon in the disease literature (e.g., Carey et al., 2006), leading to a variety of statistical methods that attempt to adjust for test accuracy (reviewed in Enøe et al., 2000). The occupancy models discussed in this book deal explicitly with detection and misclassification (or state uncertainty; Chapters 6 and 10) to provide unbiased estimates of disease effects and pathogen prevalence and dynamics across space and time.

Occupancy modeling approaches have started to appear in the wildlife disease literature. Bailey et al. (2014) emphasized the importance of study objectives and design, and corresponding model assumptions, when applying occupancy models to disease systems. These authors illustrate how different disease-related objectives result in dramatically different study designs, with applications of occupancy models to detection–nondetection data collected at the individual, population, or site levels.

The most common occupancy application involves estimating pathogen prevalence via single-season models (Chapter 4) and investigating variation in detection among different laboratory methods or individual host characteristics (e.g. age, sex). This approach has been used to estimate the prevalence of several pathogens including parasites in fish (e.g., Thompson, 2007, whirling disease) and birds (e.g., Lachish et al., 2012, avian malaria), amphibian fungal pathogens (e.g., Miller et al., 2012b; Schmidt et al., 2013, chytrid fungus), and bacteria in disease vectors (e.g., Gómez-Díaz et al., 2010, Lyme disease). Likewise, Elmore and colleagues used single-season occupancy models to evaluate detection differences among serology assays while estimating *Toxoplasma gondii* exposure

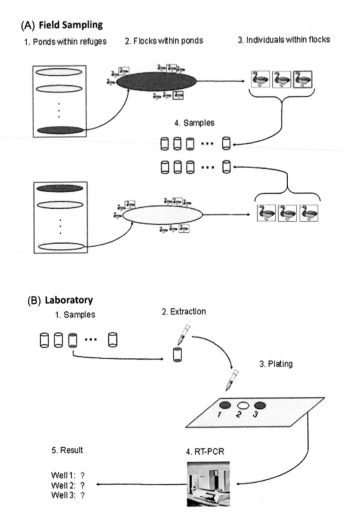

(A) Field Sampling

1. Ponds within refuges   2. Flocks within ponds   3. Individuals within flocks

4. Samples

(B) Laboratory

1. Samples   2. Extraction

3. Plating

5. Result   4. RT-PCR

Well 1: ?
Well 2: ?
Well 3: ?

FIGURE 2.6 Conceptualization of the myriad ways in which uncertainty can emerge in wildlife disease ecology (e.g., avian influenza in waterfowl populations), from the spatio-temporal allocation of field sampling effort (A) to laboratory practices (B). Red indicates infected samples and sample units. Whether or not a sample is infected, false negative or false positive test results can conceivably occur. *Source: McClintock et al. (2010c).*

(i.e., antibody prevalence) in arctic foxes (*Vulpes lagapus*; Elmore et al., 2016) and migratory geese (Elmore et al., 2014). Variation in pathogen detection is often associated with infection intensity or pathogen load (Gómez-Díaz et al., 2010; Miller et al., 2012b; Lachish et al., 2012). When quantitative measures of these metrics are not available, occupancy models presented in Chapter 7 can be used to account for intensity-induced heterogeneity in pathogen detection (e.g.,

Lachish et al., 2012). Two critical assumptions of these occupancy applications are that captured individuals are a random subset of the population of interest and that there are no false-positive detections. Elmore et al. (2014) utilized multistate models (Chapter 5) to accommodate ambiguous test results, but models in Chapters 6 and 10 should also be useful if false-positive test results are suspected. When individuals are not captured at random from the population (e.g., diseased individuals are more, or less, likely to be captured than non-diseased individuals; Jennelle et al., 2007), multi-state capture–recapture models and occupancy models could be combined to make improved population-level inferences.

Several recent studies have focused on the prevalence or occurrence of disease vectors and their dynamics over time. Eads et al. (2013, 2015) studied the occupancy dynamics of ticks on black-tailed prairie dogs (*Cynomys ludovicianus*) to better understand plague outbreaks that influence both prairie dog and black-footed ferret (*Mustela nigripes*) populations. In the Neotropics, Abad-Franch and colleagues have used occupancy models to estimate the occurrence of blood-sucking triatomine bugs ('kissing bugs', the most important vector of Chagas disease) among palm trees (e.g., Abad-Franch et al., 2010, 2015) and to better understand the dynamics of mosquito-borne vectors responsible for transmitting dengue virus (Padilla-Torres et al., 2013).

In most of the examples given above, the sample unit of interest is an individual organism (e.g., fish, bird, mammal, or plant), but sometimes investigators are interested in a pathogen's distribution among host populations. This approach was employed to investigate factors influencing the occurrence of the fungal pathogen *Batrachochytrium dendrobatidis* among amphibian populations in the Pacific northwest (Adams et al., 2010), Arizona (Schmidt et al., 2013), and elsewhere in the United States (Chestnut et al., 2014).

While many disease systems are hierarchical in nature (Fig. 2.7, from McClintock et al., 2010c), few investigators have utilized multi-scale occupancy models (but see Schmidt et al., 2013; Elmore et al., 2014). We suspect that occupancy models presented in Chapters 5 and 9 will be increasingly popular as detection–nondetection data are collected at multiple levels (e.g., host and pathogen). As both false-negative and false-positive results are possible for many disease diagnostic tests, we believe that models accounting for both types of errors (Chapters 6 and 10) will be important in future studies of disease systems. Finally, host dynamics are undoubtedly linked to pathogen dynamics, where detection of the pathogen may depend on the presence and/or detection of the target host species. In these cases, investigators may choose to use the co-occurrence or multi-species models (Chapters 14 and 15) to simultaneously investigate factors influencing host–pathogen dynamics or explore multi-species

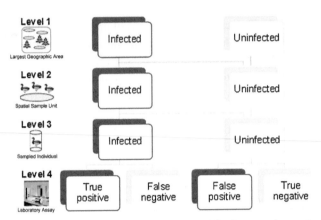

FIGURE 2.7 Hierarchical formulation of uncertainty in wildlife disease ecology under four general themes. Conditional on the disease state at the upper levels, many different sample paths can lead to a false negative or false positive result upon analysis at Level 4. Spatial subunits may be added or removed within Level 2 of the hierarchy. *Source: McClintock et al. (2010c).*

concepts, such as the dilution effect (e.g., Keesing et al., 2010), while accounting for nondetection of hosts and pathogens.

2.8   NON-ECOLOGICAL APPLICATIONS

Occupancy modeling was developed for surveys of animal populations and communities carried out at specific units using methods that do not always detect species that are present. However, these key survey features of possible nondetection at a fixed number of units are shared by other kinds of inference problems, some of which are ecological in nature and some of which are not. We will discuss some possible applications of occupancy models here, but many have not yet been implemented anywhere (to the best of our knowledge) and are thus without published examples. In addition, we do not claim this list of potential applications to be exhaustive. Indeed, it is our hope that these examples will promote further thinking of other less conventional uses.

Goswami et al. (2015) used an occupancy modeling approach to study conflicts between people and the Asian elephant (*Elephas maximus*) in the Garo Hills, India. The authors investigated potential drivers of this human–wildlife conflict, primarily crop depredation, recognizing that nondetection would be a likely occurrence in this rural landscape. Data on incidents of conflict came from a trained team of 17 informants who surveyed communal lands owned and managed by residents of 49 villages across a 172 km² study area subdivided into 4 km² grid cells. Months within a season were the basis for temporal replication and were recorded as either having at least one incident of conflict in each cell or not. Multiseason occupancy models incorporated hypotheses about detection probabilities and both spatial and seasonal dynamics of conflict. Evidence indicated that detection probabilities were a function of location remoteness (covariates were distance to road and elevation), whereas conflict was a function of seasonal rainfall and distance from forests (Goswami et al., 2015).

Typically, illegal activities are difficult to detect by design, such that inferences about them must deal with the issue of nondetection. Barber-Meyer (2010) noted the potential utility of occupancy modeling for dealing with the clandestine nature of illegal wildlife trade. She wrote generally about the potential of occupancy modeling for use in addressing questions about the availability of single species in illegal trade, co-occurrence patterns for pairs of illegally traded species, and disease occurrence in the live trade of animals. She conducted a simulation study of a wildlife trade market survey designed to estimate the proportion of towns engaged in a specific trade and concluded that occupancy modeling was superior to approaches to data analysis that ignored possible nondetection (Barber-Meyer, 2010).

An ongoing project developed by Moore et al. (2017) is focusing on illegal anthropogenic threats to wildlife in Nyungwe National Park, Rwanda. Various indicators can be used to assess human threats to a site in the park, but nondetection is always possible. Thus, the project used dynamic multistate occupancy models (Chapter 9) to estimate site-specific probabilities of different levels of threat as a function of such variables as distance to nearest roads and settlements, and ranger patrol activities. These investigators are also considering the use of single-season multistate occupancy modeling (Chapter 5) to investigate patterns of co-occurrence for illegal threats and different park mammal species, and possibly multiseason models to investigate relationships between illegal threats and local extinction and colonization probabilities of park mammals.

There is increasing recognition in a variety of disciplines extending well beyond wildlife and conservation of the pervasiveness of nondetection and the need to deal with it in inference methods. This recognition has led to increased consideration of methods developed in wildlife and animal ecology for dealing with this issue in other subject areas. For example, the United States Census now uses capture–recapture approaches to deal with the 'undercount' (Fienberg, 1992), and capture–recapture thinking has become widespread in certain types of human epidemiological research (Chao et al., 2001). One of us (JDN) attended two multidisciplinary workshops that focused on the general problem of nondetection, and below we discuss two potential uses of occupancy modeling that generated substantial interest in these workshops.

We have already outlined a number of potential uses of occupancy modeling in disease and epidemiological research. However, one example from the workshops is sufficiently different that we instead include it in this section.

Epidemiologists focusing on human diseases often base inferences about spatio-temporal dynamics of disease on reports from physicians, clinics and hospitals. However, accessibility to physicians, clinics and hospitals may be very limited in rural areas and in less developed countries, and diagnostic skills may also vary, such that the probability of a disease being detected and reported may be substantially < 1 and quite variable. Multiseason occupancy modeling might be useful in drawing inferences about dynamics in such cases. For example, for any specific location, a list of physicians and hospitals within some specified radius could be viewed as the replicate sampling 'occasions', with 1's and 0's indicating detection (report of the disease) or nondetection of the focal disease during some relevant time period (e.g., week). Detection probabilities would be modeled as functions of such covariates as distance between focal location and medical facility, and nature of the 'detector' (e.g., a village doctor versus a city hospital). 'Removal' models (Chapter 11) might be useful in some cases, as once an emerging disease is detected by a physician or facility, it is frequently more likely to be recognized in subsequent diagnoses. Modeling spatial auto-correlation in occupancy (Chapters 6 and 10) is also likely to be important, with the probability of an emerging disease entering an area likely dependent on the incidence in neighboring locations.

Yet another workshop topic was the reporting of incidents of political 'unrest', whereas the precise definition of this term (e.g., protests, riots producing at least one death) would depend on the objectives of the investigators or surveys. Variable detection probabilities < 1 are again expected, especially for remote locations in less developed areas. Different online news agencies with widely distributed reporting networks in certain regions could be used as the survey 'occasions', and occupancy models could be employed to estimate the prevalence of such incidents throughout a focal region. Reporting probabilities would be modeled as functions of covariates that included remoteness of the focal location and distance to nearest base of the news agency, with removal models possible if nonindependence of reports is suspected. Spatio-temporal dynamics and the possible influences of unrest in neighboring locations could be incorporated by accounting for spatial autocorrelation in occupancy (Chapter 6) and occupancy dynamics (Chapter 10).

Border control and biosecurity are other areas where we believe occupancy models could be applied. The intent of such operations is typically to intercept prohibited or undesirable items present in shipments or other packages at inspection stations. Hence the probability of occupancy could be regarded as the probability that a unit of interest (e.g., a container, suitcase, or package) contains a contraband item. Given the clandestine nature of illegal smuggling operations, it would be reasonable to expect that the contraband is not always detected in a unit of interest when searched (i.e., detection given presence < 1) and that the detection probability may vary by the quantity of contraband within the unit, and also due to other factors. In a biosecurity setting undesirable organisms may be small relative to the size of the shipment, thus difficult to detect (e.g., fruit-fly larvae in fruit shipments). Given the availability of data on the outcome of multiple inspections or screenings of units (e.g., X-ray and visual inspection), we foresee multiple options for the application of the modeling techniques discussed in this book as a profiling tool (i.e. to identify what types of packages are more likely to contain contraband), for comparing the efficacy of different inspection methods, or for estimating what fraction of units containing contraband do not get intercepted. Covariate modeling of occupancy and detection probabilities (e.g., Chapter 4) is likely to be very important in such applications to account for variation due to package size and type, point of origin, transit locations, etc. The approaches that allow for heterogeneous detection probabilities (Chapter 7) could be particularly useful to account for variation in the amount of contraband present within a unit, which would be equivalent to detection heterogeneity of the species due to local abundance (Royle and Nichols, 2003). When interest is in the presence of undesirable items at different scales (e.g., probability of presence in a vessel, and probability of presence within a shipping container on vessel), the multi-scale occupancy model of Chapter 6 could be used. By regarding different types of contraband as different species, species co-occurrence models could be used to investigate whether some types of contraband are more likely to occur together in the same shipments (e.g., illicit drugs and illegal immigrants) and possible predictors of that co-occurrence. Community-level models (Chapter 15) could be used to consider a suite of contraband items (e.g., different types of illicit drugs) simultaneously.

The detection of errors in computer software is uncertain, and capture–recapture thinking has long been used to estimate number of errors remaining in software (e.g., Chao and Yang, 1993). Approaches to inference can be discrete, based on different programmers looking at code, or it can be continuous, with detection of an error modeled as a function of time expended debugging. In addition to number of remaining errors, in some circumstances with very large programs, it may be useful to divide the software into subunits (e.g., subroutines) and to search for errors in a sample of these. With an occupancy framework, replicate looks at a sample of subunits (e.g., by two different programmers) would permit inference about the proportion of subunits expected to have errors, and, more importantly, covariates (sub-unit length or complexity) associated with the probability of containing at least one error.

The basic issue of detecting problems can be extended well beyond software and indeed is a very general application area for occupancy modeling. We note just one additional example, with the understanding that many more applications exist. The detection of structural damage is an important topic in engineering.

Sometimes, structural damage (e.g., to a building or a ship) is readily detectable, for example, via highly visible cracks in a material. However, many other forms of damage are not readily visible, and nondetection is very likely without application of specialized detection methods (Nichols and Murphy, 2016). It may be reasonable to view very large structures (skyscraper buildings, large ships, rockets, missiles) as sets of segments or units. A reasonable approach to structural monitoring may be to apply expensive damage detection methods to a subset of such units in order to estimate the proportion of units experiencing damage and perhaps effects of unit-specific covariates on the probability of damage.

The final application area that we will discuss entails a blurring of the definition of occupancy models. These models were initially developed for dealing with nondetection in surveys conducted on some finite and predefined number of sample units. One of us (JAR) recognized that an occupancy approach could also be taken without a fixed set of existing sample units. Royle et al. (2007a) proposed the approach of data augmentation, perhaps best exemplified by application of occupancy ideas to capture–recapture thinking. Royle et al. (2007a) noted that abundance estimation via closed-population capture–recapture models (see Royle and Dorazio, 2008; Royle et al., 2014), for example, entails counting the number of animals caught and then adding to this number the estimated number of animals present, yet not caught. They proposed augmenting existing capture–recapture data (encounter histories of detected animals) with some number of all-0 encounter histories, indicating animals never detected. If this number of extra all-0 histories is sufficiently large (larger than abundance could possibly be) then occupancy modeling can be used to estimate the proportion of these all-0 histories that represent animals that are in the population, yet were not encountered. This proportion can be multiplied by the number of all-0 histories to estimate the number of animals not detected, and hence population size. The data augmentation approach can be viewed as an application of occupancy thinking to capture–recapture problems and has proven to be very useful (e.g., Royle et al., 2014; Kéry and Royle, 2016).

In summary, the ingredients for application of basic occupancy modeling are a set of units, a focal event that categorizes units into at least two categories (e.g., presence or absence of the event), a survey method that does not always detect the event when present at a unit, and an interest in the proportion of units at which the event is present. In the work that motivated the development of occupancy modeling, the units are geographic locations and the focal event is presence of a particular species. As evidenced by the above examples, however, these features characterize a variety of problems, leading to a number of potential and realized application areas.

2.9 DISCUSSION

The purpose of this chapter has been to review a number of ecological questions that are typically addressed using occupancy or detection–nondetection data. Based on this abbreviated review, it is clear that occupancy data are widely used. For some topics (e.g., use of occupancy in large-scale monitoring), occupancy may be viewed as a surrogate for abundance. However, for other important topics including studies of range size, species–habitat relationships, metapopulation and disease dynamics, interspecific interactions and interspecific variation in extinction and colonization, occupancy is the state variable of primary interest. For many questions, the dynamics of this state variable, and the vital rates that underlie these dynamics (patch-level probabilities of extinction and colonization), are the subjects of primary interest.

It is also clear from this review that the vast majority of studies conducted prior to the development of occupancy models included no mechanism for dealing with the likelihood that the species of interest goes undetected at some locations at which it is present. The review indicates that many workers are aware of this problem, but often assume the problem away or hope that its likelihood is sufficiently small as to cause few problems. Some workers try to minimize the probability of missing a species by visiting units multiple times. What is somewhat surprising is that, conditional on the assumption that nondetection is equivalent to absence, substantial statistical rigor has been applied to the analysis of occupancy data for both single-species and multi-species problems. Thus, it cannot be claimed that those who deal with occupancy data simply shy away from statistical methods. Yet, until the early 2000's collectors and analysts of occupancy data had not followed the approach of scientists studying animal population ecology at the level of the individual animal and incorporated detection parameters directly into their inference methods.

Consider the following statements from a study of the influences of incomplete data sets (meaning missed species presence) on species–area relationships and nested subset analyses:

> *Most large data sets on species distributions and community composition will be incomplete. ... The question becomes, then, how complete the data must be to change the qualitative results and the inferences that are drawn from them.*
>
> Kodric-Brown and Brown (1993, p. 741)

The authors then recommend "additional field studies" to "help to assess the completeness of the data and to detect and correct for certain kinds of bias" (Kodric-Brown and Brown, 1993, p. 741). Our recommendation throughout this book will be to instead incorporate the collection of additional information (typically replicate surveys of units within a relatively short timeframe) directly into the sampling protocol and to then utilize models that explicitly incorporate detection probabilities. We have given several examples in the previous sections where investigators have employed occupancy models, and highlighted sections of the book we believe will be useful in future studies.

In summary, the approaches for estimation of occupancy from detection–nondetection sampling at a single season (Chapters 4–7) will be useful for most questions discussed in the present chapter. Studies of range at a fixed point in time should be based on occupancy estimates, rather than on raw detection data. Similarly, studies of occupancy dynamics and long-term monitoring studies should be based on the models of Chapters 8–10 that explicitly incorporate detection probabilities that may change over time. Efforts to model occupancy as a function of habitat covariates, are included in every chapter, as these methods separate effects of habitat on occupancy and detection probability.

Methods of Chapters 14 and 15 will be especially relevant to attempts to deduce inferences about process from occupancy patterns of multiple species in the face of detection probabilities that may vary among species and even depend on presence of certain other species. These methods should be useful in addressing various questions about occupancy dynamics for multiple species in the case of independent dynamics (Chapter 15) and dependent dynamics where the vital rates of one species are a function of presence of another species (Chapter 14). Thus, our intention with this book is to provide a beginning toolbox for addressing the ecological questions of this chapter in ways that deal adequately with the sampling reality that species are not always detected in sample units when present.

Chapter 3

Fundamental Principals of Statistical Inference

In this chapter we give an overview of the important statistical concepts that we will be using in this book. Our intent here is to provide an introduction to many basic ideas in terms that those less familiar (or comfortable) with statistics can understand. We have also included a short appendix at the end of the book to provide details on some topics that are secondary to the main thrust of this chapter, but topics that some readers may not be familiar with (e.g., Σ and Π notation for sums and products, basic matrix algebra, and differentiation and integration). In the following pages there are a number of equations that may appear daunting at first glance; however we provide ample explanation so that readers may interpret what the equations represent. After all, equations are simply a mathematical short-hand for conveying sets of ideas or concepts. Once one learns to read and interpret the mathematical language, understanding of the information being conveyed follows naturally (just like learning any other language such as French, Japanese, or Spanish).

We strongly encourage readers to take the time to fully understand the concepts discussed in this chapter. It is our experience that those with a sound grasp of the underlying statistical methods and assumptions make the best use of modeling procedures such as those included in this book. They also tend to be the people who are less likely to misuse the methods or attempt to apply them incorrectly. Sound use of statistical inference is at the heart of successful analyses of all wildlife studies. We are usually basing our inferences on a small sample from the population and obtaining only uncertain estimates of the parameters of interest. Statistical inference can be viewed as a rigorous method of studying and drawing conclusions about the quantities of interest in the face of that uncertainty, which is due to sampling variability (i.e., by not being able to census the entire population).

We begin with some background and definitions of key statistical concepts before discussing the desirable properties of parameter estimators, including discussion of methods to assess the bias and precision of estimators using simulation and related computer intensive methods. We then consider the important role of the likelihood in parameter estimation with particular emphasis on the

Occupancy Estimation and Modeling. http://dx.doi.org/10.1016/B978-0-12-407197-1.00004-1

use of maximum likelihood estimation. An overview is provided of the underlying theory of Bayesian statistical inference and computational methods that are commonly used in such analyses. We also discuss methods of including the effects of predictor variables, or potential covariates, on parameters of interest using appropriate link functions. Our attention is then turned to hypothesis testing and in particular likelihood ratio tests and goodness of fit tests. We conclude this chapter by discussing both frequentist and Bayesian methods for comparing models, outside of a hypothesis testing framework. For ecologists with little statistical background, a good introduction to many of these standard methods is Chapter 4 of Williams et al. (2002). A much more detailed treatment suitable for ecologists with a very strong background in statistics is Casella and Berger (2002). For those that become overwhelmed with some of the contents of this chapter, we encourage you to use this chapter as a reference, and revisit the concepts explained here as you encounter them later in the book.

Bayesian methods have become widely used in wildlife and ecological studies since the late 1990's, primarily due to the advent of software such as WinBUGS/OpenBUGS and JAGS that provide a relatively user-friendly framework for implementing such approaches. Aside from being an alternative philosophical approach to frequentist methods of statistical inference such as maximum likelihood estimation, some modeling applications are much easier to implement using a Bayesian approach and the associated estimation tools. In this chapter we provide a brief introduction to Bayesian methods, and utilize them for some examples throughout the book. It should be appreciated that the underlying models which are detailed in this book are neither inherently Bayesian nor frequentist, they are simply descriptions of the system of interest. Bayesian and frequentist methods are just alternative approaches to statistical inference using such models.

When applying the methods detailed in this book to real studies, we will often have to consider a suite of statistical models (representing competing hypotheses about the system or population) and identify the 'better' models from among them. One approach for doing so is the use of information-theoretic methods, such as Akaike's Information Criterion (AIC; Akaike, 1973), which rank the models according to a selected metric. Oftentimes there will not be a single model that is clearly much better than the others being considered, which will lead to *model selection uncertainty*; uncertainty about whether other highly-ranked models should be disregarded in our inferences. An honest approach to the estimation and inferential procedure would be acknowledge that the model selection procedure itself introduces a source of uncertainty that should be accounted for rather than just using the top-ranked model. We discuss the use of *model averaging* as one approach for doing so. Following Burnham and Anderson (2002), we will emphasize the use of the AIC for assessing the weight of evidence for members of the model set for the purposes of model choice and model averaging. Generalizations to allow for small sample sizes (AIC_c) and overdispersion (QAIC) will also be considered. This should not be seen as wholesale endorsement of these methods but simply a reflection of our belief that they represent an improvement over previous approaches to model selection based upon statistical hypothesis tests (e.g., forward stepwise selection).

3.1 DEFINITIONS AND KEY CONCEPTS

Statistical inference always involves the study of a population of interest, by collecting data from a selection of sampling units that are drawn from the population using a sampling procedure. For example, a sampling unit might be a site or plot where we would like to establish the presence or absence of a plant or animal species, and the population would then be all of the potential units within the region of interest. One key parameter, in the context of this book, would be the probability a unit is occupied by the species (i.e., probability of species presence), which could be estimated by selecting a sample of units and surveying them. Some important definitions are now presented.

- *Population:* The population is the complete set of sampling units that we are interested in studying.
- *Sample:* The sample is the set of sampling units that are actually surveyed. It is chosen from the population using an appropriate probabilistic sampling scheme (e.g., simple random sampling, stratified random sampling, unequal probability sampling, etc.).
- *Parameter:* A parameter is a characteristic of the population that we would like to know about. For our earlier example, the parameter would be the probability a unit is occupied by the species. Generally, a population parameter is never known exactly. In our example, the symbol ψ may be used to represent the parameter that we refer to as the probability of occupancy. A parameter will often be associated with a *model* (as discussed in Chapter 1), although the model may not be explicitly stated.
- *Estimator of a Parameter:* As we never know the value of a parameter exactly, we have to use an estimator of the parameter based on a sample. An estimator is an equation or process applied to the collected data to produce an estimate of the parameter. In our notation, $\hat{\psi}$ may be used to represent an estimator of the parameter ψ. For example, if the individuals of the species are very easy to detect, the proportion of sampled units occupied by the species can be used as the estimator for the population probability of occupancy. Clearly ψ, the parameter, and $\hat{\psi}$, the estimator, are distinct, and use of the circumflex or 'hat' to represent an estimator has become standard notation.

In general, detection probability has to be considered and this simple estimator is unsatisfactory, but we defer this discussion to the next chapter.

- *Parameter Estimate:* When an estimator is applied to a specific data set, the resulting numerical value is a parameter estimate. The 'hat' notation is also used to denote an estimate. For example, if ten units were surveyed and six were occupied by the conspicuous species (the data) then applying the above estimator our estimate would be $\hat{\psi} = 0.6$.

Importantly, note that in the above definitions the population is the complete set of sampling units of interest, however they are to be defined (e.g., ponds, grid cells, etc.), and *not* the population of individuals of the target species within the area of interest. This is a fundamental distinction in occupancy modeling that is sometimes unappreciated.

3.1.1   Random Variables, Probability Distributions, and the Likelihood Function

A *random variable* is the outcome of a stochastic process (hence the term *random* variable), collected during the sampling. Our statistical inference is based upon these random variables and the underlying stochastic process that is assumed to have generated them, i.e., the underlying assumed model and associated parameters. To illustrate let us consider again our example of estimating the probability of occupancy for a species. Suppose you randomly selected five units from the set of units within the area of interest, and the following is the outcome of surveying the five units for the presence of the species:

$$01010$$

where 0 indicates the unit is unoccupied and 1 the unit is occupied. If we assume that all the units visited have perfect detection for the species, the probability of getting this particular sequence of occupied and unoccupied units is:

$$Pr(01010|\psi) = (1 - \psi)\psi(1 - \psi)\psi(1 - \psi)$$
$$= \psi^2(1 - \psi)^3 \tag{3.1}$$

where ψ is the probability a unit is occupied. In this instance, the exact sequence is being considered as the random variable. Alternatively, rather than being concerned with the sequence of occupied and unoccupied units we could summarize this information by defining our random variable (x) to be the number of occupied units. The probability of observing exactly two occupied units out of five follows the well-known binomial distribution:

$$Pr(x = 2|\psi) = \binom{5}{2}\psi^2(1 - \psi)^3 \tag{3.2}$$

with general form

$$Pr(x|\psi) = \binom{s}{x} \psi^x (1 - \psi)^{s-x} \qquad (3.3)$$

where s is the total number of units surveyed and

$$\binom{s}{x} = \frac{s!}{x!(s-x)!}$$

is the number of ways x 'successes' could be observed from s independent samples (known as the binomial coefficient). Note that the exclamation point (!) indicates a *factorial* which is the product of all integers up to and including the indicated value, e.g., $3! = 1 \times 2 \times 3 = 6$, with $0!$ defined to $= 1$. By only considering the number of occupied units in our sample, rather than the exact sequence of observations, the difference between Eqs. (3.1) and (3.2) is the binomial coefficient. Here it equals

$$\binom{5}{2} = \frac{5!}{2!3!} = \frac{120}{2 \times 6} = \frac{120}{12} = 10.$$

That is, by sampling five units there are 10 possible sequences of 1's and 0's with two 1's (e.g., 11000, 10100, 100100, ...). In Eqs. (3.1)–(3.3), the term '$|\psi$' on the left hand side simply denotes that the probability of observing x is conditional on (or depends upon) the value for ψ. To simplify notation we do not always include this term, but it is implied.

To summarize, we denote the discrete probability distribution of the observed random variable(s) (**x**; which may be a vector, as indicated by the bold font) as a function of the parameter(s) (θ) as $Pr(\mathbf{x}|\theta)$. An alternative, more general, notation for the probability distribution is $f(\mathbf{x}|\theta)$. Random variables may be discrete as in our example or continuous. Distributions may be univariate (relate to only a single random variable) or multivariate (relate to multiple random variables).

Another important discrete distribution is the multinomial distribution that is an extension of the binomial distribution to the case of more than two categories. If there are k categories then:

$$Pr(\mathbf{x}|\theta) = \frac{n!}{x_1! x_2! \dots x_k!} \prod_{i=1}^{k} \pi_i^{x_i}.$$

This distribution can be viewed as providing a probabilistic description of possible outcomes resulting from distributing n objects into k categories with x_i indicating the number of the n entities or objects that are found in category i, and π_i indicating the probability of an entity being in category i. The π_i can also be viewed as the expected proportion of the n entities that is found in category i.

While the value of a random variable is often observed during the data collection, this is not always the case, particularly in the context of this book's subject (i.e., with imperfect detection, species presence/absence is not observed directly). When the value of the random variable is unobserved or unknown it is referred to as a *latent* random variable, or just latent variable. The concept of a latent variable is very important as models are often defined in terms of latent variables in applications that involve imperfect detection, or other sources of ambiguity, that result in the true value of the random variables not being observed completely.

The likelihood function is a fundamental statistical concept used throughout this book. It is also a very simple concept. As mentioned above, the purpose of statistical inference is to draw conclusions about population/model parameters (θ) based upon the observed data or random variables ($\mathbf{x}$), and using the likelihood function is one approach to achieving that inference. Compared to the probability statement for a random variable(s) above, in a likelihood function the role of the data and parameters is reversed. Now interest lies in the values of θ given the observed data $\mathbf{x}$; we denote this as $L(\theta|\mathbf{x})$. Once the probability statement for a random variable has been determined, calculating the likelihood function is simply a matter of changing notation; with no additional mathematics required (i.e., $L(\theta|\mathbf{x}) \equiv f(\mathbf{x}|\theta)$). Hence, from a practical perspective the likelihood function can be regarded as equivalent to determining the probability of observing a particular set of data (although interpretations of the two concepts are very different). For example, in the above case where two out of five units were occupied by our species of interest, the likelihood function would be:

$$L(\psi|x = 2) = \binom{5}{2} \psi^2(1 - \psi)^3.$$

Furthermore, as the binomial coefficient does not include any of the parameters of interest (ψ in this case), it could be removed such that the likelihood could be simplified to:

$$L(\psi|x = 2) = \psi^2(1 - \psi)^3.$$

The removal of such constant terms will have no effect on the estimation of parameters. The likelihood function can be used to obtain estimates of the parameters of interest based upon the collected data using either maximum likelihood or Bayesian methods of inference.

TABLE 3.1 Example of determining the expected value for a fair six-sided dice where x denotes the possible values on each face and $f(x|\theta)$ is the probability of each value

x	$f(x|\theta)$	$xf(x|\theta)$
1	1/6	1/6
2	1/6	2/6
3	1/6	3/6
4	1/6	4/6
5	1/6	5/6
6	1/6	6/6
Sum	1	21/6

3.1.2 Expected Values and Variance

Before we can consider statistical properties of estimators we need to define expected values. The expected value of a random variable x can simply be viewed as its long-term average. More technically, it is a weighted average of the possible values of the random variable, where the weights are the probabilities of observing x, as given by the probability density function $f(x|\theta)$ (note here we are only concerned about the expected value of a single random variable). For a discrete-valued random variable, the expected value can be calculated as:

$$E(x) = \sum_{x} xf(x|\theta) = \mu,$$

where the summation is performed over all allowable values of x. For example, consider determining the expected value for a fair six-sided dice. Table 3.1 details the steps required to calculate that the expected value is $21/6 = 3.5$.

The expected value can also be expressed in a general analytical (i.e., non-numerical) form. In our occupancy example expressed in terms of the binomial distribution, the expected number of occupied units would be:

$$E(x) = \sum_{x=0}^{s} xf(x|\theta)$$

$$= \sum_{x=0}^{s} x \binom{s}{x} \psi^{x}(1-\psi)^{s-x}$$

$$= \sum_{x=1}^{s} \frac{xs(s-1)!}{x(x-1)!(s-x)!} \psi^{x}(1-\psi)^{s-x}$$

$$= s\psi \sum_{x=1}^{s} \frac{(s-1)!}{(x-1)!(s-x)!} \psi^{x-1}(1-\psi)^{s-x}$$

$$= s\psi \sum_{y=0}^{r} \binom{r}{y} \psi^{y}(1-\psi)^{r-y}$$

$$= s\psi.$$

The simplification is possible because the summation in the second to last line $\left(\sum_{y=0}^{r} \binom{r}{y} \psi^{y}(1-\psi)^{r-y} \right)$ is equal to 1. This is because our new random variable $y \ (= x - 1)$ can take the possible values of 0, 1, ..., $r = (s - 1)$, which is the range of values that the summation is performed over (i.e., the summation is conducted across all possible values for y), and the term being summed is the probability of observing y. Therefore the summation is simply adding together the probabilities of observing all possible values of y, which must equal 1 (i.e., y must have a value between 0 and r, inclusive). The result is that the expected number of surveyed units occupied by the species is the number of units surveyed times the probability the species occupies a unit (e.g., if $s = 5$ and $\psi = 0.6$, the expected number of occupied units would be $5 \times 0.6 = 3$).

Calculating the expected value for a continuous random variable is very similar, although now integration over all possible values of x is required rather than summation:

$$E(x) = \int_{x} xf(x|\theta) \, dx = \mu.$$

Note that integration can be viewed as the summation of infinitely many, infinitely small terms (see Appendix); in fact many computer algorithms for calculating integrals use such an approach.

The expected value of a function or transformation of a random variable, $g(x)$, is:

$$E(g(x)) = \int_{x} g(x) \, f(x|\theta) \, dx,$$

or an analogous equation for a discrete random variable. However, unless $g(x)$ is a linear transformation of x the expected value of the transformed variable is not equal to applying the transformation to the expected value of x, i.e., $E(g(x)) \neq g(E(x))$. For example, $E(3x) = 3\mu$ but $E(log(x)) \neq log(\mu)$.

The variance of a random variable is defined as:

$$Var(x) = E\left((x - E(x))^2\right)$$
$$= E\left((x - \mu)^2\right)$$

$$= \int_x (x - \mu)^2 f(x|\theta) \, dx$$
$$= \sigma^2.$$

For the binomial distribution in our above example:

$$Var(x) = \sum_{x=0}^{s} (x - \mu)^2 f(x|\theta) = s\psi(1 - \psi).$$

The variance thus provides a metric for the variation or spread of a random variable about its average or expectation. An alternative measure of variation is the standard deviation (σ), which is related to the variance by:

$$\sigma = \sqrt{Var(x)}.$$

The covariance between two random variables x and y is:

$$Cov(x, y) = E\left((x - E(x))(y - E(y))\right) = \int_y \int_x (x - \mu_x)(y - \mu_y) \, dx \, dy,$$

and the correlation between two random variables is:

$$\rho(x, y) = \frac{Cov(x, y)}{\sqrt{Var(x) \, Var(y)}}.$$

3.1.3 Introduction to Methods of Estimation

The methods of estimation used primarily in this book are maximum likelihood and Bayesian inference. Maximum likelihood estimators (MLEs) are the values for the parameters that maximize the likelihood function given the observed data. The properties of MLEs are derived from the view that the observed random variables (the data) are one realization from an infinitely many number of possible realizations from the population – this is the frequentist inference paradigm. We provide details of the likelihood estimation procedure in Section 3.2.

Bayesian parameter estimation has a different philosophical basis in which inference is achieved conditional on the particular data set that was observed. This is done using the *posterior* probability distribution of the parameters which is derived from the probability of observing the random variable, or data, given the parameter value(s) (which, recall, is equivalent to the likelihood function), in combination with the *prior* probability distribution for the parameters. We provide more details of this estimation approach in Section 3.3.

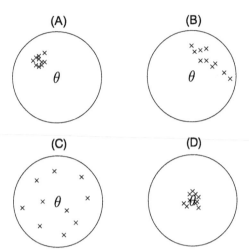

FIGURE 3.1   Illustration of estimator bias, precision and accuracy. Parameter estimates, $\hat{\theta}$ (denoted by ×), compared to true value, θ. (A) Precise and biased: inaccurate. (B) Imprecise and biased: inaccurate. (C) Imprecise and unbiased: inaccurate. (D) Precise and unbiased: accurate.

3.1.4   Properties of Point Estimators

The key properties of a point estimator are the *bias*; the *variance* (and *standard error*) that characterizes the *precision* of an estimator; and the *mean squared error* (and *root mean squared error*) that characterizes the overall *accuracy* of an estimator (Fig. 3.1). We will show that the accuracy of the estimator is a combination of both the bias and precision of the estimator. The formal statistical definitions are now presented.

Bias

$$Bias\left(\hat{\theta}\right) = E\left(\hat{\theta}\right) - \theta.$$

Expressed in words, this means the *bias* of a parameter estimator ($\hat{\theta}$) is the difference between the expected value of the estimator and the true value of the parameter. For our example, a natural estimator for the proportion of units occupied would be, $\hat{\psi} = x/s$, which is a linear function of the random variable x. Using the above results the expected value of $\hat{\psi}$ would be:

$$E\left(\hat{\psi}\right) = E\left(\frac{x}{s}\right) = \frac{s\psi}{s} = \psi,$$

and therefore

$$Bias\left(\hat{\psi}\right) = E\left(\hat{\psi}\right) - \psi = 0,$$

so we have an *unbiased* estimator. Although recall that here we have assumed the species is always detected when present at a unit, and we argue that assumption will not hold generally.

Precision (Variance and Standard Error)

$$Var\left(\hat{\theta}\right) = E\left(\left(\hat{\theta} - E\left(\hat{\theta}\right)\right)^{2}\right),$$

and in words this means the variance is the expected squared deviation of the estimator from its expected value. This is a measure of how 'close' an estimator is to its expectation. To obtain a measure that is on the original scale we define the square root of the variance of the estimator to be the standard error of the estimator.

$$SE\left(\hat{\theta}\right) = \sqrt{Var\left(\hat{\theta}\right)}.$$

For our example,

$$Var\left(\hat{\psi}\right) = E\left(\left(\frac{x}{s} - \psi\right)^{2}\right) = \frac{E\left((x - s\psi)^{2}\right)}{s^{2}}$$

$$= \frac{E\left((x - E(x))^{2}\right)}{s^{2}} = \frac{Var(x)}{s^{2}}$$

$$= \frac{s\psi(1 - \psi)}{s^{2}} = \frac{\psi(1 - \psi)}{s}$$

and therefore

$$SE\left(\hat{\psi}\right) = \sqrt{Var\left(\hat{\psi}\right)} = \sqrt{\frac{\psi(1 - \psi)}{s}}.$$

Note that the standard error decreases in proportion to $1/\sqrt{s}$, which is common in many statistical problems.

Accuracy (Mean Squared Error)

$$MSE\left(\hat{\theta}\right) = E\left(\left(\hat{\theta} - \theta\right)^{2}\right) = Var\left(\hat{\theta}\right) + Bias\left(\hat{\theta}\right)^{2}$$

and, in words this means the mean squared error denotes the expected squared deviation of the estimator from the parameter. This measure combines bias and precision in one overall measure of how 'close' an estimator is to the parameter it is estimating (note the subtle difference in the interpretation of the mean squared error and the variance). The root mean squared error just converts the measure to the original scale.

$$RMSE\left(\hat{\theta}\right) = \sqrt{MSE\left(\hat{\theta}\right)}.$$

For our example,

$$MSE\left(\hat{\psi}\right) = Var\left(\hat{\psi}\right) + Bias\left(\hat{\psi}\right)^2$$
$$= \frac{\psi(1-\psi)}{s} + 0 = \frac{\psi(1-\psi)}{s}.$$

Generally, for an unbiased estimator the mean squared error is equal to the variance of the estimator. Note that in Fig. 3.1, the situation illustrated in panel (D) will have the lowest MSE as it is both unbiased and displays less variation, and is therefore considered the most accurate.

3.1.5   Computer Intensive Methods

Monte Carlo or simulation methods can be very useful approaches to evaluating the properties of estimators, and they are widely used. In particular, they can be used to evaluate bias in estimators, either due to small samples or to the failure of assumptions.

Further, simulation may be very valuable for calculating or evaluating variances (standard errors) and confidence intervals for situations where there are questions about the validity of large sample approaches (Manly, 1997). Also, in some cases, the simulations may be used to include other sources of uncertainty in variance or interval estimation that were not included in the estimation algorithm.

Another related, widely used procedure is *bootstrapping* which is a computer intensive resampling procedure (Efron, 1979; Manly, 1997) which is often used to obtain estimates of the bias and precision of estimators. It can also be used to obtain interval estimators with better properties than those from profile likelihood or asymptotic normality approaches (see Section 3.2.4 below). *Nonparametric* bootstrapping involves direct resampling of the observations, while *parametric* bootstrapping uses the parametric model that has been fit to the data to generate resampled datasets. We direct those interested in reading more on computer intensive methods within a biological context to Manly (1997).

3.2   MAXIMUM LIKELIHOOD ESTIMATION METHODS

Here we provide greater detail on maximum likelihood estimation and related topics that follow directly from it.

3.2.1 Maximum Likelihood Estimators

A widely used method of estimation that produces estimators with desirable properties is the method of maximum likelihood. As discussed above, we specify the probability distribution of the observed data as a function of the parameters $Pr(\mathbf{x}|\boldsymbol{\theta})$, or more generally $f(\mathbf{x}|\boldsymbol{\theta})$, where $\mathbf{x}$ denotes the data and $\boldsymbol{\theta}$ denotes the parameters. This probability distribution is then viewed as a function of the parameters conditional on the data, and this is called the likelihood function $L(\boldsymbol{\theta}|\mathbf{x})$. We need to find the values of the parameters that maximize this function; that is, given the underlying model, for what values of the parameters are these data most likely? These are the maximum likelihood estimators (MLEs).

A very simple example of a likelihood function involves our example on occupancy of a conspicuous species. Recall that from a sample of five units, two were found to be occupied, and that the probability for two units being occupied was:

$$Pr(x = 2|\psi) = \binom{5}{2} \psi^2 (1 - \psi)^3$$

with the likelihood (a function of ψ) given the data

$$L(\psi|x = 2) = \binom{5}{2} \psi^2 (1 - \psi)^3$$

where ψ is the probability a unit is occupied.

Here it is easy to find the maximum by just plotting the likelihood function with respect to different values for ψ (Fig. 3.2). Regardless of whether we consider our random variable to be the sequence of occupied and unoccupied units or the summarized number of occupied units, i.e., the likelihood is based on Eq. (3.1) or (3.2), we would obtain the same MLE. The two functions are proportional, and the constant of proportionality (i.e., $\binom{5}{2}$) does not affect the resulting MLE. In practice, other methods must generally be used to obtain MLEs, as the likelihood may be very complex. One approach to finding MLEs for likelihoods that do not have many parameters is to use standard results from calculus. Set the partial derivatives of the likelihood with respect to each parameter equal to zero, and then solve the resulting equations (e.g., Williams et al., 2002, p. 47). Another general approach that is used for more complex likelihoods is to use computer software that may use a variety of numerical algorithms to find the maximum, which is know as *numerical optimization*.

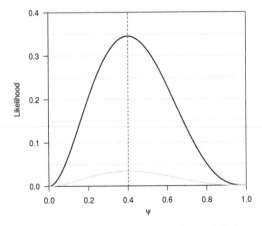

FIGURE 3.2 Plot of the likelihood functions where the random variable is considered to be either the number of occupied units (black line) or the specific sequence of occupied and unoccupied units (gray line). Note that in both cases the likelihood is maximized when $\psi = 0.4$.

3.2.2　Properties of Maximum Likelihood Estimators

The asymptotic (large sample) properties of MLEs (which were derived by R.A. Fisher in the 1920's) make the method of MLE very powerful. To quote Williams et al. (2002, p. 48):

> *The MLE $\hat{\theta}$ has an approximately normal distribution for large sample sizes. Furthermore, its distribution converges asymptotically to a normal distribution as sample sizes increase. Though the estimator may be biased, it is asymptotically unbiased in the sense that the expected value of $\hat{\theta}$ converges to the parameter θ as sample sizes increase. The variance of the estimator $\hat{\theta}$ is asymptotically minimum, in that $\hat{\theta}$ has the least variance of all unbiased estimators of θ when sample size is large.*

3.2.3　Variance, Covariance (and Standard Error) Estimation

Exact variances and covariances can be computed for some simple estimators (in our example, $Var\left(\hat{\psi}\right) = \psi(1 - \psi)/s$), however, most often in practice only large sample results are possible. From the asymptotic likelihood theory just discussed, the approximate variance–covariance matrix of a set of MLEs can be obtained from functions of the second partial derivatives evaluated at the MLEs (Williams et al., 2002, p. 735). Most computer software uses numerical algorithms based on this principle to estimate the variance–covariance matrix for all of the parameters being estimated from the likelihood.

Sometimes we will also need to obtain the MLE of a function of the original parameters. To obtain a point estimator here is straightforward, as the MLE of a function is just the function evaluated at the MLEs of the original parameters. For example, if $h(\theta)$ is the function, and $\hat{\theta}$ is the MLE of θ, then the MLE of $h(\theta)$ is $\widehat{h(\theta)} = h\left(\hat{\theta}\right)$. To obtain the estimated variance (if there is one parameter) or variance–covariance matrix of this estimated function (if, as more commonly, there is a whole set of parameters) is a little more difficult. Typically, the *delta method* (that takes advantage of a Taylor series expansion to linearize the function) is used to obtain large sample (approximate) variances and covariances (Seber, 1982, p. 8; Williams et al., 2002, p. 736). Here we present the key result first in one dimension, and then in matrix form, as these methods will be used later in the chapter. For one parameter we have:

$$Var\left(h\left(\hat{\theta}\right)\right) \approx \left(h'\left(\hat{\theta}\right)\right)^2 Var\left(\hat{\theta}\right)$$

where $h'\left(\hat{\theta}\right)$ is the derivative of $h(\theta)$ with respect to θ, evaluated at its MLE, $\hat{\theta}$. The matrix equivalent is:

$$Var\left(\mathbf{h}\left(\hat{\boldsymbol{\theta}}\right)\right) = \mathbf{h}'\left(\hat{\boldsymbol{\theta}}\right) \mathbf{V}\, \mathbf{h}'\left(\hat{\boldsymbol{\theta}}\right)^T,$$

where $\mathbf{V}$ is the estimated variance–covariance matrix for the vector of parameter estimates $\hat{\boldsymbol{\theta}}$ and $\mathbf{h}'\left(\hat{\boldsymbol{\theta}}\right)$ is the row vector of partial derivatives of $h(\boldsymbol{\theta})$ with respect to $\boldsymbol{\theta}$ evaluated at the MLEs, $\hat{\boldsymbol{\theta}}$. We illustrate the use of the delta method in Section 3.4.

3.2.4 Confidence Interval Estimators

Confidence intervals are one method of expressing the uncertainty in an estimator. While it is sometimes possible to obtain exact confidence intervals, in many cases we obtain approximate or large sample confidence intervals based either on the asymptotic normal distribution of the ML estimator or the profile likelihood method which is based on an appropriate log-likelihood ratio having an asymptotic chi-square distribution.

One common approach for constructing a $(1 - \alpha)\%$ confidence interval, also known as a Wald confidence interval, is based on the normal distribution and takes the form:

$$\hat{\theta} \pm z_{\alpha/2} SE\left(\hat{\theta}\right),$$

with $z_{\alpha/2}$ the appropriate critical value from the standard normal distribution. With 95% confidence the z value is 1.96 while with 90% confidence it is 1.645.

A profile likelihood confidence interval can be based on the following quantity (Williams et al., 2002, p. 49):

$$\varphi(\theta_0) = 2ln\left(\frac{L\left(\hat{\theta}_0, \hat{\boldsymbol{\theta}}|\mathbf{x}\right)}{L\left(\hat{\theta}_0, \hat{\boldsymbol{\theta}}^*|\mathbf{x}\right)}\right).$$

$L\left(\hat{\theta}_0, \hat{\boldsymbol{\theta}}|\mathbf{x}\right)$ is the likelihood evaluated at the MLEs for all the parameters and $L\left(\hat{\theta}_0, \hat{\boldsymbol{\theta}}^*|\mathbf{x}\right)$ is the likelihood evaluated at the MLEs of all the other model parameters, with θ_0 allowed to vary over its range (i.e., the parameter θ_0 is not estimated in $L\left(\hat{\theta}_0, \hat{\boldsymbol{\theta}}^*|\mathbf{x}\right)$ but can take any allowable value). We find the two values of θ_0 that satisfy the equation $\varphi(\theta_0) = \chi_1^2(\alpha)$ where $\chi_1^2(\alpha)$ is the $(1 - \alpha)\%$ percentile of the chi-square distribution (Williams et al., 2002, p. 728) with one degree of freedom. The profile likelihood method typically gives confidence intervals with somewhat better coverage than the confidence intervals based on the normal approximation.

Alternatively, confidence intervals can be calculated using computer intensive methods such as those discussed above (e.g., the bootstrap).

3.2.5   Multiple Maxima

One important issue associated with maximum likelihood estimation, particularly in a practical sense, is that likelihood functions may have more than one maximum, especially for more complex models. Hence while the intent is to find the global maximum (i.e., the combination of parameter values that results in the greatest value for the likelihood function), there may also be local maxima that are identified through either analytic or numeric approaches (i.e., a combination of parameter values that results in a likelihood value that is greater than that for other nearby parameter values, but is lower than the global maximum). The situation is completely analogous to a mountain range, with the topography representing the likelihood function. While one of the peaks within the mountain range will be the highest, there may be a number of peaks within the range each separated by a series of valleys.

This is not to say that there will always be multiple maxima, but practitioners of the methods described in this book should be aware of the potential for them and take appropriate steps to verify that the MLEs obtained are not the result of numerical optimization algorithms locating a local maxima. Practical options include initiating the optimization routine at alternative values to more fully explore the parameter space, or changing the order of the parameters to be estimated (i.e., alter the order of any covariates in included in a model). Some optimization algorithms are more robust to the presence of local maxima, such as simulated annealing, but tend to be slower.

We stress that this issue is relevant to any statistical analysis that uses maximum likelihood methods, which is a very large set of approaches that are commonly used with ecological data (e.g., capture–recapture, generalized linear models and distance sampling). It is not an issue that is restricted to the approaches we detail in this book.

3.2.6 Observed and Complete Data Likelihood

Recall that a latent random variable is one whose value(s) is unknown or not directly observable. There are two approaches for dealing with latent random variables within a maximum likelihood framework; the observed data likelihood and complete data likelihood. We will not get into the detail of the two approaches at this stage, but simply aim to introduce the concepts.

Using the observed data likelihood approach the likelihood function is constructed by considering all values of the latent random variables that are possible, and their associated probabilities, given the observed data. Technically this is known as integration of the latent random variables. Most standard approaches can be used for maximizing the observed data likelihood to obtain MLEs.

Using the complete data likelihood, the likelihood function is constructed as if the values of the latent random variables are known. Typically the complete data likelihood is simpler than the observed data likelihood, however most standard methods for maximizing the likelihood cannot be used as, obviously, the values for the latent random variables are unknown. Instead, alternative maximization methods must be used that can account for the unknown values of the latent random variables. One such approach is the expectation-maximization (EM) algorithm (Dempster et al., 1977).

3.3 BAYESIAN ESTIMATION

The frequentist view of statistics (e.g., when using maximum likelihood estimation) supposes that parameters are fixed, and seeks to find procedures with desirable properties for estimating those parameters. Usually, probability is used as a basis for evaluating the procedures, under a scenario in which replicate realizations of the data are imagined. This view supposes that it is sufficient to draw inferences about parameters based on what might have happened (but did not), not on what actually did happen (i.e., the observed data). Frequentist theory relies on asymptotic (large sample) arguments to assess the operating characteristics of procedures, although their properties can be assessed via simulation for small samples. For example, the frequentist approach to statistical inference would view the parameter ψ as fixed, and a 95% confidence interval constructed for ψ will contain the true value 95% of the time (in large samples). Thus, while the specific interval computed from our observed data may or may not contain the true value (we cannot know), if we were able to repeat our study a very large number of times, we would expect the stated interval to contain the true value for ψ 95% of the time. Note that the view of the parameter as fixed, and the data random, is manifest as a probability statement about the interval, and not the parameter.

Alternatively, a Bayesian view of statistics seeks to provide a direct probabilistic characterization of uncertainty about parameters given the specific data at hand. Bayesian statistical methods have become increasingly popular in recent years, partially due to the advent of fast computers and efficient methods for solving Bayesian inference problems that, typically, require solving complex integration problems. Similar to a Frequentist, the Bayesian views the data as the realization of a random variable. However, the Bayesian also views the parameters of a model as random variables and provides a probabilistic characterization of the state of knowledge of these parameters by statement of a prior distribution. Then, with both data and parameters viewed as random variables, a conceptually simple calculation known as Bayes Rule yields the probability distribution of the parameters given the data, a quantity known as the posterior distribution. The Bayesian formulates inferences for the parameters using this posterior distribution, conditional on the observed data, and not by entertaining notions of repeated realizations of the data.

3.3.1 Theory

The *prior* distribution of the parameters represents a statement about their likely range and frequency of values before consideration of the observed data, and is denoted here as $f(\theta)$. Recall from the previous section that $f(\mathbf{x}|\theta)$ is the probability distribution of the observed data (the random variables) given the parameters which is the basis of the likelihood function $L(\theta|\mathbf{x})$ used in maximum likelihood estimation. Therefore, by use of Bayes' Theorem, the *posterior* distribution of the parameters is:

$$f(\theta|\mathbf{x}) = \frac{f(\mathbf{x}|\theta)\,f(\theta)}{f(\mathbf{x})}$$

Bayesian inference is based on this posterior distribution. For example, the posterior mean of θ is commonly used as a measure of the center of the posterior distribution (the posterior median or mode may also be used), and one may construct 'Bayesian confidence intervals', known as *credible intervals*, using quantiles of the posterior distribution. For example, a central 95% Bayesian credible interval would be 2.5 and 97.5 percentiles of the posterior distribution, while a central 50% credible interval would be the 25 and 75 percentiles. Such summaries are easily attainable from the outcome of an analysis using Markov chain Monte Carlo (MCMC) methods (see below). Credible intervals need not be based upon the centralized percentiles of the posterior distribution and alternatively the so-called highest posterior density (HPD) interval could be used, which is obtained by taking the shortest interval that contains a prescribed fraction of the posterior mass. The HPD interval need not be continuous which is particularly advantageous if the posterior distribution is multi-modal (i.e., has multiple 'peaks'). In practice, one usually reports several numerical summaries of the posterior distribution of a parameter, such as the mean, median, various quantiles, and the standard deviation. A graphical summary of the posterior distribution is also often useful particularly, again, for a multi-modal posterior distribution.

When constant or uniform priors are used (i.e., priors that assume all parameter values within the defined range have equal probability, and that may be 'improper' if the prior does not integrate to 1), the posterior and the likelihood function are proportional, i.e., $f(\theta|x) \propto f(x|\theta) \propto L(\theta|x)$. Therefore, in this case, the posterior mode is equivalent to the maximum of the likelihood, thus producing an equivalence of sorts between frequentist and Bayesian point estimators based on the posterior mode. Uniform priors are widely used and are sometimes referred to as *noninformative* prior distributions although this term is not precise as all prior distributions are informative on some scale. While there are major differences in the underlying philosophies between Bayesian and likelihood approaches, what the posterior distribution being proportional to the likelihood means for applied usage is as follows. When sufficient quality data have been collected and when constant or uniform priors have been used, then resulting inferences from the Bayesian and likelihood methods tend to be very similar.

However in many situations there will be prior information available based on other similar field studies and on strong but diffuse knowledge from 'expert opinion'. This knowledge can be very helpful and used to define strong prior distributions leading to much less uncertainty in the posterior distribution of the parameters. Ignoring all this knowledge and using 'noninformative priors' in such cases seems illogical, yet this is what maximum likelihood methods are effectively doing. For example, consider a simple coin tossing experiment where we wish to estimate the probability of obtaining a head, p, when 7 out of 10 tosses of the coin resulted in a head. Assuming a noninformative prior distribution where all possible values for p are equally likely (i.e., p could be reasonably expected to have any value between 0 and 1, and all are equally likely), the resulting posterior distribution is given in Fig. 3.3A. Note the most likely value for p from this posterior distribution is 0.70, with a 95% credible interval of (0.39, 0.89). However, generally it would be reasonable to assume that prior to conducting a coin tossing experiment, p should about 0.5. Fig. 3.3B illustrates the effect of assuming a much more informative (but reasonable) prior distribution for p. Now the most likely value for p is 0.52 with a 95% credible interval of (0.43, 0.61). Assuming an informative prior distribution has clearly reduced the uncertainty in the posterior distribution, although also note that in this instance the prior and posterior distributions are very similar. This indicates that the collected data have had little impact upon our preconceived notions about p, which is partially due to the relatively small sample size. Had the coin been tossed 100 times resulting in 70 heads, then the posterior distribution in Fig. 3.3C would be obtained. While there is clear justification in this case for using an informative prior distribution for p, one should always be aware of the potential that resulting inferences based upon a posterior distribution may be sensitive to the choice of prior distribution, 'noninformative' or otherwise.

Another key advantage of Bayesian estimation methods is that the Bayesian view of parameters themselves arising from a distribution, rather than being fixed quantities, is especially useful when considering models with *random effects* (e.g., Royle and Kéry, 2007). For example, instead of assuming that all units within some area of interest have the exact same probability of being occupied by a species of interest, we might consider random effects models in which we view the occupancy probabilities for all units as coming from the same underlying distribution, but with different realized probabilities as governed by that distribution. In long-term studies, year-to-year variation in either occupancy or rates of extinction and colonization are likewise usefully viewed as random effects in many situations. Such thinking and modeling are possible from a frequentist or likelihood perspective as well, but are a very natural outcome of Bayesian thinking.

We note that as in frequentist case, the distinction between the observed and complete data likelihood is also relevant to Bayesian inference. Exactly how one applies Bayesian inferential procedures will depend on which form of the likelihood is used.

3.3.2 Computing Methods

Since the late 1990's the power of modern computation has led to an explosion of interest in Bayesian methods and an emphasis on use of the methods in a wide variety of applied problems. Modern Bayesian inference sometimes uses numerical integration methods to obtain posterior distributions if the number of parameters in the model is fairly small. More typically simulation methods based on Markov chain Monte Carlo (MCMC) methods are used.

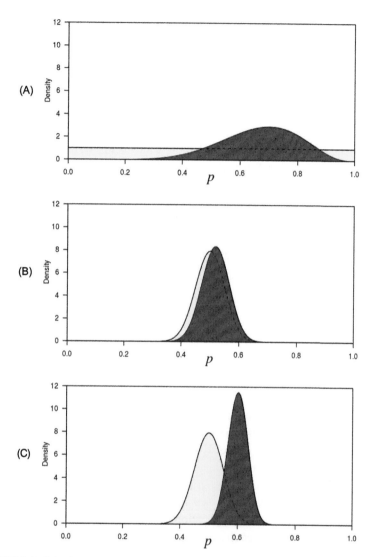

FIGURE 3.3 Prior (yellow) and posterior (red) distributions for the probability of a head, p, from a coin tossing experiment where 7 out of 10 tosses were heads with (A) a 'non-informative' prior, and (B) an informative prior. Prior and posterior distributions for p from a coin tossing experiment where 70 out of 100 tosses were heads with (C) the same informative prior used in (B).

The advent of MCMC has proved extremely important for advancing the widespread adoption of Bayesian analysis. This is due to the fact that, while the posterior distribution is easy to characterize in general terms (using Bayes' rule), it is usually not possible in practice to reduce the posterior distribution to a known distributional form. Thus, obtaining posterior summaries such as the mean, mode, variance or quantiles cannot usually be done directly. What MCMC allows us to do is simulate values of the parameter(s) in question from the correct posterior distribution without having to know its precise mathematical form. By using the prescribed prior distribution and probability distribution of the data (which is defined by the assumed data model), MCMC methods allow for simulating a sequence of parameter values $\theta^{(1)}, \theta^{(2)}, \ldots, \theta^{(M)}$ which theory dictates have the correct distribution, which is to say, they are a sample from the target posterior distribution. Given the manner in which the sequence of values $\theta^{(1)}, \theta^{(2)}, \ldots, \theta^{(M)}$ is simulated, they form a Markov chain, i.e., $\theta^{(t)}$ depends upon the previous values in the chain. Because of this dependence, often a very large posterior sample size M must be simulated in order to obtain numerically precise characterization of the posterior distribution. Summaries of the posterior distribution are obtained by calculating the desired quantities from the simulated values. For example, the posterior mean is obtained by

$$E(\theta|\text{data}) = \frac{1}{M} \sum_{m=1}^{M} \theta^{(m)}.$$

There are many specific algorithms and methods that are commonly used for developing MCMC algorithms for specific models including rejection sampling, Gibbs sampling, the Metropolis–Hastings algorithm and many others. See Link and Barker (2009) for examples. In practice, it is convenient to use popular software packages such as WinBUGS/OpenBUGS (Lunn et al., 2000; Kéry, 2010), JAGS (Plummer, 2003), or the newly developed NIMBLE package (de Valpine et al., 2017). These software packages allow for specification of the model in a type of 'pseudo-code' which the program uses to derive a suitable MCMC algorithm. We find that these software packages are suitable for the vast majority of all Bayesian occupancy model applications.

Most MCMC methods require an initial simulation period (called the burn-in period) in order to ensure that the simulated values achieve the target posterior distribution. One practical deficiency with MCMC in general is that there are no hard and fast rules for deciding whether the burn-in period has been sufficient. In practice, multiple chains are run and an assessment is made based on whether the summary statistics from the different Markov chains are similar. The most widely used convergence diagnostic is the so-called Brooks–Gelman–Rubin or 'R-hat' statistic (Gelman and Rubin, 1992).

Regardless of the investigator's views about Bayesian philosophy, our pragmatic view is that, from a practical standpoint of modeling and estimation, the two paradigms are largely consistent with each other in the vast majority of problems where both approaches can be applied. Moreover, Bayesian analysis by MCMC can readily provide estimates under models that would be extremely difficult to deal with from a strictly frequentist perspective (e.g., by using maximum likelihood estimation).

3.4 MODELING PREDICTOR VARIABLES

In many examples, important parameters such as species occupancy (ψ), or nuisance parameters such as detection probability (p), may be functions of predictor variables or covariates (e.g., habitat type, patch size, elevation, rainfall in previous 24 hours, time of day or distance to nearest road). Modeling these relationships can be viewed as a type of generalized linear regression technique. A *link function* is used as a transformation of the parameter that is more convenient for expressing the linear relationship with the covariates. Typically, here, the parameters that we wish to be functions of covariates are probabilities and the use of a special link function called the *logit* link is widely used when the parameter of interest is to be restricted to values between 0 and 1. Throughout this book we typically use the logit link function so provide some details about it below, but note that other appropriate link functions are available (e.g., probit, log–log, complementary log–log, sin) for parameters that are probabilities. Another common link function for parameters whose values must be > 0 is the *log* link.

3.4.1 The Logit Link Function

The logit link function is used to model the probability of 'success' as a function of covariates (e.g., logistic regression). The purpose of the logit link is to take a linear combination of the covariate values (which may take any value between $\pm\infty$) and convert those values to the scale of a probability, i.e., between 0 and 1. The logit link function is defined in Eq. (3.4).

$$logit(\theta_i) = ln\left(\frac{\theta_i}{1 - \theta_i}\right) = \beta_0 + \beta_1 x_{i1} + \beta_2 x_{i2} + \cdots + \beta_U x_{iU} \qquad (3.4)$$

where θ_i is the probability of interest for the ith sampling unit and $x_{i1}, x_{i2}, \ldots,$ x_{iU} are the values for the U covariates of interest measured at the ith sampling unit. The regression coefficients $\beta_1, \beta_2, \ldots, \beta_U$ determine the size of the effect of the respective covariates, and β_0 is the intercept term. Eq. (3.4) can be rearranged such that θ_i has the following relationship to the covariates:

$$\theta_i = \frac{exp(\beta_0 + \beta_1 x_{i1} + \beta_2 x_{i2} + \cdots + \beta_U x_{iU})}{1 + exp(\beta_0 + \beta_1 x_{i1} + \beta_2 x_{i2} + \cdots + \beta_U x_{iU})}. \qquad (3.5)$$

The logit of θ_i is also known as the log-odds for 'success'. The term $\frac{\theta_i}{1-\theta_i}$ is the *odds* of success (i.e., how much greater the probability of success is compared to that of a failure) and is often expressed as a ratio. For example, odds of 3:1 suggest the probability of success is 3 times that of a failure. The probability of success can be calculated from the odds as:

$$\theta_i = \frac{odds}{1 + odds}$$

so in this instance the probability of success would be, $\theta_i = 3/4 = 0.75$. Note in comparison to Eq. (3.5), the above suggests that $odds = exp(\beta_0 + \beta_1 x_{i1} + \beta_2 x_{i2} + \cdots + \beta_U x_{iU})$. In some instances, the odds may simply be stated as a single number rather than as a ratio (e.g., 3), is which case the ':1' is implied. While not commonly used in ecological applications, most people will be familiar with the concept of odds through gambling and games of chance (e.g., horse racing or roulette). Interestingly, in most gambling situations the odds are usually given in terms of the player *losing*. For example, a 'long shot' in a horse race is one that is thought unlikely to win and a bookie may offer odds of 30:1 on such a horse. Here the odds reflects the amount of money to be paid out for every $1 bet should the horse win, but undoubtedly also reflects the bookie's belief about the probability that the horse will *not* win the race (i.e., the probability of the 'long shot' *not* winning is 30/31, or in terms of winning the race 1/31). Alternatively, consider a simple raffle with ten tickets. If you bought only one ticket, then your odds of winning would be 1:9 (as there are nine other tickets purchased by others) or 0.111, whereas if you bought two tickets your odds of winning would be 2:8 or 0.25. Purchasing six tickets your odds of winning would be 6:4 or 1.5, i.e., the probability of you winning is 1.5 times greater than your probability of not winning (0.6 vs. 0.4).

We appreciate that there will often be some confusion about the concepts of *odds* and *probabilities* as the two words are often used interchangeably in everyday language, but while linked they are distinct quantities.

The logit link is appropriate for the probability associated with an observation that has only two possible outcomes (e.g., success/failure). When there are > two outcomes, there are multiple probabilities associated with the observations, and in addition to the individual probabilities having to be between 0 and 1, the sum of the set of probabilities must also be $\leqslant 1$. The *multinomial logit link* function is an extension of the above to deal with such situations.

When there are M possible outcomes, typically the $M - 1$ probabilities will be estimated $(\theta_i^{[1]}, \ldots \theta_i^{[M-1]})$ and the probability for the final outcome obtained by subtraction $(\theta_i^{[M]} = 1 - \sum_{m=1}^{M-1} \theta_i^{[m]})$. This final outcome can be considered as the reference or standard outcome. The multinomial logit link (mlogit) is defined

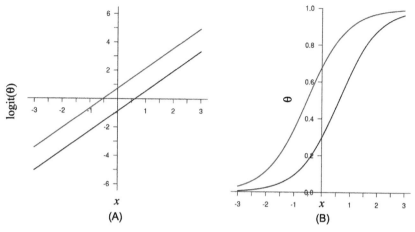

FIGURE 3.4   Example of the non-linear nature of the logit link function. Linear relationships that are parallel on the logit-scale (A) are not parallel and have different slopes for a given x value on the probability scale (B).

similarly to Eq. (3.4), except the denominator is now $\theta_i^{[M]}$ instead of 1 minus the probability of interest (although when $M = 2$, they are equivalent). That is:

$$mlogit\left(\theta_i^{[m]}\right) = ln\left(\frac{\theta_i^{[m]}}{\theta_i^{[M]}}\right) = \beta_0^{[m]} + \beta_1^{[m]}x_{i1} + \beta_2^{[m]}x_{i2} + \cdots + \beta_U^{[m]}x_{iU}.$$

Note that the regression coefficients can be different for each possible outcome, i.e., the effect of covariates can be different for different probabilities.

The probability can thus be calculated as:

$$\theta_i^{[m]} = \frac{exp\left(\beta_0^{[m]} + \beta_1^{[m]}x_{i1} + \beta_2^{[m]}x_{i2} + \cdots + \beta_U^{[m]}x_{iU}\right)}{1 + \sum_{l=1}^{M-1} exp\left(\beta_0^{[l]} + \beta_1^{[l]}x_{i1} + \beta_2^{[l]}x_{i2} + \cdots + \beta_U^{[l]}x_{iU}\right)}.$$

While the above equation indicates that the same set of covariates is considered for each probability, this does not have to occur in practice as excluding a covariate for a particular probability is equivalent to setting the associated regression coefficient to 0.

3.4.2   Interpretation

Given the nature of the logit link function, relationships that are linear on the logit-scale are non-linear when transformed to the probability scale (Fig. 3.4). This can make interpretation of covariate effect sizes on the probability of interest difficult, particularly when multiple or continuous-valued covariates are included in an analysis. For example in Fig. 3.4 note how the parallel, linear relationships on the logit-scale become non-parallel and non-linear on the probability scale. Fig. 3.4A might represent the results from an analyst fitting the following model on probability θ:

$$logit(\theta_i) = \beta_0 + \beta_1 x_i + \beta_2 h_i,$$

where x is a continuous-valued predictor variable (e.g., elevation) and h is a binary-valued covariate (e.g., to represent two different habitat types). β_1 is therefore the effect of x (the slope of the lines), and β_2 the difference between habitat types (the separation between lines) on $logit(\theta)$. While conclusions could be described in terms of $logit(\theta)$, this is not an intuitive scale that can be related to in real, practical terms, and describing the results in terms of the probability θ is not straightforward (e.g., the difference in the probability between habitats is small when x is low, large when x is near 0 and small again when x is high). This issue has long been recognized with logistic regression where the regression coefficients are typically interpreted in terms of odds and odds ratios, and we make the same suggestion here.

As the term implies the *odds ratio* is simply the ratio of two sets of odds (i.e., $OR = odds_2/odds_1$), or alternatively the value one set of odds is multiplied by to get a second set of odds (i.e., $odds_2 = OR \times odds_1$). Hence, the odds ratio is a scaling factor and a value close to 1 would suggest the two odds (and therefore the two probabilities) are similar.

Now Eq. (3.4) could be rearranged in terms of the odds of success for sampling unit i as:

$$\frac{\theta_i}{1 - \theta_i} = exp(\beta_0 + \beta_1 x_{i1} + \beta_2 x_{i2} + \cdots + \beta_U x_{iU})$$

$$= exp(\beta_0) exp(\beta_1 x_{i1}) exp(\beta_2 x_{i2}) \ldots exp(\beta_U x_{iU}).$$

Expressed in this form, $exp(\beta_0)$ is simply the odds of success at a sampling unit when the value for all covariates is zero (i.e., a baseline odds), and $exp(\beta_u x_{iu})$ is simply an odds ratio that depends upon the value of covariate x_{iu}. In fact, $exp(\beta_u)$ can be interpreted as the odds ratio for a 1-unit change in the covariate x_{iu}. While the notation may look disconcerting to some, recall that ultimately the odds ratio is just a number. For instance, suppose in our example of a conspicuous species the probability of occupancy could be expressed as:

$$logit(\psi_i) = -1.1 + 1.8 Rocks_i,$$

where $Rocks_i$ is an indicator covariate that equals 1 if unit i had a predominately rocky soil type, and 0 otherwise. At a unit without a rocky soil type (i.e., where

$Rocks_i = 0$), the odds of occupancy is $exp(-1.1) = 0.33$ (:1), which if we scale up to integer values is odds of 1:3. That is, in the non-rocky soil type for every 1 unit where the species is present there are 3 units where the species is absent; $\psi_i = 0.25$. At units with a rocky soil type, these odds are multiplied by a factor of $exp(1.8) = 6.05$; hence the odds of occupancy at a rocky unit will be $0.33 \times 6.05 = 2.00$ (:1). Hence, in the rocky soil type, there are 2 units with the species present for every 1 unit where the species is absent so $\psi_i = 0.67$.

Interpretation of odds ratios does take practice (like any skill), and the important aspect to appreciate is that the odds ratio is a scaling factor applied to the number of 'successes' in a baseline case for every 'failure', irrespective of how many 'successes' there actually were.

Note that confidence intervals for probabilities, odds, and odds-ratios can be calculated by first calculating the required interval limits on the logit-scale as normal (e.g., Section 3.2.4), then applying the appropriate transformation on the limits.

When using the multinomial logit link, similar interpretations can be made in terms of odds and odds ratios. Although rather than them being relative to the number of 'failures', they should be interpreted relative to the number of outcomes in the reference observation type (i.e., the type of observation defined as category M).

3.4.3 Estimation

Estimation of the regression parameters $(\hat{\beta}_u)$ and their standard errors is straight-forward. These parameters are in the likelihood function so computer software can be used to obtain the estimators and their standard errors using the asymptotic properties of MLEs, or their posterior distribution can be obtained when using a Bayesian analysis.

Using maximum likelihood theory, estimation of the probabilities at a unit $(\hat{\theta}_i)$ and their standard errors is a little more involved, although plenty of software and numerical routines are available to perform the following procedure. The probabilities are derived parameters that are not directly in the likelihood (i.e., their values are computed for use in the likelihood, but the likelihood is now maximized with respect to the β_u values, not θ_i directly). The derived parameters $(\hat{\theta}_i)$ are functions of the MLEs, and we can use the theorem quoted earlier that states that the MLE of a function of parameters is the function of the parameters' MLE. To obtain standard errors we need to use the delta method in matrix form. Recall that:

$$Var\left(h\left(\hat{\theta}\right)\right) = \mathbf{h}'\left(\hat{\theta}\right) \mathbf{V} \mathbf{h}'\left(\hat{\theta}\right)^T$$

and here we have:

$$h_i\left(\hat{\boldsymbol{\beta}}\right) = \hat{\theta}_i = \frac{exp\left(\hat{\beta}_0 + \hat{\beta}_1 x_{i1} + \hat{\beta}_2 x_{i2} + \cdots + \hat{\beta}_U x_{iU}\right)}{1 + exp\left(\hat{\beta}_0 + \hat{\beta}_1 x_{i1} + \hat{\beta}_2 x_{i2} + \cdots + \hat{\beta}_U x_{iU}\right)}.$$

When using the logit link function the partial derivative with respect to β_k is of the (relatively) simple form:

$$h'_{ik}\left(\hat{\boldsymbol{\beta}}\right) = x_{ik}\hat{\theta}_i\left(1 - \hat{\theta}_i\right),$$

and $\mathbf{V}$ is the estimated variance–covariance matrix of $\hat{\boldsymbol{\beta}}$. For example, consider once again the equation given above where the probability of occupancy for our conspicuous species depended upon soil type. Let us now consider that the coefficients are estimated quantities, and we wish to calculate standard errors for the estimated probabilities of occurrence, i.e., we have the equation:

$$\hat{\psi}_i = \frac{exp(-1.1 + 1.8 Rocks_i)}{1 + exp(-1.1 + 1.8 Rocks_i)}$$

where $\hat{\beta}_0 = -1.1$ and $\hat{\beta}_1 = 1.8$. For a unit without rocky soil type, the covariate $Rocks_i$ ($= x_{i1}$) will be 0, and the 'covariate' x_{i0} will be 1 (as β_0 represents an intercept term which applies to all units, though it is often omitted from equations as above for notational convenience). Hence, the estimated probability for a site with non-rocky soil type (NRST) is:

$$\begin{aligned}\hat{\psi}_{NRST} &= \frac{exp(-1.1 + 1.8 \times 0)}{1 + exp(-1.1 + 1.8 \times 0)}\\ &= \frac{exp(-1.1)}{1 + exp(-1.1)}\\ &= \frac{0.33}{1 + 0.33}\\ &= 0.25.\end{aligned}$$

To calculate the standard error we need to evaluate the vector of partial derivatives, $\mathbf{h'}\left(\hat{\boldsymbol{\beta}}\right)$, and the variance–covariance matrix for the $\hat{\beta}$'s, $\mathbf{V}$. $\mathbf{V}$ would generally be outputted (or at least obtainable) from the software used to fit the model to the data. Suppose here we obtained the variance–covariance matrix of:

$$\mathbf{V} = \begin{bmatrix} 0.30 & -0.05 \\ -0.05 & 0.40 \end{bmatrix}$$

where $Var\left(\hat{\beta}_0\right) = 0.30$, $Var\left(\hat{\beta}_1\right) = 0.40$, and $Cov\left(\hat{\beta}_0, \hat{\beta}_1\right) = -0.05$. The vector of partial derivatives would be:

$$
\begin{aligned}
h'_{NRST}\left(\hat{\beta}\right) &= \left[x_{i0}\hat{\psi}_{NRST}\left(1 - \hat{\psi}_{NRST}\right) \quad x_{i1}\hat{\psi}_{NRST}\left(1 - \hat{\psi}_{NRST}\right) \right] \\
&= \left[1 \times 0.25 \times 0.75 \quad 0 \times 0.25 \times 0.75 \right] \\
&= \left[0.1875 \quad 0 \right]
\end{aligned}
$$

which can then be used to obtain the variance of $\hat{\psi}_{NRST}$ by application of the delta method:

$$
\begin{aligned}
Var\left(\hat{\psi}_{NRST}\right) &= \left[0.1875 \quad 0 \right] \begin{bmatrix} 0.30 & -0.05 \\ -0.05 & 0.40 \end{bmatrix} \begin{bmatrix} 0.1875 \\ 0 \end{bmatrix} \\
&= 0.01
\end{aligned}
$$

and hence a standard error of 0.10 (after some allowance for rounding error). For sites with rocky soil type (RST) a similar procedure can be used where $\hat{\psi}_{RST} = 0.67$ and $h'_{RST}\left(\hat{\beta}\right) = \left[0.22 \quad 0.22 \right]$, to give $Var\left(\hat{\psi}_{RST}\right) = 0.03$ and a standard error of 0.17. We leave the mechanics of doing so as an exercise for the reader.

Estimation of the derived probabilities is, however, relatively straightforward in a Bayesian analysis when using computer intensive methods such as MCMC. Here the probabilities can be defined as additional *nodes* that the computer algorithm simply calculates and stores as part of the MCMC procedure. As such, the posterior distribution for $\hat{\theta}_i$ and $\hat{\beta}_u$ can be obtained simultaneously.

3.5 HYPOTHESIS TESTING

In this chapter we focus mainly on estimation methods and choices between models with similar parameter structures, however, we will need to discuss likelihood ratio tests for comparing models and goodness of fit tests for assessing the 'fit' of models as these techniques will be used in later chapters. Therefore, we now present a very brief introduction to hypothesis testing.

3.5.1 Background and Definitions

In hypothesis testing there is a null hypothesis and an alternative hypothesis. The idea is to develop a test statistic that has a known distribution under the null hypothesis and see if the observed value of the test statistic based on the data is unusual when compared against this known distribution.

Null Hypothesis (H_0): $\theta = \theta_0$

Alternative Hypothesis (H_A): $\theta \neq \theta_0$

The alternative hypothesis may be one-sided (e.g., $\theta > \theta_0$) or two-sided, but often takes the two-sided form presented above. The first step after one defines the null and alternative hypotheses is to define a test statistic that has a known distribution under the null hypothesis. The next step is to define a critical region of values for the test statistic where the probability of obtaining values in this region is equal to α (the size of the test).

Hypothesis testing leads to a dichotomous decision; one either rejects the null hypothesis or not. The investigator's decision (to reject or not) can be either right or wrong with respect to the truth. Therefore, statisticians provide a nomenclature for the errors that can be made, depending on the decision of whether or not to reject the null hypothesis.

- *Type 1 Error*: The probability that the null hypothesis is rejected when it is true (α).
- *Type 2 Error*: The probability that the null hypothesis is not rejected when it is false (β).
- *Power of a Test*: The probability that the null hypothesis is rejected when it is false is called the power of the test and is ($1 - \beta$). A good test will have high power even for values of θ close to, but distinct from, θ_0.

3.5.2   Likelihood Ratio Tests

To compare a model m_0 with another model m_A, when model m_0 is nested inside model m_A, we can use a likelihood ratio test. The idea of nesting simply means that the less general model m_0 (representing the null hypothesis) can be obtained by constraining some of the parameters of the more general model m_A (representing the alternative hypothesis). The likelihood ratio test statistic is:

$$LRT = -2ln\left(\frac{L\left(\hat{\theta}_0|\mathbf{x}\right)}{L\left(\hat{\theta}_A|\mathbf{x}\right)}\right)$$

where $\hat{\theta}_0$ and $\hat{\theta}_A$ are the MLEs for the parameters of the respective models. If the model m_0 is true and sample size large, then the distribution of *LRT* is chi-square with degrees of freedom equal to the number of additional parameters in model m_A vs. model m_0. This is related to the profile likelihood method used for computing approximate confidence intervals. Both are based on a likelihood ratio having an asymptotic chi-square distribution.

3.5.3 Goodness of Fit Tests

An old, common, and simple goodness of fit test to assess the fit of a model to data that is still widely used is Pearson's chi-square test. It has a test statistic of the form:

$$\chi^2 = \sum_{i=1}^{k}\left(\frac{(O_i - E_i)^2}{E_i}\right), \tag{3.6}$$

where O_i is the observed number of sample units found in class i, E_i is the number of sample units expected to appear in class i under the hypothesis of interest, and k is the number of classes being considered (e.g., these could be the cells or bins of a multinomial distribution). Under the null hypothesis, χ^2 is asymptotically distributed as a central chi-square and the degrees of freedom are $df = k - \delta$, where k is the number of classes or cells, and δ is the number of parameters fitted in the model. A statistic with similar properties is the *deviance* (as used in Chapter 7).

The use of goodness of fit tests is also important in accounting for overdispersion in model selection methods discussed and defined below. MacKenzie and Bailey (2004) used the Pearson statistic as a goodness of fit test and to estimate overdispersion for single-species, single-season occupancy problems. However, they used a parametric bootstrap procedure to estimate the exact distribution rather than the chi-square distribution. Often the expected values in some of the k classes or cells were too small for the chi-square to be a good approximation to the exact distribution (Chapter 4).

3.6 MODEL SELECTION

In many classic statistical inference books, the focus is on inference for a particular model, which can be viewed as a hypothesis (in the sense that it represents a story about how the system functions), with the parameter structure totally specified. Maximum likelihood and Bayesian estimation methods can equally be used. However, in many of the examples we consider in this book, we may have multiple hypotheses about our biological system, and the first question we have to consider is how to decide on which model (i.e., hypothesis), or models, from a set of candidate models are 'better' for a given data set. Only once we have settled that question can we present good point or interval estimates of the parameters required from the selected model(s).

Burnham and Anderson (2002) have studied model selection and multimodel inference in detail. Here we give a very brief treatment based on their approach using the Akaike Information Criterion (AIC). We agree with them that this approach is better than older approaches such as forward or backward stepwise methods, but it should not be used unthinkingly.

A key concept at the basis of their approach is *the principle of parsimony*. Statisticians view the principle of parsimony as a 'bias versus precision' trade-off. In general, bias (of parameter estimates) decreases, but variance (of the parameter estimates) increases as the dimension of the model (number of parameters) increases. The fit of any model can be improved (and hence a reduction in bias achieved) by increasing the number of parameters; however, a tradeoff or cost is increasing variance, and must be considered when selecting a model(s) for inference. Parsimonious models achieve a proper tradeoff between bias and variance. All model selection methods are based to some extent on the principle of parsimony (Burnham and Anderson, 2002, p. 31). One older method of model choice used by quantitative ecologists selecting from a set of candidate capture–recapture models, for example, is to come up with a global model that fits the data adequately based on a goodness of fit test and then use sequential likelihood ratio tests to decide whether simpler nested models can be used for the data (for examples see Pollock et al., 1990; for a more detailed description of the approach see Williams et al., 2002, p. 54). However, this method cannot handle non-nested model structures very well, and it tends to favor models with more parameters than some other approaches such as AIC (Burnham and Anderson, 2002). Burnham and Anderson (2002) emphasize that model selection is really distinct from traditional hypothesis testing and present the Akaike Information Criterion (AIC) based on the pioneering work of Akaike in the 1970's and 80's (e.g., Akaike, 1973).

3.6.1   Akaike's Information Criterion (AIC)

In information theory, the Kullback–Leibler (KL) information is a relative measure of how much information is lost by using a model to represent 'truth'. As truth is unknown, the absolute information loss is unknown, but in comparing a candidate set of models they can be ranked in terms of which models result in greater, or lesser, information loss. AIC is an approximation of the KL information and can be calculated as:

$$AIC = -2ln\left(L\left(\hat{\theta}|\mathbf{x}\right)\right) + 2\delta$$

where $ln\left(L\left(\hat{\theta}|\mathbf{x}\right)\right)$ is the log of the likelihood function evaluated at the MLEs and δ is the number of parameters estimated in the model. The first term is a relative measure of how much variation in the data is unexplained by the model (note the same term is used in the likelihood ratio test above), while the second amounts to a penalty term as models with greater numbers of parameters will be able to explain a greater amount of variation. As such, AIC encourages parsimonious models; models that explain the variation in the data well, with as few parameters as possible. We hasten to add that the derivation of AIC was based on sound theoretical work, and the simplistic nature of the calculation is fortuitous.

The absolute magnitude of AIC is not very relevant, but the differences in AIC among different models are the focus of model selection. Usually all the models are compared to the model with minimum achieved AIC by using a table of differences in AIC. Thus for a particular model (k) the difference would be:

$$\Delta \text{AIC}_k = \text{AIC}_k - \text{AIC}_{min}.$$

Burnham and Anderson (2002) note as a rough rule of thumb that all models with AIC differences of less than 2 have a substantial level of empirical support, 4–7 substantially less support, and greater than 10 essentially no support. However, through practical experience we have found situations where the normalized AIC model weights provide a fairer indication of the support for each model. The AIC model weights can be calculated from the ΔAIC values:

$$w_k = \frac{exp(-0.5\Delta \text{AIC}_k)}{\sum\limits_{r=1}^{R} exp(-0.5\Delta \text{AIC}_r)}$$

for a suite of R models, and w_k is interpreted as "the weight of evidence in favor of model k being the actual Kullback–Leibler best model for the situation at hand …" (Burnham and Anderson, 2002, p. 75). The AIC weights sum to 1 for all of the members of the model set, again emphasizing the relative nature of AIC comparisons, conditional on the model set. AIC model weights may also be interpreted (heuristically) as the probability that model k is the 'best' model in the candidate set (Burnham and Anderson, 2004), although we prefer the former interpretation.

When multiple models in the candidate set have some common feature (e.g., different formulations of the same hypothesis, or multiple models containing the same factor or covariate), one approach for determining the overall level of support for that common feature (given the model set) is to sum the model weights for each of the respective models (Burnham and Anderson, 2004). For example, suppose there were six models in the candidate set and the AIC model weights for Models 1–6 were 0.30, 0.30, 0.20, 0.10, 0.05, and 0.05, respectively. All models have a comparatively similar level of support; hence there is no clear group of models (and therefore hypotheses) that may be considered as 'better' representations of the data. However, only Models 1, 2, and 3 included habitat as a covariate for occupancy, and adding these model weights could be used to assess the total level of support that occupancy is related to a unit's habitat. This gives a combined model weight of 0.8 ($= 0.3 + 0.3 + 0.2$), which translates to substantial overall support for the habitat hypothesis.

Another way of interpreting model weights (or summed model weights), is the use of evidence ratios (Anderson, 2008). An evidence ratio (*ER*) can be calculated as:

$$ER = \frac{w}{1 - w}$$

where w may either be the model weight for a specific model, or the summed model weight for a common feature as above. The *ER* indicates how much support there is for that model (or set of models) compared to all other models, where the value of 1 indicates equal support, <1 less support, and >1 more support. For example, if the summed model weight is 0.8, then the $ER = 4$, i.e., there is four times the evidence supporting the models that include habitat in comparison to models that do not. At which point one might conclude an 'important' effect based on the value of the *ER* will depend on the context of the analysis, and Anderson (2008) suggests there is no standard value people should use, and that it is largely a personal choice they must be comfortable with.

Burnham and Anderson (2002) also suggest a small sample second order bias adjustment to AIC:

$$\text{AIC}_C = \text{AIC} + \frac{2\delta(\delta + 1)}{n - \delta - 1},$$

where n is the effective sample size. Unless the sample size is very large compared to the number of parameters, then AIC_C is generally recommended for use in practice, with Burnham and Anderson (2002) suggesting that AIC_C should be used if $n/\delta \leqslant 40$ (i.e., 40 observations per parameter). However, we note that the exact nature of sample size is not always obvious, for example in capture–recapture and occupancy modeling. In fact, the 'effective sample size' may also vary for different parameters in the model (e.g., be different for occupancy and detection probabilities). Because of this dilemma, we currently have no firm suggestions on what should be regarded as an 'effective sample size' for the occupancy models we describe in this book, hence simply use AIC. The effect of this is that more complex models are perhaps ranked higher than they should be in the examples we present. Although we note that the original derivation of AIC and AIC_C was based on normally distributed data and, to the best of our knowledge, no one has specifically explored how good these approximations to the KL information are for the type of data considered here. Although we do expect them to be reasonable approximations.

3.6.2 Goodness of Fit and Overdispersion

In models of overdispersed data, the mean or expectation structure of the model is adequate, but the variance structure (σ_θ^2) is inadequate. One approach is to think of the true variance structure as following the form ($\gamma_\theta \sigma_\theta^2$), but it is complex to fit such a form. As a simpler approach, we suppose $\gamma_\theta = c$, so that the true variance structure $c\sigma_\theta^2$ is some constant multiplier of the theoretical variance structure. For example, in capture–recapture models, a multinomial likelihood is often used, and while the expectation structure of the multinomial model may be adequate (i.e., point estimates of parameters may be valid), the variance structure may be inadequate. Values of c in the range of 2–4 would not be unreasonable due to positive correlations among individuals within a group or flock of animals (e.g., by frequently recapturing or resighting both members of a breeding pair).

A common method of estimating overdispersion is to use the observed chi-squared goodness of fit statistic for a global model, the most general model in the model set, e.g., Eq. (3.6), divided by its degrees of freedom:

$$\hat{c} = \frac{\chi^2}{df}.$$

If there is no overdispersion or lack of fit, then c should equal 1, and $\hat{c}$ should be approximately 1 (because the expected value of a chi-square statistic is equal to its degrees of freedom). In Chapter 4 we discuss one way of estimating $\hat{c}$ specifically for single-species, single-season occupancy models. In the next section we show how the AIC criterion is modified by overdispersion.

3.6.3 Quasi-AIC

It is our experience that overdispersion can be common in modeling ecological data and that there may be the need to include this in our model selection criterion. The AIC and AIC_C criterion can be modified for overdispersion ($\hat{c}$) (Burnham and Anderson, 2002, p. 70) using:

$$\text{QAIC} = \frac{-2ln\left(L\left(\hat{\theta}|\mathbf{x}\right)\right)}{\hat{c}} + 2\delta$$

and

$$\text{QAIC}_c = \text{QAIC} + \frac{2\delta(\delta + 1)}{n - \delta - 1}.$$

When $\hat{c} > 1$, the consequence of using QAIC or QAIC_C is that models with a greater number of parameters will tend to get ranked lower compared to using AIC or AIC_C.

Once QAIC or QAIC$_C$ have been used, the empirical estimates of variances and covariances can be obtained by multiplying the theoretical model-based variances and covariances by $\hat{c}$. Note that although $\hat{c}$ is estimated based on the most general model in the set, $\hat{c}$ should be used to estimate variances and covariances for parameters of all models in the set. Burnham and Anderson (2002) point out that this approach has been used at least since the 1970's.

It is important to note that the use of QAIC and QAIC$_C$ is justified by the presence of overdispersion, which may be identified through a goodness of fit test as the lack of independence between observations is a violation of model assumptions. The lack of independence leads to greater variation in the data than predicted by the model. A goodness of fit test may also indicate other violations of model assumptions, e.g., unmodeled heterogeneity or other inadequacies in the model's structure, for which use of QAIC is unsupported. In fact, for such situations using QAIC and QAIC$_C$ is the exact opposite of what is required. The poor model fit is identified by the most general model not being complex enough in some manner, yet use of QAIC over AIC will lead to the simpler models in the candidate set being ranked higher!

3.6.4 Model Averaging and Model Selection Uncertainty

In many ecological situations, multiple models in the candidate set may be reasonable, i.e., the 'best' model is not always apparent. Instead of choosing a single 'best' model to draw inferences from, we may use estimates from multiple models in the candidate set, calculating *model averaged* estimates. In this case the AIC (or similar) weights of the candidate models are used to obtain a weighted average of the individual parameter estimates. For a suite of R models, the estimator is:

$$\hat{\theta}_A = \sum_{r=1}^{R} w_r \hat{\theta}_r,$$

with the subscript indicating model r. Burnham and Anderson (2002) suggest the variance of the model-averaged estimator, allowing for model uncertainty, should be:

$$Var\left(\hat{\theta}_A\right) = \left(\sum_{r=1}^{R} w_r \sqrt{Var\left(\hat{\theta}_r | m_r\right) + \left(\hat{\theta}_r - \hat{\theta}_A\right)^2}\right)^2,$$

where $Var\left(\hat{\theta}_r | m_r\right)$ is the variance of the estimate obtained from model r. The notation indicates that this variance is conditional on the model r. The second component of the variance, $\left(\hat{\theta}_r - \hat{\theta}_A\right)^2$, then reflects model uncertainty by focusing on the difference between a model-specific estimate and the weighted mean estimate based on all models. However, Anderson (2008) presents a revised calculation arguing it is more appropriate:

$$Var\left(\hat{\theta}_A\right) = \sum_{r=1}^{R} w_r \left(Var\left(\hat{\theta}_r | m_r\right) + \left(\hat{\theta}_r - \hat{\theta}_A\right)^2 \right).$$

In this book we use the revised version recommended by Anderson (2008) and refer the interested reader there for more detail.

3.6.5 Bayesian Assessment of Model Fit

Goodness of fit in Bayesian analyses is routinely assessed using a method referred to as the 'Bayesian p-value' and posterior predictive checks (Gelman et al., 1996). This is based on comparing the values of a fit statistic computed from data sets simulated from the posterior distribution, to the value of the fit statistic for the data set in hand. For example, the Pearson chi-square statistic could be used, Eq. (3.6), which we shall denote here as FS (for fit statistic). This fit statistic is calculated for each MCMC iteration, which represents one sample from the posterior distribution, given the parameter values at that iteration of the MCMC algorithm. Similarly, for each MCMC iteration a new data set is simulated using the current model parameter values and the equivalent statistic is computed, FS^{new}. Note that as the model of interest was used to simulate the new data set, those data, and therefore FS^{new}, must represent what might be realized if the model were properly specified.

The Bayesian p-value is the fraction of MCMC iterations where $FS^{new} > FS$. If the observed data set is consistent with the model in question, then the Bayesian p-value should be close to 0.50. In practice, a p-value close to 0 or 1 indicates that the model is inadequate in some way – close to 0 suggests a lack of fit and close to 1 suggests that the model over-fits the data, which may occur when it is too complex.

Remediation in poor fitting models can be done in a similar manner to what is done in classical inference based on likelihood methods, using the lack of fit ratio which we introduced above:

$$\hat{c} = \frac{\chi^2}{df}.$$

This has expected value of 1 under the model fitting, and thus values of the observed test statistic larger than expected produce a lack of fit ratio > 1. Similarly when assessing model fit using a Bayesian p-value we can compute $\hat{c}$ by taking the mean of the posterior distribution of the fit statistic for the data divided by its expected value under the fitting model. We can multiply the variance–covariance matrix by $\hat{c}$ to obtain variances adjusted for over-dispersion, e.g., see Section 6.8 in Kéry and Royle (2016). However, as when using maximum likelihood methods, such adjustments are only justified when the lack of fit is expected to be from overdispersion and not some other violation of model assumptions.

In addition to using a more traditional fit statistic, such as Pearson's chi-square, it is also possible to use any summary of the data which is thought to characterize certain aspects of the data. For example, total number of detections, number of units with x number of detections, etc. This is known as a posterior predictive check, and these alternative fit statistics could also be used to calculate a Bayesian p-value in the same manner as above, although there is no theoretical justification for using them to calculate a value for $\hat{c}$ and adjust the variance–covariance matrix. Should a practitioner attempt to use a large number of fit statistics to assess a model, we would point out that some may indicate a problem with the model simply because of random chance with the large number used. We would therefore recommend judicious choice of the fit statistics used to those that are thought to be most informative about suspected causes of poor model fit.

3.6.6 Bayesian Model Selection

Bayesian model selection is much less procedural than model selection based on AIC. There is more diversity of options and also less agreement on their validity in any particular problem. Spiegelhalter et al. (2002) devised a Bayesian metric for model selection, analogous to AIC (in its application), called the Deviance Information Criterion (DIC), which is a function of model deviance and a measure of effective number of parameters. The model deviance is often calculated as twice the negative log-likelihood, although strictly speaking there is an additional constant representing the fit of a *saturated model* (a model that is a perfect fit to the data) that will cancel out when comparing deviance values from different models. For each MCMC iteration, the deviance is calculated for the current parameter values $\theta^{(m)}$, i.e.,

$$Dev\left(\theta^{(m)}\right) = -2ln\left(L\left(\theta^{(m)}|\mathbf{x}\right)\right).$$

The posterior mean of the deviance is calculated by:

$$\overline{Dev}(\theta) = \sum_{m=1}^{M} Dev\left(\theta^{(m)}\right),$$

where M is the number of MCMC iterations. DIC is defined as:

$$DIC = \overline{Dev}(\theta) + p_D,$$

where p_D is effective number of parameters. The standard definition of p_D is:

$$p_D = \overline{Dev}(\theta) - Dev(\bar{\theta}),$$

where $Dev(\bar{\theta})$ is the deviance evaluated at the posterior mean of the model parameters, i.e., the average of each parameter value from all MCMC iterations. Gelman et al. (2004) suggest a different version of p_D based on one-half the posterior variance of the deviance, i.e.,

$$p_D = \frac{\sum_{m=1}^{M} \left(Dev(\theta^{(m)}) - \overline{Dev}(\theta)\right)^2}{2(M-1)}.$$

DIC is widely used although its use has been called into question by several authors. Millar (2009) noted that, for hierarchical models, DIC based on the conditional likelihood was invalid. Lunn et al. (2012) noted other problems including "An inability to calculate p_D when θ contains a categorical parameter, since the posterior mean is not then meaningful. This renders the measure inapplicable to mixture models [...]." Unfortunately for us, the latent occupancy state variables in all occupancy models are exactly such categorical parameters! Thus, DIC may not be very suitable for Bayesian model selection in occupancy models.

An alternative convenient way to deal with model selection and averaging problems in Bayesian analysis by MCMC is to use the method of indicator variables (Kuo and Mallick, 1998). To implement the Kuo and Mallick approach, we include all covariates of interest in the model, then expand the model to include latent indicator variables, say w_u, associated with each covariate in the model, such that $w_u = 1$ if the linear predictor contains covariate x_u and, if $w_u = 0$, then the linear predictor does not contain covariate x_u. We assume that the indicator variables w_u are mutually independent with prior distribution:

$$w_u \sim Bernoulli(0.50),$$

for $u = 1, 2, \ldots, U$. For example, consider a simple linear regression with two covariates, x_1 and x_2, then the expanded regression model, to accommodate the indicator variables has the form:

$$y_i = \beta_0 + \beta_1 w_1 x_{1,i} + \beta_1 w_2 x_{2,i}.$$

The posterior mean of w_u is a gauge of the importance of covariate x_u, with values closer to 1 indicating stronger evidence of support for the importance of x_u being included in the model, whereas values close to 0 suggest that x_u is less important. An important consideration with this approach is that all covariates should be standardized to have a mean of zero, otherwise the MCMC chain for β_0 may have difficulty converging as its interpretation, and what might be considered a 'reasonable' value, will depend upon what combination of w_u's $= 1$ at each MCMC iteration. A further consideration is proper construction of the model should interactions between covariates be included.

In general, using this indicator variable formulation of the model selection problem, we can characterize unique models by the sequence of indicator variables. For just one covariate, there are only two models defined by $w_1 = 1$ or $w_1 = 0$. Considering a more general case, e.g., with three covariates, then each unique sequence (w_1, w_2, w_3) represents a model, and we can tabulate the posterior frequencies of each model by post-processing the MCMC histories of (w_1, w_2, w_3) just as if they were any other parameter of the model. This method then produces posterior probabilities for each of the models. This method of indicator variables to do model selection is especially useful for producing model-averaged predictions of latent variables because the MCMC output for a variable represents a posterior sample from all possible models in the set defined by combinations of the variables.

Other alternatives also exist for model selection in a Bayesian context, such as reversible jump MCMC (e.g., Link and Barker, 2006), although we do not consider such approaches in this book.

One broader, technical consideration is that posterior model probabilities are well known to be sensitive to priors on parameters (Link and Barker, 2006). What might normally be viewed as vague priors are not usually innocuous or uninformative when evaluating posterior model probabilities. There is no general resolution to this issue at the present time (that we know about). To be safe and informed, one should evaluate the sensitivity of posterior model probabilities to different prior specifications (Tenan et al., 2014).

3.7 DISCUSSION

In this chapter we gave a very brief overview of classical inference based on likelihood and also Bayesian inference based on the posterior distribution. We believe that both inference paradigms are suitable for most problems, although one or the other may have an advantage in specific instances. For example, if the model in question contains complex latent variable structure, or requires explicit inference about functions of random effects or latent variables, then a Bayesian analysis may be preferred. On the other hand, for analysis of a basic set of models that include fixed covariates, then direct analysis of the likelihood is both computationally efficient, and can produce direct and straightforward analysis of model selection and fit.

We now proceed to use many of these statistical techniques in the remainder of the book. The general approach will be to derive a general likelihood for each sampling design we will consider and then to use model selection methods and goodness of fit tests to choose a particular model for use on a data set. In Chapter 4 we consider the simplest practical case of sampling a series of units for occupancy on multiple occasions in a single season where detection probability for the species at each unit may be less than one.

Part II

Single-Species, Single-Season Occupancy Models

Chapter 4

Basic Presence/Absence Situation

In the preceding chapters we have introduced many of the basic concepts that readers should keep in mind when estimating and modeling species occurrence. It is important that these concepts are well understood so that the approaches presented in the following chapters may be used correctly, and to their fullest utility. By covering the underlying principles early in this book, we hope readers will be mindful of how these issues interrelate with the analytic techniques that we shall cover for estimating occupancy and related parameters. Further, without a firm understanding of the underlying statistical and design concepts it would be very easy to misuse these techniques, potentially resulting in erroneous conclusions.

In the remainder of this book, we develop a series of methods that can be used to estimate and model occupancy patterns and dynamics. Many of the methods we present in the latter chapters have evolved from considering the problem of estimating the proportion of area (or patches) occupied by a single species, at a single point in time. In this chapter we outline the general sampling problem; review previous two-step *ad hoc* methods; and then present a flexible, robust, and useful model-based approach. Several practical examples are given throughout to illustrate these methods. We conclude the chapter with a discussion of the important issues raised and show how this general approach leads us to more complex topics that will be covered in subsequent chapters.

4.1 THE SAMPLING SITUATION

We may wish to estimate the proportion of an area, or proportion of suitable habitat within an area, that is inhabited by a target species. Alternatively, we may be interested is estimating the probability the species is present at sampling units within an area. As noted at the beginning of Chapter 2 there is a subtle, but important, distinction between the terms proportion and probability that can often be overlooked without consequence in many practical applications. We use the term 'area' in the general sense of a statistical population; that is, a collection of sampling units about which we wish to make inference. Sampling units may constitute arbitrarily defined spatial units (such as grid cells of a specified size

Occupancy Estimation and Modeling. http://dx.doi.org/10.1016/B978-0-12-407197-1.00006-5
Copyright © 2018 Elsevier Inc. All rights reserved.

and shape) or discrete, naturally occurring sampling units (e.g., forest remnants, ponds or islands). From this population of S sampling units, s units are selected at which the intent is to establish the presence (occupied) or absence (unoccupied) of the target species. Throughout this book we generally regard that S is very large in comparison to s, and that we wish to make inference about the population of S units, not the sample of s units. This may not always be the case, and it is in this situation that the distinction between proportion and probability becomes critical. Potential modifications to the methods of this chapter are detailed in Section 6.1 for such cases. The manner in which units are selected has important consequences for the appropriateness and accuracy of resulting inferences, although at this point we shall assume units have been selected in such a manner that they are representative of the entire population (e.g., a random sample). We shall revisit issues related to unit selection in Chapter 11.

While species presence may be confirmed by detecting the species at a unit, it is usually impossible to confirm whether a species is absent. The nondetection of the species may result either from the species being genuinely absent, or from the species being present at the unit, but undetected during the surveying. Unless the species is so conspicuous that it will always be detected when present (a situation we believe to be rare), or a very intensive level of surveying has been conducted at each location, there may be a good chance that the species was indeed present but not detected. This has long been recognized by many field biologists who have used repeated surveys to minimize the possibility of declaring a species 'falsely absent' from a location.

We therefore consider a basic sampling scheme where s units are each surveyed K times for the target species (later we generalize to allow an unequal number of surveys). During each survey, appropriate methods are used to detect the species at the units. Such methods may include the visual, aural or indirect confirmation of at least one individual of the species. Indirect methods may involve scent stations, tracking tunnels, or detecting other species sign, e.g., fresh droppings. Initially, it is assumed that the target species is never falsely detected at a unit when absent (i.e., by misidentification of the species), which is likely to be a reasonable assumption in many situations (Tyre et al., 2003; Wintle et al., 2004), although extensions that relax this assumption are detailed in Section 6.2.

The K surveys are conducted over a suitable timeframe during which the s units are all closed to changes in the state of occupancy, i.e., units are either always occupied or unoccupied during the surveying period (this may be relaxed in some situations, see Section 4.4.6). We define this period of population closure as a 'sampling season' or just 'season'. The actual timeframe encompassed by a season will vary from situation to situation. For example, during a study of breeding bird colonies, a season may last for two or three months, while for small mammal studies the closure assumption may only be reasonable for a week. In effect, the concept of a 'season' enables us to consider a snapshot of the population at a given point in time, from which we can attempt to infer patterns about the level of occupancy. Importantly, the duration of the sampling season completely determines how species 'presence' should be interpreted (discussed further in Chapter 11 in terms of study design).

The sequence of detections (1) and nondetections (0) of the target species from the K surveys of unit i is recorded as a detection history ($\mathbf{h}_i$). For example, suppose three surveys were conducted at a unit with the species being detected in the first and third surveys, but not detected during survey two. The detection history for this unit could be expressed as $\mathbf{h}_i = 101$. Similarly, the detection history for a unit where the species was not detected in any of the three surveys would be $\mathbf{h}_i = 000$.

We consider that there are two processes occurring in the general sampling situation: occupancy and detectability. Occupancy is a biological quantity and relates to the presence or absence of the species from units during the sampling period (the season), which will be the quantity of prime interest in many situations. Detectability is an aspect of the sampling/surveying protocols that will generally be regarded as a nuisance parameter. However, as illustrated in Chapter 2, not accounting for the imperfect detection of the species may result in misleading conclusions about the population.

In the remainder of this chapter we first consider how occupancy could be estimated if the target species was always detected when present (i.e., no false absences) or if the probability of detection was known. We then review two-step *ad hoc* approaches for estimation when detectability is unknown, before moving to a full model-based approach for the simultaneous estimation of both occupancy and detectability.

4.2 ESTIMATION OF OCCUPANCY IF PROBABILITY OF DETECTION IS 1 OR KNOWN WITHOUT ERROR

In most practical situations, we consider it unlikely that a target species will always be detected when present at a unit. This is especially true for those rare or cryptic species that are often of special interest to managers and researchers. However, it is instructive to consider this best possible case because it provides useful insights and a 'gold standard' for determining how well an estimator may be performing in a given situation. The precision of any estimator that incorporates detectability could not do better than the estimator for which the occupancy state of units is known without error.

Let us assume that all units have a common probability of being occupied by the species, ψ. Therefore the number of units that are occupied by the target species (x) from a random sample of s units will follow a binomial distribution with $E(x) = s\psi$ and $Var(x) = s\psi(1 - \psi)$ (Chapter 3). Using the standard results for a binomial proportion, an estimate of ψ when the species is detected perfectly will be:

$$\hat{\psi} = \frac{x}{s} \tag{4.1}$$

with an associated variance of:

$$Var(\hat{\psi}) = \frac{\psi(1 - \psi)}{s}, \tag{4.2}$$

which can naturally be approximated by substituting in the estimated value for ψ. When there are covariates or predictor variables of interest for the presence/absence of the species, then standard logistic regression, or similar approaches, can be used.

Now let us assume that the target species is detected imperfectly, and that the probability of detecting the species in a single survey of an occupied unit is equal to p, which has a value that is known exactly (i.e., has no associated sampling error). Therefore, the probability of detecting the species at least once after K surveys of the unit will be $p^* = 1 - (1 - p)^K$. This is one minus the probability of the species being undetected in all K surveys. The number of units at which the target species is *detected* (s_D) from a random sample of s units will again follow a binomial distribution with $E[s_D] = s\psi p^*$ and $Var[s_D] = s\psi p^*(1 - \psi p^*)$, that is, the probability of the species being present and detected at a unit will be ψp^*. Based on this, an estimator for the proportion of units occupied when p is known would be:

$$\hat{\psi}_p = \frac{s_D}{s p^*} \tag{4.3}$$

with variance:

$$
\begin{aligned}
Var(\hat{\psi}_p) &= \frac{\psi(1 - \psi p^*)}{s p^*} \\
&= \frac{\psi(1 - \psi)(1 - \psi p^*)}{s(1 - \psi) p^*} \\
&= \frac{\psi(1 - \psi)}{s} + \frac{\psi(1 - p^*)}{s p^*}.
\end{aligned}
\tag{4.4}
$$

Eq. (4.4) highlights that this variance consists of two components. The first component corresponds to the binomial variation associated with the true underlying value of ψ. The second component is due to the imperfect detection of the species and, in effect, is the result of having to estimate the number of units that were occupied in the sample. As we shall see, the variance for most occupancy estimators can be written in a similar form.

Another important point highlighted by Eq. (4.4), is that when a species is detected imperfectly, the variance of an occupancy estimator can never be smaller than the binomial variation term. For example, if 50 units were surveyed ($s = 50$) and occupancy was 0.4 ($\psi = 0.4$), then with perfect detection the expected variance will be 0.0048 (from Eq. (4.2)), or a standard error of 0.07 (recall the standard error is the square root of the variance). Whereas, if p^* is known to be 0.75, then the expected variance will be 0.0075 (from Eq. (4.4)) or a standard error of 0.09. This is because the second component must be greater than 0, although it will get close to 0 as p^* tends to 1.0 (i.e., as it becomes almost certain that the species will be detected at least once in K surveys of an occupied unit). This is also useful information from a study design perspective, to which we shall return in Chapter 11.

However, rarely will the detection probability be exactly known prior to a study being conducted, so while the above discussion does provide some interesting insights to which we shall return later, the estimators and associated variances above are likely to be of little practical value. More often than not, both occupancy and detection probabilities will be unknown and must be estimated jointly from the collected data.

4.3 TWO-STEP *AD HOC* APPROACHES

There have been a number of independent efforts to estimate the proportion of units occupied by a target species, in the face of an unknown level of imperfect detection. These approaches can be loosely grouped into two classes. In the first class of methods, a two-step approach is taken where the probability of detecting the species is estimated in the first step, and is then used in a second step to estimate the occupancy parameter. We define such methods as two-step, *ad hoc* estimation procedures. They do provide valid estimators, but they do not provide a framework that is as flexible for comparing competing hypotheses about the system. Further they do not lend themselves to generalization to more complex situations. The second type of approach, and in our mind the superior one, involves directly modeling the biological and sampling processes in a way that enables the simultaneous estimation of both occupancy and detectability parameters in a single model-based framework. This latter class of model is considered in the next section and for the remainder of this book.

4.3.1 Geissler–Fuller Method

The earliest method for occupancy estimation of which we are aware is that proposed by Geissler and Fuller (1987). They considered a sampling situation where unit i is surveyed K_i times for the presence/absence of the species, and the species is first detected at the unit in survey t_i. A conditional probability of detection can be estimated for each unit as the proportion of surveys in which the species is detected, following the first detection of the species at the unit, that is:

$$\hat{p}_{GF,i} = \frac{\sum_{j=t_i+1}^{K_i} h_{ij}}{K_i - t_i},$$

where h_{ij} is a binary indicator for whether the species was detected (1) or not (0) in survey j of unit i. For example, if a unit was surveyed six times, with the species being detected for the first time in the second survey, and detected subsequently only once, then $K_i = 6$, $t_i = 2$ and $\hat{p}_{GF,i} = 1/(6-2) = 0.25$. The probability of detecting the species at least once during the K_i surveys can then be calculated as:

$$\hat{p}_{GF,i}^* = 1 - \left(1 - \bar{\hat{p}}_{GF}\right)^{K_i},$$

where $\bar{\hat{p}}_{GF}$ is the simple average (across all units) of the $\hat{p}_{GF,i}$'s. A Horvitz–Thompson based estimate of the occupancy probability is then given:

$$\hat{\psi}_{GF} = \frac{\sum_{i=1}^{s} \frac{w_i}{\hat{p}_{GF,i}^*}}{s},$$

where w_i is an indicator variable equaling 1 if the species was observed at the unit, and 0 otherwise.

Geissler and Fuller (1987) then suggest that the actual estimator should be the median or the mean value obtained from a large number of non-parametric bootstraps, and that the standard error for the estimators can also be approximated from a non-parametric bootstrap procedure as the sample standard deviation of the B values of $\hat{\psi}_{GF}$ obtained from the bootstrapped data sets (Manly, 1997).

One of the key assumptions of the Geissler–Fuller approach is that the probability of detecting the species at a unit is constant for all surveys, and especially that the detection probability does not change after the first detection of the species. This is a particularly important assumption in many practical situations. Once observers have detected the species at a unit, they may be more, or less, likely to detect the species in subsequent surveys. One could readily imagine a situation where once the observer has found an indication that the species occupies a unit, the observer uses that information to make detection of the species easier in future visits, e.g., by returning to an animal's den or nest that was discovered during a previous survey. In addition, while Geissler and Fuller (1987) define the detection probabilities $\hat{p}_{GF,i}$ as unit specific, by using the average of these values in the calculation of $\hat{p}^*_{GF,i}$, they effectively assume that the probability of detecting the species in any given survey is equal across all units. Another implicit assumption is the probability that a unit is occupied by the species is also constant across all units. These assumptions are not uncommon, and indeed the majority of the approaches in this book make similar assumptions. However as we shall see, the model-based approach discussed below provides a flexible framework that allows some of these factors to be incorporated directly into the method of analysis.

4.3.2 Azuma–Baldwin–Noon Method

Azuma et al. (1990) consider a slightly different sampling scheme to that given above. For the monitoring of spotted owls (*Strix occidentalis*) in Washington, Oregon, and California, they consider a situation where s units are monitored for the presence/absence of owls with up to K repeat surveys of each unit. However, the unit is not surveyed again once the presence of a pair of spotted owls has been confirmed at the location (by a detection). In later sections we refer to this general design as a removal study design, as units are 'removed' from the pool of units being actively surveyed once the species is first detected, and also because of the analogy with removal studies in mark–recapture (Otis et al., 1978).

Azuma et al. (1990) assumed a constant probability of detecting the species (given presence), p, and then modeled the number of surveys required to detect the species (Y) as a truncated geometric distribution (TGD):

$$Pr(Y = t|0 < Y \leqslant K) = \frac{p(1-p)^{t-1}}{1-(1-p)^t}, \quad t = 1, 2, 3, \ldots, K$$

$$= 0, \quad \text{otherwise.}$$

This provides a model for the time to first detection for those locations where the species was detected during the K surveys. Azuma et al. (1990) suggest a method-of-moments based estimator for p, by equating the expected value of Y to the average number of surveys required to detect the species (conditional on the species being detected) and iteratively solving for p:

$$E(Y|0 < Y \leqslant K) = \frac{1}{p} - \frac{K(1-p)^K}{1-(1-p)^K}.$$

An estimate of the proportion of units occupied would then be:

$$\hat{\psi}_{ABN} = \frac{s_D}{s\hat{p}^*_{ABN}},$$

where s_D is the number of units where the species was detected at least once and $\hat{p}^*_{ABN} = 1 - \left(1 - \hat{p}_{ABN}\right)^K$ is the estimated probability of detecting the species at least once during K surveys based upon the TGD. Azuma et al. (1990) present a number of equations for estimating the variance of their occupancy estimator, although provided that the number of units surveyed is reasonably large (i.e., $s \approx s - 1$), they should all provide similar results. Assuming that S, the total number of units in the population, was very large (i.e., an infinite population), the variance for their estimator will be approximately:

$$Var\left(\hat{\psi}_{ABN}\right) = \frac{\hat{\psi}_{ABN}}{s-1}\left(\left(1 - \hat{\psi}_{ABN}\right) + \frac{\left(1 - \hat{p}_{ABN}^*\right)}{\hat{p}_{ABN}^*}\right).$$

Note that $Var\left(\hat{\psi}_{ABN}\right)$ can be rearranged into a form similar to Eq. (4.4), the estimator variance for when p is known exactly, but with $s - 1$ rather than s in the denominator. Azuma et al. (1990) suggest this adjustment as estimated quantities for occupancy and detection probabilities have been used, though no justification for the adjustment is given. They also provide extra details on correcting for finite sampling (where s represents a substantial fraction of the total population of possible units); estimating the number of locations to sample for a given level of precision; and assessing whether the TGD assumption is reasonable (see Azuma et al., 1990, for these details).

4.3.3 Nichols–Karanth Method

Nichols and Karanth (2002) suggested a more general approach to the problem of estimating the proportion of area occupied, in the context of large-scale monitoring of tiger and their prey species (Fig. 4.1) in India. They advocated that an estimate of the number of occupied units (x) can be obtained by using closed population mark–recapture methods (e.g., Otis et al., 1978; Williams et al., 2002). By focusing on the detection histories for units at which the species was detected at least once, and considering them as the capture histories of individuals encountered during a mark–recapture experiment (e.g., White et al., 1982), estimating the number of occupied units where the species was not detected is completely analogous to estimating the number of individuals in the population that were never captured. Once x has been estimated using an appropriate technique, an occupancy estimator and its associated variance are obtained by:

$$\hat{\psi}_{NK} = \frac{\hat{x}}{s},$$

and:

$$Var\left(\hat{\psi}_{NK}\right) = \frac{\psi\left(1 - \psi\right)}{s} + \frac{Var(\hat{x})}{s^2}.$$

Note the general form of the variance equation as compared to Eq. (4.4), although here uncertainty, with respect to detection probability, is taken into account directly via $Var(\hat{x})$. One shortcoming of this approach is that there are no constraints on the value that $\hat{x}$ may take, hence the estimated number of occupied units could be greater than the number of units surveyed, implying $\hat{\psi}_{NK} > 1$.

4.4 MODEL-BASED APPROACH

An alternative to the *ad hoc* estimation methods given above is to model the probability of the observed outcomes resulting from the stochastic biological and sampling processes. Doing so, it is possible to estimate both the occupancy and detection parameters simultaneously. This framework also provides a direct means of investigating potential relationships between the probabilities of occupancy and detection, and factors such as habitat type or weather conditions at the time of surveying. That is, competing hypotheses about the system can be easily explored and compared within a model-based context, unlike in the *ad hoc* approaches given above. In addition, the flexibility of the approach enables unequal survey effort at different units, providing a great deal of flexibility for realistic study designs. In this section we describe a relatively basic model to introduce the main ideas and philosophies behind this type of modeling, which is extended in various ways throughout this book.

While we focus our presentation on the underlying theories and examples of their applications, we generally avoid detailed discussion of how to actually analyze the data with specific pieces of software as the available software is continually evolving, hence such detailed instruction may quickly become obsolete. However, the underlying methods will remain relatively unchanged. Indeed, providing detailed instruction on how to use each piece of software would likely require its own book. The ability to implement the methods in a user-friendly manner is, however, critical to enable people to apply the methods we are discussing. Table 4.1 presents a list of software options that we are aware of that are available for implementing some of these methods at the time of writing.

4.4.1 Building a Model

A model can be constructed in the manner outlined in Chapter 1: take a set of hypotheses to create a conceptual model about the system, construct a verbal model from this set of ideas, then translate the verbal descriptions into a set of mathematical equations which may be used to estimate the parameters of interest. The development of the model here is presented slightly differently to

TABLE 4.1 Readily available software options for applying model-based approaches for occupancy estimation. An internet search will provide current options for download sites and further information. Given are the software name, operating system (OS), primary input method (graphical user interface, GUI; or syntax), a brief description of the software, and whether estimation is primarily performed using maximum likelihood estimation (MLE) or Bayesian methods

Software	OS	Method	Brief description	Estimation
Program PRESENCE	Win, Mac,[a] Linux[a]	GUI	Windows-based software for fitting occupancy models	MLE
Program MARK	Win, Mac,[a] Linux[a]	GUI	Windows-based software for fitting mark–recapture and occupancy models	MLE
Unmarked	Win, Mac, Linux	Syntax	R package for fitting occupancy, N-mixture and distance sampling models	MLE
RPresence	Win, Mac, Linux	Syntax	R package that utilizes the same source code as Program PRESENCE for model fitting	MLE
RMark	Win, Mac, Linux	Syntax	R package that interfaces with Program MARK	MLE
OpenBUGS	Win, Mac,[a] Linux[a]	Syntax	Generalist software for applying Bayesian methods using Markov chain Monte Carlo	Bayesian
JAGS	Win, Mac, Linux	Syntax	Generalist software for applying Bayesian methods using Markov chain Monte Carlo	Bayesian

[a] *Requires a Windows emulator or virtual Windows PC.*

MacKenzie et al. (2002), with the various steps in the process being outlined more explicitly. The underlying principles of the model-building procedure are, however, the same.

The basis of our conceptual model is that there is a biological and a sampling (or observation) process occurring that affect the outcome of whether the species is detected at a unit (Fig. 4.1). First, the biological layer suggests a unit may be either occupied (with probability ψ) or unoccupied (with probability $1 - \psi$) by the species. In terms of the sampling process, if the unit is unoccupied, then

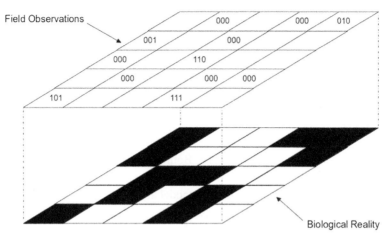

FIGURE 4.1 Illustration of the biological and sampling processes that are accounted for in the model-based approach. Dark and light squares represent occupied and unoccupied sample units, respectively. Not all units are surveyed, and even when surveyed the presence of the focal species is not always detected.

obviously the species cannot be detected there. If the location is occupied, then at each survey (j) there is some probability of detecting the species (p_j; therefore the probability of not detecting the species in the survey is $1 - p_j$); this is the second process. This conceptual model also includes the assumption that the occupancy status of units does not change between surveys (i.e., closure of the unit to changes in occupancy state), but we will discuss relaxing that assumption later.

The biological and sampling processes can also be described more formally in the following manner. Let the presence or absence of the species at unit i be a Bernoulli (i.e., binary; 0 or 1) random variable denoted as z_i, where the probability of presence is ψ. That is:

$$z_i \sim Bernoulli(\psi).$$

The outcome of the jth survey is detection or nondetection of the species, which is also a Bernoulli random variable, that we denote as h_{ij}. Given the species is present at a unit (i.e., $z_i = 1$), the probability of detection is p_j. If the species is absent (i.e., $z_i = 0$), the probability of detecting the species in a survey will be 0. This is represented below through the product $z_i p_j$, indicating the probability of detection the species at unit i depends on the presence or absence of the species (i.e., $z_i p_j = p_j$ if species present, and $z_i p_j = 0$ if species absent). That is:

$$h_{ij} | z_i \sim Bernoulli(z_i p_j).$$

Note that under this formulation, z_i (the true presence/absence of the species) is considered as a latent random variable as it is not directly observable.

Given this basic conceptual model, there are two general approaches in which a model likelihood can be developed for parameter estimation; the observed data likelihood and the complete data likelihood. The former approach is that used by MacKenzie et al. (2002) and in the first edition of this book (MacKenzie et al., 2006). Importantly, the two different approaches do not represent different modeling approaches but simply two different ways of implementing the same modeling technique. Both approaches should give the same results.

The main assumptions of this initial model are:

1. the occupancy state of the units does not change during the period of surveying;
2. the probability of occupancy is equal across all units;
3. the probability of detecting the species in a survey, given presence, is equal across all units;
4. the detection of the species in each survey of a unit is independent of detections during other surveys of the unit;
5. the detection histories observed at each location are independent;
6. there is no species misidentification resulting in false absences.

We shall discuss ways in which many of these assumptions can be relaxed in due course.

Observed Data Likelihood

Using this conceptual model we can develop a verbal description of the underlying processes that gave rise to any observed detection history. In constructing the observed data likelihood, we consider all possible processes that could have generated the observed data. For example, consider the detection history $\mathbf{h}_i = 0101$ resulting from four detection/nondetection surveys for the target species at unit i. A verbal description of this detection history is:

The unit was occupied by the species, the species was not detected in survey 1, was detected in survey 2, was not detected in survey 3 and was detected in survey 4.

For this example, there is only one possibility in terms of the process that generated the observed data (at least within the constraints of the conceptual model). The probability of observing this particular detection history can then be obtained by translating this verbal description into a mathematical description. This is simply achieved by substituting for the appropriate phrases, the parameters that represent the relevant stochastic processes. Hence the above description of the observed detection history could be translated to the following probability statement.

$$Pr(\mathbf{h}_i = 0101|\boldsymbol{\theta}) = \psi(1 - p_1)\,p_2(1 - p_3)\,p_4,$$

where $\boldsymbol{\theta}$ denotes the parameters ψ and p_1–p_4. Compare the verbal description with the probability statement for this detection history. Recognize that the phrase "the unit was occupied by the species" has been represented in the probability statement by the parameter ψ; the phrase "species detected in survey j" is represented by p_j; and the phrase "species not detected in survey j" is represented by $(1 - p_j)$. Probability statements for the observed detection histories can be developed in a similar manner for all units at which the species was detected at least once. Note that whenever the species is detected at least once, that is confirmation of the species being present (assuming no misidentification), which implies the latent variable $z_i = 1$.

For units at which the species was never detected, the same procedure of translating a verbal description into a probability statement is still used, although it is marginally more complicated as there are now two possibilities for why the species was never detected at a unit. Consider the history $\mathbf{h}_i = 0000$, for which the verbal description of the data would be:

*The unit was occupied by the species and the species was not detected in any of the 4 surveys, **OR** the unit was unoccupied by the species.*

The first possibility occurs when the species is present (hence $z_i = 1$) and could be translated to $\psi \prod_{j=1}^{4}(1 - p_j)$, while the second option occurs when the species is absent (and $z_i = 0$) which translates to $(1 - \psi)$. Because the two possibilities are indistinguishable from the data, both must be represented in the probability statement as:

$$Pr(\mathbf{h}_i = 0000|\boldsymbol{\theta}) = \psi \prod_{j=1}^{4}(1 - p_j) + (1 - \psi).$$

The fact that both options are possible for the observed data is represented by the addition of the two terms within the probability statement. This technique is used frequently throughout this book, whenever there are multiple possibilities for the same set of observed data. Technically it is referred to as integration of the latent variable. Note that $\prod$ is denoting the product of a set of terms (see Appendix).

Once the probability statement for each of the s observed detection histories is formed, assuming that the detection histories for the s units are observed independently (so the probabilities can be simply multiplied together), the model likelihood for the observed data can be constructed in the usual manner (see Chapter 3), that is:

$$ODL(\psi, \mathbf{p}|\mathbf{h}_1, \mathbf{h}_2, \ldots, \mathbf{h}_s) = \prod_{i=1}^{s} Pr(\mathbf{h}_i),$$

which reduces to:

$$ODL(\psi, \mathbf{p}|\mathbf{h}_1, \mathbf{h}_2, \ldots, \mathbf{h}_s)$$
$$= \left[\psi^{s_D} \prod_{j=1}^{K} p_j^{s_j} (1 - p_j)^{s_D - s_j} \right] \left[\psi \prod_{j=1}^{4} (1 - p_j) + (1 - \psi) \right]^{s - s_D}, \quad (4.5)$$

where s_D is the number of units where the species was detected at least once, and s_j is the number of units where the species was detected during the jth survey.

Under the assumption that the probability of detection is constant for all occasions, the above approach is equivalent to modeling the number of detections at each unit ($y_i = \sum_{j=1}^{K} h_{ij}$) as a binomial random variable with an inflated zero class, that is:

$$Pr(Y = y_i | \psi, p) = \psi \binom{K}{y_i} p^{y_i} (1 - p)^{K - y_i}, \quad y_i > 0$$
$$= \psi (1 - p)^K + (1 - \psi), \quad y_i = 0.$$

Such an approach has been used by a number of authors including Stauffer et al. (2002, 2004), Tyre et al. (2003), Wintle et al. (2004). However, by considering the sampling process more generally as a series of surveys with specific detection probabilities we maintain a greater degree of flexibility.

Complete Data Likelihood

The complete data likelihood (CDL) is constructed assuming that the value of the latent variable z_i is known. This is of course untrue, but let us pretend that we are omniscient at the moment. If the value of z_i was known, then we could define the probability of observing its value and the probability of the observed sequence of detections conditional upon the value for z_i. That is:

$$Pr(\mathbf{h}_i, z_i | \mathbf{\theta}) = Pr(\mathbf{h}_i | \mathbf{p}, z_i) Pr(z_i | \psi).$$

For example:

$$Pr(\mathbf{h}_i = 0101, z_i = 1|\boldsymbol{\theta}) = (1 - p_1) \, p_2 (1 - p_3) \, p_4 \times \psi.$$

For any unit where the species is detected at least once, this amounts to the same probability as determined using the observed data likelihood because in this situation it is known that $z_i = 1$ from the observed data. In this case, ψ is at the end of the above equation following convention given the ordering of the conditional random variables; as the resulting probability statement is a simple product term, the ordering does not matter.

When the species is not detected, then if it was known that the species was present ($z_i = 1$), the probability statement would be:

$$Pr(\mathbf{h}_i = 0000, z_i = 1|\boldsymbol{\theta}) = \left(\prod_{j=1}^{4} (1 - p_j) \right) \times \psi.$$

While if it was known the species was absent ($z_i = 0$):

$$Pr(\mathbf{h}_i = 0000, z_i = 0|\boldsymbol{\theta}) = 1 \times (1 - \psi).$$

Note that these are just the respective probabilities for the two possibilities that were added together in the observed data likelihood formulation for units where the species was never detected. However, using the complete data likelihood only one of the two options would be used to develop the overall likelihood function as it is being assumed that it is known whether the species is present or not, even for units where the species was never detected.

This simplifies the process of constructing the model likelihood as it is assumed there is no potential for false absences. As for the observed data likelihood, the complete data likelihood can be constructed by taking the product of the detection history probabilities, that is:

$$CDL(\psi, \mathbf{p}|\mathbf{h}_1, \mathbf{h}_2, \ldots, \mathbf{h}_s, \mathbf{z}) = \prod_{i=1}^{s} Pr(\mathbf{h}_i, z_i|\boldsymbol{\theta})$$

$$= \prod_{i=1}^{s} Pr(\mathbf{h}_i|\mathbf{p}, z_i) \, Pr(z_i|\psi)$$

$$= \prod_{i=1}^{s} Pr(\mathbf{h}_i|\mathbf{p}, z_i) \prod_{i=1}^{s} Pr(z_i|\psi).$$

Conceptually, this final version of the CDL amounts to having two distinct independent components; one for the probability of the species being present at each surveyed unit $\left(\prod_{i=1}^{s} Pr(z_i|\psi) \right)$ and one for the observed detection data given the true occupancy status of each unit $\left(\prod_{i=1}^{s} Pr(\mathbf{h}_i|\mathbf{p}, z_i)\right)$. This construction of the model highlights the hierarchical nature of the sampling situation, and the model. The CDL can be further simplified to:

$$CDL(\psi, \mathbf{p}|\mathbf{h}_1, \mathbf{h}_2, \ldots, \mathbf{h}_s, \mathbf{z}) = \left[\prod_{j=1}^{K} p_j^{s_j}\left(1 - p_j\right)^{x-s_j}\right]\left[\psi^x (1 - \psi)^{s-x}\right] \quad (4.6)$$

where $x = \sum_{i=1}^{s} z_i$, that is, the number of sampled units where the species was present. This is a simpler formulation than the corresponding equation for the observed data likelihood, and again note that it decomposes neatly into two components (as indicated by the square brackets); one that only contains detection probabilities and one that only contains the occupancy probabilities. This is a rather elegant result, although unfortunately the values for the latent z_i variables are unknown which means they need to be approximated as part of the estimation procedure. This necessitates the use of methods and algorithms other than those which are typically used in maximum likelihood estimation, such as Markov chain Monte Carlo (MCMC) or the expectation-maximization (EM) algorithm. While the use of the EM algorithm in this context is only starting to be explored, MCMC is regularly used to handle estimation problems involving latent variables. In fact when using software such as OpenBUGS or JAGS, it is often much more convenient to specify the estimation problem in terms of the underlying processes which leads to the software using the CDL approach.

4.4.2 Estimation

As discussed in Chapter 3, parameter estimates could be obtained using either a frequentist or a Bayesian philosophy based upon the likelihood equations defined in Eqs. (4.5) or (4.6). In this section we focus primarily on estimation using maximum likelihood techniques (i.e., a frequentist approach) for the ODL formulation. This is not a reflection of our beliefs about which statistical inference philosophy is more appropriate, but is a matter of convenience. As noted above, use of the EM algorithm for obtaining maximum likelihood estimates using the CDL is only starting to be explored. Bayesian inference for both the ODL and CDL formulations require computer intensive methods such as MCMC to numerically obtain approximations of the posterior distribution of the model parameters. This makes it more difficult to discuss general results and comparisons with other methods, without extensive computer simulation studies. By contrast, the maximum likelihood estimates (MLEs) of the model parameters can be written in a relatively simple form that makes discussion of results somewhat easier.

However, data analysis will be performed with computer software making it feasible to consider either approach in practice, and for some applications there are a number of advantages in using Bayesian approaches.

Below we consider obtaining MLEs of the model parameters using the ODL for two different situations; where detection probabilities are assumed to be either constant or survey-specific. We also present the formula for calculating the asymptotic variance of the MLE of occupancy in the former situation for comparison with the *ad hoc* occupancy estimators given above.

In most circumstances we envision that readers will use computer software to analyze their data; hence we only present relatively rudimentary estimating equations to illustrate the intuitive nature of the estimators. It should also be noted that, for some data sets, the equations given below may result in estimates of occupancy that are greater than 1. This is because the equations do not enforce the constraint that the probabilities must take values between 0 and 1, which is further encouragement for using software where this constraint can be enforced automatically.

4.4.3 Constant Detection Probability Model

Assuming a constant detection probability, the observed data likelihood becomes

$$ODL(\psi, p|\mathbf{h}_1, \mathbf{h}_2, \dots, \mathbf{h}_s)$$

$$= \left[\psi^{s_D} p^{\sum_{j=1}^{K} s_j} (1-p)^{K s_D - \sum_{j=1}^{K} s_j} \right] \left[\psi(1-p)^K + (1-\psi) \right]^{s-s_D}. \tag{4.7}$$

By taking first derivatives with respect to each parameter, and equating to zero we obtain the following equations:

$$\hat{\psi}_{MLE} = \frac{s_D}{s \hat{p}^*_{MLE}} \tag{4.8}$$

where $\hat{p}^*_{MLE} = 1 - (1 - \hat{p}_{MLE})^K$ is the estimated probability of detecting the species at least once during a survey (given the species is present); and

$$\tilde{p}_{MLE} = \frac{\hat{p}_{MLE}}{\hat{p}^*_{MLE}} = \frac{\sum_{j=1}^{K} s_j}{K s_D} \tag{4.9}$$

where $\tilde{p}_{MLE}$ is the estimated probability of detecting the species during a survey, conditional upon the species being detected at least once at a unit. Eq. (4.9) can be rearranged to obtain a closed form estimate of $\hat{p}_{MLE}$, but requires finding the roots of a $K - 1$ order polynomial (which would typically involve using numerical methods whenever $K > 2$). However, note the intuitive form of the equations as they are given. Eq. (4.8) is very similar to the estimators given in Sections 4.2 and 4.3; hence while not derived from a likelihood perspective, the other estimators should approximate the MLE whenever detection probabilities have been estimated appropriately. Eq. (4.9) shows that a conditional estimate of detection probability ($\tilde{p}_{MLE}$; from which we may numerically derive $\hat{p}_{MLE}$) is given by the ratio of total number of species detections, to the total number of surveys conducted at units where the species was detected at least once.

Likelihood theory suggests that the asymptotic variance formula for $\hat{\psi}_{MLE}$ can be obtained by inverting a quantity known as the *information matrix* (a matrix whose (i, j)th element is the expected value of the negative second partial derivative of the log-likelihood with respect to parameters i and j). We do not go through the mechanics of this here, but note that many software packages use a numerical method based on this theory to obtain variance and standard error values for parameter estimates for a wide range of likelihood-based statistical methods (e.g., mark–recapture modeling, generalized linear models).

The variance formula for $\hat{\psi}_{MLE}$, Eq. (4.10), can be expressed in two forms, with either two or three components. The two-component form has the familiar features of a component due to the binomial proportion and the second component related to uncertainty in the number of units that were actually occupied in the sample. This second component can be rearranged to express $Var\left(\hat{\psi}_{MLE}\right)$ as a function of three components. Now, the second component is the uncertainty in the number of occupied units, assuming that p is known, and the third component is the contribution to $Var\left(\hat{\psi}_{MLE}\right)$ from having to estimate p from the data simultaneously. That is, in the latter form, the first two components give the variance formula above for the situation where p is known, Eq. (4.4).

$$Var\left(\hat{\psi}_{MLE}\right) = \frac{\psi(1-\psi)}{s} + \frac{\psi(1-p^*)(1-p)}{s\left[p^*(1-p) - Kp(1-p^*)\right]}$$
$$= \frac{\psi(1-\psi)}{s} + \frac{\psi(1-p^*)}{sp^*} + \frac{\psi(1-p^*)Kp(1-p^*)}{sp^*\left[p^*(1-p) - Kp(1-p^*)\right]}.$$
(4.10)

Clearly, for a given set of data, the variance may be approximated by substituting in the estimated values for ψ and p (and p^*).

4.4.4   Survey-Specific Detection Probability Model

Based upon the model likelihood with survey-specific detection probabilities, using the same techniques as before, we obtain the following estimating equations:

$$\hat{\psi}_{MLE} = \frac{s_D}{s\,\hat{p}^*_{MLE}},$$

where now

$$\hat{p}^*_{MLE} = 1 - \prod_{j=1}^{K} \left(1 - \hat{p}_{j,MLE}\right),$$

and

$$\tilde{p}_{j,MLE} = \frac{\hat{p}_{j,MLE}}{1 - \prod_{j=1}^{K} \left(1 - \hat{p}_{j,MLE}\right)} = \frac{s_j}{s_D}.$$

Again, $\hat{p}_{j,MLE}$ cannot be simply expressed in a closed form; hence a numerical method should be used, but note the intuitive form of $\tilde{p}_{j,MLE}$. Here s_D is equivalent to the number of surveys conducted at time j at units where the species was detected at least once during the K surveys. Hence $\tilde{p}_{j,MLE}$ is just the fraction of surveys conducted at time j at units where the species was eventually detected, which resulted in a detection.

We do not present an equation for $Var\left(\hat{\psi}_{MLE}\right)$ when detection probabilities are survey-specific here, but note that similar methods to those described above (inverting the information matrix) could be used.

4.4.5 Probability of Occupancy Given Species Not Detected at a Unit

In some instances one quantity that will be of interest is the probability that the species was present at a unit given it was never detected. From Bayes theorem we have:

$$\begin{aligned}
\psi_{condl} &= Pr(\text{species present}|\text{species not detected}) \\
&= \frac{Pr(\text{species present and not detected})}{Pr(\text{species not detected})} \\
&= \frac{\psi \prod_{j=1}^{K} \left(1 - p_j\right)}{\psi \prod_{j=1}^{K} \left(1 - p_j\right) + (1 - \psi)}.
\end{aligned}$$

This can be simply calculated from the estimated parameters. For example, suppose a unit was surveyed twice and the species was never detected. Given an estimated occupancy probability of $\hat{\psi} = 0.65$ and detection probabilities estimates of $\hat{p}_1 = 0.3$ and $\hat{p}_2 = 0.6$, the estimated probability of the species being present given it was never detected at the unit would be:

$$\frac{\hat{\psi} \prod_{j=1}^{2}\left(1-\hat{p}_{j}\right)}{\left(1-\hat{\psi}\right)+\hat{\psi} \prod_{j=1}^{2}\left(1-\hat{p}_{j}\right)} = \frac{0.65 \times (1-0.3) \times (1-0.6)}{(1-0.65)+0.65 \times (1-0.3) \times (1-0.6)}$$

$$= \frac{0.65 \times 0.7 \times 0.4}{0.35 + 0.65 \times 0.7 \times 0.4}$$

$$= \frac{0.182}{0.532}$$

$$= 0.34.$$

That is, the unconditional estimated probability of the species being present was 0.65, but by taking into account the fact that the species was not detected in two surveys, the estimated probability of presence is now 0.34. Hence, the fact that the species was not detected at a unit can be incorporated into our inferential procedures. An approximate asymptotic variance for ψ_{condl} could be obtained by use of the delta method (Section 3.4.3), by differentiating the expression for ψ_{condl} with respect to ψ and the p's and given the variance–covariance matrix for ψ and the p's. The derivative of ψ_{condl} with respect to ψ is:

$$\frac{\psi_{condl}}{\psi} = \frac{1-p^{*}}{(1-\psi p^{*})^{2}}$$

where p^{*} is the probability of detecting the species at least once in the K surveys (as defined above), and when detection probability is constant:

$$\frac{\psi_{condl}}{p} = \frac{\psi(1-\psi) K (1-p)^{K-1}}{(1-\psi p^{*})^{2}},$$

or when detection probability is survey-specific:

$$\frac{\psi_{condl}}{p_{j}} = \frac{\psi(1-\psi) \prod_{k \neq j} K (1-p_{k})}{(1-\psi p^{*})^{2}}.$$

While the above derivation may appear daunting to readers who are less quantitative, calculation of ψ_{condl} is included as part of the standard output of some software packages (e.g., Program PRESENCE) or by implementing the above process using numerical algorithms that are commonly available.

For units where the species is detected at least once, $\psi_{condl} = 1$ and $Var(\psi_{condl}) = 0$, as the detection is confirmation of species presence.

4.4.6 Example: Blue-Ridge Two-Lined Salamanders

Maximum Likelihood Estimation

The Blue-Ridge two-lined salamander (*Eurycea wilderae*) is one of over 30 species of salamanders that occur within the Great Smoky Mountains National Park (Dodd, 2003). Like many amphibians, *E. wilderae* has a dual life strategy, with a 1–2 year aquatic larval period, followed by a more terrestrial adult phase, with individuals often occurring far from water (Petranka, 1998). In the late 1990s researchers, working in cooperation with the National Park Service, conducted a series of studies aimed at developing efficient long-term monitoring methods for a suite of salamander species in the southern Appalachians. Occupancy was one state variable explored, and more detailed occupancy analysis of these data can be found in Bailey et al. (2004) and MacKenzie et al. (2005). Here we use detection information from one year (2001) and one species (Blue-Ridge two-lined salamander) to illustrate simple, single season models described above.

Salamanders were sampled using two detection methods: an area-constrained natural cover transect (50×3 m) and a 50 m coverboard transect, consisting of five coverboard stations spaced at 10 m intervals. The two transects were parallel to each other and separated by approximately 10 m. Natural cover objects (wood and rock) or coverboards (pine boards) were carefully turned, and encountered salamanders were identified to species or species complexes. Together, the area sampled by these transects constituted a sample unit. Transects were located near trails, approximately 250 m apart to ensure independence among them. Thirty-nine units ($s = 39$) were sampled once every two weeks from April to mid-June ($K = 5$ surveys) when salamanders are believed to be most active and near the surface.

We explore two simple models: one model assumes that occupancy and detection probabilities are constant across units and surveys, denoted $\psi(\cdot)\, p(\cdot)$, and the second model assumes constant occupancy among units, but detection probabilities are allowed to vary among the five surveys, denoted $\psi(\cdot)\, p(Survey)$. The convention used to denote each model is to list each parameter type in the model (ψ and p) followed by the factors affecting that parameter type in parentheses. A '·' is used in the parentheses to indicate the parameter is being held constant, i.e., an 'intercept only' model. The data were analyzed using the software Program PRESENCE.

Of the 39 transects surveyed, Blue-Ridge two-lined salamanders were detected at 18 which leads to a naïve occupancy estimate (s_D/s) of 0.46. Compare this to the parameter estimates given in Table 4.2 where the two models have been ranked according to AIC and model averaged estimates have been calculated (see Chapter 3 for details of these procedures).

TABLE 4.2 Summary of models fit to Blue-Ridge two-lined salamander example data. ΔAIC is the relative difference in AIC values compared with the top-ranked model; w is the AIC model weight; $Npar$ is the number of parameters; and $-2l$ is twice the negative log-likelihood value. Estimates of occupancy ($\hat{\psi}$) and its standard error ($SE\left(\hat{\psi}\right)$) are given, along with estimates of detection probabilities ($\hat{p}$). The naïve occupancy estimate is 0.46 (18 of 39 transects had one or more detections).

Model	ΔAIC	w	$Npar$	$-2l$	$\hat{\psi}$	$SE\left(\hat{\psi}\right)$	$\hat{p}_1$	$\hat{p}_2$	$\hat{p}_3$	$\hat{p}_4$	$\hat{p}_5$
$\psi(\cdot)\,p(\cdot)$	0.00	0.73	2	161.76	0.60	0.12	0.26	0.26	0.26	0.26	0.26
$\psi(\cdot)\,p(Survey)$	1.95	0.27	6	155.71	0.58	0.12	0.18	0.13	0.40	0.35	0.27
Model averaged estimate					0.59	0.12	0.24	0.22	0.30	0.28	0.26

From the constant detection model $\psi(\cdot)\,p(\cdot)$, the estimated probability of occupancy, $\hat{\psi}$, is 0.60. This means, if a 50 m transect was randomly placed in the study area, the probability of Blue-Ridge two-lined salamanders being present on that transect is 0.60. Alternatively, if the study area was divided up into transect-sized units, the constant detection model would estimate that the proportion of transects occupied by the salamanders is 0.60. Note there is a subtle distinction between the underlying probability of occupancy for a unit, and proportion of units occupied, that we shall ignore for now, but return to in Section 6.1. The estimated detection probability of 0.26 indicates the probability of finding at least one salamander in a single survey (using the same survey methods) of a transect where Blue-Ridge two-lined salamanders are present. That is, the searchers will only find salamanders in approximately one out of every four surveys conducted at a transect where salamanders are present. Under the $\psi(\cdot)\,p(\cdot)$ model, the estimated detection probability is the same for all five surveys. Results are very similar for the $\psi(\cdot)\,p(Survey)$ model in terms of estimated occupancy, but in allowing detection probability to vary with each survey occasion, $\hat{p}$ ranges from 0.13 to 0.40. The naïve occupancy estimate (ignoring detection) is approximately 20% lower than the estimates that account for detection, suggesting there is likely to be a substantial number of false absences.

The model with constant detection probability has much greater support as indicated by the model weights, but with a model weight of 0.27 there is still some support for model $\psi(\cdot)\,p(Survey)$ suggesting the evidence that detection probabilities are equal in each survey is not overwhelming. Model averaged estimates suggest the probability of detecting Blue-Ridge two-lined salamanders during a single survey of occupied units is in the range of 0.22–0.30, thus the probability of not detecting the species when it is present at a transect from the five surveys is approximately 0.22 (i.e., $=(1-0.24)(1-0.22)(1-0.30)(1-0.28)(1-0.26)$).

Obviously, one does not have to use AIC (or similar) model selection and model averaging with these types of approaches. For example, one could perform a hypothesis test on whether the detection probabilities are equal using a likelihood ratio test (Section 3.5.2). That is, test the hypotheses:

$$H_0 = \text{detection probability equal in all surveys.}$$

$$H_A = \text{at least one detection probability is different.}$$

Our test statistic (TS) for the likelihood ratio test is calculated as the absolute difference in the $-2l$ values of the respective models, and the degrees of freedom (df) for the χ^2 distribution (for determination of the test P-value) is the difference in the number of parameters between the two models, i.e., the number of extra parameters required to allow detection probability to be survey-specific.

$$TS = -2l_{\psi(\cdot)p(\cdot)} - -2l_{\psi(\cdot)p(Survey)}$$
$$= 161.76 - 155.71$$
$$= 6.05.$$

$$df = Npar_{\psi(\cdot)p(Survey)} - Npar_{\psi(\cdot)p(\cdot)}$$
$$= 6 - 2$$
$$= 4.$$

This yields a P-value of 0.195, from which it would be reasonable to conclude there is insufficient evidence to reject the null hypothesis and accept that detection probability is equal for all surveys. While we believe there are some situations where hypothesis testing is appropriate (such as with a designed experiment), in general we have a strong preference for comparing models using AIC-type frameworks.

Finally, suppose we want to estimate the probability of a unit being occupied, given Blue-Ridge two-lined salamanders were not detected there in any of the five surveys ($\hat{\psi}_{condl}$). To illustrate the calculation we shall estimate ψ_{condl} based upon the top-ranked model, although given the model selection uncertainty, our 'best' guess would be to calculate the value from each model, then model average the estimates. We also note that Program PRESENCE provides this value as part of the output for a standard single-season, or static, occupancy model so the calculations do not have to made 'by hand'. From the model $\psi(.) p(.)$, an estimate of ψ_{condl} would be:

$$\hat{\psi}_{condl} = \frac{\hat{\psi}\left(1 - \hat{p}\right)^K}{1 - \hat{\psi}\left[1 - \left(1 - \hat{p}\right)^K\right]} = \frac{0.59(1 - 0.26)^5}{1 - 0.59\left[1 - (1 - 0.26)^5\right]}$$
$$= 0.25.$$

The standard error for $\hat{\psi}_{condl}$ can be approximated by application of the delta method, where the variance–covariance matrix for ψ and p given by Program PRESENCE is:

$$\mathbf{V} = \begin{bmatrix} 0.0150 & -0.0038 \\ -0.0038 & 0.0033 \end{bmatrix}$$

and

$$\mathbf{h}'\left(\hat{\theta}\right) = \begin{bmatrix} \frac{\partial \hat{\psi}_{condl}}{\partial \hat{\psi}} & \frac{\partial \hat{\psi}_{condl}}{\partial \hat{p}} \end{bmatrix}$$
$$= \begin{bmatrix} 0.7720 & -1.2551 \end{bmatrix},$$

giving

$$Var\left(\hat{\psi}_{condl}\right) = \begin{bmatrix} 0.7720 & -1.12551 \end{bmatrix} \begin{bmatrix} 0.0150 & -0.0038 \\ -0.0038 & 0.0033 \end{bmatrix} \begin{bmatrix} 0.7720 \\ -1.12551 \end{bmatrix}$$
$$= 0.0215$$

or $SE\left(\hat{\psi}_{condl}\right) = \sqrt{0.0215} = 0.15$. That $\hat{\psi}_{condl} = 0.25$ indicates that Blue-Ridge two-lined salamanders would be expected to be present at a quarter of the transects that yielded no detections after five surveys. If detection was higher, or a greater number of surveys conducted per transect, then ψ_{condl} would be smaller, with a value of 0 indicating no probability of occupancy given no detections, i.e., confirmed absence.

Bayesian Estimation

An alternative approach to statistical inference from maximum likelihood (as used above) is to use Bayesian methods (Chapter 3). Here we reanalyze the Blue-Ridge two-lined salamander with a Bayesian philosophy, using 'pseudo-code' that is compatible with generalist software packages such as OpenBUGS (or WinBUGS) and JAGS. We do not provide detailed instruction on how to use those software packages here, but would refer interested readers to Kéry and Schaub (2012) or Kéry and Royle (2016) for specifics of using these software packages in ecological applications, including occupancy models.

The 'pseudo-code' given in Fig. 4.2 is for fitting a $\psi(\cdot)$, $p(\cdot)$ model. Lines 2 and 3 are defining the prior distribution to use for detection (p) and occupancy (ψ), respectively, both of which are uniform distributions. The uniform prior distribution suggests that before the data were collected, the researcher's believed that detection and occupancy probabilities could be any value between 0 and 1, and all values were equally likely. Lines 4–10 define the actual model to fit to the data, expressed in terms of the latent (unobserved; z) and observed (h) random variables, as defined in Section 4.4.1. Lines 5 and 8 are defining that each of these variables is a random value from the Bernoulli distribution ($\sim$dbern(θ)) with a corresponding probability of a '1' equal to θ. The for loops are simply cycling through the different observations, firstly through each of the s units (the loop indexed by i), then within each unit, looping through each of the K surveys (the loop indexed by j). The only real 'trick' to the code appears on line 6, where for each unit the quantity mu is calculated which will

```
 1:  model {
 2:     p~dunif(0,1)
 3:     psi~dunif(0,1)
 4:     for(i in 1:s) {
 5:        z[i]~dbern(psi)
 6:        mu[i] <- p*z[i]
 7:        for(j in 1:K) {
 8:           h[i,j]~dbern(mu[i])
 9:        }
10:     }
11:  }
```

FIGURE 4.2   OpenBUGS/JAGS code for fitting an occupancy model with constant occupancy and detection probability (i.e., $\psi(\cdot)$, $p(\cdot)$). Line numbers on the left-hand side are not part of the code and are for ease of interpretation.

take the value of p if the species is present at the unit (i.e., z[i]=1), and 0 if the species is absent (i.e., z[i]=0). The quantity mu is then used as the probability of detecting the species in a survey (line 8), indicating that the species can only be detected at units where the species is present, otherwise a nondetection must be observed (presuming no misidentification of the species) with probability 1. If mu[i] was replaced by p in line 8, that would imply the species could be detected in any survey, irrespective of the species presence or absence at a unit.

The code presented in Fig. 4.2 encompasses the definition of the prior distributions for all model parameters and the model itself (in terms of random variables); the actual observed data values and other input values (e.g., number of surveyed units, s) are specified separately. This basic code can be extended to account for increased complexity (e.g., covariates). Note that this is an example of using the complete data likelihood approach, as the model is defined in terms of the 'data we wish we had'; the true presence/absence of the species at each unit (z[i]) which is a latent random variable.

Posterior distributions for ψ and p were approximated using MCMC methods. Three chains were run for a total of 11,000 iterations, with the first 1000 iterations discarded as the 'burn-in period'. Inferences are therefore based on the retained 30,000 samples. Traceplots indicated good mixing of the chains (Fig. 4.3), with a graphical summary of the posterior distributions presented in Fig. 4.4, and a numerical summary in Table 4.3. The mean of the posterior distribution for the probability of occupancy (ψ) is 0.62 with a central 95% credible interval of (0.40, 0.89), and the posterior mean for the probability of detection is estimated to be 0.26 with a central 95% credible interval of (0.16, 0.37). In this case, the posterior distributions provide similar inferences about occupancy and detection probabilities as the MLEs resulting from fitting the corresponding

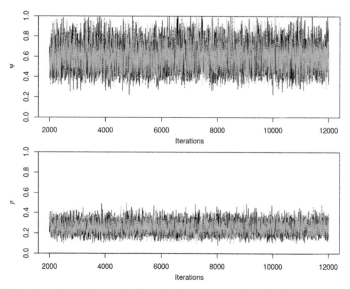

FIGURE 4.3 Traceplots of the retained iterations for ψ (top panel) and p (bottom panel) for Blue-Ridge two-lined salamander example data, from $\psi(\cdot)$, $p(\cdot)$ model. Each color indicates a different Markov chain.

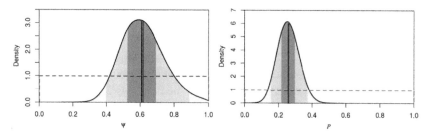

FIGURE 4.4 Posterior distributions of ψ (left panel) and p (right panel) estimated for Blue-Ridge two-lined salamander example data, from $\psi(\cdot)$, $p(\cdot)$ model. Darker shaded regions indicate the central 95% and 50% of the distribution, with the blue and purple lines indicating the median and mean of the distribution, respectively. The dashed red line indicates the uniform prior distribution.

model to the data using maximum likelihood techniques (Table 4.2), although in some situations different conclusions may be reached depending on the prior distributions assumed.

4.4.7 Missing Observations

MacKenzie et al. (2002) developed their likelihood equation by assuming that the probability of detecting the species during survey j was the same at all units. The logic behind this assumption is that for many species, detection prob-

TABLE 4.3 Numerical summary of the posterior distributions of ψ and p for Blue-Ridge two-lined salamander example data, from $\psi(\cdot)\,p(\cdot)$ model. Uniform prior distributions were assumed for all parameters, and summary is from three chains of 10,000 iterations that were retained for inference. Given are the mean, standard deviation (SD), median, and 2.5 and 97.5 percentiles of the distributions.

	Mean	SD	Median	2.5%	97.5%
ψ	0.62	0.12	0.61	0.40	0.89
p	0.26	0.06	0.26	0.16	0.37

abilities will vary with environmental conditions (e.g., level of precipitation, temperature, etc.) and that these conditions will affect all units in a similar manner at the same instant. It is, therefore, important to realize that from a biological viewpoint, the model with survey-specific detection probabilities only makes sense when units are surveyed simultaneously (or within a relatively short timeframe).

However in many situations this may not be the case. From a logistical perspective it is unlikely that all, or even a substantial fraction, of the units can be surveyed at any one time. Data are often collected by small teams of surveyors who must travel from one unit to another, precluding the simultaneous surveying of all units. We use this practicality of the sampling to introduce the concept of missing observations. For example, suppose that over a five-day period, a number of units are to be searched multiple times for the target species. Consider the data below for two of the units, where the first unit was searched on days 1, 2, 3, and 5, whereas the second unit was searched on days 2, 4, and 5. For the first unit, day 4 could be treated as a missing observation, and days 1 and 3 as missing observations for the second unit.

Unit	Day 1	Day 2	Day 3	Day 4	Day 5
1	1	0	1	–	0
2	–	0	–	1	1

By considering the general form of the likelihood above, where the stochastic processes that resulted in an observed detection history are considered to construct an associated probability statement, missing observations can be easily accommodated. In essence, if unit i was not surveyed on occasion j, then the probability of detecting the species at that occasion must be zero, that is, $p_{ij} = 0$. Effectively, by imposing this constraint, whenever a unit is not surveyed, the jth survey occasion is ignored for that unit, and neither p_j nor $(1 - p_j)$ appear in the probability statement. This fairly reflects the fact that no information regarding the detection (or non-detection) of the species has been collected from that unit, at that time. Returning to the above example, the probability of observing the respective detection histories could be expressed as:

$$Pr(h_1 = 101\text{-}0|\boldsymbol{\theta}) = \psi p_1(1 - p_2) p_3(1 - p_5)$$

and

$$Pr(h_2 = \text{-}0\text{-}11|\boldsymbol{\theta}) = \psi(1 - p_2) p_4 p_5,$$

where the '-' in the detection history denotes the missing observation(s).

While we have introduced the concept of missing observations from a design perspective, where logistical constraints prohibit units being surveyed simultaneously, missing observations may arise through a myriad of other circumstances, such as a change in weather, a vehicle or equipment breakdown, or hung-over field technicians (or principle investigators!). To date we have treated the two types of missing observations (design-based and random) in a similar manner, although we note that missing observations arising from random events could also be modeled by an additional set of parameters. Such an approach would be more in line with one of the most important tenets of statistical inference, that the method of analysis should reflect the manner in which the data were collected. However in this situation we doubt that such an extension would have an effect on the resulting inference, provided that the probability of a missing observation does not depend upon the occupancy and detection probabilities. This is due to the statistical concept of 'missingness' and the chances that the data would be missing at random.

The ability to accommodate missing observations has important ramifications for the design of occupancy studies based on this method of analysis. Equal sampling effort is not required across all units. Indeed, as one of the purposes of the repeated surveys is to collect appropriate information allowing detection probabilities to be estimated, going to all units an equal number of times may not be an efficient use of resources. This and other design considerations are explored more fully in Chapter 11.

One special design that we note here is the 'removal design' where a maximum of K surveys will be conducted at a unit, but surveying for the species halts once the species is first detected (e.g., the design used by Azuma et al., 1990). The surveys that are not conducted after the species has been detected at a unit can be regarded as a form of missing observations. In this special case, survey-specific detection probabilities can no longer be estimated (technically, they are not identifiable), although imposition of a single constraint (e.g., $p_{K-1} = p_K$) renders the remaining detection probabilities estimable, or they could be a function of measured covariates (see Section 4.4.8). However, in the case where detection probabilities are constant, the observed data likelihood can be expressed as:

$$ODL(\psi, p|\mathbf{h}_1, \mathbf{h}_2, \ldots, \mathbf{h}_s)$$

$$= \left[\psi^{s_D} p^{s_D} (1-p)^{\sum\limits_{i=1}^{s_D} t_i - s_D} \right] \left[\psi(1-p)^K + (1-\psi) \right]^{s-s_D}, \qquad (4.11)$$

where t_i is the number of surveys required until the species was detected at unit i, with the summation only applied to the s_D units where the species was detected.

4.4.8 Covariate Modeling

So far, all of the methods detailed above have assumed that the probability of a unit being occupied and the probability of detecting the species in a survey (given species presence) are equal across units. Often, this is unlikely to be a reasonable assumption and the probabilities will actually vary among units, that is, the probabilities will be heterogeneous. In fact, in many situations (such as species–habitat modeling; see Chapter 2) the manner in which these probabilities vary in accordance with characteristics of the unit may be the primary focus of the study. By using an appropriate link function, and the model-based approach of this chapter, it is possible to model the probability of a unit being occupied as a function of measured covariates.

There is some choice in the type of link function one might use (McCullagh and Nelder, 1989), but here we tend to use the logit link or logistic equation (see also Chapter 3). In doing so, we can draw parallels with logistic regression techniques. In fact the modeling we discuss in detail could be considered a form of generalized logistic regression analysis where there is some uncertainty as to whether an observed absence equates to a true absence. Therefore, simultaneous logistic regression analyses are conducted on each of the detection and occupancy components to account for this uncertainty. There are also clear parallels between our modeling and that used for capture–recapture studies, where generalized logistic regression is frequently used in the face of uncertainty (e.g., Lebreton et al., 1992; Williams et al., 2002).

Using the logit link, we can express the probability of unit i being occupied as:

$$logit(\psi_i) = \alpha_0 + \alpha_1 x_{i1} + \alpha_2 x_{i2} + \cdots + \alpha_U x_{iU},$$

which is a function of U covariates associated with unit i ($x_{i1}, x_{i2}, \ldots, x_{iU}$) and the $U + 1$ coefficients that are to be estimated: an intercept or constant term (α_0) and U regression coefficients for each covariate. While the probability of occupancy can now vary among units, the actual parameters being estimated (the α's) are still assumed to be constant across all units. In addition, note that if ψ_i is modeled only as a function of α_0 (i.e., there are no covariates in the model), then $\psi_i = \psi$ for all units.

As the units are assumed to have constant occupancy status within a season, the types of covariates that may be considered appropriate for modeling ψ_i are those that remain constant for the duration of the season. This includes most types of covariates that could be used to characterize a unit or its general locality, e.g., habitat type, unit size, unit isolation, elevation, distance from some focal point, and generalized weather conditions. This approach to modeling the occupancy probabilities of units (in this type of framework) is now commonplace.

Using the same principles, MacKenzie et al. (2002) show that the probability of detecting the target species can also be modeled as a function of covariates (this was also noted by Tyre et al., 2003). Here, however, there are two types of covariates that may be considered. The first type are those that remain constant within a season (as for occupancy probabilities) while the second type are those that may vary from survey to survey, e.g., local environmental conditions, time of day or surveyor experience. Again, using the logistic equation the probability of detecting the species at unit i during survey j could be expressed as:

$$logit(p_{ij}) = \beta_0 + \beta_1 x_{i1} + \cdots + \beta_U x_{iU} + \beta_{U+1} y_{ij1} + \cdots + \beta_{U+V} y_{ijV},$$

where $x_{i1}, \ldots, x_{iU}$ denote the U season-constant covariates associated with unit i (which may be different from those used to model occupancy probabilities), and $y_{ij1}, \ldots, y_{ijV}$ are the V survey-specific covariates associated with survey j of unit i.

Note that some of the well-known closed population, capture–recapture models can be specified by way of covariates on detection probability. For example, by specifying the covariate $y_{ij} = 1$ if the target species was detected at unit i prior to survey j, $= 0$ otherwise, one could fit analogs of models M_b and M_{tb} (Otis et al., 1978; Williams et al., 2002) which allow the probability of re-detection to be different from the probability of first detection of the species at a unit (i.e., allow for 'trap response'). This approach is also conceptually similar to the modeling of data collected under the 'removal design' just described where surveying of a unit halts after first detection of the species.

Having the ability to model both occupancy and detection probabilities as functions of covariates, enables a large range of models to be investigated with appropriately collected data. This, in conjunction with the ability to accommodate missing observations, is why we believe model-based frameworks such as that presented above, coupled with sound model selection procedures (e.g., AIC), provide a superior structure to *ad hoc* approaches for making inferences about occupancy-related parameters. Examples illustrating the use of predictor variables as potential covariates, and other features of these models are given in Section 4.4.12.

4.4.9   Violations of Model Assumptions

There are several critical assumptions for the model described above. Briefly, they include:

1. occupancy status at each unit does not change over the survey season; i.e., units are 'closed' to changes in occupancy;
2. the probability of occupancy is constant across units, or differences in occupancy probability are modeled using covariates;
3. the probability of detection is constant across all units and surveys or is a function of unit-survey covariates; there is no unmodeled heterogeneity in detection probabilities;
4. detection of species and detection histories at each location are independent;
5. species are correctly identified; i.e., no false positives.

If these assumptions are not met, estimators may be biased and inferences about factors that influence either occupancy or detection may be incorrect. A number of authors have investigated the impact of assumption violations on parameter estimators in the model described above, and insights can also be gleaned from assessments of many of these same assumptions in the context of closed population mark–recapture estimators (recall the close relationship between occupancy models and closed capture–recapture models; see Section 4.3).

The basic occupancy model extensions that are discussed in Chapters 6 and 7 were developed to address violations of the basic model assumptions. We direct readers to the extensions that might be most relevant to overcome the violations as we discuss each of the key assumptions.

Violation of Closure

Kendall (1999) explored several types of violations of the closure assumption when estimating the size of a population using mark–recapture methods. When individuals exhibited completely random movement in and out of a study area then the estimated population size was larger than the number of individuals present within the study area at any particular sample occasion, but was accurate for the total number of individuals in the 'superpopulation' surrounding the study area. For example, suppose the study area is one half of a forest. On any given day each individual of the target species flips a coin to determine which half of the forest it will reside in for the day. Only those individuals that are within the study area portion of the forest are, therefore, at risk of capture each day. Assuming that the total number of individuals within the forest does not change during the sampling (i.e., the superpopulation is closed), resulting closed-population mark–recapture estimates of abundance (and associated variances or standard errors) correspond to the total number of individuals within the forest, not the number within the study area at any occasion. Similarly, the estimated capture probabilities represent the probability that an individual is within the study area (i.e., at risk of capture) *and* captured. By analogy, we would thus expect the occupancy estimator to be unbiased if species randomly moved in and out of a sampling unit, though the occupancy estimator should now be interpreted as the probability a unit is *used* by the target species (MacKenzie, 2005a). That is, the species is present at the unit during the season (in the sense that at any point in the season the sample unit has a non-negligible probability of containing at least one member of the species), but not always present. The probability of use is likely to be greater than the instantaneous probability of occupancy; the probability the species is present at a unit in a given instant. Similar to the mark–recapture situation, given that a unit is used by the target species, the detection probability is now a combination of two different and confounded components: the probability that the species was present in the sampling unit and the probability that the species was detected, given that it was present. For example, suppose that we wish to determine whether the area sampled with a fixed-radius point count is used by male warblers during the early breeding season, and a male warbler's breeding territory overlaps, but is not completely enclosed by, that area. Therefore, the male may not be within the sampled area during a given survey. Its movement in and out of the area could be considered 'random' if the chances of the male being in the sampled area on the previous survey has no effect on whether it is present in the area for the current survey. An important point here is that the degree of randomness depends on the length of time between surveys. If the second survey was conducted immediately after the first survey, then if the male was in the area for the first survey it is more likely that it will be in the area for the second survey compared to if the male was not in the area during the first survey. This is because there has been insufficient time for the male to move to a random point in its territory. Whether the movement could be considered random would therefore be questionable. However, if surveys were spaced further apart then that would allow the male more time to move and thus movement in and out of the sampled area between surveys might appear more random. Of course, if the intent was to determine what areas are used by warblers over a much shorter time frame (e.g., in a single day) then closely spaced surveys may be reasonable if the level of movement is negligible relative to the spatial and temporal scales of the sampling.

Kendall's (1999) work also suggests that if movement in and out of the sampling unit is not random, the occupancy estimator may be biased. If the species is present during initial surveys and then permanently vacates sampling units during the surveying (i.e., all individuals leave a sampling unit for the remainder of the season; emigration-only movement), then with only two survey occasions per unit, the occupancy estimate should reflect the probability that the species was present at a unit in the first survey occasion. This does not hold, however, when more than two surveys are conducted, with occupancy at the first survey occasion now being underestimated. Similarly, for immigration-only movement (i.e., when additional units are becoming occupied by the species during the course of the surveying), with two survey occasions, the occupancy estimate will reflect the probability that the species is present at a unit in the second survey. Again this no longer holds for greater than two surveys, with occupancy in the final survey now being underestimated. We predict the direction of the bias (or lack of bias) from the work of Kendall (1999) assuming that a model with survey-specific detection probabilities has been fit to the data. The bias is less predictable if a model with constant detection probability is used; allowing detection probabilities to vary among survey occasions absorbs some of the effect of the emigration or immigration. In fact, the systematic decrease or increase of detection probability estimates during a season may indicate that emigration or immigration, respectively, is occurring.

Kendall (1999) suggests that the effect of the bias may be reduced by pooling data, such that the modified dataset consists of only two survey occasions. If emigration is suspected, then the detection data for the second survey onward could be pooled, while for immigration the detection data for all surveys prior to the final survey could be pooled (although note carefully how the occupancy estimate should be interpreted in each case). If both types of movements occur within the system (and are not random as described in the preceding paragraph), bias is likely to remain. These scenarios of entry-only or exit-only movement may occur for species that congregate at breeding or wintering or migration stopover locations (e.g., neotropical migrants; shorebirds at stopover locations; pond-breeding amphibians; spawning fish).

A number of authors have explored the effects of non-random changes within a season specifically for occupancy modeling, with mixed results in terms of potential bias. Using the multi-season occupancy modeling framework (Chapter 8), Rota et al. (2009) and Otto et al. (2013) found occupancy probability to be overestimated when there were non-random changes in occupancy within a season. Their premise was to sub-divide a longer period of time that was being considered as a single season into smaller time periods within which there were at least two surveys, and assume closure within the sub-season periods but non-random changes between sub-seasons. By comparing results from models that were fit to the data that estimated the non-random changes, and those that assumed no changes between sub-seasons (i.e., closure over the whole season), using simulation they were able to identify a positive bias. However, by assuming slightly different scenarios, Hines et al. (2010) and Kendall et al. (2013) both found a negative bias. Details of the model used to simulate their data are given in Sections 6.4 and 6.5 respectively, but the salient point is that both approaches allowed non-random changes between individual surveys, rather than between groups of surveys as in the multi-season based approach. Further investigation of these conflicting results has not been undertaken so far, although we would suggest that the different results are due to the manner in which the data were simulated. A negative bias was found when non-random changes were allowed between individual surveys, but a positive bias when changes were between groups of surveys.

Studies that provide such conflicting results are confusing to practitioners, and without clairvoyant powers it is impossible to know whether one is in a situation where biased estimates are likely to result. The best recourse is that investigators should use their knowledge about the phenology of the target species and design their study to try to minimize violations in the closure assumption. If they are concerned about potential violations once the data have been collected, then in addition to the basic single season model, they should also consider using some the extensions detailed in Chapter 6 to assess the robustness of their results.

The above warbler example raises an important point; closure is species-level not individual-level. Hence, individual animals may move in and out of the sampling unit, but provided at least one (and it does not have to be the same one) is present in the unit at the time of a survey, the closure assumption will be met. When multiple individuals may be moving in and out of a sample unit, it may not be intuitively obvious whether it is reasonable to consider the species-level changes random. If this is a concern, then it may be feasible to consider applying some of the extensions of the basic occupancy model described so far to essentially test for closure or random changes. Finally, when surveys are being conducted to detect species sign (e.g., tracks, scat, markings, etc.) as an indication of species presence in a unit, the closure assumption is in relation to the sign itself, not the individuals that left the sign. We would also point out that 'occupancy' is defined with respect to the probability units contain sign, not necessarily actual individuals of the species.

Extensions of the basic single-season occupancy model that can be used to address violations of the closure assumption include the multi-scale (Section 6.3), correlated detection (Section 6.4), and staggered entry (Section 6.5) occupancy models. If there are sufficient repeat surveys it may feasible, and biologically interesting, to break a single-season into multiple time periods and use the multi-season occupancy models (Chapter 8) to examine how occupancy changes within the season, e.g., early vs. late breeding season. These changes may be random or non-random, although note that 'random' changes do not necessarily require abandonment of the single-season model depending on desired inferences. The multi-season model may also be used as the basis for 'testing' for closure (Chapter 10).

Heterogeneity in Occupancy Probability

The impact of unmodeled variation in occupancy probability among units (occupancy heterogeneity) is relatively unknown compared to other model assumptions, and more thorough simulation studies are needed. Moreover, there is no clear parallel assumption in closed-population capture–recapture literature from which we might draw intuition about violations in this assumption. However we anticipate that in a simple case where two groups of units exist with different occupancy probabilities, but detection probabilities among all units are equal, the estimated occupancy value may appropriately reflect the average level of occupancy for the pooled groups, though the reported variance will be too large (i.e., the estimated precision of the estimator would be conservative). Initially this may seem counter intuitive, but consider the example of pooling the data from two independent binomial experiments, with n_1 and n_2 trials and probabilities of success π_1 and π_2, respectively. The true variance for the total number of successes $(x_1 + x_2 = x_T)$ will be:

$$
\begin{aligned}
Var(x_T) &= Var(x_1) + Var(x_2) \\
&= n_1\pi_1(1 - \pi_1) + n_2\pi_2(1 - \pi_2).
\end{aligned}
$$

However, if group membership was unknown then one might estimate this variance based upon the pooled estimator for the probability of a success, $\hat{\pi}_P = x_T/(n_1 + n_2)$, that is:

$$
\widehat{Var}(x_T) = (n_1 + n_2)\,\hat{\pi}_P\left(1 - \hat{\pi}_P\right).
$$

Avoiding algebraic details, it can be shown that the bias of $\widehat{Var}(x_T)$ is:

$$
bias\left(\widehat{Var}(x_T)\right) = \left[n_1 n_2(\pi_1 - \pi_2)^2\right]/(n_1 + n_2)
$$

(i.e., $\widehat{Var}(x_T)$ will be too large unless π_1 and π_2 are equal).

If there are greater than two groups, or occupancy probability differs among units as a result of an unmodeled continuous covariate, it may again be reasonable to interpret the parameter estimates as representing an average for the units from which data were collected. Whether these estimates are reasonable for the population at large depends on whether the distribution of covariates among sampled units is consistent with the distribution of covariates in the population of all sample units (something that should be achieved, on average, with the random selection of units), although variances may again need adjusting.

Other than using information about unit characteristics (e.g., habitat, vegetation cover, elevation, location, etc.) as covariates (previous section), there are few options to account for heterogeneity in occupancy probability. One option would be to include a unit-specific random effect with a specified distribution to act as a random error term; to account for other sources of variation not accounted for by any covariates. We would point out that this issue of heterogeneous occupancy probabilities is not unique to occupancy models, but also applies to regular logistic regression, and other methods that might be used to assess occupancy–habitat relationships, when detection is presumed perfect; inferences may not be as reliable as desired if an important covariate has not been identified and included the analysis.

Heterogeneity in Detection Probability

Heterogeneity in detection probability can result in negatively biased occupancy estimates. The problem has been studied extensively in closed-population capture–recapture studies and has been explored within the unit occupancy context by Royle and Nichols (2003), MacKenzie and Bailey (2004), MacKenzie et al. (2005), Royle (2006). Naturally, low detection probabilities and high levels of variation among detection probabilities (either among units or surveys) increase the potential bias in occupancy estimates (e.g., Royle and Nichols, 2003). Bias is further exaggerated in studies involving a small number of units or a few repeated surveys at each unit. Increased detection heterogeneity also causes ambiguity in determining appropriate model structure (Section 7.5; Royle, 2006).

Anticipating heterogeneity, and minimizing its effects, both through study design where possible and collecting relevant covariates to model variation in detection, are essential for good performance of models presented in this book. Detection probabilities may vary among surveys as a result of factors such as environmental variables (e.g., weather conditions), seasonal behavioral patterns, or differences among observers. Importantly, when the range of values of these factors is similar across all units, the level of bias, if any, is likely to be small and the previously described models are likely to be adequate. When the values of the factors are different at different units (e.g., some units tended to be predominately surveyed during early morning, while other units were predominately surveyed during late morning), that could result in a greater bias that should be accounted for. Detection probabilities may also vary among units because of habitat features (e.g., forest density). When information has been collected on factors that are suspected to influence detection probabilities, these can be included as potential covariates in an analysis of the data (Section 4.4.8).

One unique source of heterogeneity in detection probability among units is the size of the local population at each unit. Logically, the probability of detecting a single individual of a target species increases with increasing local population size of the species. Royle and Nichols (2003) suspect that local abundance might be the biggest source of heterogeneity in detection probabilities for many occupancy studies, and they develop a model to accommodate local abundance as the source of heterogeneity in detection. This 'abundance' model and others that deal with more generic forms of heterogeneous detection probabilities are discussed in detail in Chapter 7.

Lack of Independence

Events are not independent if the outcome of one event affects the outcome of a second event. Our discussion of violations of the independence assumption is at two levels: survey-level within a single unit and the among unit-level. However before entering that discussion, it is important to realize that at this point we are only considering how the independence assumption relates to the *observations* from the units that were selected for surveying, not independence of the units themselves. Hence even if the occurrence of the species among units in an area displays a high degree of spatial correlation (i.e., nearby units are more likely to be either occupied, or unoccupied, than units farther apart), that is not necessarily a violation of the independence assumption from a data analysis perspective. This has important ramifications in terms of study design as there is no requirement of individuals being constrained to single sampling units; it is perfectly valid to have the same individual using multiple sampling units, which may occur when sampling units are smaller than an individual's home range. As inference is at the species level, it is largely immaterial whether nearby units are occupied by the same, or different individuals of the species. We provide greater discussion of this issue in Chapter 11.

At the survey-level a lack of independence between survey outcomes can cause occupancy probability estimates to be biased (Hines et al., 2010). When the survey outcomes are not independent the probability of detecting the species in one survey depends on whether or not it was detected in another survey. This might happen, for example, in call surveys when surveys are spaced too closely together and the calling behavior of the species is likely to exceed the duration of a survey, hence the same vocalization event (which may include responses from other individuals) spans multiple surveys. Once the calling begins, the probability of detection is going to be higher in the subsequent surveys. Another example is a species where individuals are highly territorial and the sample unit being surveyed constitutes a smallish fraction of its home range. As an individual moves about its home range, when it is in the vicinity of the sample unit the species may be detected, but when the individual is elsewhere in its home range, there is no chance of detecting the species. If the frequency of surveying is high relative to the individual's movement rate, such that the duration for which the individual is near the sample unit is greater than a single survey, if the species is detected in a survey then it may be more likely to be detected in the immediately following surveys.

These are both examples where surveys are repeated temporally, but a similar effect can occur with spatial replication. A defining feature of these examples is that there are pulses of detections and non-detections. If the pulses are somewhat predictable, or thought to relate to some measurable variable (e.g., time of day), then it may be possible to include a suitable covariate. Where pulses are less predictable or suitable covariate information is not available, then the correlated detection occupancy model of Hines et al. (2010; Section 6.4) would be useful. An extreme example of detection pulses is if surveying within a sampling unit is at two scales, e.g., the unit is surveyed on multiple days, and multiple surveys each day. If the species is not always present within the unit at that first scale (that is, the unit is used by the species rather than continuously occupied), then regardless of how many surveys are conducted at the second scale, the species will never be detected. Therefore, it would be inappropriate to treat all of the surveys of the unit as independent. The multi-scale occupancy model of Nichols et al. (2008; Section 6.3) could be used in such situations. Other forms of dependence (e.g., more likely to redetect the species once it has been detected at a unit for the first time) may also create biases in the parameter estimates, but could be accommodated by defining an appropriate survey-specific covariate (e.g., $y_{ij} = 1$ if species detected at unit i prior to survey j, $= 0$ otherwise). Incorporating a covariate defined in this manner for detection probability will allow the probability of redetecting the species to be different from the probability of first detection, just as in the 'behavioral' or 'trap response' models used in capture–recapture for estimating animal abundance (Otis et al., 1978; Williams et al., 2002).

At the unit-level, it is assumed the detection histories of units are independent observations. There are two ways in which this independence assumption could be violated. The first is when multiple units are being surveyed simultaneously and similar detection events are consistently registered across the multiple units such that the detection histories are very similar across the multiple units. For example, an owl hooting at one unit is also recorded by an observer at a nearby unit, or if a bait or lure is being used and the effective range of the attractant at multiple units overlaps, resulting in an individual moving from one unit to another within a survey period. In these instances, the effective sample size (the number of independent units or detection histories) is actually smaller than the number of units surveyed, and the estimated standard errors obtained from the above model are too small (MacKenzie and Bailey, 2004). This is a form of overdispersion (Chapter 3). Investigators should attempt to address this issue, in the first instance, from a design standpoint, using their knowledge of the target species' movements to distribute sample units to minimize possible independence violations. The assessment of model fit, described next, has some power to detect problems caused by non-independence, and standard errors can be adjusted with an estimated variance inflation factor.

The second way of violating the independence assumption of unit-level observations is when units are not selected from the population independently of one another, when species occurrence is spatially correlated. Often researchers will cite logistical reasons for clustering their survey units in some manner, e.g., randomly select the intersection of grid lines and survey the four adjoining cells and consider them four independent observations, or randomly select a number of transects and place sampling units along the transects at predefined intervals. With such designs, the occupancy status of units within a cluster may be more similar than if units were independently selected from the population of interest, dependent upon the nature and scale of any spatial correlation relative to the size of, and spacing between, units. This does not imply that there should always be some minimum spacing between units when they are independently selected, as a sample that results in some neighboring units being selected is just as valid as a sample with no neighboring units selected. When a clustered design is being used then estimates of occupancy from the basic model described so far may be biased depending on the level of spatial correlation, and the true level of uncertainty in the occupancy estimate will be larger than indicated by the reported standard error, as the number of independent samples will be closer to the number of clusters surveyed rather than the number of units. This could, again, be considered as a form of overdispersion, hence model selection and standard errors could be adjusted based upon a variance inflation factor (i.e., $\hat{c}$; Chapter 3 and next section). Alternatively, the multi-scale occupancy model (Section 6.3) could be used to account for some broad-level spatial correlation where the clusters represent one spatial scale of occupancy and units within clusters a second scale of occupancy. Another option would be to explicitly account for the spatial correlation in occupancy (e.g., Pacifici et al., 2016; Section 6.6). This issue is discussed further in Chapter 11 with respect to study design although a take home message for practitioners is that deviations from a statistical ideal in the name of logistical convenience will have repercussions on the required analyses (i.e., increased complexity) and resulting inferences.

Species Misidentification

False-positive detections (i.e., incorrectly recording the target species as detected in a survey) will cause occupancy probability to be overestimated as it may lead to the species being recorded as present at the unit when it was truly absent (Royle and Link, 2006; McClintock et al., 2010a, 2010b; Miller et al., 2011). While nondetection (or false-negatives) may be more prevalent, even relatively low false-positive rates, e.g., $<5\%$, may induce substantial bias in occupancy estimates. It is therefore an important issue that should not be ignored. There is now a set of methods available for accounting for misidentification (see Section 6.2), although it is always desirable to ensure, as far as practical, that appropriate protocols are used to minimize species misidentification and false-positive detections (e.g., rigorous observer training programs). Some of the extensions described later also allow incorporation of other sources of information about misidentification rates, therefore field protocols may also require the collection of relevant auxiliary data (e.g., field samples collected for DNA confirmation of species).

While species misidentification causes occupancy probability to be overestimated when accounting for imperfect detection, it is important to realize that it has a similar impact regardless of whether detection is perfect otherwise, ignored completely, or presence-only data are being used. In any of those cases, there are likely to be too many 1's in the data set. The effects of misidentification can be particularly insidious when it does not happen at random, but when it correlates with factors that might be determinants of species occurrence, producing misleading inferences.

4.4.10 Assessing Model Fit

Whenever possible it should be demonstrated that a fitted model adequately describes the observed data, i.e., a model should be assessed for lack-of-fit (McCullagh and Nelder, 1989; Lebreton et al., 1992). It should be demonstrated that the models being considered for the data are realistic and capture the important features of the system under study. Substantial lack-of-fit in a model(s) may lead to inaccurate inferences, either in terms of bias or in terms of precision (e.g., reported standard errors are too small). In order to have some degree of confidence in the inferences resulting from an analysis of real data, it is important that the model fit be assessed.

A popular approach for analyzing ecological data is to fit a suite or candidate set of models to the data, and use a model selection technique such as AIC, or similar measures, for choosing the 'best' model(s). The selection of a 'best' model(s) does not guarantee the selection of a 'good' model. Given the popularity of using such techniques in the analysis of ecological data, it is important to realize that they assume that the candidate set contains at least one model that fits the data adequately (Burnham and Anderson, 2002), and that they are not a substitute for assessing model fit. A common approach to model selection using AIC and related metrics is to test fit of the most general model in the model set (Burnham and Anderson, 2002). If fit of that model is deemed adequate, then model selection proceeds in the usual manner based on AIC. If the fit is not adequate, then an analyst needs to assess whether lack of fit is likely due to non-independent observations, or structural inadequacies in the model (e.g., missing covariates, detection heterogeneity). In the former case, a quasi-likelihood overdispersion parameter is estimated and used to modify model selection, i.e., use QAIC, and inflate standard errors (Chapter 3), while in the latter case it would be necessary to consider different models to account for the inadequacies.

However, while we advocate that models should be assessed at every opportunity, the reality in many ecological studies is that sample sizes may often be too small to detect poor model fit, i.e., tests for model fit may have low power. This could lead to a sense of false confidence in that one may decide that a model is adequate merely because there was insufficient evidence to the contrary, not because the model structure was appropriate. This dilemma is one to which we have no solution except to suggest that if using a statistical test to assess lack-of-fit, then it may be appropriate to use larger type I error rates than are often used, thus be willing to reject a null hypothesis of adequate fit based upon weaker evidence, i.e., considering rejecting the null hypothesis even when a P-value is greater than 0.05. As noted above, if fit is not viewed strictly in a hypothesis-testing context, then a reasonable approach is to try to estimate the degree to which fit is inadequate. Overdispersion parameter estimates, $\hat{c}$, provide such estimates for use in model selection and inference (see below and Chapter 3).

For many models that are frequently used by ecologists (including those described later in this book), no adequate methods for assessing fit have been developed. For example, there are no tests for lack-of-fit for the vast majority of mark–recapture models, particularly when individual covariates are used. We believe the reason for this is that developing such techniques is often viewed as a secondary or tertiary level problem when compared to the development of more flexible and realistic methods of data analysis. Only once the new models are accepted and receive widespread use (provided they are not quickly superseded) will an impetus be created to develop techniques for assessing the fit of the models. However we believe that such techniques will be developed over time. Finally, we note that often the best 'test' of a model is to compare predictions from the model with an independent data set.

In terms of single-season occupancy models, MacKenzie and Bailey (2004) developed a method for assessing the fit of the model described in this chapter. They suggest a relatively straightforward approach that effectively tests whether the observed number of units with each particular detection history, has a reason-

TABLE 4.4 Example of detection histories and how one would consider them to be different cohorts for the purpose of assessing model fit

Unit	h_i	Cohort
1	1010	1
2	1001	1
3	1101	1
4	-110	2
5	-100	2
6	0-11	3
7	1---	4

able chance of occurring if the target model (the model which is being assessed) is assumed to be correct.

Let O_h be the number of units observed to have detection history h, and E_h be the expected number of units with history h according to the target model. For example, suppose the target model assumes that occupancy and detection probabilities are constant across units and time, i.e., the model $\psi(\cdot)\,p(\cdot)$, and parameter estimates are $\hat{\psi} = 0.82$ and $\hat{p} = 0.43$. The expected number of units with the detection history 101 would be:

$$E_{101} = s \times \widehat{Pr}(h = 101)$$
$$= s\hat{\psi}\,\hat{p}(1 - \hat{p})\,\hat{p}$$
$$= s \times 0.82 \times 0.43^2 \times 0.57$$
$$= 0.09s.$$

More generally, E_h equates to the sum of the estimated probabilities of observing h across all units, as the occupancy or detection probabilities may be unit specific depending upon the model that has been fit to the data; for example:

$$E_{101} = \sum_{i=1}^{s} \widehat{Pr}(h_i = 101)$$
$$= \sum_{i=1}^{s} \hat{\psi}_i \hat{p}_{i1}(1 - \hat{p}_{i2})\,\hat{p}_{i3}. \tag{4.12}$$

This assumes, however, there are no missing observations and equal sampling effort, as an implicit requirement is that the estimated probabilities of observing each possible detection history sum to 1 (i.e., $\sum_{h} \widehat{Pr}(\mathbf{h} = h) = 1$) so that $\sum_h O_h = \sum_h E_h$. To account for missing observations, MacKenzie and Bailey (2004) suggest that units with each unique combination of missing values be regarded as separate cohorts, e.g., Table 4.4. Therefore, for each cohort c, the expected number of units with each detection history becomes:

$$E_{h_c} = \sum_{i=1}^{s_c} \widehat{Pr}(\mathbf{h}_i = h_c),$$

where s_c is the number of units in the cohort.

Once the E_h's have been calculated, MacKenzie and Bailey (2004) recommend the use of a simple Pearson's chi-square statistic to test whether there is sufficient evidence of poor model fit.

$$X^2 = \sum_c \sum_h \frac{\left(O_{h_c} - E_{h_c}\right)^2}{E_{h_c}}. \tag{4.13}$$

As many of the E_{h_c} are likely to be relatively small (<2) for even moderate values of K (say $\geqslant 5$), the usual distributional arguments used to justify that X^2 will have a chi-square distribution with df degrees of freedom are unlikely to hold. Therefore MacKenzie and Bailey (2004) suggest using a parametric bootstrap procedure to determine whether the observed value of X^2 is unusually large.

Parametric bootstrapping involves assuming that the target model is correct (i.e., a good approximation to the process generating the data) and then generating alternative sets of data subject to the constraints that the s_c's are fixed. As the target model is known to be correct for the generated data, if the observed data appear typical in comparison, then it would seem reasonable to conclude that the target model may be adequate for the observed data. Hence parametric bootstrapping may be an ideal technique for assessing the model's structure. For the single-season occupancy model, MacKenzie and Bailey (2004) implement parametric bootstrapping as follows:

1. Fit target model to the observed data and estimate parameters $\hat{\psi}_i$ and $\hat{p}_{ij}$ (which may be functions of covariates).
2. Calculate the test statistic for the observed data, X_{Obs}^2, using the model fit in step 1, i.e., using Eqs. (4.12) and (4.13).
3. For each unit generate a pseudo-random number (r) between 0 and 1. If $r \leqslant \hat{\psi}_i$ then the unit is occupied, hence generate K further pseudo-random numbers (r_j) between 0 and 1. If $r_j \leqslant \hat{p}_{ij}$ then the species was 'detected' and the corresponding bootstrapped observation is a '1', otherwise '0'. If $r > \hat{\psi}_i$, then the unit is unoccupied and the bootstrapped observations will all be '0' for that unit.

4. Fit a model with the same structure as in step 1 to the bootstrapped data set.
5. Calculate the test statistic for the bootstrapped data, X_B^2, using the model fit in step 4, and store the result.
6. Repeat steps 3–5 a large number of times to approximate the distribution of the test statistic, given the fitted model is correct.
7. Compare X_{Obs}^2 to the bootstrap distribution of X_B^2 to determine the probability of observing a larger value (the P-value).

If the target model is found to be a poor fit for the data then the analyst needs to contemplate the likely source of any lack of fit. When the suspected source is overdispersion due to a lack of independence of observations (see previous section), then an overdispersion parameter ($\hat{c}$) may be used to inflate standard errors (McCullagh and Nelder, 1989) and adjust model selection procedures (Burnham and Anderson, 2002). Following White et al. (2002), $\hat{c}$ may be estimated as

$$\hat{c} = \frac{X_{Obs}^2}{\bar{X}_B^2},$$

where $\bar{X}_B^2$ is the average of the test statistics obtained from the parametric bootstrap. If the target model is an adequate description of the data, then $\hat{c}$ should be approximately 1. Values greater than 1 suggest that there is more variation in the observed data than expected by the model (overdispersion), while values less than 1 suggest less variation (underdispersion). Typically, corrections are only applied when there is evidence of overdispersion, and underdispersion is ignored (Burnham and Anderson, 2002).

When the analyst believes the lack of fit might be due to structural inadequacies in the model (e.g., missing covariates), then further consideration will be required to attempt to improve the model. Unfortunately, as of 2017, there has been limited development of methods that may aid diagnosing where such deficiencies may lie, although some suggestions are given below.

Note that there are also alternative calculations for $\hat{c}$, the performance of which has not been evaluated in an occupancy setting at the time of writing. Fletcher (2012) proposed a method for calculating $\hat{c}$ for sparse data that White and Cooch (2017) applied to closed population mark–recapture models as:

$$\hat{c} = \frac{\hat{c}_x}{\bar{r}},$$

where

$$\bar{r} = \frac{1}{H} \sum_{h=1}^{H} \frac{O_h}{E_h}.$$

Here $\hat{c}_x$ is a regular Pearson χ^2 statistic divided by its degrees of freedom and H is the total number of observable capture histories ($= 2^K - 1$ in mark–recapture studies). A modification of this approach may be useful for occupancy models given that sparse data are not uncommon, particularly when accounting for missing values with the above cohort approach.

When multiple models are to be considered for the data and some form of model selection procedure is to be used, it is generally recommended that the most general or 'global' model (i.e., the most complex model with the greatest number of parameters) be assessed for lack-of-fit first. The logic is that if the global model fits the data, then any reduced model that has fewer parameters, but explains a similar level of variation in the data (i.e., a more parsimonious model), should also provide an adequate description of the data. If the global model is found to be a poor fit, then any adjustments to standard errors or model selection procedures should be made on the basis of the $\hat{c}$ value calculated from the global model (Burnham and Anderson, 2002). For example, in an analysis of occupancy data collected on members of the terrestrial salamander complex *Plethodon glutinosus*, MacKenzie and Bailey (2004) found evidence that the global model considered was a poor fit to the data ($X^2 = 63.1, p$-value $= 0.056$). QAIC (Section 3.6.3; Burnham and Anderson, 2002) was, therefore, used for model selection with a value $\hat{c} = 1.43$ obtained using their approach described above. The standard errors for the parameter estimates were also inflated by a factor of $\sqrt{\hat{c}} = 1.20$.

MacKenzie and Bailey (2004) found via simulation that their procedure had some power for detecting poor model fit for the scenarios considered (at a level of $\alpha = 5\%$), particularly when $K > 5$. Interestingly, when lack-of-fit was caused by omission of an important unit-specific covariate for detection probabilities in the target model, their test had greater power when the average detection was high (>0.5), but when the poor model fit was caused by a lack of independence among units, the test performed better when detection probabilities were lower (<0.5). In terms of assessing the fit of a target model with respect to heterogeneity of occupancy probabilities, MacKenzie and Bailey (2004) found their test to have no power. They suggest this should not be unexpected, because when the model is misspecified with respect to detection probabilities, some units will have an unusually large (or small) number of species detections. However as occupancy is effectively a single binary observation for each unit, there is no such outward indication that the model may be inadequately describing the data. They go on to draw a comparison with the similar problem that exists for logistic regression which led to the development of the Hosmer–Lemeshow Test (Hosmer and Lemeshow, 1989, p. 140) that uses the predicted probabilities of a success to classify the observations into k groups, and speculate that such an approach may be modified to assess the fit of a target model with respect to occupancy probabilities. However, point estimates of the occupancy-related parameters may still be considered reasonable as an average for the units in the study (as discussed in the previous section), hence the consequences of not identifying heterogeneous occupancy probabilities will not be as serious as for heterogeneous detection probabilities.

Experience with this method of assessing model fit has identified that there are situations where it does not perform overly well, particularly if there are a large number of missing values in the detection histories and/or a large number of surveys per unit (e.g., >10), and a relatively small number of units. As the number of survey occasions increases, the number of possible detection histories increases exponentially, e.g., with five surveys there are $2^5 = 32$ possible detection histories, but with 10 surveys there are $2^{10} = 1024$ detection histories. If the number of surveyed units is relatively small (e.g., 20) the expected number of units with a particular detection history (E_{h_c}) is going to be extremely small, resulting in extremely large X^2 values, i.e., several millions. When you also factor in different combinations of missing values being treated as different cohorts for the calculation of the test statistic (i.e., more possible detection histories), the expected number of units may be extremely small. While the parametric bootstrap was used to mitigate the effect of small expected values, and the simulation results of MacKenzie and Bailey (2004) indicate it was successful in the scenarios they considered, it would seem there are situations where the parametric bootstrap also fails to provide meaningful results.

If Bayesian methods of inference are being used then the appropriateness of the model could be assessed using Bayesian p-values and posterior predictive checks (PPC; Gelman et al., 1996, 2004). As noted in Chapter 3, PPC involves generating new sets of data for the surveyed units, then comparing some aspect of the observed data to the generated datasets to determine how typical the observed data might be. There is a great deal of flexibility in what aspect of the data is being compared; it may be some form of test statistic as above, or something more general that the analyst thinks is reasonable, e.g., number of units with at least 1 detection, total number of detections, or frequency of number of detections per unit. Kéry and Royle (2016, Section 10.8) found that assessing the fit of occupancy models on the raw binary data was unsuccessful and some level of aggregation was required (e.g., by unit, occasion, or encounter history). To generate a new dataset, model parameter values are drawn from their joint posterior distribution, occupancy and detection probabilities are calculated, then random detection histories generated in the same manner as for the parametric bootstrap. In practice, a routine for conducting PPC is often incorporated into the MCMC procedure when analyzing the observed data such that new datasets are generated using the parameter values in the current iteration rather than resampling the joint posterior distribution after completing the analysis. A Bayesian p-value is calculated as the proportion of times the selected summary statistic calculated for the generated data is greater than the value calculated from the observed data.

4.4.11   Diagnostic Plots

One set of tools for assessing model fit that is largely lacking (as of 2017) for single-season occupancy models, and actually all occupancy models, is graphical diagnostics similar to residual plots in regular regression problems. These may help identify, for example, structural inadequacies in the model and whether the functional relationship for a covariate that has been included in the model is appropriate. For example, if a logit-linear relationship between occupancy and a covariate has been used, is there any indication of a systematic lack of fit that might suggest a quadratic relationship (or some other transformation of the covariate) should be considered.

A notable exception is a diagnostic plot suggested by Warton et al. (2017) using Dunn–Smyth residuals (DSR; Dunn and Smyth, 1996). DSR are based upon the cumulative distribution functions (CDF) of a standard normal variable and the observed variable x_i. The intent is to be able to produce residual plots that are analogous to those used in standard linear regression problems. For a discrete valued x_i the DSR, r_i, can be calculated as:

$$r_i = \Phi^{-1}\big((1 - u_i)F(x_i) + u_i F_-(x_i)\big),$$

where $\Phi^{-1}(\cdot)$ is the inverse standard normal CDF, $F(\cdot)$ is the CDF of x_i and $F_-(\cdot)$ is the value of the CDF preceding $F(\cdot)$, and u_i is a random value from the standard uniform distribution. The inclusion of u_i is to provide some random 'jittering' given the discrete values of x_i. Warton et al. (2017) describe methods to calculate residuals for both the occupancy and detection components in order to assess the appropriateness of each component separately.

As true presence/absence of the species is not observable, Warton et al. (2017) use apparent occupancy, z_i^*, for the basis of their occupancy residual which is defined as:

$$z_i^* = 0, \ \text{if species not detected at unit } i, \text{ and}$$

$$= 1, \ \text{if species detected at least once at unit } i.$$

The CDF for z_i^* is therefore:

$$F\left(z_i^*\right) = 1 - \psi\left(1 - \prod_{k=1}^{K}(1 - p_{ik})\right), \text{ if } z_i^* = 0$$

$$= 1, \text{ if } z_i^* = 1.$$

Given that *apparent* occupancy has been used, which is a combination of occupancy and detection, the residual may be influenced by detection to some degree.

For the detection residuals, Warton et al. (2017) use the total number of detections at a unit, d_i, conditional upon there being at least one detection at the unit (i.e., residuals are only calculated for those units where the species is detected). When detection probability is constant across repeat surveys, then the CDF is:

$$F(d_i|d_i > 0) = \frac{1}{1-(1-p)^K} \sum_{j=1}^{d_i} \binom{K}{j} p_i^j (1 - p_i)^{K-j},$$

and when detection probability varies across repeat surveys of unit i:

$$F(d_i|d_i > 0) = \frac{1}{1 - \prod_{k=1}^{K}(1 - p)} \sum_{\mathbf{h}|1 \leqslant d_h \leqslant d_i} \prod_{k=1}^{K} p_{ik}^{h_k} (1 - p_{ik})^{1-h_k}. \quad (4.14)$$

The summation term in Eq. (4.14) is calculating the probability of observing between 1 and d_i detections for all possible detection histories. This calculation becomes more computationally intensive as K and d_i increase. By calculating a detection residual based upon the total number of detections, there is only a single residual value for each unit (where the species was detected at least once), which creates a difficulty when the desire is to compare the residual to a survey-specific covariate whose value may be different for each survey. Warton et al. (2017) suggest that one can either plot r_i against the average covariate value for unit i, or replicate the value of the residual r_i for each specific covariate value.

Warton et al. (2017) suggest the use of a smoother (e.g., a generalized additive model, GAM) to aid interpretation of the residual plots, as patterns are not always obvious given the underlying discreteness of the data, and the random jittering introduced by the u_i terms when calculating the residuals (which will actually cause the residuals to have slightly different values each time they are calculated). Warton et al. (2017) demonstrate that their approach does have some ability to diagnose problems with model fit, particularly with respect to evaluating covariate relationships.

We have recently investigated an alternative diagnostic plot, a cumulative sum (CUSUM) residual plot, based upon the method developed by MacKenzie (2002) for graphically evaluating the fit of mark–recapture models. Pearson-type residuals are proposed as model diagnostic aids for both occupancy and detection components of a model. For an occupancy residual, we suggest using the expected value of z_i (the presence/absence latent variable), conditional upon the unit's detection history, and estimated parameters as the 'observed' value. That is, the occupancy residual is defined as:

$$\zeta_i = \frac{E\left(z_i | \mathbf{h}_i, \hat{\boldsymbol{\theta}}\right) - \hat{\psi}_i}{\sqrt{\hat{\psi}_i \left(1 - \hat{\psi}_i\right)}},$$

where $\hat{\boldsymbol{\theta}}$ represents all the estimated parameters and

$$E\left(z_i | \mathbf{h}_i, \hat{\boldsymbol{\theta}}\right) = 1, \text{ if } \sum_{j=1}^{K} h_{ij} > 0$$

$$= \frac{\hat{\psi}_i \prod_{j=1}^{K} \left(1 - \hat{p}_{ij}\right)}{1 - \hat{\psi}_i \left(1 - \prod_{j=1}^{K} \left(1 - \hat{p}_{ij}\right)\right)}, \text{ if } \sum_{j=1}^{K} h_{ij} = 0.$$

Note that as detection probability increases and/or the number of surveys at a unit (K) increases, then $E\left(z_i | \mathbf{h}_i, \hat{\boldsymbol{\theta}}\right)$ will tend to 0 when the species is never detected at a unit; therefore it will tend to the value of the latent variable z_i.

For detection, definition of a residual is more straightforward as h_{ij} is observed directly, although detection is only relevant at units where the species is present. Therefore, a detection residual is defined here as:

$$\delta_{ij} = E\left(z_i | \mathbf{h}_i, \hat{\boldsymbol{\theta}}\right) \frac{h_{ij} - \hat{p}_{ij}}{\sqrt{\hat{p}_{ij} \left(1 - \hat{p}_{ij}\right)}}.$$

To overcome difficulties in interpreting plots of individual residuals due to their dichotomous nature, we suggest plotting the CUSUM of the residuals, where residuals were placed in a meaningful order (e.g., according to the rank of a potential covariate). The underling logic is that any systematic pattern in the residuals (in that particular order) should result in a distinctive shape of the CUSUM curve. That is, if there is a long run of negative values in the ordered residuals, the CUSUM value would become increasingly negative, resulting in a negative trend for the corresponding portion of the CUSUM curve, and similarly

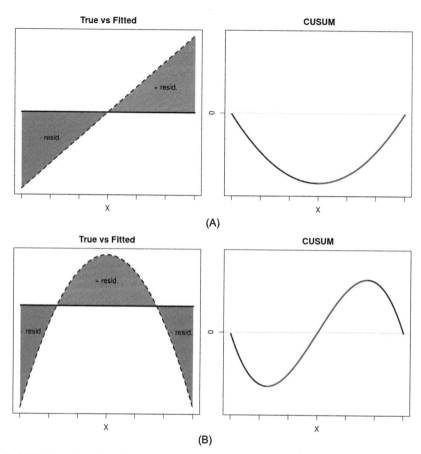

FIGURE 4.5 Illustration of how discrepancies between a true (dashed black line) and fitted (solid black) relationship appear using a CUSUM plot. (A) Linear true relationship; (B) quadratic true relationship. Blue regions indicate where residuals are negative and red regions where residuals are positive. Colors in the CUSUM plot are used to highlight the values of the corresponding residuals.

a long run of positive values would result in a positive trend. If the sign of the ordered residuals alternates or has only short runs (as would be expected if there is good agreement between the estimated parameters and observations), then the CUSUM curve will remain flat in that region. The resulting CUSUM curve can provide some insight not only about whether a potential covariate should be included in a model to be fit to the data, but also about its functional form (e.g., linear or quadratic; Fig. 4.5). To aid in the interpretation of the CUSUM plot, MacKenzie (2002) used parametric bootstrapping to produce an 'envelope' of what curves might be expected if the model was a good fit to the data, and we use the same approach here. As with the parametric bootstrap used by MacKen-

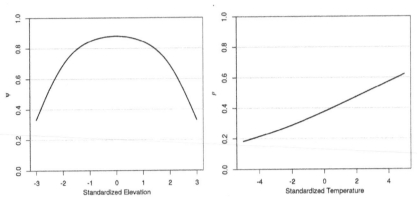

FIGURE 4.6 True scenarios for occupancy (left) and detection (right) probabilities for simulated data to illustrate diagnostic plots.

zie and Bailey (2004), the estimated parameters of the fitted model are used to simulate new detection/nondetection datasets, then treat them as observed data and analyze them with a model that has the same structure (e.g., same set of covariates) as the model that was used to create the data. In doing so, the model that is fit to the bootstrapped datasets must be a good fit, as it was used to simulate the observations. Therefore, the CUSUM curves from such datasets would be representative of what would be expected when there is no systematic lack of fit in the model. Where there are ties in the covariate values, or when a categorical covariate is being used, the residual values could be averaged for the tied cases, or a small random value could be used to provide some jittering of the tied values. This can lead to random patterns within the tied values, but the analyst should only consider the initial and final values of the CUSUM for the tied values.

To illustrate the use of these diagnostic plots we simulated detection/nondetection data for five surveys of 100 units where occupancy and detection probabilities each depended upon the value of different continuous predictor variables (Fig. 4.6). For the purpose of this example, we shall refer to the occupancy covariate as 'elevation' and the detection covariate as 'temperature', each of which has been standardized on some appropriate scale, centered at 0. Occurrence probability was defined as a quadratic function of elevation, and detection probability defined as a linear function of temperature, both on the logit-scale. The elevation value used for each unit, and the temperature value used for each survey, were randomly selected from the ranges indicated in Fig. 4.6.

Three models were fit to the simulated data; $\psi(\cdot)\,p(\cdot)$, $\psi(Elev)\,p(Temp)$, and $\psi(Elev + Elev^2)\,p(Temp)$, where the final model has the same structure as that which generated the data. Fig. 4.7 presents the DSR plots for the occupancy and detection residuals from each model, plotted against the elevation

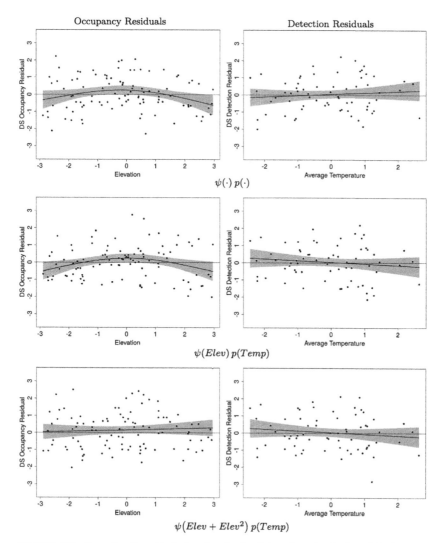

FIGURE 4.7 Dunn–Smyth residual plots for simulated data from the fitted models $\psi(\cdot)\,p(\cdot)$, $\psi(Elev)\,p(Temp)$, and $\psi\left(Elev + Elev^2\right)\,p(Temp)$, where the final model has the same structure as that which generated the data. The black line is a GAM smoother and shaded area is the 95% confidence interval for the expected value of the smoother.

and temperature covariates, respectively. A GAM smoother has been used to add interpretation (black line), along with a 95% confidence interval on the expected value of the smoother (shaded area). For the models that do not include the quadratic term for elevation, the occupancy residuals tend to be more negative for low and high elevation values, and more positive for middling elevation values. This would suggest the quadratic elevation term is required, and when it is included in the fitted model there is no indication of a systematic pattern in the residuals. The detection residual plots do not indicate any systematic patterns for any of the models in this example when plotted against the average temperature at each unit. This is most likely due to the fact that the detection residuals proposed by Warton et al. (2017) are based upon the total number of detections at a unit, not the individual survey outcomes, and temperature was a survey-specific covariate hence the relevant information gets lost within the aggregation. Note that there are fewer residual points in the detection plots as a DSR value is only calculated for units with at least one detection. CUSUM residual plots for the same three models are given in Fig. 4.8. The shape of the occupancy residuals CUSUM line for the first two models clearly indicates that a quadratic elevation term should be included for occupancy probability (see Fig. 4.5 for interpretation of CUSUM line), and when included the CUSUM residual line fluctuates around zero. The detection residual CUSUM line for the constant detection model tends to be more extreme than any of the values obtained from the 100 parametric bootstrap samples, and is obviously atypical. The run of negative residual values for low temperature values (where the slope of the CUSUM line is negative) and run of positive CUSUM values for high temperature values (where slope of CUSUM line is positive), with residual values of alternating sign in between (where CUSUM line is relatively flat), suggests temperature should be included as a linear term for detection (Fig. 4.5). When temperature is included as a covariate for detection in the last two models, the CUSUM residual plots still indicate a run a negative values for very low temperature values, followed by a run of positive residual values, but for most of the range of temperature values the CUSUM residual lines fluctuate about zero, suggesting no systematic problem with the detection component of the models. Clearly these recently developed diagnostic plots are a useful addition to an analyst's toolbox to assess model fit. Functions to calculate and plot the DSR and CUSUM residuals have been incorporated into the R package RPresence.

One of us (DM) has also briefly investigated the use of semi-variograms as a means to identify unmodeled spatial correlation in occupancy. A semi-variogram plots the degree of variation as a function of distance between the locations of observations. When spatial correlation is present, the values of observations from nearby locations tend to be more similar than those spaced farther apart. As some forms of spatial correlation could be well explained by covariates, the semi-variogram can be constructed on residual values rather than the values of the actual observations. Using the occupancy residuals ζ_i defined above does seem to provide some indication of unmodeled spatial correlation in some initial explorations of the approach, however further work is required to better evaluate the usefulness of the method.

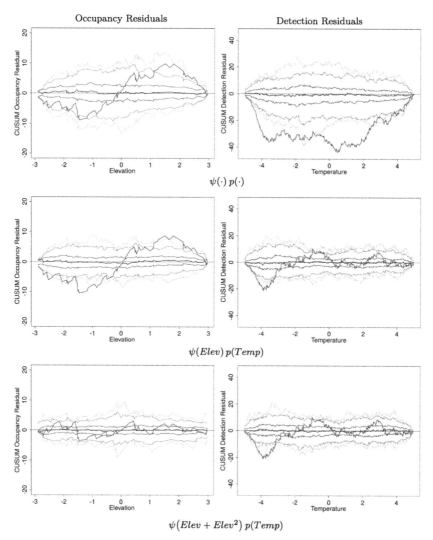

FIGURE 4.8 CUSUM residual plots for simulated data from the fitted models $\psi(\cdot)\,p(\cdot)$, $\psi(Elev)\,p(Temp)$, and $\psi\left(Elev + Elev^2\right)p(Temp)$, where the final model has the same structure as that which generated the data. The red line is the observed CUSUM residual value and gray lines indicate envelopes of expected values from 100 parametric bootstrap samples. Given are the minimum, 2.5th, 25.0th, 50.0th, 75.0th, 97.5th percentiles, and maximum value of the bootstrapped CUSUM residuals evaluated at each covariate value.

We note that in the diagnostic approaches described above, not being able to directly observe presence/absence of the species required authors to use either apparent occupancy, or the expected value of the latent variable z_i given the observed data and estimated parameters. When using the complete data likelihood with Bayesian methods, the value of each z_i will be imputed as part of the estimation procedure. That is, for each iteration of the MCMC procedure the value of z_i is assumed to be known (i.e., it will have the value of 1 or 0) which may enable other types of diagnostic procedures to be developed. That the imputed value may change with each iteration, at least at those units where the species was never detected, will result in some variation of the diagnostic as a result of both imperfect detection and parameter uncertainty. Alternatively, a non-Bayesian approach would be to use the estimated conditional occupancy probabilities (i.e., probability unit is occupied given observed detection history, ψ_{condt}; Section 4.4.5) to simulate a binary random value for each unit, and then apply the desired diagnostic procedure. Repeating the simulation and diagnostic calculation a large number of times will incorporate the uncertainty in the true values of the latent random variables.

Finally while we are aware that area under the receiver operator characteristic curve (AUROC), which is often just referred to as area under the curve (AUC), is a popular metric calculated for presence/absence data and species distribution models as a measure of model reliability, we share many of the concerns of Lobo et al. (2008) about its general appropriateness. In particular AUC is insensitive to the absolute value of the fitted probabilities from a model, and only their relative value is informative to the AUC metric, i.e., the same AUC value would be obtained whether the estimated probabilities $\hat{\theta}$ are used or $\hat{\theta}/10$. Therefore, AUC is not providing a measure of the absolute fit of the model to the data. AUC is essentially providing a metric associated with the degree of correlation (but it is not the correlation) between the *ranks* of the fitted probabilities (i.e., their ordered values) and those of the presence/absence observations. For example, AUC $= 1$ is obtained when the values of all the fitted probabilities for the presence observations (i.e., 1's) are greater than all the values for the fitted probabilities for the absence observations (i.e., 0's), that is, the values of the probabilities perfectly discriminate between the two types of observations. Importantly, note that the values of the predicted probabilities for the individual probabilities are not used otherwise. Our view is that AUC does not accurately reflect the nature of the probabilities estimated from modeling procedures such as logistic regression and those detailed in this book. We therefore choose to use other methods to assess model fit, in a relative or absolute sense.

4.4.12 Examples

To illustrate the above methods we now present four examples. The first is an example from MacKenzie (2006) on pronghorn antelope (*Antilocapra americana*) that was briefly mentioned in Chapter 2, which nicely illustrates that misleading inferences can result from not explicitly accounting for detection

TABLE 4.5 Summary of model selection procedure for pronghorn antelope example where factors affecting habitat selection have been investigated using simple logistic regression. Given are the relative differences in AIC values compared to that of the top ranked model (ΔAIC); the AIC model weights (w); twice the negative log-likelihood ($-2l$); and the number of parameters in the model ($Npar$)

Model	ΔAIC	w	$-2l$	$Npar$
$\psi(Sl + DW)$	0.00	23%	345.26	3
$\psi(DW)$	0.22	21%	347.48	2
$\psi(Sg + Sl + DW)$	0.82	16%	344.08	4
$\psi(Sg + DW)$	1.18	13%	346.44	3
$\psi(Sl + DW + A)$	2.79	6%	342.05	6
$\psi(DW + A)$	3.08	5%	344.34	5
$\psi(Sl)$	3.81	3%	351.07	2
$\psi(Sg + Sl + DW + A)$	4.05	3%	341.31	7
$\psi(Sg + DW + A)$	4.45	3%	343.71	6
$\psi(Sg + Sl)$	4.67	2%	349.93	3
$\psi(\cdot)$	5.63	1%	354.89	1
$\psi(Sl + A)$	6.11	1%	347.37	5
$\psi(Sg)$	6.65	1%	353.91	2
$\psi(Sg + Sl + A)$	7.45	1%	346.71	6
$\psi(A)$	7.67	1%	350.93	4
$\psi(Sg + A)$	9.13	0%	350.39	5

Source: MacKenzie (2006).

probability. Next we present two examples using a dataset first considered by MacKenzie et al. (2005); occupancy of Mahoenui giant weta (*Deinacrida mahoenui*) within a scientific reserve in New Zealand. One example uses maximum likelihood methods for the analysis while the other uses Bayesian methods. The final example uses the techniques discussed so far to develop a range or species distribution map for the willow tit (*Poecile montanus*) in Switzerland. As part of the examples we also introduce the idea of formally comparing competing hypotheses or models as part of the inferential procedure. Generally, different hypotheses about the system can be specified by way of covariates that are either collected in the field or that are constructed after the fact. The different hypotheses of interest can therefore be represented by a suite of candidate models. The candidate models can then be ranked according to some model selection procedure, with models (and hence hypotheses) that have a lot of support being ranked highly, with those less supported by the data having lower rankings.

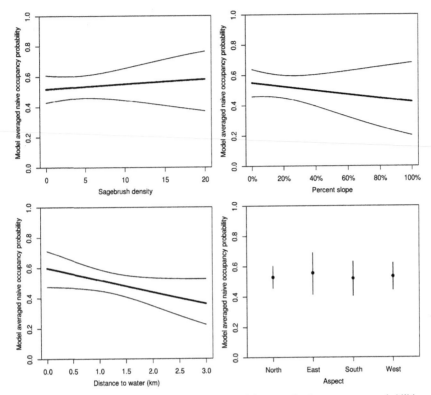

FIGURE 4.9 Illustration of relationships between model-averaged naïve occupancy probabilities using logistic regression for pronghorn antelope and the covariates sagebrush density, percent slope, distance to water, and aspect.

Here AIC has been used to rank the models, but we note other metrics can also be used.

The data files, software output files and any code used in these example analyses are available via the first author's company's website, http://www.proteus. co.nz, for free download.

Pronghorn Antelope

MacKenzie (2006) reanalyzed data on pronghorn antelope considered by Manly et al. (2002) in a resource selection context. During the northern hemisphere winters of 1980–1981 and 1981–1982, 256 locations in Wyoming, USA, were surveyed to determine whether the units where being used by pronghorn antelope (i.e., that antelope were present at a unit). A number of unit characteristics were also recorded (e.g., distance to water source and grass type) with the intent of identifying factors that affected whether pronghorn antelope used particu-

TABLE 4.6 Summary of model selection procedure for pronghorn antelope example. Factors affecting habitat selection have been investigated using the occupancy models described in this chapter, with a general model for detection probabilities, i.e., $p(Sg + Sl + DW + A)$. Given are the relative differences in AIC values compared to that of the top ranked model (ΔAIC); the AIC model weights (w); twice the negative log-likelihood ($-2l$); and the number of parameters in the model ($Npar$)

Model	ΔAIC	w	$-2l$	$Npar$
$\psi(Sl)$	0.00	23%	615.48	9
$\psi(\cdot)$	0.72	16%	618.20	8
$\psi(Sg + Sl)$	0.85	15%	614.33	10
$\psi(Sg)$	1.12	13%	616.60	9
$\psi(Sl + DW)$	1.95	9%	615.44	10
$\psi(DW)$	2.25	7%	617.73	9
$\psi(Sg + Sl + DW)$	2.85	6%	614.33	11
$\psi(Sg + DW)$	2.99	5%	616.47	10
$\psi(Sl + A)$	5.53	1%	615.01	12
$\psi(A)$	6.05	1%	617.53	11
$\psi(Sg + Sl + A)$	6.41	1%	613.89	13
$\psi(DW + A)$	6.60	1%	616.08	12
$\psi(Sl + DW + A)$	6.87	1%	614.35	13
$\psi(Sg + A)$	7.03	1%	616.51	12
$\psi(Sg + Sl + DW + A)$	8.21	0%	613.69	14
$\psi(Sg + DW + A)$	8.34	0%	615.83	13

Source: MacKenzie (2006).

lar units. Manly et al. (2002) noted that 'use' could be defined in a number of ways, and MacKenzie (2006) opted for a definition where it was assumed that there was no systematic difference in the types of units used by the antelope in both winters (i.e., the sampling season was defined to encompass both seasons). Hence the surveys of the units in each winter represent the repeated surveys of the units with the 'season'. MacKenzie (2006) analyzed the data using two approaches to illustrate how not accounting for detection probability can result in misleading inferences. The first approach was to use simple logistic regression where the nondetection of antelope in both winters is assumed to equate to pronghorn antelope not using the unit. The second approach was to use the above occupancy models to make inference about habitat selection while explicitly accounting for detection probability. Both approaches considered the effect of four covariates on habitat selection; sagebrush density (Sg), percent slope

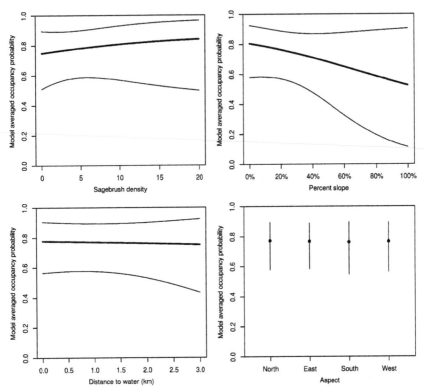

FIGURE 4.10 Illustration of relationships between model-averaged occupancy probabilities from occupancy modeling for pronghorn antelope and the covariates sagebrush density, percent slope, distance to water, and aspect.

(SI), distance to water (DW) and aspect (A; predominantly north, east, south or west facing).

The results of using logistic regression are presented in Table 4.5. Note that when used in this context, the implicit assumption made for the use of logistic regression to be valid is that either sufficient repeat surveys have been conducted such that the probability of a false absence is negligible, or that detection probability is constant across all units in which case the results should be interpreted as relative, rather than absolute, measures of occupancy (or 'use' in the present context). The summed model weights for the four variables are: distance from water: 86%; slope: 52%; sagebrush density: 35%; and aspect: 16%. Hence one would likely conclude that distance from a water source is the most important factor for determining whether a unit was used by pronghorn antelope. Because of the substantial model selection uncertainty (i.e., AIC model weights are relatively low for all models), model-averaging can be used to combine the results from all models rather than focus on the estimates from the top-ranked or a small

TABLE 4.7 Summary of model selection procedure for pronghorn antelope example. Factors affecting detection probability have been investigated using the occupancy models described in this chapter, with a general model for occupancy probabilities, i.e., $\psi(Sg + Sl + DW + A)$. Given are the relative differences in AIC values compared to that of the top ranked model (ΔAIC); the AIC model weights (w); twice the negative log-likelihood ($-2l$); and the number of parameters in the model ($Npar$)

Model	ΔAIC	w	$-2l$	$Npar$
$p(DW)$	0.00	24%	618.54	9
$p(Sl + A)$	1.37	12%	617.91	10
$p(A)$	1.75	10%	616.29	11
$p(Sg + DW)$	1.79	10%	618.33	10
$p(DW + A)$	1.82	10%	614.36	12
$p(\cdot)$	2.80	6%	623.34	8
$p(Sg + Sl + DW)$	3.17	5%	617.71	11
$p(Sl + DW + A)$	3.56	4%	614.10	13
$p(Sg + DW + A)$	3.58	4%	614.12	13
$p(Sg + A)$	3.62	4%	616.16	12
$p(Sl + A)$	3.73	4%	616.27	12
$p(Sl)$	4.76	2%	623.30	9
$p(Sg)$	4.79	2%	623.33	9
$p(Sg + Sl + DW + A)$	5.15	2%	613.69	14
$p(Sg + Sl + A)$	5.61	1%	616.15	13
$p(Sg + Sl)$	6.75	1%	623.29	10

number of models. Fig. 4.9 presents the relationship between model-averaged naïve occupancy probability estimates (from the models in Table 4.5) and each of the variables considered in the analysis. To create each univariate plot, the median value was used for the continuous-valued variables not in the plot, and the aspect was fixed to *North*. Clearly from Fig. 4.9, the strongest relationship appears to be with distance to water, with the level of use declining as distance increases and some indication of a decrease with increasing slope. Note the confidence intervals are narrower where the number of observed variable values is greater.

Quite different conclusions are reached when using the occupancy models described above (Table 4.6). Rather than attempting to perform model selection on both occupancy and detection probabilities simultaneously, MacKenzie (2006) focused on model selection of the occupancy probabilities while maintaining a very general model for detection probabilities, i.e., $p(Sg + Sl + DW + A)$. Now the summed model weights for each factor with respect to

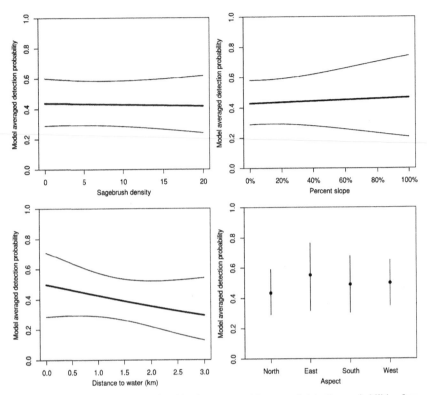

FIGURE 4.11 Illustration of relationships between model-averaged detection probabilities from occupancy modeling for pronghorn antelope and the covariates sagebrush density, percent slope, distance to water, and aspect.

use (occupancy) are: slope: 55%; sagebrush density: 41%; distance from a water source: 29%; and aspect: 6%. Distance from a water source has gone from being very strongly supported, to having only weak support. From Fig. 4.10 it would appear that occupancy decreases with increasing slope, but the width of the 95% confidence intervals indicate a large degree of uncertainty. This is consistent with the summed model weight; a value of 55% is ambiguous as half of the models fit to the data contain the slope variable.

MacKenzie (2006) suggests the difference in the conclusion is caused by these factors also affecting detection probabilities. Subsequent modeling supports this premise. The results of performing model section on the detection probabilities while maintaining a general model for occupancy, i.e., $\psi(Sg + Sl + DW + A)$, are presented in Table 4.7. The summed model weights for the factors with respect to detection probability are: distance to water source: 68%; aspect: 37%; slope: 29%; and sagebrush density: 27%. Clearly there is moderately strong support that detection probability is affected by distance to

FIGURE 4.12   Mahoenui giant weta (*Deinacrida mahoenui*) of New Zealand. (*Amanda Smale*)

water, and less support for the other factors. Interestingly, note that the model with constant detection probability, $p(\cdot)$; the de-facto assumption when using simple logistic regression in a relative sense, has only a very low model weight, suggesting little support for this hypothesis. Fig. 4.11 indicates that model-averaged detection probability declines with distance from water.

Mahoenui Giant Weta

Weta are ancient species of the order *Orthoptera* (e.g., grasshoppers, crickets and locusts) with more than 70 endemic species surviving in New Zealand today. Based upon fossil records, they have remained almost unchanged from their ancestors of 190 million years ago. In the absence of native ground-dwelling mammals, the weta of New Zealand evolved to fulfill the roles rodents play in other ecosystems. As such, the introduction of rats and other small mammals to New Zealand ecosystems with the arrival of the Maori and Europeans decimated weta populations.

The Mahoenui giant weta (*Deinacrida mahoenui*; Fig. 4.12) is endemic to the King Country on the North Island of New Zealand, with only a few surviving populations. The main naturally occurring population is restricted to a 240-ha scientific reserve at Mahoenui (near the town of Te Kuiti) maintained by the New Zealand Department of Conservation. The reserve is characterized by steep-sided gullies and is largely covered by dense gorse (*Ulex europaeus*), a perennial pest plant with sharp spiny stems and bright yellow flowers that can form dense thickets, originally introduced to New Zealand as a hedging plant by the early European settlers. The weta are believed to use the prickly gorse plants as refuge from introduced mammalian predators (e.g., rats and hedgehogs), and also as a food source. Goats and cattle are used to browse the gorse, encouraging dense foliage and providing further protection for weta.

As part of a pilot study, in March 2004, 72 circular plots of 3 m radius were surveyed for the Mahoenui giant weta within the more accessible regions of the reserve. Clearly, inference can only be made to the more accessible parts of the reserve, but this was deemed reasonable given the nature of the pilot study. Each plot was surveyed between three and five times during the five-day period ($K_{average} = 3.6$). Three different observers were used and the study was designed such that each observer surveyed each unit at least once. This was done to avoid introducing heterogeneity in detection probabilities caused by the use of multiple observers (see Chapter 11 for further discussion).

Weta were detected at 35 of the 72 plots (a naïve occupancy estimate of 0.49), however, often weta were only detected in one or two of the repeated surveys, clearly indicating that detection probabilities are less than 1. Conceivably, there may be a number of plots where weta were indeed present but were simply never detected during the surveys.

Here we wish to estimate the probability of occupancy for the weta. Detection probabilities will be allowed to vary by day and also among observers, but simpler models that do not include these effects will be included in our candidate set in the interest of parsimony. Survey day and observer are categorical survey-specific covariates with five and three levels, respectively. Software packages differ in exactly how they treat categorical covariates, and how they are incorporated into an analysis. Often, though not always, categorical covariates are converted into a series of indicator variables (an indicator variable = 1 if an observation is for that category, and = 0 otherwise) and a subset of them included in the estimating regression equation for the model. Which indicator variables are included depends on what other terms have been included in the model, which category the analyst wishes to use as a 'standard' (i.e., the contrast for the categorical covariate) and how the analyst aims to use the results. It is important to note that often the same biological model (in terms of what factors are included in the model) can be fit to the data using different estimating models with different sets of indicator variables. Overall conclusions should be identical, although specific interpretation of some parameter estimates may be different.

Using Program PRESENCE, daily variation in detection probabilities (or more generally, survey-specific detection probabilities) can be easily accommodated with the design-matrix interface used to build models rather than defining a series of 'survey day' indicator variables (although it amounts to the same thing). To allow detection probabilities to vary among the three observers, three survey occasion-specific indicator variables were defined: $Obs1$, $Obs2$, and $Obs3$. For unit i, survey j, $Obs1_{ij} = 1$ if the survey was conducted by observer 1, 0 otherwise; $Obs2_{ij} = 1$ if the survey was conducted by observer 2, 0 otherwise; and $Obs3_{ij} = 1$ if the survey was conducted by observer 3, 0 otherwise. Note that using this coding (Table 4.8), when a survey is conducted by a particular observer, only one of the three indicator variables equals 1, and the

TABLE 4.8 Coding used to define observer effects in Mahoenui giant weta example using the *Obs*1, *Obs*2, and *Obs*3 indicator variables

Survey conducted by	*Obs*1	*Obs*2	*Obs*3
Observer 1	1	0	0
Observer 2	0	1	0
Observer 3	0	0	1

other two both equal 0. Even though an indicator variable has been defined for each category in the observer categorical covariate, here we shall include only two of the three in our estimating model with the omitted indicator variable being treated as the standard or reference category (Section 3.4). Therefore, by using the *Obs*2 and *Obs*3 indicator variables the estimated coefficients represent the difference (on the logistic scale) between the respective observers and observer 1.

While the pilot study was not specifically designed for this purpose, the effect of browsing on occupancy and detection probability will also be assessed. The level of browsing at each plot was assessed by the field crew prior to the pilot study, which is a measure of the condition of the gorse bushes at each plot. A covariate *Browse* has been defined here as $= 1$ if the plot showed evidence of sustained browsing (based upon shape of bushes and foliage density; bushes are dense and compact), 0 otherwise (open and less compact bushes).

The candidate model set contains 16 models without considering interactions between factors (Table 4.9). Models are denoted with the relevant factors indicated in parentheses following each probability. For example, ψ (*Browse*) indicates the probability of occupancy was different for browsed and unbrowsed units, while $p(Day + Obs)$ indicates that detection probability varied by day with additive (on the logistic scale) observer effects.

Testing the global model from the candidate set, ψ (*Browse*) $p(Day + Obs + Browse)$, does not indicate any evidence of lack-of-fit using 10,000 bootstrap samples ($X^2 = 154.1$, P-value$= 0.999$, $\hat{c} = 0.35$), although one would perhaps be concerned that the P-value is so close to 1.0, that may indicate the model 'over-fits' the data (i.e., there may be too many parameters in the model). As such, no adjustment has been made to the model selection procedure (AIC) or the standard errors of parameter estimates. Recall that the parametric bootstrap procedure used in the lack-of-fit test has a random element to it when generating alternative datasets, therefore the P-value and $\hat{c}$ will vary with repeated runs of the test. However, the level of variation will decline as the number of bootstrap samples increases, which is why a large number of bootstrap samples is recommended.

TABLE 4.9 Summary of model selection procedure for Mahoenui giant weta example. ΔAIC is the difference in AIC value for a particular model when compared with the top ranked model; *w* is the AIC model weight; *Npar* is the number of parameters; −2*l* is twice the negative log-likelihood value; *Browse* is the value of the coefficient for the *Browse* variable with respect to its effect on occupancy probability; and SE is the associated standard error (blank entries indicate that the *Browse* variable was not included in the model)

Model	ΔAIC	*w*	*Npar*	−2*l*	*Browse*	SE
ψ (Browse) p (Day + Obs)	0.00	0.28	9	239.60	1.17	0.74
ψ (·) p (Day + Obs)	0.95	0.17	8	242.55		
ψ (·) p (Day + Obs + Browse)	1.82	0.11	9	241.41	1.24	0.75
ψ (Browse) p (Day)	1.84	0.11	7	245.44	1.17	0.89
ψ (Browse) p (Day + Obs + Browse)	2.00	0.10	10	239.60		
ψ (·) p (Day)	3.20	0.06	6	248.79		
ψ (·) p (Day + Browse)	3.58	0.05	7	247.18		
ψ (Browse) p (Day + Browse)	3.81	0.04	8	245.41	1.15	0.88
ψ (Browse) p (Obs)	4.44	0.03	5	252.04	1.18	0.70
ψ (·) p (Obs)	5.76	0.02	4	255.36		
ψ (Browse) p (Obs + Browse)	6.42	0.01	6	252.02	1.25	0.83
ψ (Browse) p (·)	6.66	0.01	3	258.26	1.23	0.72
ψ (·) p (Obs + Browse)	6.91	0.01	5	254.50		
ψ (·) p (·)	8.19	0.00	2	261.79		
ψ (Browse) p (Browse)	8.66	0.00	4	258.26	1.20	0.84

Table 4.9 presents the 16 models ranked according to AIC. The first thing to note is that no single model is demonstrably better than the others; the five top models are separated by less than 2.0 AIC units. As such, the AIC model weight is distributed across a number of models, indicating that a number of models may be reasonable for the collected data. There are, however, a number of common features among the top ranked models. The eight models where detection probability varied daily are all ranked higher than the models without daily variation. In terms of model weights, 91% is associated with models that include *Day* as a covariate for detection, providing clear evidence that *Day* is an important factor; the hypothesis that the detection probability varied among days therefore has much greater support than the hypothesis of no daily variation. Many of the top ranked models also contain the factor *Obs* for detection probability, providing evidence that the observers differed in their ability to find weta in the plots. The combined model weight for models that included observer effects on detection is 73%. There is substantially less support for the hypothesis that the level of browse affects detection probabilities, with a combined model weight of 33%.

In terms of occupancy probability, based upon rankings and AIC model weights, the results are somewhat inconclusive about the effect of browsing. The combined weight for the $\psi(Browse)$ models is 58%, i.e., similar levels of support for the hypotheses that occupancy is/is not affected by whether bushes within the plots are browsed. However, there is an important point to note that illustrates that unthinking use of model selection procedures can be misleading. Given the biology of the situation, *a priori* we would expect browsing to increase the probability of occupancy (by creating better habitat), therefore the parameter estimate associated with the factor *Browse* for occupancy should be positive. This was in fact the case (Table 4.9). From the respective models, all estimates for the *Browse* factor were very similar. Yet AIC and similar metrics (and therefore the derived model weights) do not account for the fact that one could specify *a priori* the direction of a particular relationship (i.e., one could very loosely describe these model selection procedures as 'two-sided'). Therefore, one could argue that, as the estimated effect matches our *a priori* expectations, the level of support for these models should be greater than that indicated by the model weights. Unfortunately we cannot make any firm recommendations at this time for how one might objectively incorporate this idea into an information-theoretic framework. As such, when interpreting the magnitude of the various effects below, rather than taking a model-averaging approach to account for uncertainty in which model(s) provides the most efficient representation of the data, we only consider the parameter estimates from the top ranked model. We acknowledge that our standard errors do not account for model selection uncertainty (Burnham and Anderson, 2002).

From the model $\psi(Browse)\,p(Day + Obs)$ we have the following equations for estimating occupancy and detection probabilities:

$$logit\left(\hat{\psi}_i\right) = 0.02 + 1.17 \times Browse_i \tag{4.15}$$

and

$$logit\left(\hat{p}_{ij}\right) = -1.30 \times Day1 - 1.45 \times Day2 - 2.24 \times Day3 - 1.37 \times Day4$$
$$- 0.26 \times Day5 + 0.73 \times Obs2_{ij} + 1.07 \times Obs3_{ij} \tag{4.16}$$

where $Browse_i$ is the value of the variable $Browse$ for plot i (1 or 0); $Day1–Day5$ are indicator variables for the day of the study; and $Obs2_{ij}$ and $Obs3_{ij}$ are indicator variables to denote which of the three observers surveyed plot i on day j (see Table 4.8). Recall that the 'hat' (^) over a parameter is just denoting an estimated rather than known quantity. Once the analysis has been conducted, these equations can be used separately to make inference about species occurrence and detection probabilities in exactly the same manner as regular logistic regression is used. By using the occupancy modeling framework, the regression coefficients in Eq. (4.15) have been corrected for imperfect detection, so further consideration of detection issues is unnecessary to make valid inference about species occurrence probabilities.

Below it is outlined how Eq. (4.15) can be used to calculate occupancy probabilities although typically computer software would be used to perform the calculations rather than doing it 'by hand'. It is instructive to consider a couple of examples to gain some appreciation of how the software is obtaining those values. In general, to use a logistic regression equation to estimate a probability for a particular situation, the covariate values that represent that scenario are inserted into the equation, then the required math is performed. For example, for an unbrowsed plot (where the gorse bushes are more open to predators) $Browse_i = 0$, so to determine the probability of weta being present Eq. (4.15) becomes:

$$logit\left(\hat{\psi}_{Unbrowsed}\right) = ln\left(\frac{\hat{\psi}_{Unbrowsed}}{1 - \hat{\psi}_{Unbrowsed}}\right) = 0.02 + 1.17 \times 0$$
$$= 0.02.$$

Rearranging this:

$$\hat{\psi}_{Unbrowsed} = \frac{e^{0.02}}{1 + e^{0.02}} = \frac{1.02}{1 + 1.02}$$
$$= 0.50.$$

For a browsed plot (so $Browse_i = 1$) the calculations become:

$$logit\left(\hat{\psi}_{Browsed}\right) = ln\left(\frac{\hat{\psi}_{Browsed}}{1 - \hat{\psi}_{Browsed}}\right) = 0.02 + 1.17 \times 1$$

$$= 1.19.$$

Which gives:

$$\hat{\psi}_{Browsed} = \frac{e^{1.19}}{1 + e^{1.19}} = \frac{1.02}{1 + 1.02}$$

$$= 0.77.$$

Standard errors for $\hat{\psi}_{Unbrowsed}$ and $\hat{\psi}_{Browsed}$ (obtained by application of the delta method) are 0.11 and 0.12 respectively. As a probability must be within the range of 0–1, an approximate 95% confidence interval can be initially constructed on the logit-scale, then the resulting limits transformed to the probability scale as above. This requires calculation of the standard error of $logit\left(\hat{\psi}_i\right)$, which could be derived from the variance–covariance matrix of the estimated regression coefficients, or alternatively, and somewhat easier, it can be calculated from $SE\left(\hat{\psi}_i\right)$ which will be reported by the relevant software packages (see Section 3.2.4 for further details). For example, to create a 95% confidence interval for $\hat{\psi}_{Unbrowsed}$:

$$SE\left(logit\left(\hat{\psi}_{Unbrowsed}\right)\right) = \frac{SE\left(\hat{\psi}_{Unbrowsed}\right)}{\hat{\psi}_{Unbrowsed}\left(1 - \hat{\psi}_{Unbrowsed}\right)}$$

$$= \frac{0.115}{0.506 \times 0.494}$$

$$= 0.460.$$

The calculation for the lower limit would be:

$$logit\left(\hat{\psi}_{Unbrowsed}\right) - 1.96 \times SE\left(logit\left(\hat{\psi}_{Unbrowsed}\right)\right) = 0.024 - 1.96 \times 0.460$$

$$= -0.877$$

where 1.96 is the respective critical z-value from the standard normal distribution, and

$$\hat{\psi}_{Unbrowsed,LL} = \frac{e^{-0.877}}{1 + e^{-0.877}}$$

$$= 0.29.$$

For the upper limit:

$$logit\left(\hat{\psi}_{Unbrowsed}\right) + 1.96 \times SE\left(logit\left(\hat{\psi}_{Unbrowsed}\right)\right) = 0.024 + 1.96 \times 0.460$$
$$= 0.925$$

and

$$\hat{\psi}_{Unbrowsed,UL} = \frac{e^{0.925}}{1 + e^{0.925}}$$
$$= 0.72.$$

Using similar calculations for $\hat{\psi}_{Browsed}$, an approximate 95% confidence interval is (0.47, 0.93).

Interpreting these results, $\hat{\psi}_{Unbrowsed} = 0.50$ indicates that the estimated probability of weta being present at a randomly selected plot with a low level of browsing of the gorse bushes is 0.50, while for a plot with a high level of browsing of the gorse bushes, the estimated probability is 0.77. The width of the confidence intervals suggest that these estimates are associated with a moderate degree of uncertainty due to the (statistically) small sample sizes (37 plots were classified as unbrowsed and 35 plots as browsed). That there is overlap in the 95% confidence intervals of the two estimated probabilities should not necessarily be interpreted that there is little statistical evidence of a difference between them.

If that is a question of interest, a better approach would be to calculate the difference in the probabilities along with the appropriate standard errors and confidence intervals. However, a more general approach would be to ensure that the estimating regression equation is constructed using the appropriate set of covariates such that the estimated regression coefficients can be interpreted directly. For example, by using the indicator variable *Browsed*$_i$ in Eq. (4.15), the associated regression coefficient indicates how *different* logit-occupancy is at a plot with a high-level of browsing compared to a plot with a low-level of browsing. Here it is estimated to be 1.17 with a standard error of 0.74 (from PRESENCE output), leading to an approximate 95% confidence interval of (−0.28, 2.62). Hence, from these data and this analysis, there is some chance that the effect size could be negative, i.e., occupancy probability may actually be lower at plots with a high level of browsing.

The effect of browsing on occupancy can also be interpreted in terms of an odds-ratio. Recall that because of the non-linear nature of the logit-link function, it can be difficult to interpret the effect of a covariate on a probability, as the magnitude of the change in the probability will depend on the value of other covariates in the model. Whereas the effect on the odds of the outcome (i.e., how

TABLE 4.10 Estimated daily detection probabilities for each observer in the Mahoenui giant weta pilot study from the logit-link function given in Eq. (4.16)

Observer	Day 1	Day 2	Day 3	Day 4	Day 5
1	0.21	0.19	0.10	0.20	0.43
2	0.36	0.33	0.18	0.34	0.61
3	0.44	0.41	0.24	0.43	0.69

likely a successful outcome is compared to an unsuccessful outcome) will be consistent in a multiplicative sense (Section 3.4). The odds ratio for a browsed unit being occupied by weta is simply calculated by taking the inverse-log of the regression coefficient, i.e., $e^{1.17} = 3.22$. This means that for every unoccupied plot, there are 3.22 times more occupied browsed plots than occupied unbrowsed plots. A 95% confidence interval for the odds ratio could be calculated by taking the inverse-log of each limit of the regression coefficient confidence interval. That is, $(e^{-0.28}, e^{2.62}) = (0.76, 13.74)$. As mentioned in Section 3.4, an odds ratio of 1.0 would indicate that the factor has no effect, hence as 1.0 is included in the confidence interval, we do not have strong evidence that browsing has an effect on the probability of occupancy by weta.

Although, as it was expected *a priori* that browsing should have a positive effect on occupancy, it would be more appropriate to consider a 1-sided 95% confidence interval (calculated here by taking the lower limit of a 2-sided 90% confidence interval). That is, $(1.17 - 1.65 \times 0.74, \infty) = (-0.05, \infty)$ on the logit scale, or $(0.95, \infty)$ in terms of odds ratios. As 1.0 is only just inside the confidence interval, we would be comfortable in concluding that it does appear browsing has a positive effect on occupancy, but the magnitude of the effect is only poorly known.

Table 4.10 presents the estimated detection probabilities for each observer on each day calculated from Eq. (4.16). There is clearly a reasonable level of daily variation, and substantial differences among observers. The exact consequences of the imperfect detection will depend on which observer surveyed each plot, and on which day the survey was conducted. We can quickly gain some insight however by looking at the average detection probability (0.36) and average number of surveys (3.6). The expected probability of not detecting weta at a plot where they are present (i.e., the probability of declaring a false absence) is about $(1 - 0.34)^{3.6} = 0.22$. That is, weta would not be detected at approximately 1 in 5 occupied plots.

Mahoenui Giant Weta: A Bayesian Analysis

Here we reanalyze the Mahoenui giant weta data as a brief example of incorporating covariates when using Bayesian methods. In particular, we shall examine the importance of defining appropriate prior distributions for the regression coefficient when using these techniques. We do not provide example 'pseudo-code' as there are plenty of examples available in the published literature and on the internet, but essentially one simply modifies the code in Fig. 4.2 such that occupancy and detection are unit-specific values (and possibly survey-specific in the case of detection). This is accomplished by expressing the probabilities as a function of the covariates of interest (as above; see also Chapter 15 for an example).

For simplicity, we fit the model $\psi(Browse)\,p(Obs)$ to the data where occupancy is different for browsed and unbrowsed plots, and detection is different for each observer (but the same for each day). This could be represented by the logistic regression equations:

$$logit(\psi_i) = \alpha_1 + \alpha_2 \times Browse_i$$

and

$$logit(p_{ij}) = \beta_1 + \beta_2 \times Obs2_{ij} + \beta_3 \times Obs3_{ij}$$

where the covariates $Browse$, $Obs2$, and $Obs3$ are defined as previously. We examined the effect of using three different sets of prior distributions for the regression coefficients α_1, α_2, β_1, β_2, and β_3. In all cases a normal distribution with mean equal to 0 was used and variance equal to 100, 10, or 1, representing relatively *vague*, *semi-informative*, and *informative* priors, respectively. For each analysis, three chains of 11,000 iterations were used with the first 1000 iterations being discarded as the burn-in period. The Brooks–Gelman–Rubin 'R-hat' metric was calculated to assess convergence of the chains to the same set of values, with R-hat values close to 1.0 indicating convergence. All analyses were conducted in JAGS.

A numerical summary of the posterior distributions of the regression coefficients is given in Table 4.11. While some posterior distributions have similar means and standard deviations (particularly the detection-related β_1–β_3 with vague and semi-informative priors), others are substantially different. Most notable are the results for α_2 that are all very different for each set of prior distributions, and when the vague prior is used, the R-hat metric would suggest the chains have not converged. The traceplot of α_2 (Fig. 4.13) reveals that the chain begins to 'wander' whenever the value for α_2 is greater than approximately 2.5, which in combination with the values for α_1, correspond to the occurrence probability at a browsed plot (ψ_{Browse}) being very close to 1.0. As a result the chain

TABLE 4.11 Mean and standard deviation (SD) of posterior distributions for regression coefficients in the model ψ (*Browse*) p (*Obs*) fit to the Mahoenui giant weta data using Bayesian methods. Three sets of prior distributions were assumed with different variances representing relatively vague, semi-informative, and informative priors; each was normally distributed with mean = 0. The Brooks–Gelman–Rubin convergence metric (R-hat) is also given

Coefficient	Vague ($\sigma^2 = 100$)			Semi-informative ($\sigma^2 = 10$)			Informative ($\sigma^2 = 1$)		
	Mean	SD	R-hat	Mean	SD	R-hat	Mean	SD	R-hat
α_1	0.21	0.66	1.03	0.11	0.54	1.00	0.10	0.41	1.00
α_2	4.40	5.01	1.19	1.66	1.25	1.00	0.90	0.57	1.00
β_1	−1.43	0.37	1.00	−1.31	0.36	1.00	−1.03	0.31	1.00
β_2	0.73	0.44	1.00	0.71	0.44	1.00	0.49	0.38	1.00
β_3	1.01	0.42	1.00	0.99	0.42	1.00	0.76	0.38	1.00

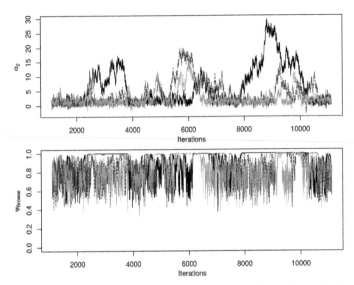

FIGURE 4.13 Traceplots of the retained iterations for α_2 (top panel) and ψ_{Browse} (bottom panel) for Mahoenui giant weta example data, from $\psi(Browse)\,p(Obs)$ model. Each color indicates a different Markov chain.

gets 'stuck' in that part of the parameter space (Fig. 4.13) as there is not enough information in the data or the prior distribution to pull α_2 back to lower values (as happens with the more informative priors, or would happen with a larger sample size). Partially, this is a consequence of using the logit-link, as on the logit-scale any value greater than 4.0 is going to result in a probability > 0.98, hence there is a wide range of values on the logit-scale that correspond to very similar probability values (close to 1.0). Note that a similar effect happens with values < -4 on the logit-scale corresponding to probabilities that are close to 0.0.

This highlights that when using Bayesian methods, resulting inferences can be sensitive to the choice of prior distributions used, just as they can also be sensitive to which predictor variables are included in a model. Therefore, the selection of appropriate prior distributions for regression coefficients is an important consideration, even when 'vague' prior distributions are to be used. Prior distributions that are too vague may result in convergence and numerical issues, particularly as the number of regression coefficients increases. Northrup and Gerber (in review) have also investigated this important issue in the context of occupancy modeling in a Bayesian context, with similar findings. As a general approach for logistic regression, Gelman et al. (2008) recommended that all predictor variables, including binary indicator variables, be standardized to have a mean value of zero, with continuous covariates also being rescaled to have a standard deviation of 0.5. After doing so, they suggest using independent

Cauchy distributions centered at 0.0 with a scale parameter of 10.0 for the intercept term, and 2.5 for all other coefficients. Attempting this approach here gave us similar results to using the semi-informative prior distributions.

Swiss Willow Tit

The willow tit (*Poecile montanus*) is a small passerine bird found throughout Europe, United Kingdom and northern Asia. Here we use data from the Swiss Ornithological Society's Swiss Survey of Common Breeding Birds (Monitoring Häufige Brutvögel; MHB) to illustrate how the single-season occupancy model can be used to develop a species range or distribution map based upon landscape covariates. The MHB has been run annually since 1999 (Schmid et al., 2004) where a systematic random sample of 267 1-km^2 quadrats (units) is surveyed three times during the breeding season (15 April–15 July), although not all units are surveyed every year. Within each unit, an irregular transect is surveyed, where the average length is approximately 5 km (Royle and Kéry, 2007). Here, the 2007 survey data from 201 units are used.

To create distribution maps, covariate information from across Switzerland is required, at the same spatial scale as the sample units (1 km^2). These data are included with the R package unmarked, which contained the average elevation and percent forest cover for each unit. Following the analysis of the survey data, the fitted model(s) can be used to calculate occurrence probability for all units based upon the covariate values.

All covariates are continuous in this example, and simply including them in a model assumes a linear relationship between the variable and the probability of interest (on the link function scale; Section 4.4.8). We hypothesized a potential quadratic relationship between occupancy and elevation, i.e., occupancy probability initially increases with elevation, then at higher elevations occupancy declines. That is, willow tit may exhibit an elevational preference where it occupies a greater proportion of units, and occupies a lesser proportion of units at lower and higher elevations. A square-root relationship was hypothesized between percent forest cover and occupancy so that small differences in the percent forest cover have a greater effect on occupancy when the percent forest cover is relatively low, and a lesser effect on occupancy when percent forest cover is relatively high. This is incorporated by including the square-root of the forest cover values as a covariate in the model. Using a log-transformation would induce a similarly shaped relationship, but the forest cover was 0 for some units and $ln(0)$ is undefined. Rather than using the log transformation after adding an arbitrary constant, the square-root transformation was used. A possible interaction between elevation and forest cover was also considered to allow the effect of percent forest cover to be different at different elevations. Therefore, our general model for the probability of occupancy at unit i takes the form:

$$logit(\psi_i) = \alpha_0 + \alpha_1 Ele_i + \alpha_2 Ele_i^2 + \alpha_3\sqrt{FC_i} + \alpha_4 Ele_i \times \sqrt{FC_i}$$

where Ele_i is the elevation value, and FC_i is the percent forest cover, for unit i. Elevation was entered in the model on the scale of 100 m increments from 1000 m (i.e., $Ele_i = 0$ equates to an elevation of 1000 m, and $Ele_i = -1.5$ equates to an elevation of 850 m), and the percent forest cover on the scale of 0–1 (i.e., $FC_i = 0.5$ equates to 50% forest cover).

For detection probability, we hypothesized similar potential relationships with elevation and forest cover as for occupancy probability, but a non-parametric curve for the effect of Julian date using natural cubic splines and log-linear effect for survey duration (i.e., log-transform the survey duration values). Parametric curves such as linear, quadratic, or logs, have well defined shapes and properties. Non-parametric curves such as splines are more flexible allowing the data to 'speak for themselves' rather than enforcing specified parametric shapes. This can be beneficial in some situations if the purpose of the modeling of the system is more descriptive or exploratory, but it is harder to incorporate specific mechanistic relationships with covariates as might be desired in a more rigorous scientific investigation. In practice, a spline is created by applying a series of piecewise polynomial transformations to the covariate values, resulting in a set of covariates that jointly represent the spline, called the basis values. Typically, the number of basis values would be one more than the number of specified *knots*, which define the piecewise intervals. The knots used for Julian date were days 32, 50, and 67 (the 25th, 50th, and 75th percentile values) resulting in a set of four basis value covariates ($JD1_{ij}$–$JD4_{ij}$). While survey duration (*Dur*) was recorded in minutes, this was converted to hours of survey effort prior to log-transforming. Note that elevation and forest cover are unit-specific covariates (so only indexed by cell number, i) while Julian date and survey duration are survey-specific covariates (indexed by cell and survey number, i and j, respectively). The general model for detection probability was therefore:

$$logit(p_{ij}) = \beta_0 + \beta_1 Ele_i + \beta_2 Ele_i^2 + \beta_3\sqrt{FC_i} + \beta_4 Ele_i \times \sqrt{FC_i}$$
$$+ \beta_5 JD1_{ij} + \beta_6 JD2_{ij} + \beta_7 JD3_{ij} + \beta_8 JD4_{ij}$$
$$+ \beta_9 ln(Dur_{ij}).$$

There is potentially a very large number of models that could be fit to the data by specifying different combinations of factors for each of the occupancy and detection components of the model. The approach taken here first compares different models for the detection component to identify which factors appear to be the most important for detection probabilities, then using that structure for detection, compares different models for occupancy probability. All models were compared by AIC, but we note other metrics and general strategies could be used.

A key point is that in the first stage, the above general model for occupancy was used while identifying which factors appear to be important for detection to provide the analysis with the flexibility to allow for variation in occupancy should the data suggest that is appropriate. A different approach would be to use a constant occupancy model (only involving the intercept term α_0) while investigating which factors may affect detection probability. However this could lead to inappropriate conclusions about detection probabilities if occupancy truly varies across different units according to a covariate that is being assessed for detection because the only way the model can express variation in the data, is in the detection component (as occupancy has been assumed constant). Note that assuming constant occupancy during this stage of model selection can not only lead to overestimation of the importance of factors on detection, but can also lead to underestimation if the true effect of the covariate on occupancy and detection are opposite.

Table 4.12 presents a summary of the detection models fit to the Swiss willow tit data. The top-ranked model includes a linear effect of elevation, the square-root of percent forest cover and survey duration as covariates, and the second-ranked model is similar but also includes the interaction between elevation and percent forest cover. These models are ranked very similarly in terms of AIC, and in fact, there are a number of models that are ranked similarly as indicated by the ΔAIC values and model weights (w). A number of the models within the nine highest-ranked models are simplifications of the top two models (obtained by removing one or more factors), while the fourth-, fifth-, and seventh-ranked models are the same as the top three models, with the addition of the elevation-squared term. However, the addition of this term explains little additional variation in the data (the -2log-likelihood values are very similar), and the estimated coefficients for the quadratic term are very close to zero. Anderson (2008) refers to this type of variable as a *pretender variable*; one that is included in relatively highly-ranked models but has little explanatory power. The models only get ranked highly by AIC because of a similar number of parameters. The highest ranked model that includes Julian date, using the spline relationship, is ranked tenth and has a low AIC model weight. Overall, these results suggest that elevation, forest cover and survey duration are the most important variables for detection.

Table 4.13 presents the estimated regression coefficients for the detection component for the two highest-ranked models by AIC, which can be used to estimate detection at a range of specified covariate values. The estimated effect

TABLE 4.12 Summary of detection models fit to Swiss willow tit example data. ΔAIC is the relative difference in AIC values compared with the top ranked model; w is the AIC model weight; $Npar$ is the number of parameters; and $-2l$ is twice the negative log-likelihood value. For all models the general occupancy model was used, $\psi\left(Ele + Ele^2 + \sqrt{FC} + Ele \times \sqrt{FC}\right)$

Model	ΔAIC	w	$Npar$	$-2l$
$p\left(Ele + \sqrt{FC} + ln(Dur)\right)$	0.00	0.25	9	254.89
$p\left(Ele + \sqrt{FC} + Ele \times \sqrt{FC} + ln(Dur)\right)$	0.41	0.21	10	253.30
$p(Ele + ln(Dur))$	1.74	0.11	8	258.63
$p\left(Ele + Ele^2 + \sqrt{FC} + ln(Dur)\right)$	2.00	0.09	10	254.89
$p\left(Ele + Ele^2 + \sqrt{FC} + Ele \times \sqrt{FC} + ln(Dur)\right)$	2.34	0.08	11	253.23
$p\left(Ele + \sqrt{FC}\right)$	3.55	0.04	8	260.44
$p\left(Ele + Ele^2 + ln(Dur)\right)$	3.70	0.04	9	258.59
$p\left(Ele + \sqrt{FC} + Ele \times \sqrt{FC}\right)$	3.90	0.04	9	258.79
$p(Ele)$	4.64	0.03	7	263.53
$p\left(Ele + \sqrt{FC} + JD + ln(Dur)\right)$	5.12	0.02	13	252.01
$p\left(Ele + \sqrt{FC} + Ele \times \sqrt{FC} + JD + ln(Dur)\right)$	5.36	0.02	14	250.25
$p\left(Ele + Ele^2 + \sqrt{FC}\right)$	5.52	0.02	9	260.41
$p\left(Ele + Ele^2 + \sqrt{FC} + Ele \times \sqrt{FC}\right)$	5.70	0.01	10	258.59
$p\left(Ele + Ele^2\right)$	6.62	0.01	8	263.51
$p(Ele + JD + ln(Dur))$	6.85	0.01	12	255.74
$p\left(Ele + Ele^2 + \sqrt{FC} + JD + ln(Dur)\right)$	7.02	0.01	14	251.91
$p\left(Ele + Ele^2 + \sqrt{FC} + Ele \times \sqrt{FC} + JD + ln(Dur)\right)$	7.03	0.01	15	249.92
$p\left(Ele + \sqrt{FC} + JD\right)$	8.71	0.00	12	257.60
$p\left(Ele + Ele^2 + JD + ln(Dur)\right)$	8.85	0.00	13	255.74
$p\left(Ele + \sqrt{FC} + Ele \times \sqrt{FC} + JD\right)$	8.98	0.00	13	255.87
$p(Ele + JD)$	9.92	0.00	11	260.80
$p\left(Ele + Ele^2 + \sqrt{FC} + Ele \times \sqrt{FC} + JD\right)$	10.51	0.00	14	255.39
$p\left(Ele + Ele^2 + \sqrt{FC} + JD\right)$	10.54	0.00	13	257.42
$p\left(Ele + Ele^2 + JD\right)$	11.80	0.00	12	260.69

continued on next page

TABLE 4.12 (*continued*)

Model	ΔAIC	w	Npar	−2l
$p\left(\sqrt{FC} + ln(Dur)\right)$	13.86	0.00	8	270.75
$p(ln(Dur))$	14.11	0.00	7	273.00
$p(\cdot)$	17.05	0.00	6	277.93
$p\left(\sqrt{FC}\right)$	17.59	0.00	7	276.48
$p\left(\sqrt{FC} + JD + ln(Dur)\right)$	18.47	0.00	12	267.36
$p(JD + ln(Dur))$	19.22	0.00	11	270.11
$p(JD)$	22.68	0.00	10	275.57
$p\left(\sqrt{FC} + JD\right)$	22.85	0.00	11	273.74

TABLE 4.13 Estimated regression coefficients (Coef.) for detection component of the two highest ranked models in Table 4.12 fit to Swiss willow tit example data, 1st ranked: $p\left(Ele + \sqrt{FC} + ln(Dur)\right)$; 2nd ranked: $p\left(Ele + \sqrt{FC} + Ele \times \sqrt{FC} + ln(Dur)\right)$. Given are the estimated values (Est.) and associated standard errors (SE) from each model for each potential covariate (Cov.) that was considered. The scale for each covariate is: *Ele*: deviations from 1000 m in 100 m increments; *FC*: percent forest cover, expressed as decimal value (0–1); *Dur*: hours

Coef.	Cov.	1st ranked		2nd ranked	
		Est.	SE	Est.	SE
β_0	Intercept	−5.30	2.14	−7.32	2.84
β_1	Ele	0.24	0.07	0.81	0.57
β_3	$\sqrt{FC}$	3.38	1.72	6.34	3.16
β_4	$Ele \times \sqrt{FC}$	–	–	−0.78	0.75
β_9	$ln(Dur)$	2.71	1.14	2.60	1.09

for *ln(Dur)* is similar in both models and indicates that detection probability is higher for longer surveys. Fig. 4.14 presents a univariate plot of these estimated relationships between detection probability and survey duration (on the as-recorded scale of minutes), where elevation and forest cover were set at the median values of 1150 m and 35% respectively (*Ele* = 1.5 and *FC* = 0.35). The curved nature of the estimated relationship is partially due to the use of the logit-link, and also the inclusion of log-survey duration as a predictor variable. In the top-ranked model, the effects of elevation and forest cover on detection probability are both estimated to be positive (Table 4.13), indicating detection increases with higher values of both variables. However in the second-ranked

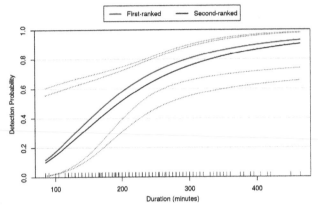

FIGURE 4.14   Estimated probability of detection of Swiss willow tits as a function of survey duration, calculated at the median values for elevation and forest cover of 1150 m and 35%, respectively (*Ele* $= 1.5$ and *FC* $= 0.35$), for the two highest-ranked models in Table 4.12. Given are the estimated values (solid lines) and limits of a 95% confidence interval (dotted lines). The recorded survey durations are indicated by tick marks along the x-axis.

model, while the estimated coefficients for the elevation and forest cover main effects are positive, the interaction term is estimated to be negative. The interaction term causes the magnitude of the effect for one variable to depend upon the value of the other variable, that is, as elevation increases, the effect of forest cover on detection decreases, and vice versa.

Fig. 4.15 illustrates the estimated effect of elevation and forest cover on detection with bivariate plots. Note that in the upper plot, the contour lines are parallel, indicating that the effect of each variable is independent of the value of the other (i.e., no interaction); the curvature in the contour lines is due to $\sqrt{FC}$ being used in the model rather than a straight linear effect of forest cover. In the lower plot, the contour lines are not parallel due to the inclusion of the interaction term. When the percent forest cover is near 0, detection increases rapidly as elevation increases (as contour lines are close together), while when forest cover is near 100% detection changes little as elevation increases (as contour lines are widely spaced). Similarly, at low elevations detection increases as forest cover increases, but at high elevations detection is estimated to decrease as forest cover increases.

For the purpose of this example, the top-ranked detection model was selected to be used when fitting different models for the occupancy component. In a full analysis, however, one should assess the robustness of the final conclusions to the detection component used by also fitting models using other highly-ranked detection models. We also note that other strategies are reasonable, e.g., use a general detection component first while performing model selection on the

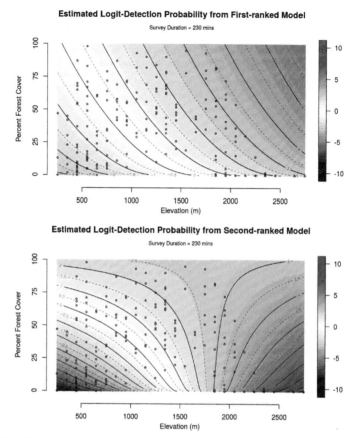

FIGURE 4.15 Bivariate plots of the estimated effect of elevation and forest cover on detection from top two models for Swiss willow tit example. Survey duration was set at the median value of 230 minutes when calculating these values.

occupancy component, then seek more parsimonious detection components. The key, we believe, is to have a clear *a priori* set of candidate sub-models for each component, a clear objective strategy for determining which sub-models from each component are combined to define the models that are fit to the data, and an approach to assess the robustness of the final inferences to any arbitrary choices that are made as part of the model fitting strategy (e.g., when using the top detection model to perform model selection on the occupancy component, or the order in which different components are assessed). As part of that process, we recommend against having very simple sub-models for the component(s) that are not being focused on, as we discuss elsewhere.

Eight occurrence models with different combinations of the elevation and forest cover covariates were fit to the data, with a summary of the fitting pro-

TABLE 4.14 Summary of occupancy models fit to Swiss willow tit example data. ΔAIC is the relative difference in AIC values compared with the top ranked model; w is the AIC model weight; $Npar$ is the number of parameters; and $-2l$ is twice the negative log-likelihood value. For all models the detection model $p\left(Ele + \sqrt{FC} + ln(Dur)\right)$ was used

Model	ΔAIC	w	$Npar$	$-2l$
$\psi\left(Ele + Ele^2 + \sqrt{FC} + Ele \times \sqrt{FC}\right)$	0.00	0.49	9	254.89
$\psi\left(Ele + \sqrt{FC} + Ele \times \sqrt{FC}\right)$	0.01	0.49	8	256.90
$\psi\left(Ele + Ele^2 + \sqrt{FC}\right)$	6.41	0.02	8	263.30
$\psi(Ele)$	17.17	0.00	6	278.05
$\psi\left(Ele + \sqrt{FC}\right)$	19.29	0.00	7	278.18
$\psi\left(Ele + Ele^2\right)$	19.70	0.00	7	278.59
$\psi\left(\sqrt{FC}\right)$	27.97	0.00	6	288.86
$\psi(\cdot)$	59.94	0.00	5	322.83

cedure given in Table 4.14. For occupancy, the top two models account for 98% of the AIC model weight, and both include the linear effect of elevation, square-root of forest cover and the interaction term. The only difference is the first-ranked model also includes the elevation-squared term. Unlike for detection, Ele^2 explains some additional variation in the data, while the interaction term ($Ele \times \sqrt{FC}$) appears to explain a lot of additional variation (e.g., compare the -2log-likelihood values of the first- and second-ranked models, and first- and third-ranked models, respectively).

The estimated regression coefficients for the occurrence component of the eight models of Table 4.14 are given in Table 4.15. When included, the coefficient for the Ele^2 term is negative indicating an inverted parabola as predicted (e.g., occurrence probability peaks at a certain elevation, and is lower at smaller or greater elevations), however the magnitude of the coefficient is close to zero indicating there is not a large degree of curvature. As the estimated effect size is close to zero, the model that excludes the Ele^2 term, but is the same in all other respects, is ranked very similarly in terms of AIC; excluding the quadratic term is equivalent to forcing the associate regression coefficient $\alpha_2 = 0$.

The fitted regression for the top-ranked model is therefore:

$$logit\left(\hat{\psi}_i\right) = -2.237 - 0.039Ele_i - 0.019Ele_i^2$$
$$+ 2.805\sqrt{FC_i} + 0.845Ele_i \times \sqrt{FC_i}.$$

TABLE 4.15 Estimated regression coefficients for occupancy models fit to Swiss willow tit example data. Models are ordered by AIC (Table 4.14), and w is the AIC model weight. For all models the detection model $p\left(Ele + \sqrt{FC} + ln(Dur)\right)$ was used

Model	w	Intercept $\hat{\alpha}_0$	Ele $\hat{\alpha}_1$	Ele^2 $\hat{\alpha}_2$	$\sqrt{FC}$ $\hat{\alpha}_3$	$Ele \times \sqrt{FC}$ $\hat{\alpha}_4$
$\psi\left(Ele + Ele^2 + \sqrt{FC} + Ele \times \sqrt{FC}\right)$	0.49	−2.237	−0.039	−0.019	2.805	0.845
		(1.274)	(0.244)	(0.015)	(1.767)	(0.328)
$\psi\left(Ele + \sqrt{FC} + Ele \times \sqrt{FC}\right)$	0.49	−2.562	−0.257	–	3.106	1.032
		(1.338)	(0.182)	–	(1.837)	(0.304)
$\psi\left(Ele + Ele^2 + \sqrt{FC}\right)$	0.02	−4.632	0.593	−0.038	6.322	–
		(1.334)	(0.226)	(0.027)	(1.751)	–
$\psi(Ele)$	0.00	−0.278	0.541	–	–	–
		(0.297)	(0.105)	–	–	–
$\psi\left(Ele + \sqrt{FC}\right)$	0.00	−6.305	0.309	–	8.562	–
		(1.074)	(0.076)	–	(1.496)	–
$\psi\left(Ele + Ele^2\right)$	0.00	0.094	0.607	−0.049	–	–
		(0.326)	(0.084)	(0.010)	–	–
$\psi\left(\sqrt{FC}\right)$	0.00	−3.762	–	–	1.083	–
		(8.141)	–	–	(2.178)	–
$\psi(\cdot)$	0.00	0.714	–	–	–	–
		(0.281)	–	–	–	–

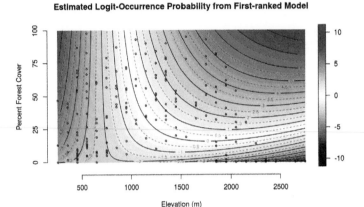

FIGURE 4.16 Bivariate plot of estimated logit-occupancy as a function of elevation and forest cover for willow tit in Switzerland, from the top-ranked model in Table 4.14. Small circles indicate observed elevation and forest cover combinations at surveyed units.

A bivariate plot (Fig. 4.16) can again be used to visualize how occupancy is estimated to be jointly affected by elevation and forest cover under this model. Occupancy is estimated to be highest where elevation and forest cover are both high, but there were very few units that had this combination of covariate values. The inclusion of the interaction term causes the effect of one variable, to be different depending on the value of the other variable, producing the non-parallel contour lines. Thus, when elevation is low, occupancy is estimated to decrease as forest cover increases, but at higher elevations occupancy is estimated to increase with higher levels of forest cover. The quadratic effect of elevation is subtle, given the small estimated value for α_2, with the elevation at which occupancy is highest changing for different values of forest cover. The peak in occupancy for a specific percent forest cover occurs in the lower, right-hand portion of the plot, at, approximately, the location of the contour line-labels '−2' to '5.5'. While we have not done so here, one could also create a bivariate plot of the standard error to examine how it differs at different combinations of covariate values and examine where uncertainty is greatest. Similar plots could also be created for the other models considered.

Another use of the fitted regression equation obtained from each model is to create species distribution maps for willow tits in Switzerland, based upon Swiss landscape data. Given the elevation and forest cover values for each 1 km² unit, the estimated regression equation from each model can be used to predict the probability of willow tit presence in each unit. Importantly, the elevation and forest cover variables from the landscape data must be on the same scale as that used in the analysis. Once calculated, these values can be plotted to create the map. An advantage of using techniques such as occupancy models for data analysis and subsequent prediction (in addition to accounting for imperfect detection) is that not only are we able to create a map of the estimated distribution, but also illustrate the levels of precision in the estimated probabilities with maps of standard errors or confidence limits. This important aspect is often ignored by many species distribution modeling techniques.

Fig. 4.17 presents the resulting species distribution maps (and maps of the standard errors; Fig. 4.18) from each of the eight occupancy models fit to the data, ordered according to the rankings in Table 4.14. Clearly, the distribution maps from each model can be quite different depending on what set of covariates is included in the estimating model, although the maps from the two highest-ranked models are markedly similar. The map from the lowest ranked model (bottom-right panel) is a single color as the estimating model did not include any covariates for species occurrence, hence all units have the same probability of willow tit being present. Maps of the standard error for the estimated willow tit occurrence probability indicate where there is greater uncertainty about the expected distribution according to the models used. There is a range of other metrics that could be used in a similar manner (e.g., coefficient of variation, i.e., CV, confidence interval limits or width, etc.), but note that in general, the standard error for a probability will always be smaller as the probability gets closer to 0 or 1.

A single distribution map based upon the set of models fit to the data could be created, if desired, using model averaging to calculate the model-averaged occurrence probability for each 1 km^2 unit. Fig. 4.19 presents the result of doing so. In this case the model averaged distribution map appears very similar to the top two maps in Fig. 4.17 as those two maps themselves were similar, and the associated models had equal AIC model weight values, however that will not always be the case.

A single distribution map could also be created by model averaging the estimated regression coefficients in Table 4.15 and using the resulting model averaged regression equation to derive predicted occurrence probabilities for each unit, based upon the characteristics of the unit. While this may initially seem to be a reasonable approach, we advise against doing so as the exact interpretation of each coefficient depends upon what factors are included in each particular model and as such, the estimated value may fluctuate widely. Combining the different estimates with different interpretations using model averaging, or other methods, seems ill advised in this case. Furthermore, regression coefficients will often exhibit some degree of correlation, particularly when some factors or covariates within a model are partially correlated or interactions of factors are included. Therefore determination of the standard error for the model averaged regression coefficient should require consideration of

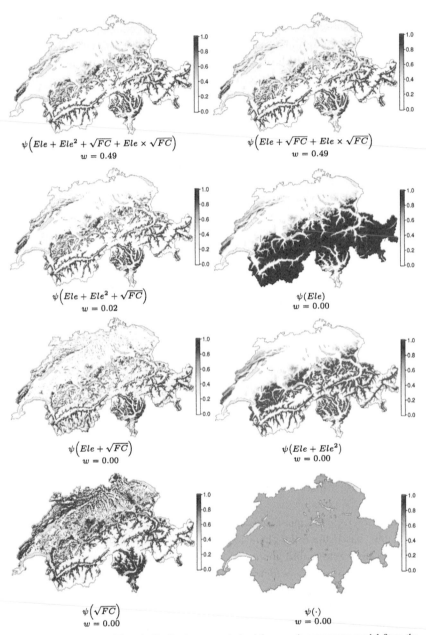

FIGURE 4.17 Swiss willow tit distribution maps derived from each occupancy model fit to the data. Distribution is expressed in terms of the estimated probability of willow tit being present in a 1 km^2 unit. The name and AIC weight (w) of the corresponding model is indicated below each map.

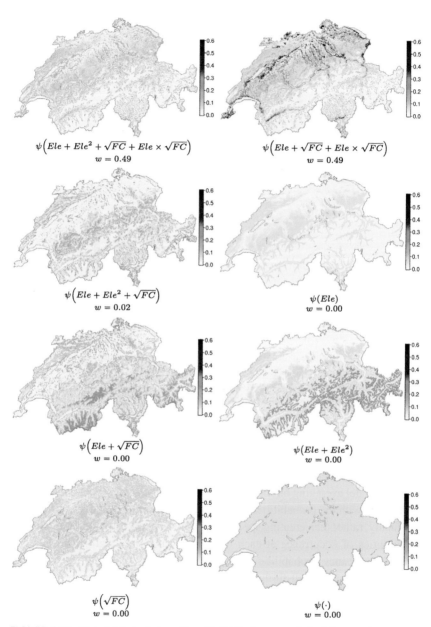

FIGURE 4.18 Uncertainty in Swiss willow tit distribution maps derived from each occupancy model fit to the data. Uncertainty is expressed in terms of the standard error of the estimated probability of willow tit being present in a 1 km^2 unit. The name and AIC weight (w) of the corresponding model is indicated below each map.

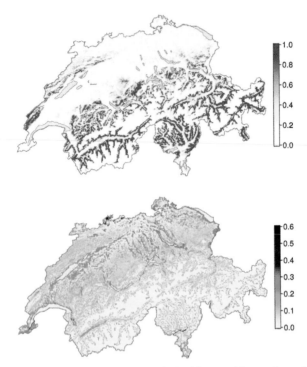

FIGURE 4.19   Swiss willow tit distribution map obtained from model averaging estimates derived from each occupancy model fit to the data. Upper panel is the predicted distribution map expressed in terms of the model averaged estimated probability of willow tit being present in a 1 km^2 unit, while the lower panel is the corresponding map of standard errors.

the full variance–covariance matrix for the coefficients from each model. This is possible, but in our experience not always appreciated by practitioners, and does not solve the previous point about differing coefficient interpretations. Performing model averaging on the estimated probabilities from each model avoids such issues with the probabilities having a consistent interpretation.

Finally, rather than creating species distribution maps based on the estimated probabilities of occurrence at each unit (as in Figs. 4.17 and 4.19), the conditional occupancy probabilities (ψ_{condl}; Section 4.4.5) could be used for the surveyed units. This will further incorporate the outcome of the field surveys into the distribution maps.

4.5   CASE STUDY: TROLL DISTRIBUTION IN MIDDLE EARTH

We finish this chapter with a case study to highlight some of the common pitfalls we have experienced with users of the methods detailed in this chapter. It is a hypothetical case study of a manuscript that has been submitted for publication on the distribution of trolls in Middle Earth, detailing the researchers' study design, analysis and findings. A critique of the manuscript is provided in the form of a reviewer's comments, noting problems with the methods used and recommendations on changes that should be made. While the setting is fictitious, the issues identified by the reviewer are not, and are ones that we believe any researcher should carefully address in his or her own work.

On the Distribution of Trolls in Middle Earth

J.H. Took, B.A. Proudfoot, and F.S. Bolger

Introduction

Despite their livestock depredation and the dangers that they pose to hobbits and other civilized creatures of Middle Earth, relatively little is known about the ecology and distribution of trolls (*Troglodyta tolkiensis*). Our objective in this study was to investigate the distribution of trolls and create a species distribution model (SDM) for the regions of Eriador and Rhovanion, Middle Earth. We surveyed for troll sign at a number of sample units and used occupancy modeling to estimate probabilities of troll occupancy throughout the study area. We investigated the relevance to troll occupancy of a number of site-specific covariates, as a means of learning about determinants of troll distribution.

Methods

Field sampling covered the general regions of Eriador and Rhovanion, bounded approximately by the Shire in the east, the Iron Hills in the west, the Grey Mountains in the north, and the White and Ash Mountains in the south. A grid of 5x5 km cells (sample units) was superimposed on the entire area. Sampling was conducted along roads for reasons of safety and logistics, and 22 locations were selected from the passable roads within the region. Each location was in a different 5x5 km sample unit. Most locations (18) were along the Great East Road between the Shire and the Misty Mountains, and the other four were along the Old Forest Road between the Misty Mountains and Mirkwood.

Troll surveys were conducted within squares of 1 km on each side, each square falling within a different 5x5 km sample unit. The side nearest the road was located 10m from the road and oriented parallel to it. Each square was surveyed by a team of the same two hobbits. The 1 km^2 area was searched for troll sign independently (different survey starting points within the square and no communication) by each team member for a period of two hours. Sign was usually either troll footprints (distinguished from those of other creatures by their large size and absence of toes) or fecal deposits (distinguished by large size and cylindrical shape). At one location, two stone trolls (trolls turn to stone when exposed to sunlight) were actually found. Each search by one hobbit was viewed as a survey occasion, with one individual always denoted as survey 1 and the other individual survey 2. Therefore, the detection histories for all units had two surveys, with a '1' denoting detection of any sign in a survey, regardless of type, and a '0' indicating no detection of sign.

Several site covariates were collected for use in the modeling of troll distribution. Geographic covariates were obtained for the 5x5 km sample units and included average elevation (*elev*) and elevation squared (*elev*^{2}), as well as distance (and distance squared) from nearest (1) mountain (denoted as *dmt* and *dmt*^{2}), (2) river (*driv* and *driv*^{2}), and (3) forest (*dfor* and *dfor*^{2}). Climatic covariates included average summer temperature (*sumtemp*) and average summer temperature squared (*sumtemp*^{2}), average winter temperature (*wintemp*) and average winter temperature squared (*wintemp*^{2}), average days below freezing (*dbf*) and average days below freezing squared (*dbf*^{2}), average rainfall for the entire year (*rain*), average summer rainfall (*rainsum*), average winter precipitation (*pptwin*) and average number of days with snow cover (*snow*). All of the above covariates were considered in the modeling of both detection probability and occupancy. In addition, observer identity (or occasion; denoted as *obs*) was included as a covariate for detection probability.

Single-season models of MacKenzie et al. (2002) were fit to the data. We first focused on modeling detection probability, holding occupancy constant, $\psi(.)$, and modeling p as constant, as a function of one covariate, two covariates, three covariates and four covariates for all possible covariates and covariate combinations. Model selection based on Akaike's Information Criterion (Burnham and Anderson 2002) was used to select the most appropriate model for detection, and this model was then used in the next set of models to explore variation in occupancy. So with the single model of detection probability selected in the previous set, a model set was created in which ψ was constant, a function of one covariate, then two, three and four covariates in all possible combinations. The low-AIC model from this step was selected as the most appropriate for the data.

All model fitting and computation of estimates was carried out by program PRESENCE (Hines 2006). In order to develop a global estimate of occupancy for the entire Eriador–Rhovanion region, we computed unconditional occupancy using the low-AIC model for ψ in conjunction with mean values of every covariate that appeared in this model.

Results

Troll sign was encountered at 8 of the 22 sites for a naïve occupancy estimate of 0.36. The low-AIC model in the model set used to investigate detection probability was $\psi(.)p(obs, elev, dmt, dfor^2)$, indicating the influence on detection of observer identity, site elevation, distance from nearest mountain, and distance from nearest forest squared (Table 1).

Table 1: AIC values, difference between AIC of focal model and top model (ΔAIC), and number of parameters for the top five models in the model set used to investigate sources of variation in detection probability.

Model	AIC	ΔAIC	#pars
$\psi(\cdot)p(obs, elev, dmt, dfor^2)$	121.20	0.00	6
$\psi(\cdot)p(obs, elev, driv, wintemp)$	122.24	1.04	6
$\psi(\cdot)p(obs, dmt, driv, wintemp)$	123.17	1.97	6
$\psi(\cdot)p(obs, dmt, dfor^2, wintemp)$	125.53	4.33	6
$\psi(\cdot)p(obs, elev, dmt, dbf)$	127.71	6.51	6

Variation among sample units in probability of occupancy was then investigated using the above model for detection probability and the described set of models for occupancy. The low-AIC model in this set was $\psi(elev, snow, dmt, dbf)$, $p(obs, elev, dmt, dfor^2)$, providing evidence of effects of elevation, days of snow cover, distance from nearest mountain and days below freezing (Table 2). The overall estimate of occupancy for the entire region, based on the top-ranked occupancy model and covariates for all potential sample units, was 0.54, substantially higher than the naïve occupancy of 0.36.

Table 2: AIC values, difference between AIC of focal model and top model (ΔAIC), and number of parameters for the top five models in the model set used to investigate sources of variation in probability of occupancy.

Model	AIC	ΔAIC	#pars
$\psi(elev, snow, dmt, dbf)$, $p(obs, elev, dmt, dfor^2)$	102.78	0.00	10
$\psi(elev, snow, dmt, wintemp)$, $p(obs, elev, dmt, dfor^2)$	103.95	1.17	10
$\psi(elev, snow, dbf, rain)$, $p(obs, elev, dmt, dfor^2)$	105.67	2.89	10
$\psi(elev, dmt, wintemp, dbf)$, $p(obs, elev, dmt, dfor^2)$	108.91	6.13	10
$\psi(elev, snow, dbf)$, $p(obs, elev, dmt, dfor^2)$	110.35	7.57	9

Discussion

We developed a species distribution model for trolls in a large region of Middle Earth and identified site-specific variables important to this distribution. Elevation appeared in each of the top five models for occupancy and appeared to be an important determinant of troll distribution. Average days with snow cover, distance from the nearest mountain, and days below freezing also appeared in the low-AIC model. The substantial difference between naïve and estimated occupancy demonstrates the importance of dealing with detection probability in species distribution modeling.

Detection probability was not a focus of the study, but its modeling was important in developing the species distribution model. Observer identity was an important determinant of detection probability and appeared in all of the top five models. Other determinants of detection appearing in the top model were elevation, distance from the nearest mountain, and squared distance from the nearest forest.

We believe this species distribution modeling effort to be an important contribution to our knowledge of troll distribution and ecology in the regions of Eriador and Rhovanion, Middle Earth. This modeling permits us to predict the probability of troll occupancy for any sample unit within these regions. Such probabilities will permit us to identify areas that are more and less likely to experience troll attacks and depredation.

Acknowledgements

We wish to thank G.T. Grey for comments on a previous version of this manuscript. Fieldwork was conducted in accordance with the Shire Council for Ethical Studies of Non-hobbit Creatures guidelines; no trolls were harmed during the course of this research.

References

Burnham, K. P., and D. R. Anderson. 2002. *Model selection and multi-model inference: a practical information-theoretic approach.* Second edition. Springer, New York, New York, USA.

Hines, J. E. (2006). *PRESENCE- Software to estimate patch occupancy and related parameters.* USGS-PWRC. http://www.mbr-pwrc.usgs.gov/software/presence.html.

MacKenzie, D.I., J.D. Nichols, G.B. Lachman, S. Droege, J.A. Royle, and C.A. Langtimm. 2002. *Estimating site occupancy when detection probabilities are less than one.* Ecology 83:2248-2255.

REVIEW OF MANUSCRIPT, ON THE DISTRIBUTION OF TROLLS IN MIDDLE EARTH, BY J.H. TOOK, B.A. PROUDFOOT, AND F.S. BOLGER

I have reviewed this manuscript and believe it to be an interesting piece of work in many ways. However, I also have a number of problematic issues to raise. These issues vary in severity, and I simply list them below and provide my rationale.

Sampling issues

Field sampling covered the general regions of Eriador and Rhovanion, with a grid of 5x5 km cells (sample units) superimposed on the entire area. Sampling was conducted at 22 locations selected from two passable roads within the region. Each location was in a different 5x5 km sample unit. This is an extremely large area containing nearly 100,000 sample units. It would have been nice to sample more than 22 units, but I recognize that such sampling is expensive of time and effort. Random selection of sampling units would have at least permitted weak inference (large variances on predicted occupancy probabilities, etc.) about the entire region. However, all samples were located along two major east-west roads within the center of the focal region, representing a form of convenience sampling. I recommend that the authors state that their inferences are restricted to the rectangular area along these roads, rather than imply that they characterize the entire focal region. I also note that this restrictive sampling did not include the full range of many covariate values because of the central location of the roads, the fact that the roads did not cross mountains, etc.

The field sampling for this paper was based on surveys of sign, primarily footprints, scat and, in one case, stone trolls. I believe that some statement should have been made about the 'freshness' of sign in order to provide the reader with some idea of the time frame to which the concept of occupancy applies. For example, in some cases, scat deposited in the open may be baked by the sun, become very hard, and may exist for substantial time periods (e.g., many months). If this had been the case in this study, the possibility should have been noted. There is no indication of how long the stone trolls had been present in the sample unit. I don't claim that there is anything wrong with the conduct of sign surveys, and I recognize the difficulties of dealing with the issue of sign freshness and time, but at a minimum, I believe that this issue merits some discussion. Of course if sign freshness can be detected in the field, one approach to establishing a time frame for occupancy is to restrict analyses to sample units at which fresh sign is found.

Hypotheses and predictions

I was disappointed at the absence of any hypotheses and predictions about covariate effects. This absence precludes use of analytic results to assess hypotheses about troll distribution and relegates work to an exercise in hypothesis generation at best. If the exercise was intended to be exploratory from the beginning, then this is fine, but should have been stated by the authors (my opinion). It is my experience that authors almost always have some hypotheses about determinants of distribution, and indeed these hypotheses guide the selection of covariates. I simply believe that these hypotheses should be made explicit to the reader, as this adds substantially to the value of the exercise.

Absence of predictions also precludes the use of β estimates to check the reasonableness of selected models. I have encountered cases where I believe that blind reliance on AIC can lead to useless models with little predictive ability (see below), and signs and magnitudes of β estimates relative to predictions can be used as an additional way to assess model utility.

Modeling strategy

The 2-step approach of first focusing on detection probability and then on occupancy merits several comments. The first point is that such a sequential approach has not been shown to be as useful as an approach that includes all considered models in a single model set. However, I recognize that the sequential approach is a pragmatic attempt to deal with large numbers of possible models, and I believe this to be reasonable.

The initial model set was designed to explore sources of variation in detection probability, and the authors used a $\psi(\cdot)$ model for this exploration. I strongly disagree with this approach, as any variation in occupancy is now forced into the modeling of detection probability. As a result, covariates that affect occupancy and not detection may appear to be important in modeling detection. The alternative that I favor would be to model occupancy with as flexible a model as possible in order to reduce the possibility of these sorts of confounding effects. I recognize that it is not always obvious what the most general or most flexible model should be, but my main point is that a 'dot' model is not a wise choice as a baseline for this sort of sequential model selection.

A point related to the last one is that I favor a reordering of the sequential model fitting. This may not be very important, but I believe it may be wise to model the focal parameter (the one of ecological interest) first using a very general model for detection probability. This permits inferences about effects on occupancy using models with as few constraints as possible, hopefully permitting the strongest inferences. This approach is an attempt to guard against constraints imposed on detection probability (based on an initial modeling step) being confounded with variation in occupancy. I have used the detection-first approach many times myself and hope that it usually performs well, but for future work I plan to reverse the sequential ordering of modeling, unless there is a good reason not to.

A final point with respect to this kind of sequential modeling deals with checking of models after the two steps are completed. For example, Table 1 shows strong support for at least 3 of the models of detection probability. One approach would be to develop model sets for exploring sources of variation in occupancy using each of these 3 models of detection probability. If this approach is viewed as too consumptive of time and effort (although I note that the required time and effort would be orders of magnitude less than that used to collect the data in the first place), then at a minimum I would recommend fitting models with the top occupancy model(s) (e.g., as determined by results of Table 2) and all 3 detection models, in order to serve as a minimal check of the selection process.

The above 3 points deal with the sequential modeling process used by the authors. The selection of covariates represented another component of modeling that merits comment. The number of covariates tested (18) was very large relative to the number of sample units (22). This insures that some sets of covariates are very likely to fit the data very well by chance alone, leading to the issue sometimes referred to as 'prediction bias'. Although such models will fit data sets very well in some cases, they are of little use for prediction, in the author's case for predicting occupancy of unsurveyed sample units. My recommendation is thus to survey many more sample units and/or to include in models only those covariates most strongly believed to influence model parameters.

A comment on model development and covariate selection related to the above point concerns the number of models fit. Even with the sequential approach to model selection, the authors fit a very large number of models. I have noted the likelihood of models appearing to fit well by chance alone just because of the large number of covariates relative to the number of sample units. However, even if the number of covariates had not been so large, the fitting of large numbers of models will also lead to some models fitting well by chance alone. The recommendation is to limit the number of tested models via hard thinking rather than relying solely on model selection algorithms.

Another issue relates somewhat to the previous discussion of hypotheses and involves some uses of quadratic terms, such as $elev^2$. Quadratic expressions, such as $\text{logit}(\psi_i) = \beta_0 + \beta_1 * elev + \beta_2 * elev^2$, are very sensible, hypothesizing a non-monotonic relationship, for example, in which probability of occupancy is highest at intermediate elevations rather than at extreme high or low elevations. However, I question modeling in which quadratic terms appear alone, e.g., as $\text{logit}(\psi_i) = \beta_0 + \beta_1 * elev^2$, as this forces the maximum (or minimum) value of the relationship to be exactly at the point where the covariate of interest equals 0, which will typically be a very strong assumption. Of course if the investigator has an *a priori* hypothesis favoring such a relationship, then this is fine. I raise this issue as a general concern, and I note the inclusion of $dfor^2$ without $dfor$ in the authors' modeling of detection probability.

As an aside, the notation the authors have chosen to denote models is somewhat ambiguous and if an additive model was used (i.e., no interactions between covariates), then I suggest they replace the ',' with '+' to be more explicit. It was also not initially obvious that when including the $dfor^2$ term that $dfor$ was omitted, as some people notate their models based upon the highest-order polynomial term. I suggest the authors add a sentence or two explaining their model notation.

Note that all of the problems listed for the covariate selection material in *Modeling strategy* could have been listed under *Hypotheses and predictions* in the sense that adequate thought devoted to *a priori* hypotheses would eliminate all of these issues. My central point is again that hypotheses should be specified up front, and model-fitting results used to test them, as opposed to mounting fishing expeditions that include a shopping list of covariates without adequate thought. Such thinking can be used to reduce the number of covariates, and the number of models, to those few with the highest probabilities of being useful. In this particular manuscript, authors considered only additive models, but if an interaction is hypothesized between 2 covariates, then a specific interaction term should be included in the modeling. I believe that the selection of link functions should be governed by hypotheses as well.

Computations

The authors indicated that they used program PRESENCE for all computations, and this is fine. However, they provided no information about checks to insure that model fitting was accomplished in a manner that led to maximum likelihood estimates. For example, there was no mention of checks (provided by PRESENCE) that estimation was successful to an acceptable number of significant digits. But this is frequently omitted from papers, so I will simply assume that such checks were conducted. There is also no indication that models were fit with multiple sets of starting values for model parameters. This practice can help analysts avoid the possibility that numerical algorithms converged on local, rather than global, maxima. I believe increasingly that such checks for local maxima should become standard practice.

Presentation/interpretation of results

Not only did the authors of this manuscript fail to present hypotheses and predictions, but they also failed to report estimates of β parameters or their standard errors, providing information about the sign, magnitude, and precision of estimated covariate relationships. The model selection information provided by the authors reports which covariates were useful in the modeling of the data sets. But without β estimates, there is no information about the nature of these relationships. Even though the authors presented no hypotheses or predictions, readers frequently have their own hypotheses about covariate relationships, but there is no way to assess whether the data are consistent with them or not. If the analyses are viewed as useful only for hypothesis generation, then the nature of generated hypotheses is unknown to the reader. In addition to providing information useful in testing or generating hypotheses, the relative magnitudes of estimated β's and their standard errors provide additional (to model selection) information about the importance of specific covariates in modeling occupancy.

Results of Table 2 indicate substantial uncertainty about the most appropriate model for these data. In such a case, it could be argued that β estimates from the low-AIC model do not provide all of the requisite information about key covariate relationships. One approach might be to compute model-averaged estimates of occupancy (not of β estimates) and to then plot these estimates against individual covariates. This approach accounts for the fact that some covariates are not included in all models and provides a reasonable overall assessment of the utility of a specific covariate for prediction of occupancy.

It can be useful to readers to include the $-2\log$-likelihood value (which is output by PRESENCE) in the tables as another means of checking results or enabling them to make additional/alternative calculations. For example when a covariate is added to a simpler model, the models are said to be 'nested' and the $-2\log$-likelihood value for the more complex model should be smaller. Checking that they are can be one way of diagnosing possible local maxima (as mentioned above). Using the -2log-likelihood values readers may also chose to calculate likelihood ratio tests (where appropriate) or alternative model selection metrics. These values can be obtained from the AIC values presented, but presenting them does make things slightly simpler for the readers.

The authors presented a summary statistic intended to reflect average occupancy within the focal region. I have already noted the mismatch between surveyed sample units and the region about which inferences are sought. But even if I profess interest in only the area surrounding the 2 sampled roads, I still dislike the approach for obtaining an average occupancy estimate for the region. The authors computed their average occupancy estimate using the low-AIC model for ψ, with its associated estimates of β, in conjunction with mean values of every covariate that appeared in this model. The resulting predicted occupancy probability reflects the probability of occupancy for a site that is characterized by average values of all relevant covariates. But this value does not represent an average in other senses. For example, if I wanted to compute an average occupancy for the sampled area, I might compute unconditional occupancy for all potential sample units in the focal area and then compute their mean. This would be an average value in the sense that if I could select a sample unit at random from the entire set of sample units in the focal area, this mean would be my best estimate of occupancy for that unit. There is also a mathematical theorem known as Jensen's inequality that is relevant on this point which states that the function of the average of a set of values, does not equal the average of the function of the values (i.e., $g(E[X]) \neq E[g(X)]$), with direction of the inequality depending upon the shape of the function. Hence, if the summary statistics the authors present are intended to reflect 'average' occupancy in the area sampled (for comparison with the naïve estimate), then I suggest they replace the predicted probability at the average covariate values with the average predicted probability calculated across all sample units and their respective covariate values.

4.6 DISCUSSION

In this chapter we have outlined a general sampling protocol that is used in many occupancy studies and introduced the concept that to explicitly account for detection probability, repeated surveys of units are required within a season (the period during which the occupancy state of units does not change, or changes in a completely random manner). We have also reviewed a number of methods for estimating the proportion of area occupied by a target species, and then detailed a model-based framework that provides a great deal of flexibility because of its ability to incorporate missing observations and covariate values into an analysis.

As reviewed in Chapter 2, many investigators have quantified occupancy for use in various kinds of ecological investigations. Such investigations include metapopulation studies (e.g., Hanski, 1999), species–habitat modeling (e.g., Scott et al., 2002, including resource selection probability functions; Manly et al., 2002) and monitoring. This large body of literature generally does not adequately address what we believe to be one of the fundamental characteristics of sampling animal populations; that in most cases a target species will not always be detected at a sampling unit when it is present. By not accounting for imperfect detection within the methods of analysis, the resulting inferences will be less robust because any apparent change or difference in 'occupancy' may actually be due to differences in the detectability of the species (e.g., the pronghorn antelope example).

In the following chapters we provide a number of useful extensions of the basic framework described above. Chapter 5 considers the case where instead of having two possible occupancy states (e.g., species presence/absence), there may be a greater number of possible states. For example, not only species presence/absence, but when the species is present, is it breeding (or not) within the unit. The multi-state occupancy model provides the framework for a number of the more advanced modeling approaches described in this book. In Chapter 6 we present a series of extensions and specialist topics for single-season occupancy applications, namely: estimating the proportion of occupied units in situations where the units that were surveyed constitute a substantial fraction of the total area of interest; accounting for false-positive detections; estimating occupancy at multiple scales; dealing with correlated detections; and incorporating spatially-correlated occupancy. A final topic covered in the single-season part of the book, is extending the above model-based estimation methods to allow for heterogeneous detection probabilities among units, beyond that accounted for by covariates (Chapter 7). In Chapter 11 study design issues for single season occupancy studies are discussed. Appropriate study design is of vital importance as it has an overriding influence on the reliability of resulting estimates and how the estimates should be interpreted. No amount of statistical magic will ever improve poorly collected data.

Chapter 5

Beyond Two Occupancy States

So far, occupancy has been treated as a dichotomous state; that is a sample unit is either occupied or it is not. This is most commonly regarded in terms of species presence/absence, although we note that a practitioner does not have to equate 'occupancy' with 'at least one member of the species is present'. 'Occupancy' may be defined in terms of a specific age-class of the species (e.g., juveniles), hence a detection is only recorded when evidence of that age-class is observed during a survey. If no evidence of that particular age-class is observed, even if there is evidence of other age-classes (e.g., adults), a nondetection would be recorded for the survey. Similarly, it may be that 'occupancy' is defined in terms of a minimum number of individuals that is >1 for areas where the species is more numerous (rather than just present), or as an attempt to avoid the effects of transient individuals that may occasionally pass through a unit, but do no occupy it in any biologically meaningful sense. In this case, the minimum number of individuals must be observed during a survey before a detection is recorded, and observing fewer individuals is regarded as a nondetection. For example, at least five individuals have to be sighted in a survey before the species is 'detected', otherwise the outcome of the survey is a 'nondetection'.

The desire to obtain greater insight about a species leads to situations where a unit might be characterized as belonging to one of multiple (greater than two) classes or states. A situation that is often considered involves reproduction or breeding. For example, we might want to differentiate between occupied units that support breeding or reproduction and those that only have non-breeders. One motivation for these considerations involves ideas about variation in the contributions of occupied units to metapopulation dynamics (e.g., Pulliam, 1988), with generally greater value and importance associated with units at which reproduction does occur (e.g., a potential population source; Fig. 5.1). Another situation where multiple occupancy states might be considered is as a measure of relative abundance, where states could be defined as: none, few, a moderate number, or lots of individuals (for example). Such an approach utilizes information on relative abundance that will often be collected in the field rather than reducing the data down to a simple detection/nondetection observation, and explicitly allows for species-level detection probabilities to depend on relative abundance, without defining a specific functional form

Occupancy Estimation and Modeling. http://dx.doi.org/10.1016/B978-0-12-407197-1.00007-7

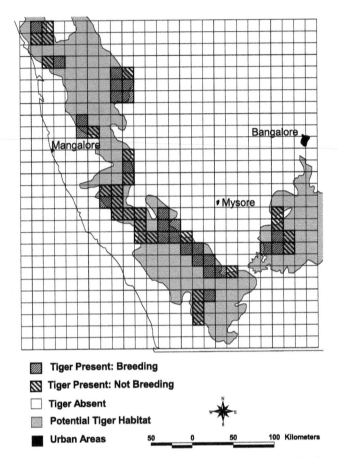

FIGURE 5.1 A hypothetical map of tiger distribution in an area of southern India. Cells are coded to show presence of tigers with and without breeding. Only cells that contain potential tiger habitat are included in the sampling frame. Such information can be viewed as a first step in identifying source and sink habitat for the purpose of making land management decisions. *Source: Stith and Kumar (2002).*

(e.g., Royle and Nichols, 2003). Therefore, even if there is little interest in the specific occupancy states, the methods detailed in this chapter accounted for abundance-induced detection heterogeneity (in a similar manner to the finite-mixture models detailed in Chapter 7), leading to a more accurate estimate of overall species occupancy. A final motivational example we have encountered that lends itself to a multi-state occupancy framework is to determine where a pathogen, or disease, may occur within the distribution of a species (noting that the pathogen may not always be detected when present in the host population). In this case, units may be unoccupied (having neither the host nor pathogen species), occupied by the host species without the pathogen, or occupied by the host species with the pathogen. Indeed, we view the multi-state occupancy framework as a very flexible and useful approach that could be applied to many ecologically interesting problems.

In this chapter we present an approach to estimate single-season occupancy with multiple states. Primarily we develop the model using species presence with breeding or without breeding, and species absence as the three occupancy states, but the reader can see that the development can deal with other occupancy states and can be extended to deal with any number of states. In Chapter 9 we extend the modeling to the multi-season situation allowing investigation of the underlying dynamics, which is utilized for joint modeling of occupancy and habitat dynamics and multi-species interactions (Chapters 13 and 14, respectively).

5.1 THE SAMPLING SITUATION

The sampling situation we consider is identical to that presented in Chapter 4, with the exception that the intent is to estimate the proportion of sampling units in each defined occupancy state, or the underlying probability of a unit being in each state. For example, in the case of assessing the presence of breeding at occupied units, the true occupancy states could be defined as: 0 = species absent; 1 = species present without breeding; and 2 = species present with breeding. Typically, additional assessments (i.e., beyond simply detecting the species) are required to determine the evidence for the type of occupied unit. For example, for pond-breeding amphibians, finding egg masses, new metamorphs, or animals in amplexus might constitute evidence of breeding in the pond. In studies of the spotted owl, investigators routinely establish different sets of criteria that constitute evidence of occupancy and of reproduction (e.g., Franklin et al., 2004). Evidence for occupancy typically involves simply hearing or seeing owls. Evidence of reproduction is provided by various types of information, including adult behavior when adults are presented with prey items (mice). Reproducing adults typically transport mice to their nest to feed their young, hence investigators will attempt to follow the adults to their nests and observe young, which is evidence of reproduction. In relative abundance studies, the unique number of individuals observed or intensity of calls could be used to assign an abundance category. For disease studies, individuals may be observed displaying symptoms of the disease, or samples may be collected to confirm disease/pathogen presence via laboratory assays.

A common characteristic of the above sampling situations, and of most other scenarios involving multiple types of occupied units, is uncertainty. In the usual situation some kinds of information provide unambiguous evidence (proof) of the true state of a unit, whereas other states are inferred through the absence of evidence and thus cannot provide certain state assignment. For example, finding egg masses of a pond-breeding amphibian confirms breeding occurred in a pond, but not finding egg masses does not confirm the absence of breeding, as the egg masses may be present but undetected. Similarly, observing a large number of individuals in a survey may confirm high abundance, but if only a few individuals are observed there may have been more present, but undetected (although it does preclude the absence of the species from the unit). This sampling problem is similar to that involved in the estimation of occupancy itself, with evidence of species presence (1 in the detection history) being unambiguous, but evidence of absence (0 in the detection history) being ambiguous (Chapter 4). The same problem arises in multi-state capture–recapture modeling where misclassification can occur in the assignment of an individual to state (Fujiwara and Caswell, 2002; Lebreton and Pradel, 2002; Kendall et al., 2003; Nichols et al., 2004; Pradel, 2005).

During each survey, records are made of the observed category or state, which may or may not be the true state for a unit. For example, if the species occupies a unit but there is no breeding (true state $= 1$), the results of a survey could be the species is detected without breeding (observed state $= 1$), or the species is not detected at all (observed state $= 0$), but the survey result cannot be detection of the species and evidence of breeding (observed state $= 2$). Therefore, the sampling situation presumes that there may be one-way misclassification of the true state of a unit through the imperfect observation, or detection, processes (i.e., observed state must $\leqslant$ true state; Fig. 5.2). This is just a generalization of the sampling situation assumed for the basic occupancy model described in the last chapter (presuming no misidentification of the species).

Given multiple surveys are conducted at a unit within a season, a detection history can be formed from the sequence of observed states, with the highest observed state within a season confirming the lowest possible true state for a unit (as it is assumed that within a season, the true occupancy state does not change, equivalent to the closure assumption of the basic occupancy model). For example, the detection history $\mathbf{h}_i = 1021$ indicates a unit at which the species was detected during survey 1 with no definitive evidence of breeding at that time, not detected in survey 2, detected in survey 3 with definitive evidence of breeding, and detected in survey 4 with no definitive evidence of breeding. Observing a '2' during the season confirms breeding at the unit, therefore observing '0' or '1' is the result of not detecting the species or the required evidence of breeding in a particular survey. Whereas the detection $\mathbf{h}_i = 0101$ indicates a unit at which the species was not detected in survey 1, detected in survey 2 with no evidence of breeding, not detected in survey 3, and detected in survey 4 but again with no

Observed State

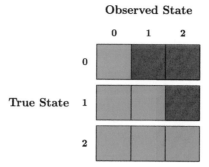

FIGURE 5.2 Illustration of the presumed relationship between observed and true states for a multi-state occupancy model. Green cells indicate possible combinations and red cells impossible combinations.

evidence of breeding. As the species was detected at least once, that precludes the possibility that the species is absent from the unit, but the failure to find evidence of breeding does not preclude that breeding may have occurred at the unit.

The basic sampling situation could also be defined in terms of the underlying latent and observed random variables, as was done for the dichotomous occupancy model in the last chapter (i.e., Section 4.4.1). That is, let z_i be a categorical-valued latent random variable of the true occupancy state of unit i, and h_{ij} be a categorical-valued random variable of the observed state of unit i in survey j which depends on the value of z_i. These variables and the associated probabilities are defined more formally in the next section.

5.2 MODEL BASED APPROACH

Single-season multi-state occupancy models have been previously developed by Royle (2004b), Royle and Link (2005), MacKenzie et al. (2006), and Nichols et al. (2007a), but here we largely follow the description of the model given by MacKenzie et al. (2009). For simplicity, the modeling is developed in terms of three possible states (unoccupied and two occupied states), but could be extended to a greater number of states in an obvious manner.

Let $\varphi^{[m]}$ be the probability that a unit is in occupancy state m ($m = 0, 1, 2, \ldots$). The states are mutually exclusive, thus there is a natural constraint that these probabilities must sum to 1; hence, for a 3-state system two of the probabilities may be independently estimated, with the final one obtained by subtraction. Often $\varphi^{[0]}$ will be obtained by subtraction (e.g., $\varphi^{[0]} = 1 - \varphi^{[1]} - \varphi^{[2]}$), but in practice it may be any of the three. Furthermore, let $\boldsymbol{\phi}_0$ be the row vector defining the probability of a unit being in each state immediately before the first survey of the season. That is:

$$\boldsymbol{\phi}_0 = \left[\begin{array}{ccc} 1 - \varphi^{[1]} - \varphi^{[2]} & \varphi^{[1]} & \varphi^{[2]} \end{array} \right].$$

The underlying true occupancy state, can therefore be defined as the latent variable:

$$z_i \sim categorical(\boldsymbol{\phi}_0).$$

Next, let $p_j^{[m,l]}$ be the probability of observing the evidence of occupancy state l in survey j, given the true occupancy state is m. Some types of observations are not possible given the true occupancy state (Fig. 5.2; e.g., if the species is truly absent, it cannot be detected, assuming no species misidentification); in these cases $p_j^{[m,l]}$ will equal 0. Furthermore, one of the possible states must be observed in any given survey, thus there is the constraint $\sum_{l=0}^{m} p_j^{[m,l]} = 1$ (where l are the possible observable states given true state m). Therefore, one of the detection probabilities for each true state must be obtained by subtraction. For the case with a total of three possible occupancy states, the detection probability structure could be summarized as follows:

True state	Observed state		
	0	**1**	**2**
0	1	0	0
1	$1 - p_j^{[1,1]}$	$p_j^{[1,1]}$	0
2	$1 - p_j^{[2,1]} - p_j^{[2,2]}$	$p_j^{[2,1]}$	$p_j^{[2,2]}$

where the probability of non-detection (i.e., observed state = 0), is obtained by subtraction. Note that ambiguity in the observed state is indicated by multiple non-zero values in a column, while the non-zero values along each row indicate possible observations given the true occupancy state of a unit. This is an extension of the detection structure for the simpler dichotomous situation, which is represented by the upper left portion of the above table (where true and observed states are 0 or 1). Readers should take the time to ensure they understand this detection probability structure before proceeding.

With this structure, let $\mathbf{p}_j$ be the matrix of detection probabilities for survey j, where each row relates to the possible true states and columns relate to observed states:

$$\mathbf{p}_j = \left[\begin{array}{ccc} 1 & 0 & 0 \\ 1 - p_j^{[1,1]} & p_j^{[1,1]} & 0 \\ 1 - p_j^{[2,1]} - p_j^{[2,2]} & p_j^{[2,1]} & p_j^{[2,2]} \end{array} \right].$$

Further, let $\mathbf{p}_j^{[m\bullet]}$ denote the row of $\mathbf{p}_j$ for true state m, and $\mathbf{p}_j^{[\bullet l]}$ denote the column of $\mathbf{p}_j$ for observed state l. For example, $\mathbf{p}_j^{[1\bullet]}$ would be a row vector defining the probability of the possible observations if the species was present without breeding (say):

$$\mathbf{p}_j^{[1\bullet]} = \left[\begin{array}{ccc} 1 - p_j^{[1,1]} & p_j^{[1,1]} & 0 \end{array} \right].$$

While $\mathbf{p}_j^{[\bullet 1]}$ would be a column vector defining the probability of observing the species but no evidence of breeding, conditional upon each true state:

$$\mathbf{p}_j^{[\bullet 1]} = \left[\begin{array}{c} 0 \\ p_j^{[1,1]} \\ p_j^{[2,1]} \end{array} \right].$$

The observed occupancy state of unit i in survey j, given the true occupancy state of the unit is m (i.e., $z_i = m$), can therefore be defined as the categorical random variable:

$$h_{ij} | z_i = m \sim categorical\left(\mathbf{p}_j^{[m\bullet]}\right).$$

As for the simpler dichotomous situation, parameters can be estimated from the collected data using either an observed data likelihood (ODL) or complete data likelihood (CDL; Section 4.4.1) approach. Here we focus on using the ODL approach, but note the CDL can be constructed in an analogous manner to the situation with only two occupancy states using the above definitions of the latent and observed random variables (z_i and h_{ij}).

5.2.1 Observed Data Likelihood

Consider again the detection history $\mathbf{h}_i = 1021$. The probability statement for this history, given the parameters defined above, could be expressed as:

$$Pr(\mathbf{h}_i = 1021 | \boldsymbol{\theta}) = \varphi^{[2]} p_1^{[2,1]} \left(1 - p_2^{[2,1]} - p_2^{[2,2]}\right) p_3^{[2,2]} p_4^{[2,1]},$$

where $\boldsymbol{\theta}$ denotes the set of model parameters. This probability statement indicates the species was present with breeding at the unit, was detected without evidence of breeding in survey 1, not detected in survey 2, detected with evidence of breeding in survey 3 and detected without evidence of breeding in survey 4. Recall from Chapters 3 and 4 that the probability statement is formed by taking the specific phrases from a verbal description of the data, and replacing them with the respective parameters that have been defined for the model, e.g., "the species was present with breeding at the unit" is replaced by $\varphi^{[2]}$; "detected without evidence of breeding in survey 1" by $p_1^{[2,1]}$; etc. As there is no ambiguity associated with this detection history (due to observing breeding during the season), the probability statement consists of only a single product term.

Compare that result to the probability statement for the detection history $\mathbf{h}_i = 0101$ which contains two terms added together because the true state is uncertain:

$$Pr(\mathbf{h}_i = 0101|\boldsymbol{\theta}) = \varphi^{[2]}\left(1 - p_1^{[2,1]} - p_1^{[2,2]}\right) p_2^{[2,1]}\left(1 - p_3^{[2,1]} - p_3^{[2,2]}\right) p_4^{[2,1]}$$
$$+ \varphi^{[1]}\left(1 - p_1^{[1,1]}\right) p_2^{[1,1]}\left(1 - p_3^{[1,1]}\right) p_4^{[1,1]}.$$

The verbal description for this probability statement would be that the species was present with breeding at the unit, was not detected in survey 1, detected without evidence of breeding in survey 2, not detected in survey 3 and detected without evidence of breeding in survey 4 (the first term above), *OR* the species was present without breeding at the unit, not detected in survey 1, detected in survey 2, not detected in survey 3 and detected in survey 4 (the second term). Note that the structure of the detection component of each term is quite different, reflecting whether the species is thought to be breeding or not, and that the two terms explicitly account for our uncertainty due to never detecting any evidence of breeding.

For completeness, when the species is never detected in any of the surveys of a unit within a season we have the most severe ambiguity, and the probability statement involves three terms, representing the three possible true occupancy states. That is:

$$Pr(\mathbf{h}_i = 0000|\boldsymbol{\theta}) = \varphi^{[2]} \prod_{j=1}^{4}\left(1 - p_j^{[2,1]} - p_j^{[2,2]}\right)$$
$$+ \varphi^{[1]} \prod_{j=1}^{4}\left(1 - p_j^{[1,1]}\right)$$
$$+ \left(1 - \varphi^{[1]} - \varphi^{[2]}\right).$$

Recall that the product term ($\prod$) is just a convenient notation for multiplying a series of terms together (Appendix).

The observed data likelihood is the product of the probability statements for all surveyed units, as detailed in Chapters 3 and 4, i.e.:

$$ODL(\boldsymbol{\theta}|\mathbf{h}) = \prod_{i=1}^{s} Pr(\mathbf{h}_i|\boldsymbol{\theta}),$$

where $\boldsymbol{\theta}$ denotes the set of parameters in the model.

5.2.2 Matrix Formulation

In general, the ODL for any observed detection history can be efficiently calculated by defining occupancy and detection probability vectors, and using matrix multiplication to sum up the relevant terms for the detection history (MacKenzie et al., 2009). The occupancy probability vector $\boldsymbol{\phi}_0$ is defined above, and let $\mathbf{p_h}$ be a state-specific column vector where each element is the probability of observing the detection history given the respective true state. For example, for the detection history $\mathbf{h}_i = 1021$:

$$\mathbf{p}_{1021} = \begin{bmatrix} 0 \\ 0 \\ p_1^{[2,1]}\left(1 - p_2^{[2,1]} - p_2^{[2,2]}\right) p_3^{[2,2]} p_4^{[2,1]} \end{bmatrix}.$$

The first and second elements are both 0 because the detection history $\mathbf{h}_i = 1021$ cannot be observed if the species was absent, or present without breeding at the unit. That is, as the evidence of breeding was detected in at least one survey, it is known that the species must be present with breeding at the unit (true state $= 2$). That there is only one non-zero element in $\mathbf{p}_{1021}$ indicates there is no ambiguity about the true occupancy state associated with that detection history. When there is ambiguity about the true state from the detection history, $\mathbf{p_h}$ will contain more than one non-zero element, for example:

$$\mathbf{p}_{0101} = \begin{bmatrix} 0 \\ \left(1 - p_1^{[1,1]}\right) p_2^{[1,1]}\left(1 - p_3^{[1,1]}\right) p_4^{[1,1]} \\ \left(1 - p_1^{[2,1]} - p_1^{[2,2]}\right) p_2^{[2,1]}\left(1 - p_3^{[2,1]} - p_3^{[2,2]}\right) p_4^{[2,1]} \end{bmatrix}$$

or,

$$\mathbf{p}_{0000} = \begin{bmatrix} 1 \\ \prod_{j=1}^{4}\left(1 - p_j^{[1,1]}\right) \\ \prod_{j=1}^{4}\left(1 - p_j^{[2,1]} - p_j^{[2,2]}\right) \end{bmatrix}.$$

Note that $\mathbf{p_h}$ is formed by taking the element-wise product of the relevant columns of the detection matrix $\mathbf{p}_j$. For example, let $\odot$ denote the element-wise product of two vectors or matrices (see Appendix for details), then:

$$\mathbf{p}_{1021} = \mathbf{p}_1^{[\bullet 1]} \odot \mathbf{p}_2^{[\bullet 0]} \odot \mathbf{p}_3^{[\bullet 2]} \odot \mathbf{p}_4^{[\bullet 1]}$$

$$= \begin{bmatrix} 0 \\ p_1^{[1,1]} \\ p_1^{[2,1]} \end{bmatrix} \odot \begin{bmatrix} 1 \\ 1 - p_2^{[1,1]} \\ 1 - p_2^{[2,1]} - p_2^{[2,2]} \end{bmatrix} \odot \begin{bmatrix} 0 \\ 0 \\ p_3^{[2,2]} \end{bmatrix} \odot \begin{bmatrix} 0 \\ p_4^{[1,1]} \\ p_4^{[2,1]} \end{bmatrix}$$

$$= \begin{bmatrix} 0 \\ 0 \\ p_1^{[2,1]} \left(1 - p_2^{[2,1]} - p_2^{[2,2]}\right) p_3^{[2,2]} p_4^{[2,1]} \end{bmatrix},$$

where $\mathbf{p}_j^{[\bullet/]}$ was defined previously.

The probability statement for any particular detection history can then be conveniently expressed as the product of the vectors using matrix multiplication, i.e.,

$$Pr(\mathbf{h}|\boldsymbol{\theta}) = \boldsymbol{\phi}_0 \mathbf{p_h}.$$

The matrix multiplication is simply summing the product of the respective pairs of elements (i.e., product of first elements, product of second elements, etc.).

Readers are again encouraged to ensure they have a good understanding of the above development before proceeding, as matrix-based formulations of the models are used throughout the latter chapters of this book because it is a notationally convenient method for defining more complex models.

5.3 ALTERNATIVE PARAMETERIZATIONS

The above parameterization has a multinomial structure and the associated probabilities are estimated separately, although with the constraints that the probabilities of a unit being in each occupancy state must sum to 1 (i.e., $\sum_{m=0}^{M} \varphi^{[m]} = 1$, where M is the greatest true state) and the probabilities of detecting each observed state, given the true state of the unit, also sum to 1 (i.e., $\sum_{l=0}^{m} p_j^{[m,l]} = 1$ where l are the possible observable states given true state m). In some applications it may be reasonable to express the multinomial structure as a series of conditional dichotomous outcomes. Using the same example, where the possible states are species absent, present without breeding, and present with breeding, this could be expressed in terms of species present/absent, and breeding/not breeding given species present. Hence the multi-state occupancy model could be parameterized in terms of a series of conditional binomial probabilities. Nichols et al. (2007a) suggested the following:

$\psi = Pr$(unit is occupied by the species),
$R = Pr$(breeding present at the unit | unit occupied),

in which case the state probability vector would be:

$$\boldsymbol{\phi}_0 = \begin{bmatrix} 1 - \psi & \psi(1-R) & \psi R \end{bmatrix}.$$

Detection probabilities can also be redefined as:

$p_j^{[1]} = Pr(\text{detect species in survey } j \mid \text{unit occupied without breeding}),$

$p_j^{[2]} = Pr(\text{detect species in survey } j \mid \text{unit occupied with breeding}),$

$\delta_j = Pr(\text{detect evidence of breeding in survey } j \mid \text{unit occupied with breeding and species detected in survey } j),$

hence the detection probability matrix becomes:

$$\mathbf{P}_j = \begin{bmatrix} 1 & 0 & 0 \\ 1 - p_j^{[1]} & p_j^{[1]} & 0 \\ 1 - p_j^{[2]} & p_j^{[2]}(1 - \delta_j) & p_j^{[2]}\delta_j \end{bmatrix}.$$

Expressed in terms of the probabilities from the multinomial parameterization, the conditional binomial parameterization is:

$$\psi = \varphi^{[1]} + \varphi^{[2]},$$

$$R = \frac{\varphi^{[2]}}{\varphi^{[1]} + \varphi^{[2]}} = \frac{\varphi^{[2]}}{\psi},$$

$$p_j^{[1]} = p_j^{[1,1]},$$

$$p_j^{[2]} = p_j^{[1,2]} + p_j^{[2,2]},$$

$$\delta_j = \frac{p_j^{[2,2]}}{p_j^{[1,2]} + p_j^{[2,2]}} = \frac{p_j^{[2,2]}}{p_j^{[2]}}.$$

An example of a conditional binomial parameterization is graphically illustrated in Fig. 5.3.

Other parameterizations are possible, for example Royle and Link (2005) used a multinomial parameterization for their occupancy component, but defined detection probabilities in terms of:

$p_j^{[m]} = Pr(\text{correctly detect species as true state } m \text{ in survey } j),$

$\beta_j^{[l,m]} = Pr(\text{observing state } l \text{ in survey } j \mid \text{observed state is } \leqslant l \text{ and true state } m),$

therefore the detection probability matrix becomes:

$$\mathbf{P}_j = \begin{bmatrix} 1 & 0 & 0 \\ 1 - p_j^{[1]} & p_j^{[1]} & 0 \\ (1 - \beta_j^{[1,2]})(1 - p_j^{[2]}) & \beta_j^{[1,2]}(1 - p_j^{[2]}) & p_j^{[2]} \end{bmatrix}.$$

We also note that one could use a combination of conditional binomial and conditional multinomial probabilities. For example, suppose four true states are defined as categories of relative abundance (e.g., none, few, some, and lots of individuals), the states could be modeled in terms of species presence/absence,

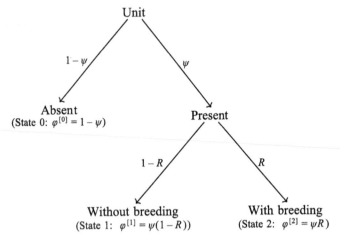

FIGURE 5.3 Example of expressing the multinomial structure as a series of conditional dichotomous outcomes in terms of a unit's true occupancy state. The terminus of each series of dichotomous outcomes represents the possible true states for a unit. Values beside each arrow are the probabilities associated with each dichotomous outcome. This is also known as a conditional binomial parameterization.

then given presence, the relative abundance of the species (few, some, or lots) could be modeled as a conditional multinomial outcome. One reason for doing so is that the results of such an analysis in terms of factors influencing the distribution of the species (i.e., presence/absence) would be directly comparable to the results of a simpler occupancy analysis where the relative abundance information is ignored and only the detection/nondetection information is utilized (i.e., by using the methods from the previous chapter). Another more ecological reason is that, in many cases, the processes (and relevant covariates) governing occupancy are expected to differ from those governing abundance of occupied sites.

The choice of parameterization an analyst should use primarily depends upon the nature of the biological or sampling questions being addressed. In many cases we have found that it is natural to consider the multiple true states as the result of a series of dichotomous outcomes, hence use of the conditional binomial parameterization is an obvious choice. Although in other cases, particularly with more than three relative abundance states, the multinomial parameterization is more sensible. However, when an analysis involves investigating the effects of a number of covariates on occupancy- or detection-related probabilities, in our experience the conditional binomial parameterization can be more numerically stable due to its use of the logit link function. When using the multinomial parameterization with the multinomial logit link function, the numerical optimization algorithms that are typically employed to obtain maximum likelihood estimates, appear to be more prone to converging upon local maxima in the likelihood function (Chapter 3). For this reason we tend to favor the use of the conditional binomial parameterization when potential covariates are being investigated.

5.4 MISSING OBSERVATIONS

As with the simpler situation discussed in the previous chapter, multi-state occupancy models do not require an equal number of surveys at all units, and allow for missing survey information in the same manner. That is, no information about detection (or nondetection) is included in the probability statements when survey j was not conducted at a unit. For example, consider the detection history $\mathbf{h}_i = 1\text{-}21$ where the '-' indicates survey 2 was not conducted at the unit. The associated probability statement (using the multinomial parameterization) is:

$$Pr(\mathbf{h}_i = 1\text{-}21|\boldsymbol{\theta}) = \varphi^{[2]} p_1^{[2,1]} p_3^{[2,2]} p_4^{[2,1]}.$$

Essentially, the survey with the missing observation is skipped, hence no detection probabilities associated with survey 2 appear in the probability statement. Allowing for unequal survey effort is certainly a strength of these approaches; in fact, the ability to account for unequal survey effort at different units within a probabilistic framework was one the original motivations for the models developed by Nichols et al. (2007a).

In some situations the data analyst may choose to regard the outcome of surveys as 'missing' even though a value was actually recorded. One such case might be when survey outcomes are not independent. For instance, once the location of an active bird nest at a unit is determined, it will likely be checked at each subsequent visit to the unit. In such a situation, the occupancy state (species present and breeding) is essentially known once the nest is found. The probability of re-observing the breeding behavior is likely to be higher as the surveyors know where to look for the nest once the nest has been found. An analyst could choose to model that as a 'trap response' effect (Section 4.4.9), or recode the actual observations following the discovery of the active nest as 'missing' observations in recognition that the subsequent surveys are not independent, e.g., recode $\mathbf{h}_i = 0222$ as $\mathbf{h}_i = 02\text{-}\text{-}$. We have not explored the relative benefits of using a 'trap response' approach to dealing with non-independence versus recoding observations as missing, but based on knowledge of trap-response in capture–recapture modeling, we expect that the two approaches should be equivalent when detection probability following first detection is completely independent of detection probabilities for subsequent occasions. However, if there is some relationship between occasion-specific detection probabilities before initial detection and those following initial detection, then the trap response modeling provides a means of taking advantage of such a relationship. Perhaps most important, we certainly believe that failing to address the issue will lead to biased parameter estimates.

5.5   COVARIATES AND PREDICTOR VARIABLES

In the development above, all probabilities have been assumed to be equal across all units. Oftentimes there is interest in determining whether probabilities are different across space (units) or time (surveys). These factors of interest can be included as predictor variables or potential covariates in a very similar manner to the simpler situation (Section 4.4.8). When using a conditional binomial parameterization, probabilities may be allowed to vary between units (and survey occasions for detection probabilities) using an appropriate link function, e.g., logit link (Section 3.4.1).

When using the multinomial parameterization, the standard logit (or similar) link function should not be used as it does not enforce the requirement that the sum of the probabilities equals 1. An option is to use the multinomial logit link function, which is an extension of the regular logit link, i.e., multinomial logistic regression (Section 3.4.1), although, as noted above, we have found the multinomial logit link function to be less numerically reliable. That said, it is still a viable and appropriate approach to use, but the analyst must be more vigilant about numerical issues such as converging to local maxima (Chapter 3).

5.6   MODEL ASSUMPTIONS

The main assumptions underlying the single-season multi-state occupancy model are virtually the same as those of the simpler models of Chapter 4, with a few additions. Namely, it is assumed that:

- occupancy state of a unit is static during the period of repeat surveys (closure);
- there is no unmodeled heterogeneity in occupancy or detection probabilities;
- the outcomes of surveys conducted at a unit are independent;
- detection histories observed at each unit are independent;
- there is no species/state misidentification; the observed state of unit cannot be greater than the true state.

The effects of assumption violations have not been well explored for multi-state occupancy models, although we would expect consequences to be similar to those for the simpler occupancy models (Section 4.4.9). With respect to the closure assumption, we expect that the true state may not be observable during all surveys within a season for some scenarios. For example, definitive evidence of successful breeding may not be observable until later in the breeding season. While this could be regarded as a violation of the closure assumption, provided that the detection process is modeled appropriately (i.e., allowed to vary within a season), we do not anticipate such situations to result in a major bias in parameter estimates. We also point out that in such cases, the parameter estimates must be interpreted accordingly.

Many of the extensions that have been developed for violations of assumptions required by the basic single-season occupancy model (detailed in Chapters 6 and 7), could also be applied to the multi-state occupancy model. Indeed, some of the extensions utilize the multi-state framework (e.g., as one approach to account for species misidentification; Miller et al., 2011), hence such extensions may be as simple as defining additional true and/or observed states. Therefore, there are practical options that could be implemented when a multi-state occupancy approach is desired, but some of the assumptions of the model described above may be violated.

5.7 EXAMPLES

We now provide two example applications of the multi-state model. While in both cases the states have been defined relative to the presence of breeding, we would stress that there is a wide range of options for how states could be defined depending on the biological questions of interest. The first example illustrates the effect accounting for imperfect detection and classification probabilities has on the biologically relevant parameters of interest, while the second example illustrates the incorporation of covariates into multi-state occupancy models. Files associated with these example analyses can be downloaded from http://www.proteus.co.nz.

5.7.1 California Spotted Owl Reproduction

One of the motivating examples for Nichols et al. (2007a) was estimation of the reproductive rate of California spotted owls (*Strix occidentalis*) in the central Sierra Nevada (California, USA), while allowing for state classification uncertainty (e.g., whether territories are unoccupied, occupied without breeding, or occupied with breeding) and variable sampling protocols. For the California spotted owl, reproductive rate was defined as the proportion of breeding pairs that produce young, and was equated to the proportion of occupied owl territories with young, as territories will only be inhabited by a single breeding pair (Nichols et al., 2007a).

Potential spotted owl territories would be initially surveyed for the presence of adult owls, and, if detected, the adult would be offered a live mouse as a prey item. Breeding adults will tend to, but not always, return to their nest site with the mouse for their young, and investigators would follow the owls through the forest with definitive evidence of reproduction being sighting of young owlets capable of leaving the nest (but not necessarily having fledged; R.J. Gutiérrez, *pers. commun.*). Non-reproductive adults will tend to consume or cache mice for themselves, but reproductive adults may also display this behavior on occasion. Therefore, the state of a potential territory (i.e., the sampling unit) may be misclassified in a number of ways:

- nondetection of adults when present at a territory;
- reproductive adults consuming or caching mice;
- investigators failing to track an adult to its nest site or otherwise failing to observe young.

Investigators thus develop criteria (usually based on a required minimum number of visits to a territory) to conclude whether reproduction has occurred at a territory or not.

Seamans (2005) analyzed data from five large study areas in Sierra Nevada conducted by different California spotted owl research groups and concluded that due to variation in the criteria used by each group, particularly with respect to the number of surveys conducted per territory, the estimates of reproduction from each group were not comparable. This precluded a meta-analysis of all groups' data to determine a Sierra Nevada-wide reproductive rate and investigate sources of variation in this rate. Nichols et al. (2007a) analyzed the data from one of the groups, but the approach could be readily used in a meta-analysis due to the ability to accommodate unequal sampling effort with missing values.

Survey data collected during April to mid-August 2004 from 54 potential territories in the Eldorado study area were analyzed using a multi-state occupancy modeling framework, where the true states of interest were defined to be:

0 = owls absent from potential territory.
1 = owls present without young at territory.
2 = owls present with young at territory.

The outcome of each survey of a territory was therefore recorded as:

0 = owls not detected.
1 = owls detected without observing young.
2 = owls present with young observed.

Each territory was surveyed between 1 and 5 times, with one survey per month. Young owlets are unlikely to be detected early in the breeding season (i.e., April and May) as eggs are typically still being incubated or would have only recently hatched. Detecting evidence of reproduction is therefore much more likely in the latter three surveys. Territories that were suspected to be occupied were selected for surveying, hence Nichols et al. (2007a) expected overall occupancy, ψ, to be close to 1; the probability of an occupied territory having reproduction, R, was the prime parameter of interest in this case. Nichols et al. (2007a) did not consider the effect of any covariates on occupancy- or detection-related probabilities, but fit a range of models primarily focused on temporal and true state effects on detection. For the purpose of this example we shall consider the results from just one of their highly-ranked models to illustrate the methods described in this chapter.

From the raw survey data, spotted owls were detected at 47 of the 54 territories, with owlets detected at 19. Naïve estimates, that do not account for detection or misclassification probabilities, would be $\tilde{\psi} = 47/54 \approx 0.87$ and $\tilde{R} = 19/47 \approx 0.40$. A naïve estimate of the overall probability of a territory being occupied with reproduction would be $\tilde{\psi}\tilde{R} = \tilde{\varphi}^{[2]} = 19/54 \approx 0.35$. The model used here for inference (using the conditional binomial parameterization) specified that the probability of detecting spotted owls ($p^{[1]}$ and $p^{[2]}$) depended upon the true state of the territory (i.e., whether reproduction occurred there or not), but was constant across surveys. The probability of detecting owlets in a survey, given owlets present and adults were detected (δ), differed between the first two surveys (early breeding season) and the last three surveys. We denote this model as $\psi(\cdot) R(\cdot) p(State) \delta(Stage)$, where *State* indicates an effect of true occupancy state on detection and *Stage* indicates an effect for stage of breeding season (early vs. late) on the conditional probability of detecting young.

Parameter estimates from this model are given in Table 5.1. As expected, after accounting for imperfect detection and state misclassification, $\hat{\psi}$ and $\hat{R}$ (0.98 and 0.44, respectively) are substantially greater than their naïve counterparts (approximately 13% and 10% greater, respectively), indicating that the overall level of occupancy, and proportion of occupied territories with reproduction, are greater than the raw data would suggest. The estimated overall probability of a territory being occupied with reproduction is 0.43 ($= \hat{\psi}\hat{R}$), which is approximately 20% greater than the naïve estimate. The probability of detecting spotted owls in a survey given no reproduction at a territory ($p^{[1]}$) and detection probability given there is reproduction at a territory ($p^{[2]}$) are very similar, close to 0.75. As expected, the probability of detecting evidence of young in a survey (given owls are detected) was estimated to be very low in the early breeding season but 0.87 later in the season. This indicates that once adult owls have been detected in a survey of a territory that has young present, the young will not always be detected, which may be due, for example, to adults consuming or caching mice for themselves, not returning to the nest site, or investigators being unable to track the adult through the forest. Hence while the detection

TABLE 5.1 Parameter estimates (and associated estimated standard errors; SE) from the multi-state occupancy model $\psi(\cdot)\,R(\cdot)\,p(State)\,\delta(Stage)$ fit to the California spotted owl reproductive success data from 54 potential territories surveyed during the 2004 breeding season at the Eldorado study area, central Sierra Nevada, California, USA. *NA* indicates a standard error was not calculated, as an estimate was on the boundary of allowable values

Parameter	Estimate	SE
ψ	0.98	0.038
R	0.44	0.082
$p^{[1]}$	0.73	0.053
$p^{[2]}$	0.76	0.058
δ_{1-2}	0.00	*NA*
δ_{3-5}	0.87	0.073

and classification probabilities are high, they are not very close to 1.0, and for those territories that are only surveyed once or twice during the season, the probability of never detecting California spotted owls is relatively high, as is the probability of not finding young even when present. Note that for investigators to detect young during a late season survey of a territory where young are present, adults and young must both be detected. The estimated probability would be $\hat{p}^{[2]}\hat{\delta} = 0.76 \times 0.87 = 0.66$. Therefore, young will *not* be detected in one of every three late season surveys on average.

5.7.2   Breeding Success of Grizzly Bears

Fisher et al. (2014) used the multi-state occupancy model to investigate spatial patterns in the breeding success of grizzly bears (*Ursus arctos*) within the Willmore Wilderness Park in the Rocky Mountains of west-central Alberta, Canada. Infrared remote cameras (i.e., camera traps) were placed at 60 locations throughout the study area to survey for grizzly bears, with detections being of single bears (possibly male or female as gender can not be reliably determined from photographs), or bears with cubs (Fig. 5.4). Therefore, detection of cubs was considered evidence of breeding success in the general vicinity of the camera trap location. Cameras were deployed for six weeks during June–August, and each week was considered a 'survey' for the purpose of the analysis. A scent lure was used to attract animals that were near the camera trap into a position where they could be photographed. There were insufficient cameras to survey all 60 locations in one summer; hence 30 locations were randomly selected to be surveyed in 2009 and another 30 in 2010. Locations were selected using a

FIGURE 5.4   Camera trap photograph of female grizzly bear (*Ursus arctos*) with cubs in Willmore Wilderness Park, Alberta, Canada. (*Jason T. Fisher, InnoTech Alberta, and Matthew Wheatley, Alberta Environment and Parks*)

mix of systematic and stratified random sampling (see Fisher et al., 2014, for further details). Fisher et al. (2014) conducted two separate analyses of the data; a dichotomous occupancy analysis that ignored breeding status (i.e., using just detection/nondetection data of grizzly bears) and a multi-state occupancy analysis which we focus on here.

The dominant land cover type at each camera trap location was classified into one of three categories: herbaceous areas, wetlands, and other (primarily conifer forest). This was considered as a potential covariate for the probability grizzly bears were present at a camera trap site (occupancy; ψ) and for the probability of breeding success given presence of grizzly bears (R). Models were fit to the data with different combinations of including land cover type as a covariate for each of these parameters. Fisher et al. (2014) did not make any specific *a priori* predictions about direction of any effects of land cover type, noting "Breeding success is expected to vary across heterogeneous landscapes, but how it varies remains unknown." We would, therefore, regard their analysis as exploratory.

The probability of detecting grizzly bears was estimated separately for each occupancy state (bears present without cubs vs. bears present with cubs), with the possible influence of two effects considered for each. A possible trend in detection probability during the six weeks was hypothesized (particularly a negative trend), reflecting the decreasing effectiveness of the scent lure over time, and the manner in which the camera trap locations were selected (stratified

TABLE 5.2 AIC comparison of the multi-state occupancy models fit to the grizzly bear camera trap data collected in Willmore Wilderness Park, Alberta, Canada, during June–August in 2009 and 2010. Given is the relative difference in AIC (ΔAIC), AIC model weight (w), number of parameters ($Npar$), and twice the negative log-likelihood value ($-2l$)

Model	ΔAIC	w	$Npar$	$-2l$
$\psi(LC)\,R(\cdot)\,p(State \times Trend)\,\delta(\cdot)$	0.00	0.44	9	236.20
$\psi(\cdot)\,R(LC)\,p(State \times Trend)\,\delta(\cdot)$	0.66	0.32	9	236.86
$\psi(\cdot)\,R(\cdot)\,p(State \times Trend)\,\delta(\cdot)$	2.35	0.14	7	242.54
$\psi(LC)\,R(LC)\,p(State \times Trend)\,\delta(\cdot)$	2.96	0.10	11	235.15

vs. systematic sampling). No interaction between these effects was considered. Fisher et al. (2014) did not investigate the potential effects of any predictor variables on the probability of detecting evidence of breeding success, given detection of grizzly bears in a survey (δ).

Data were not collected at six locations due to camera malfunctions and logistical reasons. Of the remaining 54 locations, grizzly bears were photographed at 21 camera sites (39%), with cubs also photographed at six of those sites (29% of the 21 sites, or 11% overall). Fisher et al. (2014) presented results from a range of models fit to the data, although for the purpose of this example we examine the results from a subset of four models; those that always included only a trend in detection (models ranked 2–5 in Table 2 of Fisher et al., 2014). We thank the authors of Fisher et al. (2014) for supplying the data enabling us to perform a reanalysis.

Table 5.2 presents a comparison of the four models of interest. On the basis of AIC, the top-ranked model suggests the probability of grizzly bear presence differs among land cover types and probability of successful breeding within the vicinity of a camera trap (given presence of grizzly bears) is equal across land cover types. The second-ranked model suggests the opposite, that the probability of presence is equal with the probability of successful breeding varying across land cover types. There is actually a very similar level of support for these two models (ΔAIC is relatively small and AIC model weights are similar) that have very different biological interpretations. However, both models are ranked above the model with constant probabilities of presence and successful reproduction (third-ranked), and the model with land cover included for both probabilities (fourth-ranked), although all models have non-negligible AIC weights (i.e., no models have w near 0.00). This suggests that there is some variation in the data that correlates with land cover type, but the exact nature of the variation is difficult to determine. In this case, small sample sizes are likely the primary cause of the ambiguity, as indicated by the relatively large standard

TABLE 5.3 Estimated probability of grizzly bears being present at a camera trap site during the six-week survey period (ψ) for each land cover type, from each model fit to the data and the model-averaged value. Standard errors are given in parentheses

Model	Herbaceous	Wetlands	Other
$\psi(LC)\,R(\cdot)\,p(State \times Trend)\,\delta(\cdot)$	0.78 (0.20)	0.43 (0.19)	0.30 (0.13)
$\psi(\cdot)\,R(LC)\,p(State \times Trend)\,\delta(\cdot)$	0.88 (0.45)	0.88 (0.45)	0.88 (0.45)
$\psi(\cdot)\,R(\cdot)\,p(State \times Trend)\,\delta(\cdot)$	0.51 (0.12)	0.51 (0.12)	0.51 (0.12)
$\psi(LC)\,R(LC)\,p(State \times Trend)\,\delta(\cdot)$	0.84 (0.26)	0.57 (0.35)	0.40 (0.23)
Model-averaged	**0.78 (0.32)**	**0.60 (0.37)**	**0.52 (0.38)**

errors of the parameter estimates (Tables 5.3 and 5.4). We believe that *a priori* predictions could have been somewhat useful in reducing the ambiguity in the model selection results. In addition to their use in testing scientific hypotheses, predictions sometimes help the investigator decide which competing model seems more reasonable or plausible.

Of the 54 camera trap locations, 20 were placed in herbaceous areas, 12 in wetlands, and 22 in other land cover types. Essentially, when land cover type is included as a covariate on both ψ and R, presence is estimated for each cover type individually, and successful breeding from the data for the occupied locations within that cover type. For example in this case, ψ for herbaceous areas is primarily estimated from data collected at just 20 relevant locations (we say "primarily" as some information from the other locations is used to estimate common detection probabilities), giving a value of 0.84 (fourth row of Table 5.3). The expected number of locations with grizzly bears would be 16.8 (20×0.84), which is the sample size for estimating R in herbaceous areas. Similar calculations could be made for the other land cover types. Clearly, these sample sizes are (statistically) small, especially when uncertainty due to imperfect detection is taken into account. In combination, these issues are responsible for the large standard errors reflecting the uncertainty in the parameter estimates, and therefore the ambiguity in the model selection process.

Note that similar ambiguity in our inferences would result if models were compared using a series of likelihood ratio tests (Table 5.5; Chapter 3). Individually, there would appear to be good evidence that including land cover as a covariate on ψ and R significantly improves the model, however there is substantially less evidence that including land cover type on both probabilities significantly improves the model (LRT_1). When land cover is added as a covariate to each probability, after it has already been included as a covariate for the other probability (LRT_2), there is no significant improvement in the amount of variation in the data explained by the model.

TABLE 5.4 Estimated probability of grizzly bear cubs present at a camera trap site, given presence of grizzly bears, (*R*) for each land cover type, from each model fit to the data and the model-averaged value. Standard errors are given in parentheses

Model	Herbaceous	Wetlands	Other
$\psi(LC)\,R(\cdot)\,p(State \times Trend)\,\delta(\cdot)$	0.33 (0.16)	0.33 (0.16)	0.33 (0.16)
$\psi(\cdot)\,R(LC)\,p(State \times Trend)\,\delta(\cdot)$	0.58 (0.38)	0.13 (0.15)	0.11 (0.13)
$\psi(\cdot)\,R(\cdot)\,p(State \times Trend)\,\delta(\cdot)$	0.33 (0.16)	0.33 (0.16)	0.33 (0.16)
$\psi(LC)\,R(LC)\,p(State \times Trend)\,\delta(\cdot)$	0.48 (0.24)	0.20 (0.22)	0.18 (0.20)
Model-averaged	**0.42 (0.29)**	**0.25 (0.19)**	**0.24 (0.19)**

TABLE 5.5 Likelihood-ratio test comparisons of the multi-state occupancy models fit to the grizzly bear camera trap data collected in Willmore Wilderness Park, Alberta, Canada, during June–August in 2009 and 2010. Given are the likelihood ratio test statistic for each model compared to the *null* hypothesis represented by the model $\psi(\cdot)\,R(\cdot)\,p(State \times Trend)\,\delta(\cdot)$ (LRT_1) and the test statistic for each model compared to the *alternative* hypothesis represented by the model $\psi(LC)\,R(LC)\,p(State \times Trend)\,\delta(\cdot)$ (LRT_2) and associated *P*-values. Degrees of freedom for χ^2 distribution are the difference in the number of parameters between the respective models

Model	LRT_1	P-value$_1$	LRT_2	P-value$_2$
$\psi(LC)\,R(\cdot)\,p(State \times Trend)\,\delta(\cdot)$	5.68	0.06	1.05	0.59
$\psi(\cdot)\,R(LC)\,p(State \times Trend)\,\delta(\cdot)$	6.34	0.04	1.71	0.43
$\psi(\cdot)\,R(\cdot)\,p(State \times Trend)\,\delta(\cdot)$				
$\psi(LC)\,R(LC)\,p(State \times Trend)\,\delta(\cdot)$	7.39	0.12		

As in Fisher et al. (2014), we have used model averaging to obtain parameter estimates that account for this ambiguity (although for a different set of models; Tables 5.3 and 5.4). These estimates would suggest that both presence of grizzly bears and breeding given presence is highest in herbaceous areas and at similar levels for the other land cover types, but the estimates are extremely imprecise due to the uncertainty in the estimates from each model, and estimated variation among models. The model-averaged detection probabilities given in Fig. 5.5 demonstrate a declining trend in detection over time, presumably due to the decreasing attractiveness of the lure over time or habituation of the bears to the lure (Fisher et al., 2014). Grizzly bears appear to be much more detectable at locations where cubs are also present, and the per-week detection probability is estimated to be very low where grizzly bears are present without cubs. The probability of detecting cubs in a survey week, given detection of grizzly bears $(\hat{\delta})$, is estimated to be 0.43 (SE = 0.19, 95% CI = 0.14–0.78).

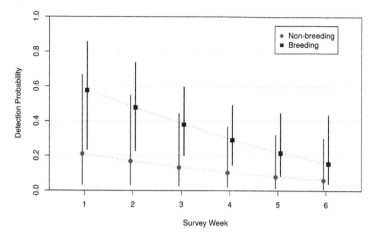

FIGURE 5.5 Model-averaged probability of detecting grizzly bears for camera trap locations with and without successful breeding (presence of cubs), estimated from data collected during a six-week survey in Willmore Wilderness Park, Alberta, Canada. All models included a trend in detection over time.

The probability of detecting grizzly bears is estimated to be different depending upon breeding success highlights an important aspect of multi-state occupancy models. If the camera trapping data were reduced to the detection/nondetection of any grizzly bear and the simpler models of the preceding chapter used, breeding success would be an unmodeled source of detection heterogeneity that would lead to occupancy (i.e., grizzly bear distribution) being underestimated. By accounting for this spatial heterogeneity in detection, the degree of bias in estimated occupancy should be reduced. Furthermore, in this case breeding success will also correlate with local abundance (sows and cubs are present at locations with successful breeding), hence the higher detection probability may not be due to any behavioral difference but to the fact that there are more bears to be detected around locations where cubs are present.

Clearly there is a high degree of uncertainty with these results, but to demonstrate some of the potential uses of these models Fig. 5.6 is a map of the estimated probability of grizzly bears being present and successfully breeding (i.e., ψR) at a camera trap location, similar to the map present by Fisher et al. (2014). This is a graphical representation of which areas are expected to contain grizzly bears with cubs according to the analysis, based upon the land cover information available for Willmore Wilderness Park. Ideally one should also present a similar plot (or plots) that represent the uncertainty in the estimates across the landscape, as we did for the willow tit example in the previous chapter. We stress, however, that unlike the willow tit example, the effective area being sampled at a camera trap location has not been clearly defined, therefore

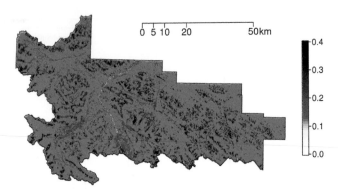

FIGURE 5.6 Map of the estimated probability of grizzly bears being present with cubs (i.e., ψR) at a camera trap location based upon model-averaged estimates of ψ and R for each land cover type across Willmore Wilderness Park, Alberta, Canada.

the results presented (including the map in Fig. 5.6) should not be interpreted as an areal measure (e.g., km^2 occupied by grizzly bears), but as the probability of occupancy (and/or reproduction) at a camera trap location if one was placed there.

To improve the reliability of the results in a future, similar study we would recommend that the number of trap locations should be increased and that methods should be explored to increase detection probability of grizzly bears in locations without successful reproduction, possibly by using an alternative lure or incorporating other information (e.g., conduct sign surveys in the vicinity of camera traps upon deployment and retrieval). Increasing the number of survey weeks would be another alternative to increase overall detection probabilities (i.e., detected at least once during survey season), however doing so will lengthen the duration of the season which would also likely influence the occupancy probabilities for a low-density, wide-ranging carnivore such as a grizzly bear. Therefore doing so may provide little overall benefit. At this stage, we would recommend simulation studies be conducted to evaluate different options for future designs.

5.8 DISCUSSION

The multi-state occupancy modeling framework is a very useful extension of the original occupancy model for a binary response (e.g., species presence/absence). It enables a greater degree of resolution to investigate interesting ecological questions at the landscape level, while allowing for the possibility that the required evidence of the activities of interest may not always be detected, or correctly classified, by the employed field methods. We believe there is a wide array of applications that would be amenable to this general modeling framework, and its use will become more widespread in the future. The key is to cast the ecological question into one of multiple states, then parameterize the model accordingly to enable estimation of biologically relevant parameters. For example, one approach to modeling patterns of species co-occurrence would be to define the status of a sampling unit in terms of which combination of species is present there (e.g., Chapter 14).

The single-season model described in this chapter should only be used to investigate patterns in the occupancy status of units at the time of surveying. In Chapter 9 we focus on a multi-season multi-state occupancy model that was developed to estimate parameters associated with the underlying dynamic processes of change in occupancy.

Little work has been done on study design issues for the multi-state model, although many of the fundamental principles discussed in Chapter 11 are likely to be relevant. A key study design consideration is how many states should be used given the available resources, as the required sample size to produce reliable parameter estimates will increase as the number of states increases. At present we do not have any firm, general advice to offer on such matters although simulation studies could provide some insights. We strongly recommend simulation studies be conducted before going into the field so researchers have realistic expectations about the outcome of their study.

Chapter 6

Extensions to Basic Approaches

In Chapters 4 and 5 we have described model-based methods for investigating patterns in species occurrence (in both the simpler dichotomous and multi-state situations), while accounting for imperfect detection of the true occupancy state at a sampled sampling unit from the field surveys. These methods have a number of assumptions associated with them that are required for valid inferences to be made, and other limitations on the types of inferences that can be drawn from them in some cases. In this chapter we focus on some extensions of these basic methods that have been developed to specifically address such assumption violations or other limitations.

In particular, we present extensions on estimating the proportion of units occupied when the number of units in the population of interest is finite (especially, small and finite), false positive detections or species misidentification, estimating occupancy at multiple scales, autocorrelated detections within a sampling unit, gradual arrival or departure of the species at a unit within a season (staggered entry/exit), and accounting for spatial autocorrelation in species occurrence. Many of these extensions have been developed to accommodate specific types of violations of the closure and independence assumptions of the previously described approaches (multi-scale occupancy, autocorrelated detections, and staggered entry). One specific set of models not discussed in this chapter is those that allow for heterogeneous detection probabilities (beyond the use of covariates); that is the topic of the next chapter.

6.1 ESTIMATING OCCUPANCY FOR A FINITE POPULATION OR SMALL AREA

As mentioned in Section 4.1, the occupancy models developed above (and elsewhere in this book) are based on the view that the sampled locations constitute a random sample from a large (theoretically infinite) population, and the occupancy parameter ψ is the probability that a unit in that infinite population is occupied. While this conceptual formulation may be appropriate, or at least adequate, for many problems, there are inference problems for which this view is insufficient. For example, suppose occupancy metrics are adopted for summarizing metapopulation status of anurans on a small number of wetland basins

Occupancy Estimation and Modeling. http://dx.doi.org/10.1016/B978-0-12-407197-1.00008-9

in a refuge or park, and interest is in what proportion of wetlands was occupied by the species at the time of surveying. Conceivably one could sample most or even all of the available wetlands, and inferences could be desired for just the particular wetlands that were surveyed or the larger set of known available wetlands (including some that were not surveyed). In such a situation, the distinction between probability of occupancy and the proportion of units occupied (Chapter 2) becomes important. While the modeling approach described in Chapter 4 can still be used to analyze the data collected from a finite population for inferences about the probability of occupancy, there are additional considerations when interest is in the proportion of occupied units in finite population. For example, we know that x (the number of occupied units in the sample) can only take on integer values between s_D and s. Consequently, the proportion of units occupied in a sample can only have a discrete set of possible values which has not been considered previously (e.g., if $s = 20$, the proportion of units occupied can only take values of 0, 0.05, 0.1, 0.15, ...). In the remainder of this section, we discuss methods for estimating the number of occupied units in the population, but note that an estimate of the proportion of units occupied can be simply derived by dividing the estimated number of units occupied by the size of the population of interest (s or S). To avoid confusion, we denote the number of occupied units in the sample as x, and the number of occupied units in the larger, but finite, population as x_{pop}. To distinguish between the probability of occupancy and proportion of units occupied, we use ψ and Ψ, respectively.

6.1.1   Prediction of Unobserved Occupancy State

The essence of estimating the number of occupied units in a population is predicting the unknown occupancy state of specific units. The exact occupancy state of a unit may be unknown either because of imperfect detection, or because the unit was not included in the sample where surveys of the species were conducted. In Chapter 4 we introduced the latent variable z_i for the occupancy state of unit i, which is a Bernoulli (i.e., binary) random variable, with probability ψ_i of 'success' (i.e., $z_i = 1$, species presence). The methods presented so far have focused on estimation of the underlying probability ψ_i, while accounting for imperfect detection. However, when inference is directed at the number or proportion of occupied units in the population, attention must be directed at the values of the actual random variables. As a precursor to the following development, we would remind readers that for a Bernoulli random variable y, its expected value is:

$$E(y) = \theta,$$

and has variance:

$$Var(y) = \theta(1 - \theta),$$

where θ is the probability that $y = 1$.

If the latent occupancy variable z_i were observed for every unit then the number of occupied units in the population would be the quantity:

$$x_{pop} = \sum_{i=1}^{S} z_i$$

$$= \sum_{i=1}^{s} z_i + \sum_{i=s+1}^{S} z_i$$

$$= x + \sum_{i=s+1}^{S} z_i.$$

That is, the number of occupied units in the population is the number of surveyed units that were occupied plus the number of unsurveyed occupied units. To distinguish that some units have been surveyed, denote the occupancy random variable for the sampled units as $z_{i|h}$ where h is the detection history for the corresponding unit and $h = 0$ indicates the species was never detected at the unit while $h \neq 0$ indicates the species was detected at least once during the surveys of the unit.

By definition, $z_{i|h \neq 0} = 1$ (the species must be present if detected at the unit, assuming no species misidentification) for the s_D units where the species was detected, however due to imperfect detection, the occupancy state is not known for the remainder of the sampled units. The logical estimator of x is therefore:

$$\hat{x} = s_D + \sum_{i=s_D+1}^{s} \hat{z}_{i|h=0},$$

i.e., the number of occupied units where the species was detected plus the estimated occupancy status of each unit where no detections occurred. Hence, an estimator for the number of occupied units in the population would be:

$$\hat{x}_{pop} = s_D + \sum_{i=s_D+1}^{s} \hat{z}_{i|h=0} + \sum_{i=s+1}^{S} \hat{z}_i. \tag{6.1}$$

It therefore follows that an estimator for proportion of sampled units occupied will be:

$$\hat{\Psi} = \frac{\hat{x}}{s},$$

and an estimator for the whole population:

$$\hat{\Psi}_{pop} = \frac{\hat{x}_{pop}}{S}.$$

A Non-Bayesian Approach

One approach to the estimation of x_{pop} is to replace the z_i values with their expected values (Chapter 3). For those units that were surveyed, but the species was never detected, i.e., the second term in Eq. (6.1), it can be shown that the expected value of $z_{i|h=0}$ is $\psi_{condl,i}$; the probability unit i is occupied conditional upon the species not being detected at the unit. Recall from Chapter 4 that $\psi_{condl,i}$ is related to p and ψ by Bayes' Theorem:

$$\psi_{condl,i} = \frac{\psi_i \prod_{j=1}^{K}(1 - p_{ij})}{(1 - \psi_i) + \psi_i \prod_{j=1}^{K}(1 - p_{ij})}.$$

For the remaining units in the population of interest, which were not surveyed, i.e., the third component in Eq. (6.1), the expected value of z_i is ψ_i. Therefore, after substituting in the estimated quantities, the estimator for x_{pop} is:

$$\hat{x}_{pop} = s_D + \sum_{i=s_D+1}^{s} \hat{\psi}_{condl,i} + \sum_{i=s+1}^{S} \hat{\psi}_i. \tag{6.2}$$

When no covariates are included in the model and all sample units are surveyed on K occasions, such that $\hat{\psi}_i$ and $\hat{\psi}_{condl,i}$ are the same for all respective units, Eq. (6.2) becomes:

$$\hat{x}_{pop} = s_D + (s - s_D)\,\hat{\psi}_{condl} + (S - s)\,\hat{\psi}.$$

If the occupancy and detection probabilities were known, so no 'hats' on them in Eq. (6.3), then the variance of $\hat{x}_{pop}$ would be:

$$Var(\hat{x}_{pop}) = \sum_{i=s_D+1}^{s} \psi_{condl,i}(1 - \psi_{condl,i}) + \sum_{i=s+1}^{S} \psi_i(1 - \psi_i), \tag{6.3}$$

assuming the occupancy state at each unit is independent of the state at other units. While we do not advocate the use of Eq. (6.3) in practice because it does not account for uncertainty in the parameter estimates, it is informative to highlight two issues. First, note that Eq. (6.3) does not include a term associated with s_D, the number of units where the species was detected, because this is an observed quantity, the value of which is known without error. Secondly, as detection probability and/or K increases, $\psi_{condl,i}$ will approach 0 and therefore the associated term will also disappear from Eq. (6.3). This behavior should be expected as when the overall probability of detection is close to 1 (i.e., near perfect detection), then if the species has not been detected at a unit, it must be almost certainly absent (i.e., $z_{i|h=0} = 0$).

To account for the fact that occupancy and detection probabilities are not known, and have been estimated, an asymptotic approximation for the variance of $\hat{x}_{pop}$ could be obtained from:

$$Var(\hat{x}_{pop}) = Var\left(E\left(\hat{x}_{pop}|\theta\right)\right) + E\left(Var(\hat{x}_{pop}|\theta)\right).$$

Typically this would be approximated numerically, especially when the model contains covariates, but in the case where no covariates are included in the model, this evaluates to:

$$
\begin{aligned}
Var(\hat{x}_{pop}) = {}&(s - s_D)\left[\hat{\psi}_{condl}\left(1 - \hat{\psi}_{condl}\right) + (s - s_D - 1)\,Var\left(\hat{\psi}_{condl}\right)\right] \\
&+ (S - s)\left[\hat{\psi}\left(1 - \hat{\psi}\right) + (S - s - 1)\,Var\left(\hat{\psi}\right)\right] \\
&+ (s - s_D)(S - s)\,Cov\left(\hat{\psi}_{condl}, \hat{\psi}\right).
\end{aligned}
\tag{6.4}
$$

Again, this would typically be evaluated using appropriate software, but it is instructive to note that $Var(\hat{x}_{pop})$ consists of three components that are related to: (1) uncertainty in the number of *sampled* units that were occupied; (2) uncertainty in the number of *unsampled* units that were occupied; and (3) correlation between $\hat{\psi}_{condl}$ and $\hat{\psi}$ as the quantities are estimated from the same data and $\hat{\psi}_{condl}$ is derived from $\hat{\psi}$.

A further point about $Var(\hat{x}_{pop})$ is that as the number of units in the population of interest (i.e., S) becomes larger relative to s (the number of sampled units), then:

$$Var(\hat{x}_{pop}) \approx S^2 Var\left(\hat{\psi}\right).$$

This indicates that $Var(\hat{x}_{pop})$ will also become very large, which is not particularly useful, but if interest was in the proportion of occupied units in the population ($\hat{\Psi}_{pop}$) of interest rather than the number of occupied units ($\hat{x}_{pop}$), then when S is very large:

$$
\begin{aligned}
Var\left(\hat{\Psi}_{pop}\right) &= Var\left(\frac{\hat{x}_{pop}}{S^2}\right) \\
&= \frac{Var(\hat{x}_{pop})}{S^2}
\end{aligned}
$$

$$\approx \frac{S^2 Var\left(\hat{\psi}\right)}{S^2}$$

$$\approx Var\left(\hat{\psi}\right).$$

Therefore, when there is a very large number of units in the population of interest, the uncertainty in the estimated proportion of units occupied is (approximately) the same as the uncertainty in the estimated probability of occupancy. Which brings us back to a point made at the beginning of Chapter 2 that the distinction between *probability* of occupancy and *proportion* of units occupied is of less importance when the population of interest is very large.

A Bayesian Approach

In Chapter 4 we discussed estimation of occupancy and detection probabilities using the complete data likelihood and Bayesian methods of inference using MCMC. As part of the estimation procedure, the values of the latent occupancy state variables are predicted, or imputed, at each iteration of the MCMC-based analysis. Therefore at the completion of the MCMC run, not only are approximations of the posterior distributions for the model parameters obtained, but also posterior distributions for the occupancy state latent variable (z_i) of each unit. Hence direct inference about the z_i's, or summaries thereof (e.g., $\hat{x}$), is possible rather than calculating expected values and using asymptotic approximations for determining estimator variances or standard errors. The resulting measures of uncertainty from a Bayesian analysis are not asymptotic and are valid for any sample size, provided that the number of MCMC iterations is sufficient to provide a good approximation of the posterior distributions. However users should be mindful that inference from a Bayesian analysis may be sensitive to the choice of prior distributions used for the model parameters, particularly with small sample sizes.

Recall that in Fig. 4.2 we presented OpenBUGS/JAGS code for analyzing a data set using a single-season occupancy model with no covariates, and constant detection. In Fig. 6.1 we present code for fitting the same model (lines 2–10), but this time with minor modifications to estimate the number and proportion of sampled units occupied (lines 11–12; i.e., $\hat{x}$ and $\hat{\Psi}$), predict occupancy in the unsampled units within the population of interest (lines 13–15), then estimate the number and proportion of occupied units in the population (lines 16–17; i.e., $\hat{x}_{pop}$ and $\hat{\Psi}_{pop}$). Clearly, using the complete data likelihood and Bayesian methods, inference about finite populations and small areas can be obtained with very little additional effort. The code can also be generalized to incorporate covariates with a key consideration being that the covariate values must be known for both the sampled and unsampled units, just as in the non-Bayesian approach.

```
 1:  model {
 2:    p~dunif(0,1)
 3:    psi~dunif(0,1)
 4:    for(i in 1:s) {
 5:      z[i]~dbern(psi)
 6:      mu[i] <- p*z[i]
 7:      for(j in 1:k) {
 8:        h[i,j]~dbern(mu[i])
 9:      }
10:    }
11:    x <- sum(z[1:s])
12:    prop <- x/s
13:    for(i in (s+1):S) {
14:      z[i]~dbern(psi)
15:    }
16:    x_pop <- sum(z[1:S])
17:    prop_pop <- x_pop/S
18:  }
```

FIGURE 6.1 OpenBUGS/JAGS code for fitting an occupancy model with constant occupancy and detection probability (i.e., $\psi(\cdot)\,p(\cdot)$), estimation of the number and proportion of occupied units in the sample (x and prop), prediction of occupancy in unsampled units, and estimation of number and proportion of occupied units in the population of interest (x_pop and prop_pop). The quantities denoted as x, prop, x_pop, and prop_pop in the code are quantities defined in the text as $\hat{x}$, $\hat{\Psi}$, $\hat{x}_{pop}$, and $\hat{\Psi}_{pop}$, respectively. Line numbers on the left-hand side are not part of the code and are for ease of interpretation.

6.1.2 Example: Blue Ridge Two-Lined Salamanders Revisited

We now return to the example on Blue Ridge two-lined salamanders considered in Section 4.4.6. Suppose it is of interest to estimate the *proportion* of the 39 surveyed transects that was occupied by the salamanders, rather than the probability of occupancy. In this case $S = s = 39$, i.e., the 39 surveyed transects are the population of interest, and Blue Ridge two-lined salamanders were detected at 18 transects, hence $s_D = 18$. For the purpose of the example we shall use the simple $\psi(\cdot)\,p(\cdot)$ model for inferences, but more complex models could be used in practice.

Using maximum likelihood methods, as inference is only desired for the 39 surveyed transects, the proportion of occupied transects can be calculated as:

$$\hat{\Psi} = \frac{s_D + (s - s_D)\,\hat{\Psi}_{condl}}{s}$$
$$= \frac{18 + 21 \times 0.247}{39}$$
$$= 0.59,$$

where the value for $\hat{\psi}_{condl}$ was given in Section 4.4.6 (although here we have reported an additional decimal place). From Eq. (6.4), calculation of the variance for $\hat{\psi}$ simplifies to:

$$Var\left(\hat{\psi}\right) = \frac{(s - s_D)\left[\hat{\psi}_{condl}\left(1 - \hat{\psi}_{condl}\right) + (s - s_D - 1)\,Var\left(\hat{\psi}_{condl}\right)\right]}{s^2}$$
$$= \frac{21(\times 0.247 \times 0.753 + 20 \times 0.022)}{39^2}$$
$$= 0.009,$$

or $SE\left(\hat{\psi}\right) = \sqrt{0.009} = 0.09$. Compare this to the estimate and standard error for the *probability* of occupancy using this model; $\hat{\psi} = 0.59$, $SE\left(\hat{\psi}\right) = 0.12$ (Section 4.4.6). While the estimated proportion and probability are the same (to two decimal places), the standard error for the proportion is approximately 25% smaller. This highlights that a primary consequence of reporting the estimated probability of occupancy as the proportion of units occupied, is the standard error for the estimated probability will be larger than the standard error for the estimated proportion, in the case of a finite population.

Similar results are obtained using Bayesian inferential methods. Code based on that given in Fig. 6.1 was used to estimate the proportion of the 39 transects occupied (the quantity `prop` in the code), and also the proportion of occupied units in the population (`prop_pop`) assuming the total number of transects in the population of interest was 50, 75, 100, 200, 500, or 1000. The analysis was conducted with the software JAGS, using three MCMC chains with 11,000 iterations, of which the first 1000 iterations were discarded as the 'burn-in period'. Therefore, the posterior distribution of each quantity is approximated from a sample of 30,000 values. Uniform priori distributions were assumed for both ψ and p.

The posterior distribution for the estimated proportion is given in Fig. 6.2. Note that it is discrete valued, asymmetric, and that the minimum value is $0.46 = 18/39$; the proportion of transects where Blue Ridge two-lined salamanders were detected. Numerical summaries of the posterior distributions for the estimated proportions are given in Table 6.1, along with a summary of the estimated probability of occupancy for comparison. These results would suggest the mean of the posterior distribution for $\hat{\psi} = 0.61$, with a median (50th percentile) of 0.59, both of which are similar to the maximum likelihood-based estimate of ψ. The posterior standard deviation for $\hat{\psi} = 0.10$ which is slightly larger, but comparable to the standard error reported above. Note that we should not expect exact agreement between the maximum likelihood-based results and those obtained using Bayesian methods. The 95% credible interval for $\hat{\psi}$ is 0.46–0.87,

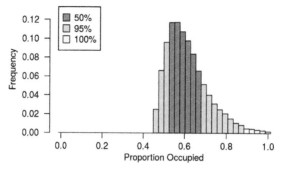

FIGURE 6.2 Posterior distribution of the proportion of the 39 surveyed transects occupied by the Blue Ridge two-lined salamander. Colors represent various central percentages of the distribution.

TABLE 6.1 Posterior distribution summaries for the estimated proportion of the 39 surveyed transects occupied by Blue Ridge two-lined salamanders ($\hat{\Psi}$), estimated proportion of transects occupied presuming total number of transects of interest (S) is 50, 75, 100, 200, 500, or 1000 ($\hat{\Psi}_S$), and the estimated probability of salamanders being present on a transect ($\hat{\psi}$). The $\psi(\cdot)\,p(\cdot)$ model was fit to the detection data

Quantity	Mean	SD	Percentile		
			2.5	50.0	97.5
$\hat{\Psi}$	0.614	0.101	0.462	0.590	0.872
$\hat{\Psi}_{50}$	0.613	0.106	0.460	0.600	0.860
$\hat{\Psi}_{75}$	0.612	0.111	0.427	0.600	0.867
$\hat{\Psi}_{100}$	0.610	0.114	0.420	0.600	0.870
$\hat{\Psi}_{200}$	0.609	0.117	0.405	0.600	0.870
$\hat{\Psi}_{500}$	0.609	0.119	0.398	0.600	0.870
$\hat{\Psi}_{1000}$	0.608	0.120	0.395	0.600	0.871
$\hat{\psi}$	0.608	0.121	0.393	0.599	0.872

the lower limit being the observed proportion of transects where salamanders were detected. This is understandable when the desired inference is about the proportion of surveyed transects that were occupied, as the observed proportion must represent a minimum, and the uncertainty is associated with potential false absences at surveyed transects (i.e., at how many surveyed units was the species present but undetected).

The posterior distribution for $\hat{\psi}$ is notably different, in particular with respect to having a larger standard deviation and the 2.5th percentile being less than the observed proportion of transects with detections. Some people have queried how the lower limit on the probability of occupancy could be less than the observed proportion, and the reason is simple. Sometimes the observed proportion in a sample will be greater than the underlying probability of 'success', particularly with a smaller sample size. For example, the probability of a head with a fair coin will be 0.5, but it would not be unreasonable to observe 6 or 7 heads from 10 coin flips, i.e., observed proportions of 0.6 or 0.7.

The posterior distributions for the proportion of occupied transects in the population converge to that of the posterior distribution of $\hat{\psi}$ as S increases from 50 to 1000. This again highlights that the importance of distinguishing between proportion of occupied units, and probability of a unit being occupied, diminishes as the total number of units in the population increases (given s is fixed).

6.1.3 Consequences of a Finite Population

As stated previously, the main issue associated with a small area or finite population surrounds the distinction between the probability of occupancy and proportion of units occupied, particularly with respect to the uncertainty associated with estimates of each quantity. Table 6.2 illustrates the difference in the expected standard error for the proportion of occupied units compared to probability of occupancy, over a range of values for the total number of units in the population. In all cases, the number of units surveyed for the species was $s = 50$. When there are only 50 units in the population of interest (i.e., the entire population is surveyed) $SE\left(\hat{\Psi}_{pop}\right)$ is almost 40% smaller than $SE\left(\hat{\psi}\right)$. As the number of units in the population increases, the fraction of the population being sampled decreases and the difference in the two standard errors reduces to nearly zero when 5% of the population is sampled. While the magnitude of the difference and rate of convergence of the standard errors will depend on a number of factors, the example given in Table 6.2 does suggest that there can be a substantial difference if sampling $>20\%$ of the population. In such situations, if the proportion of occupied units was truly the quantity of interest (e.g., as a state variable in a monitoring program), there would appear to be clear advantages in taking the necessary steps to properly determine the correct standard error. This issue is conceptually similar to the use of a finite population correction when estimating the variance or standard error of a population mean in regular sampling problems (e.g., Cochran, 1977; Thompson, 2002).

6.1.4 A Related Issue

Once an appropriate set of models is identified (some of which may include covariates), a common question is, "what is the overall level of occupancy?" Presumably, the 'level of occupancy' is the proportion of occupied units in the population of interest. While the techniques outlined in this section could be

TABLE 6.2 Comparison of the expected standard error of the proportion of occupied units in the population, $SE\left(\hat{\Psi}_{pop}\right)$, and standard error of the occupancy probability, $SE\left(\hat{\psi}\right)$, for different numbers of units in the population (S). The fraction of the population surveyed is also given (s/S). In all cases, the sample size (s) equals 50, 4 surveys per unit (K) with $\psi = 0.6$ and $p = 0.4$

S	s/S	$SE\left(\hat{\Psi}_{pop}\right)$	$SE\left(\hat{\psi}\right)$
50	1.00	0.054	0.088
100	0.50	0.073	0.088
150	0.33	0.078	0.088
250	0.20	0.082	0.088
500	0.10	0.085	0.088
1000	0.05	0.087	0.088

used even if there is a very large number of units of interest (e.g., the population of interest is the set of all 1 km^2 grid cells across the USA), the investigator needs to consider a number of very important points.

Most importantly, what is the population of interest? Is it just the surveyed locations, or is there a broader area of interest that includes unsurveyed locations? In the latter case, if certain covariates have been identified as useful predictors of occupancy, then the covariate values for all unsurveyed units will be required to predict occupancy at those units. When an ultimate goal of a project is to estimate overall occupancy, this necessitates two things: (1) the unsurveyed areas must be defined in terms of discrete units to make the required predictions; and (2) the occupancy-related covariates must be available for both the surveyed and unsurveyed units. In practice, this second consideration will often restrict the types of covariates to those that can be obtained without a physical visit to each unit (e.g., remotely-sensed data).

Do the covariates need to be included in the data analysis? Covariates can be used to explore whether certain factors explain variation in the data, but that does not change what the level of occupancy is for the units that were surveyed. For example, if one was to measure the heights of everyone in a room, a regression analysis could be conducted to identify what factors are useful predictors of a person's height (e.g., age, gender, ethnicity, parent's heights, etc.). However if the question is "what is the average height of people in the room?", such an analysis is unnecessary. The same is true here, more complex models involving covariates might be 'better' models, but may result in very similar estimates of the proportion of occupied units in the sample. As such, valid inferences about the overall level of occupancy could be made using simpler models. That is true, at least, for occupancy-related covariates; the inclusion of detection-related covariates can change the estimated level of occupancy in the surveyed units by accounting for detection heterogeneity (Chapters 4 and 7). Models that include covariates are more useful for predicting occupancy at unsurveyed units (i.e., out of sample prediction), particularly if the joint distribution of covariate values for the surveyed units is not a good approximation of the joint-distribution of covariate values in the population. Any discrepancy is more likely when a non-probabilistic sampling scheme is used to select the units to survey. When the joint distribution of covariate values in the sample is a good approximation of that in the general population of interest, then it is reasonable to expect $\hat{\psi}$ from a $\psi(\cdot)$ model to be a good estimate of the average probability of occupancy in the population. Therefore, when a relatively small fraction of the population is surveyed, it would be appropriate to use the $\hat{\psi}$ from a $\psi(\cdot)$ model and its corresponding standard error for inferences about the estimated proportion of occupied units in the population (i.e., in this case it would be suitable to assume that $\hat{\psi} = \hat{\Psi}_{pop}$). This highlights the importance of a good sampling scheme and how it can simplify final inferences. We return to study design questions in Part III.

6.2 ACCOUNTING FOR FALSE POSITIVE DETECTIONS

The classical occupancy models described thus far are based on two kinds of observations, detections and nondetections. Under these models, one of these kinds of observations, detections, are considered to be unambiguous. That is, any detection of the focal species during a survey is assumed to mean that the species is present at the sample unit with probability 1. The other kind of observation, nondetection, is ambiguous in the sense that it admits two possibilities: the species is not present in the sample unit, or it is present but not detected (a so-called 'false negative' or 'false absence'). For many kinds of animal and plant surveys, the assumption of detections providing unambiguous evidence of species presence is completely reasonable. However, there is increasing evidence that some surveys do admit the possibility of misidentification or 'false positives', creating a need for models that can deal with this source of uncertainty.

Misidentification and misclassification are frequently encountered in auditory surveys, during which one species is mistaken for another and recorded incorrectly. Although long suspected, conclusive evidence of such misclassification has been provided by a series of studies conducted using an experimental playback system developed by T.R. Simons. Experiments using this system with recorded vocalizations of both avian (Simons et al., 2007; Alldredge et al., 2008) and anuran (McClintock et al., 2010a, 2010b; Miller et al., 2012c, 2015) species have demonstrated that misidentification is relatively common (also see Bart, 1985; Genet and Sargent, 2003; Lotz and Allen, 2007; Campbell and Francis, 2011; Farmer et al., 2012). Acoustic surveys of bats using automated detection devices are being increasingly advocated, but species identification errors are reported for these detections as well (Preatoni et al., 2005; Redgwell et al., 2009; Armitage and Ober, 2010; Britzke et al., 2011; Walters et al., 2012) and can be common (Clement et al., 2014). In addition to auditory surveys, efforts to draw inferences about distribution patterns of various elusive mammalian species (e.g., carnivores) are frequently based on large-scale field surveys of animal sign (e.g., feces and tracks; Karanth et al., 2011; Mondol et al., 2015) and even reports of detection by members of the general public (e.g., Molinari-Jobin et al., 2012; Miller et al., 2013; Rich et al., 2013; Pillay et al., 2014). Such efforts based on animal sign and observations from the general public are susceptible to problems with misclassification as well. Studies of disease and pathogen prevalence can also be affected by misclassification or false positive results of epidemiological assays (e.g., Elmore et al., 2014).

Various approaches have been used to minimize misclassification errors and false positives. Training in which observers were instructed to record a species only when completely certain of species identity was successful in reducing, but not eliminating, false positive errors in one study of auditory anuran surveys (Miller et al., 2012c). Other *ad hoc* approaches to dealing with misclassification in large-scale sign surveys include use of only the most reliable data sources (e.g., dead animals, tracks verified by an expert; Molinari-Jobin et al., 2012) and requirement of two or three independent species detections to record the species as detected (Rich et al., 2013). Studies that rely on DNA amplification use various methods to classify whether a sample is positive for the target DNA (e.g. two of three wells must be positive), or samples are 're-run' if the results from the first well are positive (e.g., Kriger et al., 2006). However, complete elimination of false positives does not seem possible under any of these approaches. The fact that even small probabilities of false positives can produce substantial errors in occupancy estimates (McClintock et al., 2010b; Miller et al., 2015) motivates the need to develop models that incorporate probabilities of species misclassification.

6.2.1 Modeling Misclassification for a Single Season

A General Approach

Here we focus on the standard occupancy setting for a single season in which we conduct K surveys over s units (sample units) and record the detections ($h_{ij} = 1$) and nondetections ($h_{ij} = 0$) of a focal species for survey j at unit i.

TABLE 6.3 Classification probabilities for a general occupancy model permitting misclassification[a]; p_{lm} is the probability of recording the detection observation l given the true occupancy state m

z_i	h_{ij}	
	0 (not detected)	1 (detected)
0 (unoccupied)	p_{00}	p_{10}
1 (occupied)	p_{01}	p_{11}

[a] Modified from Royle and Link (2006, Table 1).

The true state for each unit i is either occupied, $z_i = 1$, or not occupied, $z_i = 0$, and is assumed to persist for all surveys within the season. Estimation is focused on parameter $\psi_i = Pr(z_i = 1)$. Under the standard occupancy models described previously, only one of the observations, $h_{ij} = 0$, is ambiguous with respect to true occupancy state of a unit. Failure to detect the species at a unit ($h_{ij} = 0$ for all j) could mean either that the unit was not occupied by the species or that the species was present and just not detected. However, the other observation, $h_{ij} = 1$, is unambiguous in the sense that a detection always indicates that the unit was occupied. The key distinction between misclassification models and the standard models is that under misclassification both observation states are ambiguous and admit the possibilities of both focal species occupancy and absence.

Royle and Link (2006) developed an initial misclassification model by defining the following probabilities associated with the survey and classification process, $p_{lm} = Pr(h_{ij} = l | z_i = m)$. Table 6.3 shows the classification probabilities corresponding to each possible combination of observation and information states. These probabilities sum to 1 for each value of z_i, that is $p_{00} = 1 - p_{10}$ and $p_{01} = 1 - p_{11}$. Note that these classification probabilities can also be expressed in terms of a detection probability matrix $\mathbf{p}$, as in the multi-state models of Chapter 5, but with only two possible observed and true occupancy states:

$$\mathbf{p}_j = \begin{bmatrix} 1 - p_{10} & p_{10} \\ 1 - p_{11} & p_{11} \end{bmatrix}.$$

The prime difference between the above and the detection probability matrix defined in Chapter 5 is that in allowing for potential false positive observations, all elements of $\mathbf{p}$ are non-zero values. This admits the possible two-way misclassification of units from the field observations rather than the one-way misclassification of units in the multi-state occupancy model of the previous chapter.

If classification probabilities are the same for each sampling occasion as implied above (no j subscripts on the classification parameters), then the sufficient statistics are the number of detections at each unit, $y_i = \sum_{j=1}^{K} h_{ij}$. For an occupied unit (where $z_i = 1$) y_i follows a binomial distribution with probability p_{11}, while for an unoccupied unit (where $z_i = 0$) y_i will follow a binomial distribution with probability p_{10}. Furthermore, let $\mathbf{y} = \{y_i\}$ denote the vector of the number of detections at each unit. When ψ is modeled as a constant value across all units, rather than as unit-specific, and when all units are visited on K occasions, then the observed data likelihood for p_{11}, p_{10}, and ψ can be written as:

$$ODL(\psi, p_{11}, p_{10}|\mathbf{y}) \propto \prod_{i=1}^{s} \left\{ \psi \left[p_{11}^{y_i}(1 - p_{11})^{K-y_i} \right] \right.$$
$$\left. + (1 - \psi) \left[p_{10}^{y_i}(1 - p_{10})^{K-y_i} \right] \right\}. \qquad (6.5)$$

Royle and Link (2006) note that this expression reduces to the traditional occupancy likelihood under the constraint of no misclassification of unoccupied units, $p_{10} = 0$. Maximum likelihood can be used to obtain estimates based on this expression. In addition, parameters can be constrained or modeled as functions of unit- and time-specific covariates as with standard occupancy models.

Royle and Link (2006) note that this misclassification model is a finite mixture model (Chapter 7; also see Norris and Pollock, 1996; Link, 2003; Royle, 2006) and can show identical support for multiple, distinct sets of parameter values (i.e., multiple maxima of the likelihood function, with the same value). They suggest that a reasonable approach to selecting among equally supported alternative sets of parameter estimates is to assume that detection probability will be larger than misclassification probability, $p_{11} > p_{10}$. A small simulation study showed that occupancy estimates, $\hat{\psi}$, were approximately unbiased over the conditions of the simulation (Royle and Link, 2006).

One approach to dealing with heterogeneous detection probabilities in standard occupancy modeling is to model them as finite mixtures (Chapter 7; Royle, 2006). Briefly, a finite mixture model allows the probability of detection to be different at a finite (and often small) number of different types of occupied units (e.g., units with few vs. lots of individuals), where the mixture probabilities determine the relative frequency of the different types of units. Royle and Link (2006) note that their misclassification model, with $p_{10} > 0$, is equivalent to a standard occupancy model ($p_{10} = 0$) with $\psi = 1$ and detection modeled with a finite mixture. The consequence of this equivalence is that detection history data alone cannot be used to distinguish between these alternatives of misclassification and heterogeneous detection probabilities with high occupancy. Instead, the decision of which class of model to use must be based on other information. For example, if misclassification is strongly suspected, with little belief of substantial heterogeneity in detection probabilities among units, then the misclassification model of Royle and Link (2006) would be the model of choice. Identifiability issues and the absence of information about whether variation in detection probabilities results from misclassification or heterogeneous detection probabilities have limited the practical application of the general misclassification model of Royle and Link (2006) (see Fitzpatrick et al., 2009; McClintock et al., 2010a; Miller et al., 2015).

These limitations led to the recommendation to use different survey methods at units in an effort to resolve ambiguity about sources of variation in detection (McClintock et al., 2010a). This recommendation motivated the development of misclassification models that use multiple kinds of detections (Miller et al., 2011, 2013, 2015; detailed below). Chambert et al. (2015a) then considered additional approaches that can directly use data on assessments of the accuracy of the detection methods themselves. Specifically, Chambert et al. (2015a) described three different designs, and we retain their terminology in the descriptions below. Although these approaches require different designs for data collection, they can all be viewed as attempts to provide prior information about parameters p_{11} and p_{10}. This prior information is useful in resolving the ambiguity of the Royle and Link (2006) model with respect to both the issue of identical support for different sets of parameters and the issue of distinguishing misclassification and heterogeneous detection probabilities.

Terminology

Most of the initial work on false positives defined such observations at the level of the sample unit, such that a false positive represents "the occurrence of a detection at an unoccupied unit". We follow Chambert et al. (2015a) and use the term "unit-level false positive" to describe this type of false detection, as it is explicitly conditional on the unit being unoccupied. However, we can also consider false positive outcomes at the level of individual observations, which we denote "observation-level false positive". A distinction between these two types of false positives is that the occurrence of observation-level false positives is not conditional on unit occupancy status. Thus, the occurrence of an observation-level false positive is a necessary, but not sufficient, condition for a unit-level false positive to occur. If an observation-level false positive occurs (e.g., a closely related species is mistaken for the focal species) in an occupied unit (i.e., the focal species also happens to be present at that unit), then it becomes a correct detection at the unit level. In addition, observation-level false positives can sometimes be confirmed as such (e.g., when *a posteriori* confirmation methods are available, such as DNA analyses of animal scats; Karanth et al., 2011; Mondol et al., 2015), whereas unit-level false positives cannot be confirmed, as we never know with certainty that a unit is unoccupied, because of imperfect detection. For some model extensions (the observation confirmation design), these distinctions become important and may lead to improvements in identifiability of model parameters. Below we outline three potential designs for data collection directed at inferences about occupancy in the presence of general species misclassification.

Unit Confirmation Design

The unit confirmation design corresponds to the designs and associated models developed by Miller et al. (2011, 2013). This design typically includes collection of two types of detections, 'unambiguous' (detection of the focal species at a unit means that the unit is occupied by that focal species) and 'ambiguous' (detections can represent either true or false positives, therefore the unit may be occupied or unoccupied by the focal species). The existence of a subset of units for which focal species occupancy is known (those units with unambiguous detections) resolves the issue of identifiability in the likelihood of Royle and Link (2006) (Eq. (6.5)). The two types of detections can arise in at least two different ways (Miller et al., 2011). In some cases, a single survey or detection method admits the possibility of both kinds of detections. For example, surveys of a unit can include indirect detections that are ambiguous, such as animal scat or tracks (e.g., Karanth et al., 2011), and direct observations (e.g., a visual encounter) that are unambiguous. In other cases, all units in the sample are surveyed using a detection method yielding ambiguous detections, and at least a subset of the units is also surveyed using a method that provides unambiguous detections (Miller et al., 2011). Note that Chambert et al. (2015a) use the term "site confirmation design", but here we use "unit confirmation design" for consistency throughout the book.

Two Detection Types. First, we develop the likelihood for the initial case in which a single survey method can yield either ambiguous or unambiguous detections. As in the Royle–Link model, s units are surveyed K times in the season for a focal species. True occupancy state of unit i is denoted z_i, where $z_i = 1$ for occupied units and $z_i = 0$ for unoccupied units. True occupancy state of each unit is assumed to remain constant over all survey periods within the season (the closure assumption). The probability that unit i is occupied by the focal species is denoted as ψ_i. These models admit three types of observations that can be made at any survey, where h_{ij} denotes the observation at unit i, survey j: no detection, $h_{ij} = 0$; ambiguous detection, $h_{ij} = 1$; and unambiguous detection, $h_{ij} = 2$. Hence the detection history (**h**) for a unit representing the outcome of the K surveys is composed of a series of 0's, 1's, and 2's. For example, $\mathbf{h}_i = 021$ represents the outcome of three surveys of unit i where the species was not detected in survey 1, there was an unambiguous detection of the species in survey 2, and an ambiguous detection of the species in survey 3. Parameters associated with the detection and classification process are defined as follows, dropping unit and survey subscripts for notational simplicity: $p_{11} =$ probability of detecting the focal species, given that the unit is occupied by the species; $p_{10} =$ probability of erroneously detecting the focal species, given that the unit is not occupied; $b =$ probability that a detection is classified as unambiguous, given that the unit is occupied. The conditional (on true state) probabilities associated with each observation are presented in Table 6.4, which could be represented in terms of the detection probability matrix:

$$
\mathbf{p}_j =
\begin{bmatrix}
1 - p_{10} & p_{10} & 0 \\
1 - p_{11} & (1-b)\,p_{11} & b p_{11}
\end{bmatrix}.
$$

The hierarchical structure of the observations can be formally defined as:

$$
z_i \sim Bernoulli(\psi),
$$
$$
h_{ij} | z_i = m \sim Categorical\left(\mathbf{p}_j^{[m\bullet]}\right).
$$

Recall from Chapter 5 that $\mathbf{p}_j^{[m\bullet]}$ denotes the row of $\mathbf{p}_j$ for the case $z_i = m$.

Unlike the multi-state occupancy models considered so far, in this situation the number of possible observed states is greater than the number of true states, but the modeling framework presented in Chapter 5 can still be used. Using that framework, the vector of probabilities for a unit being in each possible occupancy state before the first survey can be defined as:

$$
\boldsymbol{\phi}_0 =
\begin{bmatrix}
1 - \psi & \psi
\end{bmatrix},
$$

and the state-specific column vector for the probability of observing the detection history ($\mathbf{p_h}$) is formed by taking the element-wise product of the respective columns of $\mathbf{p}_j$ (Chapter 5). For example, for $\mathbf{h}_i = 021$:

$$
\mathbf{p}_{021} = \mathbf{p}_1^{[\bullet 0]} \odot \mathbf{p}_2^{[\bullet 2]} \odot \mathbf{p}_3^{[\bullet 1]}
$$
$$
=
\begin{bmatrix}
1 - p_{10} \\
1 - p_{11}
\end{bmatrix}
\odot
\begin{bmatrix}
0 \\
b p_{11}
\end{bmatrix}
\odot
\begin{bmatrix}
p_{10} \\
(1-b)\,p_{11}
\end{bmatrix}
$$
$$
=
\begin{bmatrix}
0 \\
(1-p_{11})\,b p_{11}(1-b)\,p_{11}
\end{bmatrix}.
$$

The probability statement for any detection history $\mathbf{h}$ can therefore be expressed as:

$$
Pr(\mathbf{h}|\boldsymbol{\theta}) = \boldsymbol{\phi}_0 \mathbf{p_h},
$$

TABLE 6.4 Probabilities associated with different observations for the unit confirmation design with two detection types

True state	Observation		
	$h = 0$	$h = 1$	$h = 2$
$z = 0$	$1 - p_{10}$	p_{10}	0
$z = 1$	$1 - p_{11}$	$(1 - b) p_{11}$	$b p_{11}$

where θ is the vector of parameters to estimate. The observed data likelihood would then be:

$$ODL(\theta|\mathbf{h}) = \prod_{i=1}^{s} Pr(\mathbf{h}_i|\theta).$$

While we have omitted unit- and time-specific subscripts on the various probabilities above for the sake of clarity, this could be achieved in practice through the inclusion of covariates with appropriate link functions, as with the other modeling we have detailed.

For completeness, the observed data likelihood can also be written as:

$$ODL(\psi, p_{11}, p_{10}, b|\mathbf{h}) \propto \prod_{i=1}^{s} \left\{ \begin{array}{l} \psi_i \left[\prod_{j=1}^{K} \left(\prod_{l=0}^{2} \pi_{l,z=1}^{I(h_{ij}=l)} \right) \right] \\ + (1 - \psi_i) \left[\prod_{j=1}^{K} \left(\prod_{l=0}^{2} \pi_{l,z=0}^{I(h_{ij}=l)} \right) \right] \end{array} \right\}$$

where $I(h_{ij} = l)$ is an indicator variable assuming value 1 when $h_{ij} = l$ and 0 otherwise, and where $\pi_{l,z=1}$ and $\pi_{l,z=0}$ are the probabilities of observation l when $z = 1$ and $z = 0$, respectively (Table 6.4).

Two Detection Methods. In this sampling approach, one survey method (denoted as M1) yields only ambiguous detections and a second survey method (M2) provides only unambiguous detections. In the usual situation, we expect method M1 to be a primary method that is applied during most sampling occasions and method M2 a secondary method that typically requires more effort, but that is applied during one or more sampling occasions as a means of dealing with false positives (Miller et al., 2011, 2015; Chambert et al., 2015a). Here, we follow Miller et al. (2011) and Chambert et al. (2015a) and describe the situation in which all units are sampled via both methods. However, inference is also possible if M2 is applied to a randomly selected subset of all units to which M1 is applied (Miller et al., 2013).

Under this design, the s units are sampled on K occasions using method M1 and on K' occasions using method M2. Denote h_{ij} as the observation ob-

TABLE 6.5 Probabilities associated with different observations for the site confirmation design with two detection methods

True state	Observation			
	$h = 0$	$h = 1$	$w = 0$	$w = 1$
$z = 0$	$1 - p_{10}$	p_{10}	1	0
$z = 1$	$1 - p_{11}$	p_{11}	$1 - r_{11}$	r_{11}

tained using method M1 at unit i on occasion j, and $w_{ij'}$ as the observation obtained using method M2 at unit i on occasion j'. Both y_{ij} and $w_{ij'}$ are binary random variables assuming value 1 when the focal species is detected and 0 otherwise. Three detection/classification parameters are required under this approach (again dropping unit and survey subscripts for ease of presentation): (1) p_{11} is the probability of detecting the focal species with method M1, given that the unit is occupied; (2) p_{10} is the probability of erroneously detecting the focal species with method M1, given that the unit is unoccupied; and (3) r_{11} is the probability of detecting the focal species using method M2, given that the unit is occupied. Detection probabilities associated with each observation state are presented in Table 6.5, and detection probability matrices can be defined similarly to above, although separately for each detection method. That is:

$$\mathbf{p}_{1,j} = \begin{bmatrix} 1 - p_{10} & p_{10} \\ 1 - p_{11} & p_{11} \end{bmatrix},$$

and

$$\mathbf{p}_{2,j'} = \begin{bmatrix} 1 & 0 \\ 1 - r_{11} & r_{11} \end{bmatrix}.$$

The random variables associated with this model can be formally defined as:

$$z_i \sim Bernoulli(\psi),$$
$$h_{ij}|z_i = m \sim Bernoulli\left(\mathbf{p}_{1,j}^{[m\bullet]}\right),$$
$$w_{ij'}|z_i = m \sim Bernoulli\left(\mathbf{p}_{2,j'}^{[m\bullet]}\right).$$

Detection histories now consist of a combination of outcomes from ambiguous and unambiguous survey methods, therefore we introduce a '/' to distinguish between the two methods with values preceding the '/' being the survey outcomes from the primary method (that yields ambiguous detections). For example $\mathbf{h}_i = 01/1$ represents the outcome of two surveys being conducted at unit i using M1, resulting in a nondetection in survey 1 and detection of the focal species in survey 2, and a single survey using M2 which resulted in a detection of the species. The state-specific detection probability vector $\mathbf{p}_h$ can be formed in a similar manner to above, but taking the element-wise product of the respective columns of $\mathbf{p}_{1,j}$ and $\mathbf{p}_{2,j'}$. Otherwise, construction of the observed data likelihood proceeds as described previously using the matrix notation.

The observed data likelihood can also be written as:

$$ODL(\psi, p_{11}, p_{10}, r_{11} | \mathbf{h}, \mathbf{w})$$

$$\propto \prod_{i=1}^{s} \left(\begin{array}{c} \psi \prod_{j=1}^{K} p_{11}^{h_{ij}} (1-p_{11})^{1-h_{ij}} \times \prod_{j'=1}^{K'} r_{11}^{w_{ij'}} (1-r_{11})^{1-w_{ij'}} \\ + (1-\psi) \prod_{j=1}^{K} p_{10}^{h_{ij}} (1-p_{10})^{1-h_{ij}} \times I\left(w_{ij'}=0\right) \end{array} \right)$$

where $I\left(w_{ij'}=0\right)$ is an indicator function that assumes the value of 1 when $w_{ij'}=0$, and 0 otherwise.

Inference for the models corresponding to both of these unit confirmation designs can be obtained using maximum likelihood or Bayesian methods. While we have not detailed it specifically here, the complete data likelihood (CDL) for these models can also be defined as previously, based upon the random variables (see Section 4.4.1). These models have been implemented in Programs PRESENCE and MARK. Miller et al. (2011) report results of a simulation study directed at the efficacy of these two kinds of unit confirmation models. In the presence of false positives, the models performed well, yielding estimates of occupancy with smaller bias and mean squared error than estimates from standard occupancy models (MacKenzie et al., 2002). Standard approaches to model selection (e.g., using AIC; Burnham and Anderson, 2002) were effective in distinguishing data for which false positive models were needed.

Example Analyses. Miller et al. (2011) reported results of two analyses using these unit confirmation models. The first example utilized data collected from an experimental study of auditory surveys for calling anurans in which true species identities for all vocalizations were known and observers did report false positive observations. When small portions of the observational data were treated as unambiguous under both the 'two detection type' and 'two detection method' designs, resulting occupancy estimates were highly accurate, whereas occupancy estimates under the standard model of MacKenzie et al. (2002) were positively biased by as much as 80%. Miller et al. (2015) conducted a more comprehensive analysis of results from this experimental study of auditory anuran surveys. Although they focused on both heterogeneity of detection probabilities and false positive detections, the unit confirmation models of Miller et al. (2011, 2013) performed best in terms of estimator bias and precision (Miller et al., 2015).

A second example from Miller et al. (2011) used data for three frog species (American bullfrogs, *Lithobates catesbiana*; green frogs, *L. clamitans*; pickerel frogs, *L. palustris*) obtained via auditory surveys of units near the Chesapeake and Ohio Canal National Historic Park (CHOH), USA. The auditory surveys occurred on 11 routes, each consisting of 10 units, on three occasions during March–June 2005 as part of the North American Amphibian Monitoring Program (NAAMP; Weir and Mossman, 2005). An additional route, consisting of 14 units, was located within CHOH, and these units were surveyed for calling anurans on 14–30 occasions. Detections from call surveys were classified based on NAAMP calling intensity categories: low (individuals can be counted; there is space between the calls), medium (calls of individuals can be distinguished, but there is some overlapping of calls), and high (full chorus, calls are constant, continuous, and overlapping). In addition to these auditory data, 34 randomly selected wetlands located 250 m from CHOH call count locations were directly (visual and dip net) sampled eight times (i.e., by two independent observers over four survey occasions) during the same time period. Observers recorded observations of individuals in all life history phases (adults, egg masses, and tadpoles) from visual encounters and captures by dip net. Data for all ponds associated with an auditory survey unit were combined to yield certain detection data for each of the eight sampling occasions.

Miller et al. (2011) analyzed these data using a model that included both multiple detection states and methods. Specifically, auditory surveys were treated as the first detection method (M1), admitting possible false positive detections, and the pond surveys were treated as the second detection method (M2), where species were identified in the hand and species identification was assumed to be certain. In addition, medium and high intensity auditory detections from the first survey method were assumed to be certain ($h_{ij} = 2$) and low intensity detections were uncertain ($h_{ij} = 1$). The parameterization for this hybrid model combining detection types and methods is given by replacing the parameterization for the uncertain detection method ($h_{ij} = 0, 1$) in Table 6.5 with that for the multiple detection types ($h_{ij} = 0, 1, 2$) in Table 6.4.

Separate occupancy probabilities were estimated for NAAMP and CHOH units. Ambient air temperature was suspected as an influence on calling frequency and intensity (e.g., Mazerolle et al., 2007). Therefore, p_{11} and b were allowed to vary as a quadratic function of air temperature at the time of the survey (Miller et al., 2011). To use model selection to provide evidence for the presence of false positives in the data, the full parameterization described above was compared to a model in which false positive detections were assumed not to occur (model with the constraint, $p_{10} = 0$). The overall occupancy probability was computed as a weighted mean, based on the relative proportions of NAAMP and CHOH units.

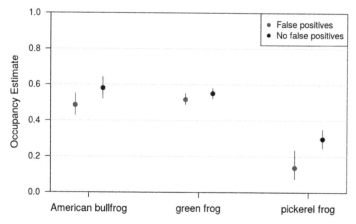

FIGURE 6.3 Occupancy estimates for three frog species at wetland sites in Maryland, in the general area of the Chesapeake and Ohio Canal National Historic Park, spring 2005. Estimates from models that do (Miller et al., 2011), and do not (MacKenzie et al., 2002), account for false positives. *Reproduced from: Miller et al. (2011).*

The estimated false positive probabilities were $p_{10} = 0.030$ for bullfrog, 0.008 for the green frog, and 0.027 for the pickerel frog. For the green and pickerel frogs the model accounting for false positives had lower AIC scores than the model where false positives were assumed not to occur (ΔAIC = 3.2 and 0.3, respectively). The model with no false positives had lower AIC for the bullfrog (ΔAIC = 0.8). As expected, for all species the estimates of occupancy were lower for the model where false positive detections were accounted for (Fig. 6.3). For the pickerel frog, the occupancy estimate based on the unit confirmation model was less than half the estimate from the standard model assuming no false positive detections.

Calibration Design

The calibration design (Chambert et al., 2015a) is based on reference units at which true occupancy state is under investigator control and thus known with certainty. This permits the collection of independent data on probabilities of false positive (p_{10}) and true positive (p_{11}) detection probabilities. For example, when using occupancy modeling to draw inferences about disease dynamics, laboratory assays can be used on reference samples of known immunological status to assess the specificity and sensitivity of the assay method (McClintock et al., 2010a). For animal surveys based on auditory detections (e.g., for birds or anurans), the experimental system developed by T.R. Simons using pre-recorded vocalizations of known species identity can be played under field survey conditions to observers (Simons et al., 2007). This system has been used to investigate probabilities of observers recording false positives and false negatives at the

TABLE 6.6 Probabilities associated with different observations at reference units with known true occupancy status, for the calibration design

True state	Observations		
	No detection	False positive	False negative
$z = 0$	$1 - p_{10}$	p_{10}	0
$z = 1$	$1 - p_{11}$	0	p_{11}

level of both the unit and the specific observation (McClintock et al., 2010b; Miller et al., 2012c, 2015). An important distinction between this calibration design and the unit confirmation design is that calibration does not provide direct information about the occupancy status of survey units. Rather the direct information supplied by calibration is about the observation process (inferences about p_{10} and p_{11}).

The sampling design for field-surveyed units is identical to that described for the unit confirmation design and yields the same observation data, h_{ij}. These data are then analogous to ambiguous data of the unit confirmation design, in that they admit the possibility of false positives and negatives. The probability structure for detections at the reference units is specified in Table 6.6, and at the sampled units it is the same as in Table 6.3, and the detection probability matrix is the same as $\mathbf{p}_{1,j}$ defined above. Define n_1 and n_0 as the numbers of reference units that respectively are, and are not, occupied by the focal species (the selection of which is under the investigator's control, hence these are known values). Further define x_1 and x_0 as the numbers of observations that represent unit-level true and false positives, respectively, that occur at these reference units. The random variables associated with these different processes are defined as:

$$z_i \sim Bernoulli(\psi),$$
$$h_{ij} | z_i = m \sim Bernoulli\left(\mathbf{p}_{1,j}^{[m\bullet]}\right),$$
$$x_1 \sim Binomial(n_1, p_{11}),$$
$$x_0 \sim Binomial(n_0, p_{10}).$$

The information from the calibration units is incorporated into the estimation by construction of a joint observed data likelihood. This can be written as:

$$ODL(\psi, p_{11}, p_{10} | \mathbf{y}, \mathbf{x}, \mathbf{n}) \propto p_{11}^{x_1} (1 - p_{11})^{n_1 - x_1} p_{10}^{x_0} (1 - p_{10})^{n_0 - x_0}$$
$$\times \prod_{i=1}^{s} \left[\begin{array}{l} \psi \prod_{j=1}^{K} p_{11}^{h_{ij}} (1 - p_{11})^{1 - h_{ij}} \\ + (1 - \psi) \prod_{j=1}^{K} p_{10}^{h_{ij}} (1 - p_{10})^{1 - h_{ij}} \end{array} \right].$$

The terms of the first line of the likelihood (to the right of the '$\propto$' symbol) incorporate the information on the known number of true positive and false positive detections from the reference units; this is the observed data likelihood for the reference unit data (note the binomial coefficients have not been included as they are constants with respect to the parameters being estimated). The second two lines of the likelihood correspond to the observed data likelihood from the sampled units, and is the finite mixture of Royle and Link (2006); also see Eq. (6.5). The identifiability issue is resolved by the extra information from the reference units about p_{10} and p_{11}. Chambert et al. (2015a) note that the confirmation units of the unit confirmation design (i.e., those units reported as occupied based on unambiguous data) serve a similar role to that of reference units of the calibration design in providing direct information about p_{11}. However, important distinctions between the two designs are that (1) the unit confirmation design has no set of units known to be unoccupied (to directly inform p_{10}), and (2) confirmation units under the unit confirmation design directly inform the occupancy parameters, ψ.

Observation Confirmation Design

This design corresponds to the sampling situation in which field observations are ambiguous in admitting false positives, but *a posteriori* confirmation of species identification is possible for at least a subset of these observations. A sampling situation for which this design will be common is surveys based on indirect animal sign, such as scats (e.g., Jeffress et al., 2011; Karanth et al., 2011; Mondol et al., 2015). For example, Karanth et al. (2011) surveyed the Malenad–Mysore tiger landscape of the Western Ghats, India, for sign of tigers (*Panthera tigris*). Field identification in the absence of other sign (e.g., pugmarks) is based primarily on scat size, and large scat from leopards (*Panthera pardus*) may be confused with small tiger scat, leading to false positive detections. However, if samples of fresh scat are transported to the laboratory, molecular techniques can be used to identify scat by species with certainty based on DNA analysis (Mondol et al., 2015).

The confirmation occurs at the observation level (e.g., an individual scat), rather than the unit level, and multiple observations of sign may be made at any unit. Thus, at the level of the individual observation, both true and false positive observations may be made at a single unit. Chambert et al. (2015a) developed their original model for this sampling design assuming an omnibus source of false positive observations that is independent of the presence or detection of the focal species. A 'true positive observation' is defined as occurring when sign from the focal species is present, detected, and correctly assigned to the focal species. A 'false positive observation' occurs when sign from a non-focal species is present, detected and mis-assigned to the focal species. Model assumptions are that (1) the occurrence of a false positive observation at a unit is independent of focal species presence at the unit, z_i, and (2) occurrences of true and false positive observations are independent events.

Modeling details can vary somewhat depending on the specific details of the sampling situation. In the following development (based upon Chambert et al., 2015a), we assume that field observations are of scats and that species confirmation is obtained via DNA analysis of selected scats. The data are divided into two groups, the first group corresponding to only field observations of scats obtained during K replicate surveys at each of s units. None of these observations is confirmed via laboratory analysis, and all are thus ambiguous. At the unit level, data for survey j can again be denoted by h_{ij}, taking a value of 0 when no observations are assigned to the focal species and 1 when at least one observation is assigned to the focal species. The second group of data consists of field observations that are then taken to the laboratory for DNA analysis and confirmation. These observations occur during K' replicate surveys of s' units. At the unit level, data from this second group are denoted v_{ij} and observations can take one of four possible values:

- $0 =$ no detections of any sign at unit i during sampling occasion j.
- $1 =$ all detections that were assigned in the field as belonging to the focal species were determined in the lab to be false positives (all incorrect).
- $2 =$ all detections that were assigned in the field as belonging to the focal species were determined in the lab to be true positives (all correct).
- $3 =$ detections assigned to the focal species were found in the lab to represent a mix of true and false positive observations.

We therefore have two different types of detection histories that we shall denote as $\mathbf{h}_i$ for the s units with the unconfirmed data and $\mathbf{v}_i$ for the s' units with the confirmed data.

Define u_0 as the unconditional probability of at least one false positive observation at a unit during a survey replicate. Similarly, define u_1 as the unconditional probability of at least one true positive observation at a unit during a survey replicate. These probabilities can then be used to derive the conditional probabilities associated with each observation state for both confirmed and unconfirmed data (Table 6.7). A tree diagram corresponding to the observation confirmation model is presented in Fig. 6.4 to illustrate the rationale underlying the probability structures presented in Table 6.7. Note that the original Royle and Link (2006) notation has been used for unconfirmed data in Table 6.7, but that we can write these probabilities in terms of the observation probabilities, u_0 and u_1. Specifically, $p_{10} = u_0$ and $p_{11} = u_0 + u_1 - u_0 u_1$, where the latter value can be obtained by summing the unconfirmed data probabilities associ-

TABLE 6.7 Probabilities associated with confirmed survey observations (v_{ij}) for the observation confirmation design

True state	Observations			
	$v_{ij} = 0$	$v_{ij} = 1$	$v_{ij} = 2$	$v_{ij} = 3$
$z = 0$	$1 - u_0$	u_0	0	0
$z = 1$	$(1 - u_1)(1 - u_0)$	$(1 - u_1)u_0$	$u_1(1 - u_0)$	$u_1 u_0$

ated with observation states 1, 2, and 3. The detection probability matrix for the unconfirmed data is again $\mathbf{p}_{1,j}$, and for the confirmed data observations it can be defined as (from Table 6.7):

$$
\mathbf{q}_j = \begin{bmatrix} 1 - u_0 & u_0 & 0 & 0 \\ (1 - u_1)(1 - u_0) & (1 - u_1)u_0 & u_1(1 - u_0) & u_1 u_0 \end{bmatrix},
$$

where rows represent the true occupancy state, and columns the possible observations for v_{ij}. The state-specific detection probability vector for the detection history $\mathbf{v}$, $\mathbf{q_v}$ can be calculated in the same manner as $\mathbf{p_h}$ is for the detection history $\mathbf{h}$; the element-wise product of the relevant columns of $\mathbf{q}_j$ (for $j = 1, \ldots, K'$).

The random variables associated with the observation confirmation design can be defined as:

$$
z_i \sim Bernoulli(\psi)
$$
$$
h_{ij} | z_i = m \sim Bernoulli\left(\mathbf{p}_{1,j}^{[m\bullet]}\right)
$$
$$
v_{ij} | z_i = m \sim Categorical\left(\mathbf{q}_j^{[m\bullet]}\right)
$$

where $\mathbf{q}_j^{[m\bullet]}$ is defined in a similar manner to $\mathbf{p}_{1,j}^{[m\bullet]}$, i.e., it denotes the row of $\mathbf{q}_j$ corresponding to the true occupancy state m. The probability statement for each detection history can be expressed using the matrix notation of the multi-state model, and the observed data likelihood for both sets of data becomes:

$$
ODL(\theta | \mathbf{h}, \mathbf{v}) = \prod_{i=1}^{s} (\phi_0 \mathbf{p}_{\mathbf{h}_i}) \prod_{i=1}^{s'} (\phi_0 \mathbf{q}_{\mathbf{v}_i}).
$$

This can be expressed as:

$$
ODL(\psi, u_0, u_1 | \mathbf{h}, \mathbf{v})
$$
$$
\propto \prod_{i=1}^{s} \left(\psi_i \prod_{j=1}^{K} p_{11}^{h_{ij}} (1 - p_{11})^{1 - h_{ij}} + (1 - \psi_i) \prod_{j=1}^{K} p_{10}^{h_{ij}} (1 - p_{10})^{1 - h_{ij}} \right)
$$
$$
\times \prod_{i=1}^{s'} \left(\psi_i \prod_{j=1}^{K'} \tau_{v_{ij}, z_i = 1} + (1 - \psi_i) \prod_{j'=1}^{K'} \tau_{v_{ij}, z_i = 0} \right)
$$

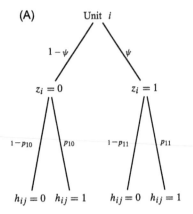

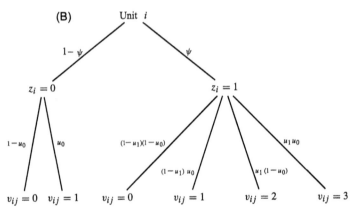

FIGURE 6.4 Diagrammed structure of the *observation confirmation* occupancy model for dealing with false positives. "A" denotes a unit with standard (non-confirmed) survey data (h), whereas "B" denotes a unit with confirmed data (v). Non-confirmed data indicate either detection ($h_{ij} = 1$) or not ($h_{ij} = 0$), whereas confirmed observations can be no detection ($v_{ij} = 0$), only false positive observations ($v_{ij} = 1$), only true positive observations ($v_{ij} = 2$), or both true and false positive observations ($v_{ij} = 3$). Both kinds of units can be either occupied ($z_i = 1$) or not ($z_i = 0$), with occupancy probability denoted as ψ. Detection probabilities for non-confirmed unit data are p_{11} and p_{10} for true and false positives, respectively. Detection probabilities for confirmed data observations are denoted as u_1 and u_0 for true and false positive detections, respectively. Note that $p_{10} = u_0$ and $p_{11} = u_1 + u_0 - u_1u_0$. *From Chambert et al. (2015a).*

where $\tau_{l,z_i=1} = Pr(v_{it} = l | z_i = 1)$, the probability of observation state l when the unit is occupied, and $\tau_{l,z_i=0} = Pr(v_{it} = l | z_i = 0)$, the probability of observation state l when the unit is unoccupied. Recall that under the observation confirmation model, the probabilities p_{11} and p_{10} are derived from the values of u_1 and u_0.

Selecting an Approach

As noted above, the three designs described in this section represent three different approaches to providing extra information to resolve identifiability issues with the general Royle and Link (2006) false positive model. All three designs accomplish that resolution, so selection among them for a specific investigation will depend on logistical constraints associated with the particular study system. The unit confirmation approach of Miller et al. (2011, 2013) may be the most generally applicable approach, as it requires no experimental systems or laboratory assays. Rather, it simply requires that one kind of observation or survey method yield observations that are assigned to species with certainty.

The calibration design should be useful in cases where experimental systems or laboratory assays can be used to directly investigate the likelihood of false positive observations. We expect this design to be useful for epidemiological studies in which molecular assays can provide unambiguous observations (see McClintock et al., 2010a; Lachish et al., 2012). The calibration design can be readily used with data from experimental systems such as that developed by T.R. Simons (see Simons et al., 2007; McClintock et al., 2010b; Miller et al., 2012c, 2015). Observer assessments such as online tests and surveys can sometimes provide calibration data on probabilities of false positives, as well as on possible individual heterogeneity in these probabilities. Such observer assessments may be especially useful in citizen science studies, with assessment scores perhaps used as individual observer covariates or as the basis for random effects models of misclassification (Miller et al., 2012c; Chambert et al., 2015a).

The observation confirmation design is well-suited for surveys based on observations, at least some of which can be somehow collected and re-examined at a later time for unambiguous identification. These include acoustic signatures from remote recording devices, photographic images from remote cameras, voucher specimens, and some forms of animal sign (e.g., scats that can be collected for laboratory DNA analysis). This approach may sometimes entail added expense, but it provides a viable approach to dealing with false positives and has potential to be extended.

One possible extension of the observation confirmation design entails better use of multiple observations for each survey sample. As described here, the four potential observation states are defined by whether there is at least one observation characterized as false positive or false negative. However, there is no need to restrict data to binary responses in this manner. Define r_0 as the probability of occurrence of a single false positive at the level of the individual observation. The likelihood of a unit-level false positive detection increases with the number of observations (n) as: $p_{10} = 1 - (1 - r_0)^n$. Such direct modeling of false positive probabilities associated with individual observations represents a model extension that may lead to increased precision.

Miller et al. (2011, 2013) note that false positive models can be extended to deal with increased numbers of observations and occupancy states. For example, rather than observations or survey methods being either certain or uncertain, uncertain observations may easily vary in degree of certainty. This possibility could be modeled using different probabilities of false positives for multiple discrete classes of observations. Degree of certainty can even be modeled as a function of observation-specific covariates as well. In addition to these other types of observations, occupancy states could be defined beyond the simplest case of occupied or not (Chapter 5; Royle, 2004b; Royle and Link, 2005; Nichols et al., 2007a; MacKenzie et al., 2009). In this case of multiple occupancy states, the concept of false positives is extended to the more general problem of misclassification, in which a unit in one state is misclassified as being in another. Miller et al. (2013) provided details on implementing such an extension, but fundamentally it just involves increasing the number of columns (for additional types of observations) and rows (for additional true occupancy states) in the detection probability matrix $\mathbf{p}$, and number of elements in the occupancy probability vector $\boldsymbol{\phi}_0$ (for additional true occupancy states), defined above. Once done, calculation of probability statements for detection histories and calculation of the observed data likelihood, or use of the complete data likelihood, proceeds exactly as for the simpler case considered here using the multi-state framework.

Another possible extension of the above models is possible when potential sources of false positive observations can be explicitly identified. The Royle and Link (2006) modeling approach on which the above designs are based is extremely general. In fact we refer to this approach as 'omnibus', in the sense that there is no specific precursor of misclassification. Thus, misclassification does not depend on presence of individuals of another species with which the focal species may be easily confused. We believe that such omnibus models should be widely applicable, as they require no identification of factors associated with misclassification. However, we have also begun to explore two-species models in which each species may be confused with the other, but this is the only source of misclassification (e.g., see Stolen et al., 2014). In such models misclassification of one species can only occur when the other species is present, so the full modeling must include inference about occupancy status of both species at surveyed units. We do not believe that these models will be as generally applicable as those we have described here and in Chapter 10, but they may permit more precise estimation when conditions for their application are met, and they will likely be essential for studying two-species interactions in the face of misclassification.

6.2.2 Discussion

Occupancy models that deal simultaneously with false positives and nondetection play a role similar to that of so-called multi-event models (Pradel, 2005) in capture–recapture. The models are very general, and virtually all more traditional models can be obtained by constraining parameters. For example, in the false positive model for two detection types, imposing the constraints of false positive probability of 0 ($p_{10} = 0$) and probability of a detection being certain of 1 ($b = 1$) yields a standard occupancy model that deals only with nondetection. The prominent role of these models leads us to believe that they will be a focus of substantial methodological development over the next decade.

As noted above, the model of Royle and Link (2006) provides a general structure for models incorporating false positives. All of the subsequent developments in false positive modeling can be viewed as ways to incorporate extra information about p_{11} and p_{10} to help resolve potential ambiguities in the original general model. The models of Miller et al. (2011) resolve these ambiguities with detection types or methods that are certain, in the sense of being unambiguous (they admit no false positives).

Chambert et al. (2015a) considered multiple designs and information sources that could be used to resolve ambiguity associated with false positives. They referred to the initial models of Miller et al. (2011, 2013) as based on a *unit confirmation* design, highlighting designs that produce a subset of units that are known to be occupied, i.e., there is a group of units for which $p_{10} = 0$ with certainty. Chambert et al. (2015a) also described a *calibration* design, relying on the independent collection of data on the probabilities of false positives, p_{10}, and possibly nondetection, p_{11}, at a set of reference units of known occupancy status. Finally, Chambert et al. (2015a) described an *observation confirmation* design, in which a set of detections is subjected to *a posteriori* confirmation. The basic likelihoods for these single-season designs were presented by Chambert et al. (2015a) with full recognition that much work remains in terms of implementing these methods, tailoring them to specific survey types, testing their efficacy, providing computer code, etc. In addition, there is no claim whatsoever that these three design types exhaust the possibilities of providing extra information for use in dealing with simultaneous non-zero probabilities of nondetection and false positives.

In this section we have focused on methods to account for false positive detections in the single-season case. Extensions of some of these approaches have been developed for the multi-season case as well, which we shall return to in Chapter 10.

Finally, all of these approaches were developed as omnibus methods, in the sense that there is no restriction on the sources of false positives. But as noted above, in some cases it will be known *a priori* that any false positives for a focal species will entail confusion with a similar species of known identity (e.g., Stolen et al., 2014). In such cases, the presence or absence of the non-focal species will provide information that can be used to resolve uncertainty in probabilities of nondetection and false positives. Development of models that fully exploit this extra information will likely require false positive co-occurrence models for both single seasons (MacKenzie et al., 2004a; Richmond et al., 2010; Chapter 14) and multiple seasons (Miller et al., 2012a; Yackulic et al., 2014).

6.3 MULTI-SCALE OCCUPANCY

So far in this book, we have only considered situations where information on a species is collected at two levels, units and surveys within units (which may be spatially or temporally replicated). Accordingly, we have focused on estimation of occupancy at the unit level, and detection at the survey level. In some applications, there may be an intermediate-level process that is operating that investigators are interested in, or need to account for to make improved inferences about species occurrence or the detection process. We refer to this intermediate process as a secondary-scale occupancy process, which may be either spatial or temporal in nature depending upon the biological questions of interest and nature of the sampling. For example, in Chapter 4 we noted that one potential violation of the closure assumption for the basic occupancy model is when the species is not always physically present within a unit during a season to be detected. When the species is randomly available to be detected at each survey, then ψ should be interpreted as a probability of 'use' (the species may be present at the unit at some stage during the season, but not always) and p is the combined probability of the species being available for detection at survey occasion j, and detected in the survey given the species was available. The two components of p in this situation cannot be separated with the type of designs considered thus far. However, with additional information collected at the appropriate level, it would be possible to make inference about this intermediate process of random availability. In this section we consider cases where information is collected at three levels, which enables estimation of occupancy at multiple scales.

There are a number of practical ways in which researchers may have access to three levels of information. Nichols et al. (2008) considered scenarios where data were collected from multiple units, with K temporally replicated surveys, and during each survey occasion multiple detection devices were deployed simultaneously to detect the species (camera traps, hair snares and track plate). Therefore, at each survey occasion there are multiple detection sources. The three levels considered by Nichols et al. (2008) were units, surveys within units, and devices within surveys. Mordecai et al. (2011) took a similar approach in their study of Louisiana waterthrush (*Seiurus motacilla*) occupancy, except instead of using different detection devices within a survey, they used replications of the same survey method within each survey occasion to obtain the third level of information. That is, they surveyed 140 point-count stations (units; first level), on four survey occasions (second level), with five successive 1-minute counts (third level). Hence each unit was surveyed with a total of 20 1-minute count surveys. While ideally there would be an interval between surveys to improve assumptions of independence (although see Section 6.4), this example highlights that investigators may often have extra information in their recorded data that they can exploit to make additional inferences about their system and species of interest. Pavlacky et al. (2012) surveyed birds via point counts conducted in the Black Hills of North Dakota and Wyoming. They sampled 56 1-km^2 units and located 16 sampling stations within each unit. The sampling stations were viewed as spatially replicated surveys of the units, and each station was surveyed with a single 5-minute point count which was divided into three intervals: (1) minutes 1 and 2; (2) minutes 3 and 4; and (3) minute 5. In this last example, the intermediate process is of a spatial rather than a temporal nature.

A defining feature of the above examples is that at each level there is nested replication (e.g., Fig. 6.5) which provides the necessary information allowing the associated probabilities to be estimated without more restrictive assumptions. Note that the sub-surveys referred to in Fig. 6.5A could be from multiple devices or methods that have been deployed simultaneously (as in Nichols et al., 2008), or the same method being replicated within the same survey (e.g., Mordecai et al., 2011).

This hierarchal nestedness, or clustering, can also create dependencies within the structure of the data that need to be accounted for during analysis. In the presence of the intermediate 'availability' process, it would be inappropriate to treat all of the surveys as independent observations. When the species is available at a particular sub-unit or survey occasion, all surveys within that occasion will have a non-zero probability of registering a detection of the species, but when the species is unavailable, none of the surveys will have any chance of detecting the species during that occasion. We believe the multi-scale occupancy model we are about to detail is a very useful approach, particularly when the field protocols used by researchers exhibit some form of nestedness or clustering, which may be spatial or temporal in nature. This includes cases in which such clustering may be for "logistical" reasons, e.g., defining sampling units as 1 ha areas, but rather than independently select 90 units from the area of interest, 30 transects are selected and three 1 ha units are repeatedly surveyed along each transect. Such clustering of the sample units may introduce a lack

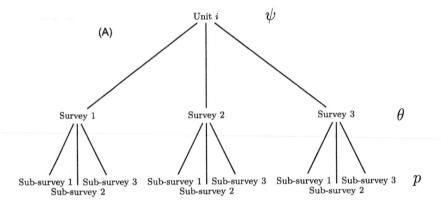

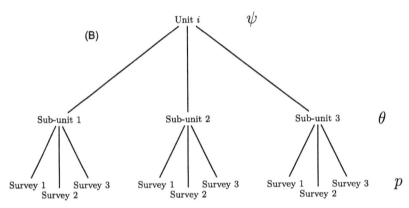

FIGURE 6.5 Examples of the nested replication associated with the multi-scale occupancy model. Panel A has replicated surveys of units, and within each survey, multiple sub-surveys of some nature. Panel B has spatial replication at the intermediate level within a unit, and repeat surveys conducted at each sub-unit. The parameter type associated with each level is indicated.

of independence in the data, particularly when species occurrence is spatially correlated.

During the last decade we have encountered applications that are well suited to the multi-scale occupancy model framework with increasing frequency. It has become increasingly common for investigators to deploy multiple devices or to use multiple methods for detecting animals. Emphasis of many recent monitoring programs has been on communities, rather than populations of a single focal species (e.g., Barrows et al., 2005; Buckland et al., 2005), with multiple detection methods deployed within a sample unit to detect multiple species of a given taxon (Sørensen et al., 2002; Manley et al., 2004, 2005; O'Connell et al., 2006). In some cases, researchers may deploy multiple detection methods to detect individuals of various sizes or life history phases for a single focal species (Mattfeldt and Grant, 2007; Smith et al., 2006). Technological advances are continually reducing the cost of many remote detection devices such as camera traps, enabling a greater number to be deployed in the field. Often a small number may be deployed within relatively close proximity in a sample unit to provide a more complete cross-section of the species present in the area, and to overcome local differences in the species assemblages due to micro-habitat effects.

Researchers are also sometimes interested in the relative effectiveness of different detection methods for a focal species (Bailey et al., 2004; O'Connell et al., 2006; Mattfeldt and Grant, 2007; Pepper et al., 2017). Studies directed at such tests might use separate analyses for each detection method and then compare resulting estimates of detection and occupancy probabilities (e.g., Bailey et al., 2004). This approach should not be misleading, but may be inefficient, as each analysis ignores information from all but the focal detection method. For example, if units have been surveyed using repeated sign and visual encounter surveys, there may be some units where individuals of the focal species were observed (so a visual encounter detection), but no sign of the species was ever found. In an analysis of just the sign survey data, for such units the modeling will assume that the species may have been present but undetected or genuinely absent, even though it is known the species was present from the visual encounter data. Using all of the data in a single analysis, as we outline here, properly incorporates such information into our inferences.

Finally, we note that this multi-scale occupancy model is sometimes referred to as the "multi-method" model, due to the applications considered by Nichols et al. (2008) where different methods were used simultaneously in each survey occasion. This terminology is somewhat restrictive and misleading, and we currently prefer to use "multi-scale model".

6.3.1 Model Definition

The multi-scale model of Nichols et al. (2008) was initially developed for use with remote sampling devices placed at approximately the same location within a sample unit. Our description here follows that motivation (replicated surveys of a unit, multiple detection methods deployed within each survey), although the approach applies more generally to any situation in which there are three levels of information available, as developed above. Consider a sampling design in which D different detection or sampling devices are deployed at each of s sample units for K survey occasions. The devices are placed at a location designated as a *sampling station* within each sample unit. Each station samples a unit, which will often be larger in area than the station itself. In this section we only consider single-season models and assume closure over the K occasions. The sampling scheme can be viewed as a robust design (Pollock, 1982; MacKenzie et al., 2003) in the sense that we have multiple possibilities for detection (multiple devices) at each sampling occasion. However, the modeling differs from that of multi-season models (Chapter 8; MacKenzie et al., 2003) in that we do not model seasonal colonizations and extinctions, but simply presence or absence of the focal species at the sample unit.

Consider a study with $D = 3$ detection devices, $K = 2$ survey occasions and a detection history of $\mathbf{h}_i = 011\ 010$. This detection history indicates that the species of interest was detected by devices 2 and 3 at survey occasion 1 and only by device 2 at survey occasion 2. Importantly, note how the detection history has been structured here, with the detection outcomes from the three devices being grouped by the two survey occasions. This grouping of the detection data within the detection history indicates the intermediate level of the sampling (at survey level in this case).

Define the following parameters:

$\psi = Pr$(sample unit occupied);

$\theta_j = Pr$(species present at immediate sampling station at occasion j | sample unit occupied);

$p_j^{[d]} = Pr$(detection at occasion j by device d | sample unit occupied and species present at immediate sampling station at occasion j).

The two occupancy parameters, ψ and θ_j, permit the modeling of occupancy at two different scales. The basic occupancy parameter, ψ, corresponds to species occurrence at the larger scale of the sample unit, where member(s) of the species have some non-negligible probability of being present at the sample unit. The occupancy parameters for the intermediate scale, θ_j, refer to the presence of member(s) of the target species at the sample station location at the time of survey occasion j, conditional on species presence in the sample unit. This probability accounts for situations in which member(s) of the species may be present at the immediate sample unit at some times and sample occasions, j, but not at others (i.e., changes in species availability for detection due to temporary absences within the season). Because of the closure assumption, the large scale occupancy probability, ψ, applies to all of the survey occasions. In addition to this larger scale occupancy, the product $\psi\theta_j$ represents the unconditional probability of small-scale occupancy, indicating presence of individual(s) of the species at the local sampling station (hence exposed to detection devices) at survey occasion j. The product $\psi(1 - \theta_j)$ indicates occupancy at the large scale, but not at the small scale, for sampling occasion j. In this latter case, the species may be temporarily unavailable for detection because members are not in the immediate vicinity of the detection devices.

When the intermediate level of sampling relates to spatial replication within a unit, θ_j is the probability of the species being present at sub-unit j, given the focal species is present at the unit during the season. In this case, θ_j is a secondary spatial-scale occupancy parameter, and $\psi\theta_j$ is the unconditional probability of a randomly selected sub-unit being occupied, irrespective of unit-level occupancy.

To illustrate the modeling, consider the following detection history: $\mathbf{h}_i =$ 010 000. This history again indicates a study with $D = 3$ detection devices and $K = 2$ survey occasions, and represents that the focal species was detected by device 2 at survey occasion 1 and was undetected by all devices at survey occasion 2. The probability of this history can be written as:

$$Pr(010\ 000|\boldsymbol{\theta}) = \psi\left[\theta_1\left(1-p_1^{[1]}\right)p_1^{[2]}\left(1-p_1^{[3]}\right)\left[(1-\theta_2)+\theta_2\prod_{d=1}^{3}\left(1-p_2^{[d]}\right)\right]\right],$$

where *uptheta* denotes the set of all model parameters (note just the θ_j's). The sample unit is known to be occupied (associated probability ψ) because the species was detected at the unit at some time during the surveying. Similarly, the species was known to have been present at the local sampling station at survey occasion 1, because it was detected by one of the devices. The product $\left(1-p_1^{[1]}\right)p_1^{[2]}\left(1-p_1^{[3]}\right)$ indicates detection by device 2, but not by devices 1 and 3 at occasion 1, conditional on local presence. The sum within the interior brackets specifies the two possible ways of obtaining no detections (000) during survey occasion 2. The first term, $(1-\theta_2)$, is the probability that the species was not locally present at the local sampling station at occasion 2. The second term, $\theta_2\prod_{d=1}^{3}\left(1-p_2^{[d]}\right)$, denotes the probability that member(s) of the species were present at the local sampling station at occasion 2, but simply not detected by any of the devices.

Now consider the detection history for a unit with no detections by any device at either sample occasion, $\mathbf{h}_i = 000\ 000$. The probability associated with this history can be written as:

$$Pr(000\ 000|\boldsymbol{\theta}) = (1-\psi) + \psi(1-\theta_1)(1-\theta_2)$$

$$+ \psi\left[\theta_1\prod_{d=1}^{3}\left(1-p_1^{[d]}\right)\right](1-\theta_2)$$

$$+ \psi(1-\theta_1)\left[\theta_2\prod_{d=1}^{3}\left(1-p_2^{[d]}\right)\right]$$

$$+ \psi\left[\theta_1\prod_{d=1}^{3}\left(1-p_1^{[d]}\right)\right]\left[\theta_2\prod_{d=1}^{3}\left(1-p_2^{[d]}\right)\right].$$

The first of these five additive terms reflects the probability that the sample unit was not occupied (focal species absent at the larger scale during the season). Hence there are no terms for small-scale occupancy or detection probabilities. The second additive term is the probability that the sample unit was occupied, but that members of the species were not present at the local sampling station during either occasion 1 or 2. The third additive term denotes the probability that the species was present at the local sampling station at occasion 1, but simply not detected then, and not locally present at occasion 2. The fourth additive term denotes presence at the local sampling station at occasion 2, but no detection then, and no local presence at occasion 1. The final term is the probability that members of the species were present at the local sampling station during both occasions 1 and 2, but went undetected by all devices at both occasions. These five additive terms represent all the possible options, according to the defined model structure, which result in the species never being detected at a unit.

Formally, the underlying random variables of this model can be defined as:

$$z_i \sim Bernoulli(\psi),$$
$$y_{ij}|z_i \sim Bernoulli(\theta_j),$$
$$h_{ijd}|y_{ij}, z_i \sim Bernoulli(p_j^{[d]}),$$

where z_i is the latent random variable for presence/absence of the species at unit i, y_{ij} is the latent random variable for the intermediate occupancy process (presently, availability of the species for detection at survey occasion j, but may represent sub-unit occupancy with spatial intermediate sampling), and h_{ijd} is the detection/nondetection of the species in each survey at the lowest level of the sampling.

The observed data likelihood is the product of probabilities corresponding to the detection histories for the set of sample units surveyed in the study:

$$ODL(\psi, \theta, \mathbf{p}|\mathbf{h}) = \prod_{i=1}^{s} Pr(\mathbf{h}_i),$$

where bold font denotes vectors and matrices of parameters and detection data. The complete data likelihood can be constructed in terms of the latent and observed random variables (Chapters 3 and 4).

All probabilities can be modeled as functions of unit-specific covariates via appropriate link functions. Probabilities of presence at the intermediate sampling scale (θ) and detection can also be modeled as functions of covariates associated with the intermediate scale (i.e., varying over the index j) that potentially vary across units. For example, in the context within which the multi-scale model has been developed above, rainfall in the 24 hours preceding survey occasion j might be considered as a covariate on both the θ and p parameter types.

Where the intermediate sampling is of a spatial nature (e.g., Pavlacky et al., 2012), micro-habitat features around each sub-unit could be potential covariates for θ and p. Finally, detection probabilities could also be modeled with covariates whose values vary at that lowest scale of the sampling, e.g., device type, observer, date, etc. The efficacy of different models can be assessed via likelihood ratio tests or model selection (Chapter 3). For example, inference about the utility of different detection devices might be based on one model with method-specific detection probabilities ($p_j^{[d]}$) and another with detection probabilities constant for all methods (p_j), with formal evaluation via model selection or a likelihood ratio test.

A useful aspect of the multi-scale occupancy model described above is the ability to account for missing observations or unequal sampling effort. For example, suppose that due to budgetary constraints it was not possible to deploy all device types at all units. At units where all device types were deployed, complete detection histories (for all survey occasions and all devices) would be obtained. However, at a unit where device 2 was not deployed, the resulting detection history could be represented as $\mathbf{h}_i = 0\text{-}1 \ 0\text{-}0$ with the associated probability statement being modified as for the basic occupancy model (i.e., $p_1^{[2]}$ and $p_2^{[2]}$ would be omitted from the probability statement; Chapter 4). Furthermore, a device may not be deployed for the entire season at a unit (e.g., the battery may fail on a camera trap after the first survey) so some detection histories may be similar to $\mathbf{h}_i = 101 \ \text{-}01$, indicating that device 1 was not active during survey occasion 2. This ability provides a great deal of flexibility with this modeling to accommodate many of the practical realities of field work.

This model described above has been incorporated into Programs PRESENCE (Hines, 2006) and MARK (White and Burnham, 1999) for easy use. We note that the model can also be viewed as a special kind of multi-season model (Chapter 8) in which each survey period is viewed as a season and each device is viewed as a survey within season, although unit-level occupancy is unchanged across all seasons with random changes in seasonal occupancy.

6.3.2   Example: Striped Skunks

The above modeling was applied to data on striped skunks (*Mephitis mephitis*) collected via surveys for carnivores conducted in 2004 at eight National Parks in the northeastern United States (Fig. 6.6; see Nichols et al., 2008). The parks range in size from 24 to 1378 ha, and sample units were selected using a stratified (based on vegetation categories), systematic sampling design with two levels of randomization (Gilbert et al., 2008; Nichols et al., 2008). Three kinds of detection device ($D = 3$) were used to detect mammals at a ran-

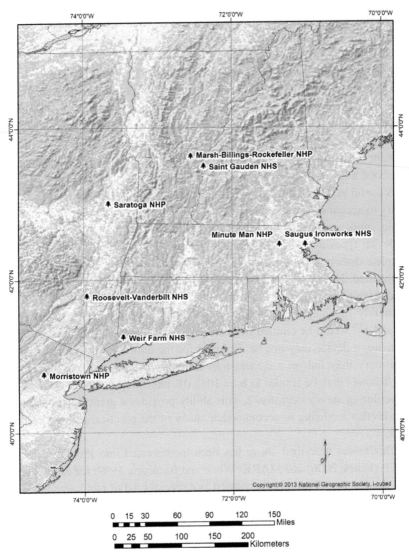

FIGURE 6.6 Map showing eight national parks in the northeastern United States where striped skunk data were collected. *Source: Nichols et al. (2008).*

domly selected sampling station within each sample unit: remote cameras with infrared sensors (Trailmaster®), hair removal traps (Mowat and Paetkau, 2002), and enclosed track plates (i.e., cubby boxes; Barrett, 1983; Zielinski and Kucera, 1995). The detection devices were arranged in a circular array at the sampling station (Fig. 6.7), and a small amount of bait and a generic scent lure (Cronks Outdoor Supplies, Wiscasset, ME) were applied at each station. Detections from

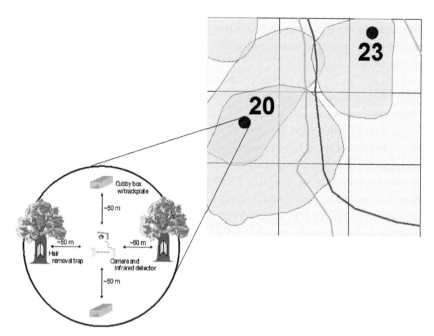

FIGURE 6.7 Diagram representing hypothetical ranges for four skunks distributed among nine sample units. Two of the sample units (units 20 and 23) were selected for sampling and a detection array (sample station location) was randomly placed within each of these sample units. Each detection array consisted of a single camera at the center and two track plates and two hair removal traps placed equidistant (~50 m) from the camera in the four cardinal directions. *From Nichols et al. (2008).*

the two track plates were combined, as were detections from the two hair traps at each station.

Each of 47 sample units was sampled continuously for approximately two weeks in each of two seasons, the winter/spring and the summer/fall. Sampling stations were visited every three days (on average) during each 2-week period, comprising five survey occasions. For any given sample unit, the two sampling seasons were separated by 4–5 months, with all sampling occurring between January–November 2004. Striped skunks were detected at six units in the winter/spring and at seven units in the summer/fall.

Initial modeling provided evidence of changes in occupancy between the two seasons. Here, we apply the above single-season, multi-scale model separately to winter/spring and summer/fall detection data. A single unit-specific covariate, percent overstory cover, was investigated as a possible determinant of sample unit occupancy (ψ). These occupancy probabilities were expected to be either constant across sample units, denoted as $\psi(\cdot)$, or negatively associated with forest overstory cover, $\psi(over)$. Detection probability was modeled as constant over survey occasions and detection methods, $p(\cdot)$, as survey-independent but different among device type, $p(dev.)$, as survey-dependent but constant among device types, $p(survey)$, or with device type as an additive effect (on the logit link scale) with survey occasion, $p(dev. + survey)$. Small-scale occupancy, θ, was modeled as either survey-independent, $\theta(\cdot)$, or survey-dependent, $\theta(survey)$. In general, skunks are more easily detected or captured during the colder months when food is scarce (Bailey, 1971; Hackett et al., 2007). Thus, we predicted that $\hat{\theta}$ might be lower in the winter/spring than in the summer/fall, but relatively constant within seasons. Models were fit and maximum likelihood estimates obtained using Program PRESENCE (Hines, 2006). In cases where there was no clear 'best' model, model-averaged estimates (Buckland et al., 1997) were computed for parameters of interest.

Estimated large-scale probabilities of sample unit occupancy, ψ, were low and consistent between the two seasons, $\hat{\psi} = 0.16$. Model selection statistics (Table 6.8) provided only limited evidence that occupancy probabilities were influenced by percent overstory cover during winter/spring (summed weights for $\psi(over)$ models $= 0.18$), but somewhat more evidence in the summer/fall (combined model weight $= 0.41$). As predicted, models that did contain the overstory covariate suggested a negative relationship with striped skunk occupancy. Point estimates of conditional (on large-scale occupancy) small-scale occupancy, $\hat{\theta}_j$, were consistent within seasons (little evidence of time variation) and higher in the summer/fall (model averaged $\hat{\theta} = 0.67$, $\hat{SE}(\hat{\theta}) = 0.31$) than winter/spring (model averaged $\hat{\theta} = 0.52$, $\hat{SE}(\hat{\theta}) = 0.20$), consistent with *a priori* expectations of greater local movement during the winter/spring. However, the standard error estimates are large and do not permit strong inferences about seasonal variation in θ.

In both seasons there was strong evidence of detection probability differences among the three detection methods (Tables 6.8 and 6.9). Hair removal traps were consistently poor at detecting skunks with model-averaged detection probabilities ≤ 0.10 in both seasons (Table 6.9). Cameras performed better, with slightly higher point estimates in the summer/fall (model averaged $\hat{p} = 0.35$, $\hat{SE}(\hat{p}) = 0.19$) than the winter/spring (model averaged $\hat{p} = 0.25$, $\hat{SE}(\hat{p}) = 0.14$, Table 6.9). Cubby boxes were 3.5 times more likely to detect striped skunks in the winter/spring than in the summer/fall (model averaged: winter/spring $\hat{p} = 0.73$, $\hat{SE}(\hat{p}) = 0.23$; summer/fall $\hat{p} = 0.20$, $\hat{SE}(\hat{p}) = 0.10$).

We point out that in this example, Nichols et al. (2008) used AIC with the small-sample correction (AIC_C) for model selection, where the effective sample size used in the correction was the number of units surveyed. While this may be reasonable with respect to occupancy estimation, there are additional data that have been collected in order to estimate the other parameters in the model. Therefore, as we have noted previously with respect to AIC_C, there is

TABLE 6.8 Summary of model selection statistics for the top five models for striped skunk data (*Mephitis mephitis*) from winter/spring and summer/fall seasons 2004. *Npar* represents the number of parameters in the model and −2*l* is twice the negative log-likelihood value. Small sample Akaike Information Criteria were calculated for each model, conservatively using the number of sample units as the effective sample size ($s = 47$). Relative AIC_c values (ΔAIC_c) and AIC model weight, *w*, are reported for each model. Detection probabilities may vary among device type (*dev*) or survey occasion (*survey*); occupancy may be modeled as a function of percentage of overstory cover (*over*) associated with each sample unit

Season	Model	*Npar*	−2*l*	ΔAIC_c	*w*
Winter/Spring	$\psi(\cdot)\,\theta(\cdot)\,p(dev)$	5	94.93	0.00	0.67
	$\psi(over)\,\theta(\cdot)\,p(dev)$	6	94.92	2.63	0.18
	$\psi(\cdot)\,\theta(\cdot)\,p(dev+survey)$	9	88.46	4.92	0.06
	$\psi(\cdot)\,\theta(survey)\,p(dev)$	9	89.04	5.50	[a]0.04
	$\psi(\cdot)\,\theta(\cdot)\,p(\cdot)$	3	106.74	6.90	0.02
Summer/Fall	$\psi(\cdot)\,\theta(\cdot)\,p(dev)$	5	123.63	0.00	0.44
	$\psi(over)\,\theta(\cdot)\,p(dev)$	6	121.69	0.71	0.31
	$\psi(\cdot)\,\theta(\cdot)\,p(\cdot)$	3	131.09	2.56	0.12
	$\psi(over)\,\theta(\cdot)\,p(\cdot)$	4	129.15	3.01	0.10
	$\psi(\cdot)\,\theta(survey)\,p(dev)$	9	120.05	7.82	[a]0.01

[a] *When reporting model averaged estimates of θ, the medians of the $\hat{\theta}_j$ estimates were used ($j = 5$ survey occasions).*
Source: Nichols et al. (2008).

uncertainty for occupancy models about how one should exactly determine the effective sample size. The number of surveyed units is one option, but likely the most conservative option in the sense that it will result in a greater penalty on more complex models (hence they will be ranked lower) than using a larger value for the effective sample size.

6.3.3 Discussion

While we have found the multi-scale occupancy model to be a useful framework for making inferences about a broad variety of ecological systems when data have been collected with three levels of information, we do not advocate that this modeling framework should always be used in such cases. Depending on the goals of the study it may be perfectly acceptable to aggregate the data in some manner to have only two levels of information and base our inferences on simpler types of models. For example, if the rationale for deployment of multiple devices is simply to detect multiple species or life stages that cannot all

TABLE 6.9 Striped skunk detection probability estimates $\hat{p}$ and associated standard errors (in parentheses) for the top multi-scale models for winter/spring and summer/fall seasons. The three detection methods include: cameras (Cam), hair removal traps (Hair), and cubby boxes with track plates (Cub). w is the Akaike weight for each model

Season	Model	w	$\hat{p}$(Cam) (SE)	$\hat{p}$(Hair) (SE)	$\hat{p}$(Cub) (SE)
Winter/ Spring	$\psi(\cdot)\,\theta(\cdot)\,p(dev)$	0.67	0.24 (0.14)	0.08 (0.08)	0.73 (0.23)
	$\psi(over)\,\theta(\cdot)\,p(dev)$	0.18	0.24 (0.14)	0.08 (0.08)	0.73 (0.23)
	$\psi(\cdot)\,\theta(\cdot)\,p(dev + survey)$[a]	0.06	0.29 (0.22)	0.10 (0.12)	0.80 (0.23)
	$\psi(\cdot)\,\theta(survey)\,p(dev)$	0.04	0.24 (0.14)	0.08 (0.08)	0.73 (0.23)
	$\psi(\cdot)\,\theta(\cdot)\,p(\cdot)$	0.02	0.25 (0.11)	0.25 (0.11)	0.25 (0.11)
Summer/ Fall	$\psi(\cdot)\,\theta(\cdot)\,p(dev)$	0.44	0.39 (0.18)	0.08 (0.06)	0.20 (0.11)
	$\psi(over)\,\theta(\cdot)\,p(dev)$	0.31	0.39 (0.18)	0.08 (0.06)	0.20 (0.11)
	$\psi(\cdot)\,\theta(\cdot)\,p(\cdot)$	0.12	0.19 (0.09)	0.19 (0.09)	0.19 (0.09)
	$\psi(over)\,\theta(\cdot)\,p(\cdot)$	0.10	0.19 (0.09)	0.19 (0.09)	0.19 (0.09)
	$\psi(\cdot)\,\theta(survey)\,p(dev)$	0.01	0.44 (0.14)	0.09 (0.06)	0.22 (0.10)

[a] *Parameter estimate reported is the median value among the five detection probability estimates for each device.*
Source: Nichols et al. (2008).

be sampled well with a single device, then detection data can simply be aggregated across all devices at a sampling station, and sample unit occupancy can be estimated in the usual manner. Detection histories resulting from such aggregation would simply indicate detection by at least one device or nondetection by all devices, e.g., the multi-scale detection history, $\mathbf{h}_i = 000\ 010$, would simply be rewritten as: $\mathbf{h}_i = 01$. The first three device-specific detection entries are collapsed to yield '0' for the first survey occasion (detected by no devices), whereas the second three entries are collapsed to yield '1' for the second survey occasion (detected by at least one device).

Analysis of data collapsed in this manner using standard single-season models (MacKenzie et al., 2002) provides unbiased estimates of ψ and p_j, where detection probability is in terms of the combination of methods or devices used (i.e., detected by at least one of the deployed methods). Detection probability obtained from such an analysis of aggregated device data can be viewed as the product of the probabilities of presence at the local sampling station and detection, conditional on presence at the local sampling station, $p_j = \theta_j \left(1 - \prod\limits_{d=1}^{D}\left(1 - p_j^{[d]}\right)\right)$. However, unlike the modeling described above, this approach using aggregated data does not readily permit the decomposition of overall detection probability into these two components (i.e., availability and detection given availability). Note that the above aggregation is not the only option for doing so. An alternative would be to aggregate across survey occasions rather than device type such that the simplified detection history has one value for each device type, e.g., simplify $\mathbf{h}_i = 000\ 010$ to $\mathbf{h}_i = 010$. Estimates of p_j from this form of aggregated data would be interpreted as the probability of detecting the species by each device type at some point during the season, and would therefore allow inferences about the efficacy of different methods to be made. Regardless of whether the multi-scale occupancy model is used with the full data, or simpler occupancy models with aggregated data, estimates of ψ and its associated standard error should be similar from all analyses.

Aggregation may not be advisable when there is substantial variation in the number of missing observations in the data set, as there will be variation in the amount of effort that is represented by the aggregated data. That is, some aggregated observations may represent five sub-survey observations, while others represent only two due to missing observations. Using the data at the finer resolution with the multi-scale model will better account for the varying levels of effort.

In a similar vein to there being multiple options for data aggregation, we point out that often there will be different options for how the full data may be formatted with each option allowing different sets of questions about the intermediate level of the sampling to be addressed. For example, consider a study where four camera traps are placed within each unit, and cameras are deployed for three weeks. Investigators have decided to regard each week as a survey occasion (hence a total of 12 surveys per unit) and wish to use the multi-scale occupancy model to analyze the data. The question is how should the data be formatted? One option would be to group the data by camera station, e.g., $\mathbf{h}_i = 010\ 110\ 000\ 111$. In this format the θ parameters would be interpreted as the probability of the focal species being present in the vicinity of a camera trap over the three-week period, given the species is present in the unit (i.e., a secondary spatial-scale occupancy), and p is the probability of detecting the species in a week at a camera trap location given it's present around the camera trap. Alternatively, the data could be formatted such that they are grouped by survey occasion, e.g., $\mathbf{h}_i = 0101 \ 1101 \ 0001$. Now θ should be interpreted as the probability the species is present somewhere in the unit during a week, given the focal species in present in the unit (i.e., weekly availability; a temporal-scale occupancy parameter), and p is the probability of detecting the species at a camera trap given it is present in the unit during that week. Note the subtle differences in the interpretation of the detection probabilities under the different formats. Clearly, the investigators should choose the format that corresponds to their questions of interest.

The discussion thus far has focused on occupancy at two scales, one at the level of the sample unit and one at the intermediate level within the sample unit. But of course scale-dependence need not be restricted to two scales but rather can be extended to several scales as dictated by sampling design and inference needs. McClintock et al. (2010c) focused on the use of occupancy modeling to inform spatial epidemiological problems and outlined a hierarchical approach to disease modeling. They considered the case of avian influenza in waterfowl populations. Their hierarchical consideration of scale included interest in infection within national wildlife refuges (the largest sample units), ponds within refuges, flocks of birds within ponds, individual birds within flocks, and even replicate blood or tissue samples within an individual (Figs. 6.8, 6.9). At each of these scales, they acknowledged the possibility of both false negatives (detection probability <1) and false positives (Figs. 6.8, 6.9). The modeling that McClintock et al. (2010c) outlined to deal with the multiple scales was relatively complex, but otherwise posed no real problems for implementation. We believe that occupancy modeling at multiple spatial and temporal scales will become increasingly important for a variety of ecological and epidemiological problems and predict this is a research area that will grow substantially in the next decade.

Finally, we note that $\theta_j = 1$ for all j is a limiting case where the multi-scale model with a total of $K \times D$ surveys is equivalent to the basic occupancy model detailed in Chapter 4 with $K \times D$ surveys. In this case, the species is everywhere at the intermediate level (i.e., always available for detection or in every spatial sub-unit depending on the application) hence the closure assumption is being met and all surveys could be considered as independent observations.

6.4 AUTOCORRELATED SURVEYS

In the previous section we introduced the concept of an intermediate random process between occupancy and detection (i.e., species availability for detec-

(A) **Field Sampling**

1. Ponds within refuges   2. Flocks within ponds   3. Individuals within flocks

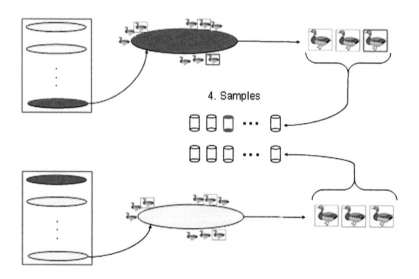

4. Samples

(B) **Laboratory**

1. Samples   2. Extraction

3. Plating

5. Result   4. RT-PCR

Well 1: ?
Well 2: ?
Well 3: ?

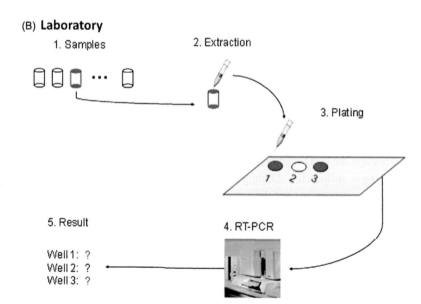

FIGURE 6.8 Conceptual diagram depicting different ways in which uncertainty can emerge in wildlife disease ecology (e.g., avian influenza in waterfowl populations), from the spatio-temporal allocation of field sampling effort (A) to laboratory practices (B). Red indicates infected samples and sample units. Whether or not a sample is infected, false negative or false positive test results can conceivably occur. *From McClintock et al. (2010c).*

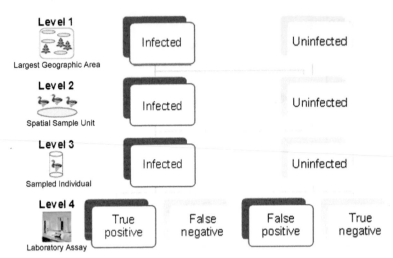

FIGURE 6.9 Hierarchical formulation of uncertainty in wildlife disease ecology under four levels or scales. Conditional on the disease state at the upper levels, many different sample paths can lead to a false negative or false positive result upon analysis at Level 4. Spatial subunits may be added or removed within Level 2 of the hierarchy. *From McClintock et al. (2010c).*

tion) and detailed how, with three levels of sampling, it was possible to estimate parameters associated with this secondary-scale occupancy. In this section we turn our attention back to situations with only two levels of sampling, but with an intermediate non-random process. For simplicity we refer to this intermediate process as *availability*, which may be of a spatial or temporal nature depending on the field methods used.

We have noted that when a free-ranging species is randomly available for detection, in the sense that there is a non-negligible probability that the species will occupy the unit on any particular survey occasion, single-season occupancy modeling is still useful. Although the closure assumption is violated in a strict sense, the occupancy modeling approaches outlined in Chapter 4 still provide useful inferences if we simply reinterpret the parameters. In the case of availability, ψ now becomes the probability that a unit is *used* by the focal species. The detection probability, p, should be interpreted as the product:

Pr(sample unit occupied by focal species at time of survey)

$\times\ Pr$(focal species detected | occupancy at time of survey).

If species availability is not random but Markovian, for example, then we may expect bias even in the reinterpreted parameter estimates. This non-random availability process is not only a violation of the strict closure assumption, but also causes a lack of independence between successive survey occasions that is another violation of the basic occupancy models.

Hines et al. (2010) developed a model to account for non-random availability motivated by a large-scale occupancy survey of tigers in Karnataka State, southwestern India. They modeled availability as a first-order Markov process, which simply means the probability of the species being available for detection in survey j, depends on whether the species was available (or not) for detection in survey $j - 1$. In the case of random availability, the probability of species availability in survey j is independent of its availability in survey $j - 1$. Hines et al. (2014) generalized the model of Hines et al. (2010) to the multi-season situation (Chapter 10), and presented a slightly modified version of the single-season model. The following is based on that modification, although using notation that is more consistent with that used elsewhere in the book.

6.4.1 Model Description

Let ψ be the probability of species presence at a unit (as previously) and p_j be the conditional probability of detecting the species in survey j given the species is available for detection at survey occasion j. Note the conditional definition of p_j which is slightly different than its definition in the simpler occupancy models. The following probabilities can be defined to model the availability process:

$\theta_1 = Pr$(species available in first survey);

$\theta_j^{[0]} = Pr$(species available in survey j | species not available in survey $j - 1$) for $j = 2, \ldots, K$;

$\theta_j^{[1]} = Pr$(species available in survey j | species available in survey $j - 1$) for $j = 2, \ldots, K$.

Note that the superscript terms on $\theta_j^{[0]}$ and $\theta_j^{[1]}$ indicate the availability status of the species in the previous survey occasion.

Consider the detection history $\mathbf{h}_i = 01110$ indicating that at unit i the focal species was not detected during survey 1, detected during surveys in 2–4, and not detected during survey 5. Using the simpler occupancy models (e.g., MacKenzie et al., 2002), there would be no ambiguity associated with this detection history as the only processes being modeled are unit-level occupancy and survey-level detection. In allowing the intermediate availability process there is now ambiguity associated with the first and last survey occasions; is the nondetection of the species due to the species being available but not detected, or was the species not available to be detected? This can be simply resolved within the probability statement for the history by summing together the probabilities of observing the various outcomes, as in the previous modeling. That is:

$$Pr(\mathbf{h}_i = 01110 \mid \psi, \boldsymbol{\theta}, \mathbf{p}) = \psi \begin{bmatrix} \theta_1(1-p_1)\,\theta_2^{[1]}\,p_2\theta_3^{[1]}\,p_3\theta_4^{[1]}\,p_4\theta_5^{[1]}(1-p_5) \\ +(1-\theta_1)\,\theta_2^{[0]}\,p_2\theta_3^{[1]}\,p_3\theta_4^{[1]}\,p_4\theta_5^{[1]}(1-p_5) \\ +\theta_1(1-p_1)\,\theta_2^{[1]}\,p_2\theta_3^{[1]}\,p_3\theta_4^{[1]}\,p_4\left(1-\theta_5^{[1]}\right) \\ +(1-\theta_1)\,\theta_2^{[0]}\,p_2\theta_3^{[1]}\,p_3\theta_4^{[1]}\,p_4\left(1-\theta_5^{[1]}\right) \end{bmatrix},$$

where parameters in bold font indicate vectors for that parameter type. Interpreting the probability statement, as the species was detected at least once during the surveys, the unit is known to be occupied, with probability ψ. The terms inside the square brackets represent the four possible options with respect to whether the species was available or not in the first and last surveys, conditional on the unit being occupied. For example, for the first term the species is available for detection in the first survey (θ_1) but not detected ($1-p_1$), and the species is always available in the remaining surveys given it was available in the previous survey ($\theta_j^{[1]}$). In the second term, the species is unavailable in the first survey ($1-\theta_1$; so no chance of being detected) and then becomes available for detection in survey occasion 2 given it was not available in survey 1 ($\theta_2^{[0]}$). The species then remains available for detection in the remaining surveys. The third and fourth terms are similar to the first two, although representing the possibility of the species being unavailable for detection in survey 5.

Expressions for the probability of any detection history can be determined in a similar manner, recognizing that any nondetection of the species may be due to a nondetection given the species was available during a survey occasion, or to the species being unavailable for detection in that survey. The probability statements are complicated by consecutive nondetections, as changes in availability must also be considered, but the expressions are tractable.

The model can also be expressed in terms of the underlying random variables:

$$z_i \sim Bernoulli(\psi),$$
$$y_{i1}|z_i \sim Bernoulli(z_i\theta_1),$$
$$y_{ij}|y_{ij-1} = m, z_i \sim Bernoulli\left(z_i\theta_j^{[m]}\right), \text{ for } j = 2,\ldots,K,$$
$$h_{ij}|y_{ij}, z_i \sim Bernoulli\left(y_{ij}z_i\,p_j\right).$$

The complete data likelihood could be derived from the defined random variables, or the observed data likelihood can be defined as:

$$ODL(\psi, \boldsymbol{\theta}, \mathbf{p}|\mathbf{h}) = \prod_{i=1}^{s} Pr(\mathbf{h}_i \mid \psi, \boldsymbol{\theta}, \mathbf{p}),$$

which can be used to obtain maximum likelihood estimates of parameters, or used within a Bayesian inferential framework to obtain posterior distributions of parameters.

The above development of the correlated detections occupancy model is very general in terms of both θ and p parameters being survey-specific. In practice, as there is only one piece of information per survey occasion, a constraint must be applied to at least one of these parameters to ensure the parameters are identifiable, e.g., constrain the parameter to be equal for all surveys or only vary by survey occasion according to a covariate relationship. If there were multiple pieces of information on the detection of the species each survey occasion (i.e., three levels of sampling as in the previous section), it would be possible to fit models with Markovian changes in availability, which would be a special case of the multi-season model of MacKenzie et al. (2003) (Chapters 8 and 10).

There is a range of constraint options associated with the estimation of θ_1, availability in the first survey. If p_1 is constrained to be equal to at least one other detection probability, then it should be possible to estimate θ_1 directly (Hines et al., 2014), although it may not always be estimable for a specific data set. θ_1 could be set equal to at least one other value of either $\theta_j^{[0]}$ or $\theta_j^{[1]}$ indicating that the analyst believes the probability of availability in the first survey is the same as either the probability of availability given the species was not available in the previous survey, or the probability of availability given the species was available in the previous survey. This option would be most reasonable when $\theta_j^{[0]}$ and $\theta_j^{[1]}$ are constrained to be time-invariant. A third option, which again would be most reasonable when $\theta_j^{[0]}$ and $\theta_j^{[1]}$ are time-invariant, is to apply the constraint:

$$\theta_1 = \frac{\theta^{[0]}}{\theta^{[0]} + (1 - \theta^{[1]})},$$

which uses the properties of Markov processes and represents a scenario where the first survey is randomly timed (or placed if using spatial replication) relative to the changes in availability. This would be reasonable assumption to make in many situations.

One constraint that cannot be made with the correlated detection occupancy model is to set $\theta_j^{[0]} = \theta_j^{[1]}$, indicating that the probability of the focal species being available in each survey is the same regardless of whether the species was available in the previous survey or not. That is, the species is available to be detected at random. While biologically reasonable, as noted above, it is not possible to estimate θ_j with just two levels of sampling. While it may seem strange that the parameters from a more complex model can be estimated while the parameters from a simpler model cannot, no one ever said that statistics had to be completely intuitive. In fact, many people have remarked to us over the years that statistics does not make any sense to them!

As with the previous modeling, the effect of potential covariates on the various probabilities can be estimated with this modeling framework through use of appropriate link functions. Missing observations and unequal sampling effort can also be easily accommodated using the same approach as detailed elsewhere in this book.

6.4.2 Example: Tigers on Trails

As mentioned above, a motivating application for Hines et al. (2010) was the monitoring of tigers in the 22,000 km^2 Malenad–Mysore Tiger Landscape in Karnataka State, southwestern India. For this monitoring program, a regular grid of 205 188-km^2 cells was overlaid across the region, and each cell was regarded as a sample unit. Tigers are known to use forest roads and trails as travel routes and to mark them intensively with tracks, scent, and scats (Karanth and Sunquist, 2000); therefore searches for tiger sign were conducted along such trails. The length of trail searched was made proportional to the amount of forest cover within each cell, with a target of 40 km to be searched in fully covered cells. Cells with less than 10% forest cover were not searched as they were deemed unlikely to shelter tigers. Therefore, between 4 and 42 km of trail were searched in each surveyed cell. The searched trails were divided into 1 km segments which were regarded as 'surveys' for detecting tiger presence in a cell, i.e., the monitoring program utilized spatial, rather than temporal, survey replication.

The field protocols were designed such that a survey team must pass through a randomly chosen point within each surveyed cell, and the starting point was typically within the forest interior (i.e., not the start of the trail on the forest edge). Given this design, Hines et al. (2010) decided to apply the third constraint listed above to θ_1 representing random tiger availability in the first segment searched (Markovian equilibrium).

Hines et al. (2010) fitted six different models to the tiger survey data representing different hypotheses about the nature of the availability and detection process, but for the purpose of this example we shall focus on the results of the correlated detection occupancy model with space-invariant θ's and p's, and no covariates on any of the parameters. Table 6.10 contains the estimated parameters for this model, which would suggest that the probability tiger are present in a randomly selected cell in this region ($\hat{\psi}$) is estimated to be 0.57. Compare this to the naive estimate (proportion of cells where tiger sign was detected) of 0.36, and the estimate from a regular occupancy model (with no covariates on occupancy or detection) of 0.41. Given tigers are present in the cell, the estimated probability of tigers being available for detection in the first 1 km segment ($\hat{\theta}_1$) is 0.26. For subsequent segments, if tigers were not available in the preceding

TABLE 6.10 Estimated probabilities from the correlated detection occupancy model fit to the tiger (*Panthera tigris*) survey data collected in Karnataka State, southwestern India. Associated standard errors (SE) are also given. Standard error is for $\hat{\theta}_1$ was calculated from the original analyses. Insufficient detail was included in Hines et al. (2010) for its calculation

Probability	Estimate	SE
ψ	0.57	0.077
θ_1	0.25	0.037
$\theta^{[0]}$	0.07	0.017
$\theta^{[1]}$	0.80	0.057
p	0.42	0.059

segment, the estimated probability of them becoming available for detection ($\hat{\theta}^{[0]}$) is 0.07, while the estimated probability of tigers being available given they were also available in the preceding segment ($\hat{\theta}^{[1]}$) is 0.80. That $\hat{\theta}^{[0]}$ and $\hat{\theta}^{[1]}$ were estimated to be so different is a clear indication of a non-random availability process, as equal values would indicate random changes in availability. Given tigers were available to be detected in a survey, the estimated probability of detecting tiger sign ($\hat{p}$) was 0.42. Detection probability from the regular occupancy model was estimated to be 0.13, although here 'detection' is a combination of tigers being available and being detected. As the correlated detection model separates out the availability component, the conditional (on being available) detection probability will be greater.

6.4.3 Discussion

In this section we have detailed an occupancy model that can be used to account for a correlated detection process, which will most likely be an issue when surveys are spaced close together (temporally or spatially) relative to the biology of the focal species. This correlation may be apparent in the detection history data as consistent pulses or groupings of 1's and 0's. The approach presented here models the correlation through Markovian changes in the availability of the species, although other approaches are possible. In addition to the Markovian model presented above, Hines et al. (2010) also explored approximating this process using a survey-specific covariate to allow the probability of detecting the species in survey j to be different if the species was, or was not, detected in survey $j - 1$. They found this approximation to perform better than the regular occupancy models, but occupancy estimates displayed a moderate negative bias when data was simulated with a Markovian availability process.

A different approach to addressing correlated detections is to aggregate surveys so that at the aggregated scale, survey outcomes are more independent of one another. In many practical situations where data are collected in an almost continuous manner (e.g., camera traps, sign surveys along a transect), and are then discretized into 'surveys' of an arbitrary length, the length used will affect the level of correlation in the detection data with shorter lengths typically resulting in a higher correlation for many species. Therefore aggregating shorter surveys, or using longer lengths to define a discrete 'survey' (e.g., in a 30-day camera-trap deployment define three 10-day 'surveys' rather than ten 3-day 'surveys') should reduce the level of correlation in the data and therefore reduce any bias in the estimates of occupancy from regular occupancy models. While there is some loss of information about the detection process, the pieces of information are not independent, so the loss may be outweighed by the benefits of more independent data. When continuous data have been collected, another alternative is to not discretize them and use the approach of Guillera-Arroita et al. (2011) which can also account for fine scale correlations or clusterings of the detections.

Successful identification of correlated detections will require a sufficient number of surveys per unit. We are not aware of any studies that have specifically addressed this issue, and what would be deemed 'sufficient' will depend on the parameter values, but we suspect that typically at least five surveys (if not more) will be required to reliably model any underlying correlation structure. Fewer than that and there will not be a sufficient number of surveys to observe 'pulses' of detections and nondetections. In fact, when there is a smaller number of surveys, and those surveys are correlated, that correlation may manifest itself in the data as a form of detection heterogeneity. There will be a greater number of detections at units that are surveyed during a period that coincides with a period when the species tends to be available, and fewer detections at units that are surveyed during a period when the species tends to be unavailable. Hence, the heterogeneity models discussed in the next chapter would be another option to reduce the bias in occupancy estimates without explicitly modeling the underlying correlation structure of the availability and detection process.

6.5   STAGGERED ENTRY-DEPARTURE MODEL

Another kind of violation of the closure assumption can occur when the focal species does not occupy the sample unit at either the beginning or the end of the season (i.e., either the early or late sample occasions). Consider migratory birds that are surveyed on the breeding grounds, for example. If we begin our surveys before the focal species has arrived on the breeding grounds, then the species will be absent from the sample unit for some number of occasions at the beginning of the study. Similarly, we may extend our sampling past the time when the species departs for nonbreeding areas. Both of these situations defy the single season model structure, as detection probability will be 0 for these early and late occasions, whereas the single-season model structure will model these nondetections instead as $(1-p_j)$. Kendall et al. (2013) developed a model to deal with this exact situation of possible species absences at the beginning and/or end of the study. Their model is based on the super-population approach of capture–recapture modeling (Crosbie and Manly, 1985; Schwarz and Arnason, 1996; Kendall and Bjorkland, 2001).

6.5.1 Model Description

Assume the standard sampling situation for single-season models with K survey occasions. Again, let ψ_i denote the probability that sample unit i is occupied at some time during the season or period of surveys. Let p_{ij} denote the detection probability for sample unit i and survey occasion j, conditional on the unit being occupied during occasion j. The staggered entry-departure model of Kendall et al. (2013) requires two new sets of parameters. For sample units that are occupied at some time during the season, let β_{ij} denote the probability that individuals of the focal species first enter sample unit i, between sampling occasions j and $j + 1$. Define β_{i0} as the probability that individuals of the species enter the sample unit prior to sampling occasion 1, and are thus available for detection during occasion 1. Thus:

$$\sum_{j=0}^{K-1} \beta_{ij} = 1.$$

The other new parameters are the d_{ij}, departure probabilities, each defined as the probability that unit i, that is occupied by the focal species at sample occasion j, is no longer occupied at sample occasion $j + 1$ (i.e., the species permanently departs from the unit).

The model structure can be illustrated by considering the probabilities associated with specific detection histories, again depicted as vectors of 1's and 0's indicating species detection and nondetection, respectively. Consider the probabilities associated with the following detection histories under this model.

$$Pr(\mathbf{h}_i = 101|\boldsymbol{\theta}) = \psi_i \beta_{i0} p_{i1} (1-d_{i1})(1-p_{i2})(1-d_{i2}) p_{i3},$$

$$Pr(\mathbf{h}_i = 011|\boldsymbol{\theta}) = \psi_i \left(\left[\beta_{i0}(1-p_{i1})(1-d_{i1}) + \beta_{i1} \right] p_{i2}(1-d_{i2}) p_{i3} \right).$$

The history $\mathbf{h}_i = 101$ admits no uncertainty, as we know that unit i was occupied at the initial sampling occasion, the species did not (permanently) depart from the unit between surveys 1 and 2, with no detection in survey occasion 2, did not depart the unit between surveys 2 and 3, and was detected in survey occasion 3. Detection history $\mathbf{h}_i = 011$ indicates that unit i was occupied at least during occasions 2 and 3 and detected at both occasions, but there is uncertainty associated with occasion 1, as the species could have been present yet undetected in occasion 1, or it entered between occasions 1 and 2. These possibilities are reflected by the two additive terms in square brackets.

Consider one additional history:

$$Pr(\mathbf{h}_i = 000|\mathbf{\theta}) = \psi_i\left(1 - p'_i\right) + (1 - \psi_i),$$

where

$$p'_i = \beta_{i0}p_{i1} + \left[\beta_{i0}(1 - p_{i1})(1 - d_{i1})p_{i2} + \beta_{i1}p_{i2}\right]$$
$$+ \left[\begin{array}{l}\beta_{i0}(1 - p_{i1})(1 - d_{i1})(1 - p_{i2})(1 - d_{i2})p_{i3} \\ +\beta_{i1}(1 - p_{i2})(1 - d_{i2})p_{i3} + \beta_{i2}p_{i3}\end{array}\right].$$

The derived parameter, p'_i, denotes the probability of being detected during at least one of the three sampling occasions at unit i, conditional on occupancy during at least one occasion. The initial term in the expression for p'_i is the probability of being detected at occasion 1. The term enclosed by the smaller set of square brackets corresponds to the probability of being first detected at occasion 2, and the term enclosed by the larger set of square brackets corresponds to the probability of being first detected at occasion 3.

Computer implementation, and perhaps understanding, are facilitated by reparameterizing the model as reflecting transitions from one occasion to the next among three possible states that define availability for detection: state 1 corresponds to the species not yet having arrived, hence unavailable for detection; state 2 indicates species present and available for detection; state 3 indicates that the unit was occupied at one time during the season but the species has now departed, and is hence unavailable for detection. Transition probabilities among states can be used to define a first order Markov process, with state transitions between occasions j and $j + 1$ conditional on state at occasion j. Departure probabilities, d_{ij}, were defined above as state-dependent transition probabilities, but the β_{ij} entry parameters were not. Thus, define state-dependent transition probabilities, e_{ij}, as the probability that a species enters and occupies the unit at occasion $j + 1$, given that the species has not occupied the unit at any sampling occasion prior to $j + 1$. These conditional entry probabilities are thus defined as:

$$e_{ij} = \frac{\beta_{ij}}{1 - \sum\limits_{v=0}^{j-1} \beta_{iv}},$$

for $j = 1, 2, \ldots K - 1$, such that $e_{iK-1} = 1$. For just prior to the initial sampling occasion, we define $e_{i0} = \beta_{i0}$ and $d_{i0} = 0$. These transition probabilities can be used to construct a state transition matrix, $\mathbf{D}_{ij}$, depicting transitions occurring from row state r to column state s:

$$\mathbf{D}_{ij} = \begin{bmatrix} 1-e_{ij} & e_{ij} & 0 \\ 0 & 1-d_{ij} & d_{ij} \\ 0 & 0 & 1 \end{bmatrix}.$$

State-specific diagonal matrices denoting probabilities of detection and non-detection can be defined as follows:

$$\mathbf{P}_{ij} = \begin{bmatrix} 0 & 0 & 0 \\ 0 & p_{ij} & 0 \\ 0 & 0 & 0 \end{bmatrix},$$

$$\mathbf{Q}_{ij} = \begin{bmatrix} 1 & 0 & 0 \\ 0 & 1-p_{ij} & 0 \\ 0 & 0 & 1 \end{bmatrix}.$$

Finally, an initial state vector indicates that prior to the survey the unit is always in state 1, not yet arrived:

$$\mathbf{n}_0 = \begin{bmatrix} 1 & 0 & 0 \end{bmatrix}.$$

Kendall et al. (2013) provided matrix notation for computing the probability corresponding to any detection history $\mathbf{h}_i$. For histories containing at least one detection, we can write:

$$Pr(\mathbf{h}_i | \boldsymbol{\theta}) = \psi_i \mathbf{n}_0 \left[\sum_{j=1}^{K} \mathbf{D}_{ij-1} (\mathbf{P}_{ij})^{h_{ij}} (\mathbf{Q}_{ij})^{(1-h_{ij})} \right] \mathbf{1}_3,$$

where h_{ij} is the detection/nondetection data for unit i, survey j. Note that $(\mathbf{P}_{ij})^0$ and $(\mathbf{Q}_{ij})^0$ are both equal to the identity matrix, and $\mathbf{1}_3$ is a column vector of 1's of dimension 3. For detection histories of all 0's:

$$Pr(\mathbf{h}_i | \boldsymbol{\theta}) = \psi_i \mathbf{n}_0 \left[\prod_{j=1}^{K} \mathbf{D}_{ij-1} \mathbf{Q}_{ij} \right] \mathbf{1}_3 + (1-\psi_i).$$

Estimation can be accomplished using maximum likelihood, as in Programs PRESENCE (Hines, 2006) or MARK (White and Burnham, 1999), or using Bayesian approaches. However, all parameters are not separately identifiable under the full model with time-specific parameters (also see Schwarz and Arnason, 1996; Kendall and Bjorkland, 2001), and some constraints are thus required for parameter estimation. β_{i0} and p_{i1} are confounded, but this can be resolved by imposing a constraint such as: $p_{i1} = p_{i2}$. Similarly, p_{iK} and d_{iK-1} can only be estimated as a product, leaving β_{iK-1} inestimable as well. This can be resolved by constraint also, for example, by setting $p_{iK-1} = p_{iK}$. Modeling these parameters as functions of covariates also resolves these identifiability problems. Finally, as with standard single-season models, unit-specific parameters cannot be estimated without unit-specific covariates or constraints (e.g., $\psi_i = \psi$). Other model assumptions are similar to those of standard single-season models.

The model of Kendall et al. (2013) permits only a single entry and exit. If a unit is randomly used by the focal species during the period between entry and exit, then the occupancy parameter reflects use during this period and the detection parameter now reflects the product of species availability and detection given availability. A small simulation study showed reasonable performance of this model, with fair power to reject the model of MacKenzie et al. (2002) (Chapter 4) in favor of this staggered entry-departure model (Kendall et al., 2013).

6.5.2   Example: Maryland Amphibians

Kendall et al. (2013) illustrated this model using surveys of amphibians for 33 wetland sites (units) located in the Chesapeake and Ohio Canal National Historic Park, Maryland. Wetland sites were sampled in April, May, June, and July, 2005. Two observers surveyed each wetland using visual encounter surveys in April and May, and dip nets in June and July. All surveys were designed to detect all life history stages: eggs, larvae, juveniles, and adults (see Mattfeldt et al., 2009, for more details). Data from both observers were combined such that detection could result from either or both observers detecting a species in any life stage during a survey. Kendall et al. (2013) fit their model to data from the spotted salamander (*Ambystoma maculatum*), American (*Bufo americanus*) and Fowler's (*Bufo fowleri*) toads, and the green frog (*Rana clamitans*). Detections of the toad species were combined (*Bufo* spp.), as field identification of tadpoles to species is difficult. Kendall et al. (2013) expected the arrival of each species to be staggered through the sampling period, in the order: *A. maculatum*, *Bufo* sp., and *R. clamitans*.

Data for each species were initially analyzed separately in Program PRESENCE using nine models, including seven open models (i.e., staggered entry-

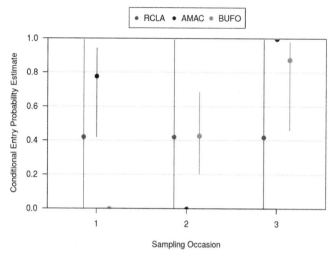

FIGURE 6.10 Estimates of probability of entry just prior to sampling occasion j, given no entry as of period $j-1$, $\hat{e}_{ij}$ (with 95% confidence intervals), from the top ranked model for each of three amphibian taxa: spotted salamanders (*Ambystoma maculatum*, AMAC), American (*Bufo americanus*) and Fowler's toads (*Bufo fowleri*) (BUFO), and green frogs (*Rana clamitans*, RCLA). Estimates on a boundary (0 or 1.0) do not permit estimation of standard error. Detection/nondetection was recorded during surveys of 33 wetland sites located in the Chesapeake and Ohio Canal National Historic Park, Maryland, during April, May, June, and July 2005. *Reproduced from: Kendall et al. (2013).*

departure models) and two closed models (i.e., simpler occupancy models described in Chapter 4) as a means of testing for closure. The highest ranked models for all species were open, and model selection statistics provided strong evidence for lack of closure for *R. clamitans* and *A. maculatum*. For *A. maculatum*, the top open model and the closed model both yielded, $\hat{\psi} = 0.27$ (SE $= 0.08$). For *Bufo* sp., the occupancy estimate for the staggered entry-departure model ($\hat{\psi} = 0.42$; SE $= 0.09$) was lower that the estimate from the closed model ($\hat{\psi} = 0.46$; SE $= 0.10$). For *R. clamitans*, estimates of occupancy were $\hat{\psi} = 0.61$ (SE $= 0.09$) for the open model and $\hat{\psi} = 0.62$ (SE $= 0.09$) for the closed model. For each species and survey occasion, detection probability from the staggered entry-departure model was estimated at 1.0, and these high detection probabilities were likely responsible for the relatively small differences between occupancy estimates based on open and closed models. For conditional entry probabilities, the temporal pattern was consistent with predictions by Kendall et al. (2013) for *A. maculatum* and *Bufo* sp. (Fig. 6.10). For *A. maculatum*, the estimated probability that individuals were available in the first survey occasion at a given unit was 0.78 (SE $= 0.14$). For *Bufo* sp., no individuals were available in the first survey occasion, but the probability that

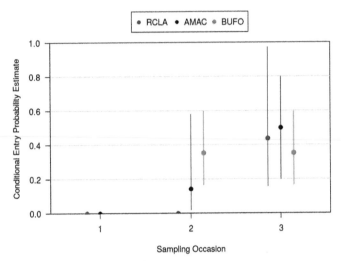

FIGURE 6.11 Departure probabilities, $\hat{d}_{ij}$ (with 95% confidence intervals) from the top ranked model for each of three amphibian taxa: spotted salamanders (*Ambystoma maculatum*, AMAC), American (*Bufo americanus*) and Fowler's toads (*Bufo fowleri*) (BUFO), and green frogs (*Rana clamitans*, RCLA). Detection/nondetection of these taxa was recorded during surveys of 33 wetland sites located in the Chesapeake and Ohio Canal National Historic Park, Maryland, during April, May, June, and July 2005. *Reproduced from: Kendall et al. (2013).*

they were available by the second occasion was estimated to be 0.43 (SE = 0.13), with the remainder available in the third occasion. Contrary to predictions, *R. clamitans* did not appear to be the last of the three species to become available, but instead showed a constant conditional probability of availability estimated at 0.42 (SE = 0.74) for survey occasions 1–3, but note the poor precision in these estimates (Fig. 6.10). For *A. maculatum* and *R. clamitans*, departure probability was low early, but subsequently rose (Fig. 6.11). Departure probability for *Bufo* sp. following initial entry at survey occasion 2 was higher than that of other species, indicating shorter residence time for *Bufo* sp. than for the other taxa.

6.5.3 Discussion

The staggered entry-departure model of Kendall et al. (2013) was initially motivated by a desire to relax the closure assumption of single season occupancy models. Stated differently, it deals with a specific kind of nonrandom pattern of temporal availability for detection within single-season designs. In this sense, the model can be viewed as analogous to the correlated detections model of Hines et al. (2010) (previous section) that deals with a specific kind of nonrandom pattern of spatial availability. In addition to relaxing the closure assumption, the model of Kendall et al. (2013) is well-suited to addressing questions about phenology. For example, under some climate change scenarios, we may expect earlier and earlier arrival times of birds on breeding grounds, and estimated entry probabilities over time can be used to formally test these ideas. Direct modeling of entry probabilities as functions of environmental covariates is also possible.

Use of the staggered entry-departure model (Kendall et al., 2013) requires some design considerations. For example, with this model, a minimum of three survey occasions is required, as opposed to two for standard closed models. To model time variation in parameters, at least four survey occasions are required. If study focus is on phenology, then for some questions the field season would need to start sufficiently early and cease sufficiently late to enable detection of the first arrivals and last departures at all units.

6.6 SPATIAL AUTOCORRELATION IN OCCURRENCE

Spatial autocorrelation occurs when the occupancy probability of a focal unit depends upon the occupancy status of 'nearby' units, where 'nearby' is defined relative to the scale at which units are defined. This creates a pattern where the occupancy status at nearby units tends to be similar. This is undoubtedly a biological reality in many situations, particularly in a contiguous environment, but one we have largely ignored so far, as do many other methods and computer programs used for modeling species distributions (e.g., MaxEnt; Phillips et al., 2006). As noted in Chapter 4, spatial autocorrelation in itself does not necessarily invalidate these methods, as the independence assumption relates to the observation process rather than the underlying occupancy status of units, which can be accounted for through appropriate design of the study (Chapter 11).

This can be easily demonstrated with a simple thought experiment. Consider the population of interest in Fig. 6.12, which clearly has a high degree of spatial autocorrelation, and suppose that we wish to estimate the proportion of occupied units. Ignoring detection issues for a moment, if units are randomly sampled from the population, then sampling theory guarantees that using the regular estimator for a binomial proportion will provide an unbiased estimate with an appropriate standard error. The reason for this is that by sampling randomly, the observation from each sampled unit is independent of the observation from all other sampled units because each unit was selected independently of the others. Essentially, by using random sampling the population of interest is broken up into individual units, placed into a (large) metaphorical hat, mixed, then units are drawn out one at a time. This process means that it is not necessary to account for spatial correlation in our inference about the proportion of occupied units. Compare this to a scenario where instead of complete random sampling

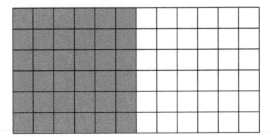

FIGURE 6.12   A hypothetical population of sample units (grid cells) exhibiting extreme spatial autocorrelation of species presence. Shaded cells indicate cells occupied by the target species.

of units, vertices on the grid would be randomly selected and observations taken from the four adjoining units. Such a design might be considered in practice for convenience, to reduce travel time between sampled units. However, now the observations from each unit are not independent of the other surveyed units and spatial autocorrelation is likely to mean that the observations from the four units in each cluster are very similar. While the regular binomial proportion estimator would still provide unbiased estimates, the standard error would overstate the precision of the estimate, as the effective number of independent samples will be closer to the number of vertices selected rather than the number of units surveyed.

Hence, study design is vitally important in deciding whether we need to account for spatial autocorrelation with our inferential methods. When a design is used such that units are selected from the population independently of other units (e.g., random sampling) then no adjustments are required to obtain appropriate estimates of the proportion of occupied units in the population. If a design is used where units are not selected independently of each other (e.g., cluster sampling) then modifications will be needed to account for the spatial autocorrelation (e.g., Pacifici et al., 2016).

Conversely, when interest lies at the level of individual units, e.g., to predict the occupancy status of individual units or to create maps of the species distribution, then it may be advantageous to exploit any spatial autocorrelation to provide more accurate unit-level inferences, even if surveyed units were selected independently of each other. In the remainder of this section we discuss some options that could be used, although we point out that using the complete data likelihood formulation of the model and MCMC, it should be possible to implement any statistical method that has been developed for modeling spatial autocorrelation. While we focus on accounting for spatial autocorrelation in occupancy, we note that there could also be spatial correlation in detection that could be dealt with using similar techniques. Finally, while in the following topics we tend to refer to units defined on a regular grid, this is not a fundamental requirement for these methods to be applied. However, the location of all units within the population of interest (surveyed and unsurveyed) must be known.

6.6.1 Covariates

One of the simplest approaches to dealing with spatial autocorrelation in occupancy is through covariates; the reason the species aggregates in some areas, and not in others, is that those areas contain a preferred habitat type, a particular substrate, or lie within a certain elevation range. Therefore, some forms of spatial autocorrelation can be suitably addressed through inclusion of predictor variables that are based upon the characteristics of each sample unit, in a model (Chapter 4). That the covariate values themselves are spatially autocorrelated is largely immaterial as they are treated as fixed, known values in an analysis.

In addition to predictor variables that are based on the physical characteristics of a unit that have a more functional biological relationship with species occurrence, other types of variables could also be considered as a general 'mopping up' of spatial autocorrelation. For example, distance from areas of high species activity (e.g., roosting sites) or distance from center of species known range. Cartesian coordinates (e.g., longitude and latitude, or UTM coordinates) could also be used to allow for a gradual change in occurrence probabilities across the area of interest. While it is a general consideration whenever covariates are being included in a model, particularly in this case polynomial or other transformations of covariate values may be appropriate. For coordinate-like covariates, bivariate splines and similar techniques allow the fitting of a flexible surface for occupancy probabilities as in a GAM (e.g., Webb et al., 2014).

6.6.2 Conditional Auto-Regressive Model

Magoun et al. (2007) used a conditional auto-regressive (CAR) model to account for spatial autocorrelation of wolverine (*Gulo gulo*) occurrence in boreal forest in northwestern Ontario, Canada, from aerial surveys for wolverine tracks. The underlying conceptual model relating occupancy and detection is the same as described elsewhere (e.g., Chapter 4), so here we focus on the details of incorporating the CAR structure on occurrence probabilities.

The probability of the target species occurring in unit i could be modeled as:

$$logit(\psi_i) = X_i \boldsymbol{\beta} + \sigma g_i,$$

where X_i is a vector of covariate values for the unit (likely including a 1 for the intercept term), $\boldsymbol{\beta}$ is the vector of regression coefficients to be estimated, σ is a scaling factor for the effect of the CAR component, and g_i is a latent (i.e., unknown) continuously-valued random variable, with $\boldsymbol{g}$ denoting the vector of values for all units. While we have defined ψ with the logit-link above, Johnson et al. (2013) found that using the probit-link provided several benefits in this situation, including computational efficiency and greater flexibility. The probit-link function uses the cumulative distribution of the standard normal distribution to transform values from the 0–1 scale to $\pm\infty$ (although most transformed values will be between ±3). Importantly, use of the probit-link does not require any additional assumptions over use of the logit-link; it is simply a different function for mapping values from the real scale to the probability scale. The CAR model is defined in terms of the conditional normal distribution:

$$g_i | \boldsymbol{g}_{-i} \sim N(\mu_i, \tau_i^2),$$

where $\boldsymbol{g}_{-i}$ denotes the vector of the CAR terms *excluding* the value for unit i, with:

$$\mu_i = \frac{\rho}{|n(i)|} \sum_{l \in n(i)} w_{il} g_l$$

and

$$\tau_i^2 = \frac{1}{|n(i)|}.$$

The term $n(i)$ denotes the set of neighboring units to unit i, and $|n(i)|$ is the number of neighbors. Note that the neighborhood of a unit could be defined in a variety of ways. Magoun et al. (2007) used a grid of hexagonal cells and defined a neighbor as any unit that shared a common border with the focal unit. When using a grid with square cells, a neighborhood could be defined as the units in the four cardinal directions adjoining the focal unit, or as the surrounding eight cells. Alternatively, all units within a certain radius of a focal cell might be considered its neighbors. Given this flexibility, the essential point is that a neighborhood should be defined in a biologically reasonable manner. One relevant point about neighborhood definition is that on the edge of the area of interest the number of units in the neighborhood will be smaller. The contribution of each g_l value to μ_i may be weighted (w_{il}; e.g., as a function of distance between units i and l), but $w_{il} = 1$ is common. The ρ parameter is a measure of spatial correlation, with a value between -1 and 1. Typically positive correlation would be expected so it may often be reasonable to restrict the values of ρ to between 0 and 1. One can also fix $\rho = 1$ which is known as an *intrinsic* CAR (e.g., Rue and Held, 2005; Aing et al., 2011; Johnson et al., 2013) which can be more computationally efficient (Johnson et al., 2013). Finally, note that a value of $\rho = 0$ implies that the values of the random variable $\boldsymbol{g}$ are independent for all units, i.e., a simple random error term. This highlights the conceptual basis of g, that it is a combination of several spatially correlated factors that influence species occurrence that have not been included in the model.

The CAR model is easily implemented with general-purposed MCMC software such as OpenBUGS. Magoun et al. (2007) used the `car.proper` distribution although an alternative is to use an intrinsic CAR distribution (`car.normal` or `car.l1` in OpenBUGS). The CAR model is also implemented within a specific occupancy modeling framework in the R packages `stocc` and `hSDM`. Examples of applying the CAR model with single-season occupancy models, while accounting for detection, include Magoun et al. (2007), Gardner et al. (2010), Aing et al. (2011), Johnson et al. (2013), Broms et al. (2014).

6.6.3 Autologistic Model

The autologistic model can be considered as a type of CAR model and described as:

$$logit(\psi_i) = X_i\beta + \beta_g g_i,$$

where X_i and β are defined as above, β_g is the coefficient associated with the spatial term, and g_i is a spatial term that depends upon the occupancy state of neighboring unit. Autologistic models can be considered as somewhat more mechanistic than more general CAR models in their focus on the occupancy status (e.g., as sources of potential immigrants) of neighborhood units. The autologistic covariate, g_i, could be defined in a number of different ways, depending on the context of the application, e.g.,

$$g_i = \frac{1}{|n(i)|} \sum_{l \in n(i)} z_l,$$

or

$$g_i = \min_l d_{il} z_l,$$

where d_{il} is the distance between units i and l. In this latter case g_i is defined as the distance to the nearest occupied neighbor, as is common with metapopulation models (e.g., Hanski, 1994a). Note that the neighbor occupancy statistic is not a standard covariate in the sense that occupancy of neighbor patches is frequently unknown, just as for the focal patch. The main distinction between the CAR and autologistic models is that in the former, g_i is a continuously-valued, spatially correlated random effect, while in the latter g_i is a spatial-correlated value that depends directly on the occupancy status of neighboring units.

Sargeant et al. (2005) used the autologistic model, in the context of an image restoration model, to examine the distribution of the swift fox (*Vulpes velox*).

While Sargeant et al. (2005) drew a distinction between their approach and "detection-history approaches", i.e., the focus of this book, the image restoration model used by Sargeant et al. (2005) included separate components for the occurrence of the species and the number of detections in a unit with allowance that swift fox may be undetected even when present, in exactly the manner as "detection-history" models.

6.6.4 Kriging

Kriging is a widely-used geostatistical method for creating smooth surfaces by interpolating the value of the quantity of interest between points at which it has been measured, based upon the distance between points. Kriging could be used in an occupancy model to account for spatial autocorrelation in a similar manner to the CAR model. That is, let

$$logit(\psi_i) = X_i\boldsymbol{\beta} + g_i,$$

although now g are random values from a multivariate normal distribution with mean values of zero and covariance matrix $\boldsymbol{\Sigma}$. Formally,

$$g \sim MVN(\mathbf{0}, \boldsymbol{\Sigma}),$$

where $\boldsymbol{\Sigma} = \sigma^2\boldsymbol{\eta}$, with σ^2 being the variance and $\boldsymbol{\eta}$ the correlation matrix of the random effect values. In Kriging the correlation between random effect values is assumed to be a decaying function of distance, e.g.,

$$\eta_{il} = e^{-\theta d_{il}}$$

where θ would be the exponential decay rate. Note the Kriging model could also be expressed as:

$$logit(\psi_i) = X_i\boldsymbol{\beta} + \sigma g_i,$$

with

$$g \sim MVN(\mathbf{0}, \boldsymbol{\eta}),$$

which is more similar with how the CAR model is defined above. As for the CAR model, the probit-link function could be used as an alternative to the logit-link. One of the main drawbacks of using Kriging is that it is computationally intensive because of the number of calculations required compared to using a CAR, thus MCMC analyses can take a much longer time to run.

Pacifici et al. (2016) used Kriging to interpolate occupancy at unsurveyed locations in the study of a rare plant species (*Tamarix ramosissima*) found in

China. This study also used an adaptive cluster sampling design (Thompson, 2002). The Matern correlation function (Cressie, 1993) was used to model the correlation between the random effects at different units, and like Johnson et al. (2013), Pacifici et al. (2016) used the probit link function. They demonstrated by simulation that with adaptive cluster sampling (where an initial set of units is surveyed and if the species is detected, neighboring units are also surveyed), not accounting for spatial autocorrelation in the analysis led to a positive bias in estimated occupancy. This is due to the increased survey effort around units where the species was detected, which with a spatially correlated population, are also more likely to be occupied. Accounting for spatial autocorrelation greatly reduced the level of bias particularly for moderate sample sizes ($s > 50$) for the scenarios they considered. Pacifici et al. (2016) also demonstrated that with a spatially autocorrelated population, applying the basic occupancy models to data simulated from a random sample from the population provided generally unbiased estimates, except when occupancy and/or detection were low (0.1 and 0.25 respectively in their scenarios), and sample size was 20. In these scenarios the data are likely to be very sparse (when $s = 20$ and $\psi = 0.1$ then the expected number of occupied units is 2!) so it should not be expected for occupancy models to work well. These results do, however, confirm our earlier claim that unbiased occupancy estimates can be obtained with random sampling from a population that is spatially autocorrelated.

6.6.5 Restricted Spatial Regression

One issue that is common to all of the above approaches is the potential confounding of the spatial random effect with fixed effect covariates that themselves are spatially correlated (e.g., vegetation or elevation). While the confounding may not be problematic if the goal of the modeling is to provide a good descriptive model, if the intent is to gain an understanding of the system ecology, then such confounding is problematic, as estimated effect sizes may vary widely depending on whether spatial autocorrelation is included in the model. Restricted spatial regression (RSR) is a technique whereby the spatial random effect is constructed such that it does not use the spatial information in the covariate value, therefore removing the confounding.

Johnson et al. (2013) applied RSR within an occupancy framework with an *intrinsic* CAR (ICAR) spatial model, based on the approach detailed by Hughes and Haran (2013). The following formal description is not for the statistical faint-hearted, but essentially, RSR involves reducing the dimension of the spatial random process down from an s-dimension process (one dimension for each unit) to a q-dimension process that is independent of the variation in the data that is explained by the predictor variables included in the model. The reduced dimension process is then mapped back into s-dimensions with a specially defined matrix U.

Formally, Johnson et al. (2013) express the ICAR model in matrix form as:

$$y = X\beta + \eta + \epsilon$$
$$= X\beta + U\alpha + \epsilon$$

where

$y = s \times 1$ vector of occurrence probabilities on a suitably transformed scale (e.g., using logit- or probit-link functions),

$X =$ the $s \times r$ matrix of fixed-effect covariate values (typically including a column of 1's for the intercept term),

$\beta =$ the $r \times 1$ vector of fixed-effect coefficients (i.e., covariate effect sizes),

$\epsilon =$ an $s \times 1$ vector of standard normal random error terms, $\sim N(\mathbf{0}, \mathbf{I})$,

$I =$ the $s \times s$ identity matrix,

$\eta =$ an $s \times 1$ vector of random effects with an (I)CAR distribution,

$U =$ the $s \times q$ spatial transformation matrix,

$\alpha =$ a $q \times 1$ vector of spatial random effects, $\sim N(\mathbf{0}, \tau U^T QU)$.

The matrix U has the property that $UU^T = I$, and Johnson et al. (2013) note that U could be constructed from the Moran operator matrix:

$$\Omega = \frac{s P^\perp A P^\perp}{\mathbf{1}^T A \mathbf{1}}$$

where $P^\perp = I - X(X^T X)^{-1} X^T$ is the projection matrix onto the residual space of X, A is the association matrix where the value in row i and column $l = 1$ if unit i and l are neighbors and $= 0$ otherwise, and $\mathbf{1}$ is an $s \times 1$ vector of 1's. Q is the CAR precision matrix defined as $Q = D - \rho A$, with D being a diagonal matrix where $D_{ii} = \sum_l A_{il}$ and ρ defined as above.

The Ω matrix has some special properties. The eigenvalues, λ_i, represent all the possible values for Moran's I statistic, a common measure of spatial autocorrelation, for a spatial process that is independent of X, given the association matrix. The corresponding eigenvectors, u_i, represent all the possible patterns of spatial dependence residual to X for an ICAR model with the association matrix. Therefore, the matrix U can be created from the set of eigenvectors $u_1, \ldots, u_q$ such that λ_q is greater than a user-specified level of autocorrelation. This results in a restricted ICAR model that is independent of the predictor variable information included in X. Note that if all of the eigenvectors of Ω were included in U, then the original ICAR model is obtained.

Johnson et al. (2013) demonstrated that using RSR, posterior distributions of the covariate effect sizes were similar to those obtained from a non-spatial model (i.e., using the methods of Chapter 4), but the posterior distributions for the effect sizes from an ICAR model were quite different, in particular exhibiting a much greater degree of uncertainty.

RSR can be performed in the R package stocc.

6.7 DISCUSSION

In this chapter we have presented a number of important extensions of the basic occupancy model detailed in Chapter 4 that were primarily developed to address assumption violations of these initial occupancy models. We make no claim that these extensions are the only, or even the best, way to address these assumption violations, and in all likelihood these extensions themselves will be extended to provide additional flexibility in the type of models that can be fit to appropriately collected data. These extensions enable researchers to investigate a broader range of questions about species occurrence patterns. However, in some cases these extensions have been developed, or applied, to overcome deficiencies in study designs where researchers have deviated from statistical ideals in the pursuit of 'logistically convenient' designs. While study designs always need to be logistically feasible, researchers need to appreciate that there are always consequences and associated tradeoffs whenever decisions on the design are made in the name of convenience. Frequently that tradeoff is a limitation on the scope of inference and often requires a more complex set of methods to provide suitable analysis of the data. In some cases, more complex modeling can be avoided through sound study design (Chapter 11).

In the next chapter methods are detailed to account for heterogeneous detection probabilities, which completes Part II of the book on single-season models, before moving into Part III of the book which focuses on multiple-season occupancy models.

Chapter 7

Modeling Heterogeneous Detection Probabilities

In previous chapters, we have formulated occupancy models under the assumption that detection probability (p) is constant among units for each occupancy state, or only varies in response to measurable covariates. Here, we consider the generalization of occupancy models that allows for heterogeneity in detection probability among units. It is natural to consider heterogeneous detection probability models because factors that influence detectability are many and varied, and it may not be possible to identify, much less control for, all of them. For example, variation in detection probability may be induced by covariates that affect detection but that were not measured and hence were omitted from the detection probability model. A crucial factor is that the data upon which occupancy models are based are typically observations of detection/nondetection of the species, and variation in unit-specific abundance of the species must surely affect the probability of detecting the species (i.e., detecting at least one individual). Detection of at least one member of the species will tend to be higher at units where abundance is high and low at units where abundance is low. Thus, one could only reasonably rule out abundance induced heterogeneity if abundance could be viewed as being relatively constant across occupied units, which we believe is unlikely to be the case in many animal sampling problems. The phenomenon of abundance-induced heterogeneity in detection probability is more likely to be important when sampled populations are small (e.g., < 10) and will diminish in importance as average population size becomes larger, in which case detection probability may be sufficiently well approximated as being constant. Nevertheless, heterogeneity may arise from other sources that vary at the level of the unit, and which are not accounted for explicitly in the model. In such cases, we might consider general models that accommodate latent heterogeneity.

The problem of unaccounted for heterogeneity in detection probabilities has been considered extensively in conventional capture–recapture models. Analogous to the effect of heterogeneity on closed population abundance estimators, unaccounted for heterogeneity yields biased estimates of unit occupancy (Royle and Nichols, 2003). The literature on modeling heterogeneity in the context of

Occupancy Estimation and Modeling. http://dx.doi.org/10.1016/B978-0-12-407197-1.00009-0

estimating the size of a closed population is vast (e.g., Burnham and Overton, 1978; Norris and Pollock, 1996; Coull and Agresti, 1999; Pledger, 2000; Fienberg et al., 1999; Dorazio and Royle, 2003; Pledger et al., 2003; Link, 2003). These existing approaches suppose that p varies by individual (p_i; where i denotes individuals in the capture–recapture context) and that each p_i is a realization of a random value from some distribution, the mixture or random effects distribution. Mixture distributions that have been used for p include discrete distributions (i.e., there is only a finite, usually small, number of values p could take; also known as a finite mixture), and continuous distributions such as the beta and the logit-normal (i.e., values are normally distributed on the logistic scale). These mixture models can be extended directly to the occupancy modeling framework.

Royle and Nichols (2003) suggested an alternative formulation of models for heterogeneity by exploiting the relationship between detection probability and abundance under binomial sampling. While this model is appealing because it admits an explicit linkage between abundance and occupancy, other (more standard) mixture distributions might also be considered such as those commonly employed in conventional capture–recapture studies. Such models do not require that parametric assumptions be made about abundance, and so might be viewed as being more general, as they encompass generic heterogeneity due to all mechanisms.

We provide (Section 7.1) a general formulation of occupancy models with heterogeneous detection probabilities that is based on the conventional view that p is a random variable, endowed with a probability distribution. We begin by considering the discrete or finite mixture distribution, then advance to continuous or infinite mixtures such as the beta and logit-normal. We also consider the Royle and Nichols (2003) model that arises by considering that heterogeneity in detection probability is derived from variation in abundance. We provide an example of fitting and interpreting these models to avian detection/nondetection data in Section 7.2. In Section 7.3 we generalize these basic heterogeneity models to allow for covariates that are thought to influence detection probability, and we provide an illustration using anuran calling survey data. Finally, we discuss the issue raised by Link (2003) in the context of estimating the population size of closed populations using analogous mixture models. He noted that different mixture distributions may produce equivalent (or nearly so) fits to the observed data, but vary widely in their estimates of population size. That is population size is not identifiable across mixture distributions, and we consider this problem within the context of occupancy models with heterogeneous detection probabilities.

7.1 OCCUPANCY MODELS WITH HETEROGENEOUS DETECTION

In this section we consider a class of occupancy model where detection probability may vary among units, but is constant in all other respects (i.e., detection probabilities do not vary in time, or in response to measurable covariates). Such models are analogs of 'Model M_h' used in classical abundance estimation problems (Otis et al., 1978; Williams et al., 2002). The incorporation of time or covariate effects is considered in Section 7.3. A key concept is that now there may be a (potentially infinite) number of possible values for the detection probability at each unit, and the likely value for a particular unit is not known *a priori*. Therefore, when formulating an expression for the probability of observing a particular detection history we must account for the fact that the detection probability could take a number of possible values.

7.1.1 General Formulation

Recall from Section 4.4, under the assumption that detection probability is constant for each survey, the number of detections at each unit (Y) is a zero-inflated binomial random variable. The probability of observing y_i ($= \sum_{j=1}^{K} h_{ij}$) detections at unit i for K surveys is therefore

$$Pr(Y = y_i | \psi, p) = \psi \binom{K}{y_i} p^{y_i} (1-p)^{K-y_i} \text{ if } y_i > 0$$

$$= \psi (1-p)^K + (1-\psi) \text{ if } y_i = 0.$$

Here we use an alternative form for expressing the zero-inflated binomial distribution that is more convenient,

$$Pr(Y = y_i | \psi, p) = \psi \binom{K}{y_i} p^{y_i} (1-p)^{K-y_i} + (1-\psi) I(y_i = 0) \qquad (7.1)$$

where $I(y_i = 0)$ is an indicator function that equals 1 if the species was not detected at the unit (i.e., $y_i = 0$), 0 otherwise. Expressed in this manner the probability of observing y_i detections can be considered as the combination of two components; one conditional on occupancy and the other conditional on the unit being unoccupied. Conditional on the unit being occupied, the probability of observing y_i detections is simply given by the binomial distribution (i.e., $\binom{K}{y_i} p^{y_i} (1-p)^{K-y_i}$). Given the unit is unoccupied, the species will never be detected; hence the probability that $y_i = 0$ is 1, and 0 otherwise (as represented by the indicator function). The zero-inflated binomial distribution is then obtained by multiplying each of the conditional probabilities by the probability that a unit is either occupied or unoccupied (ψ and $1 - \psi$ respectively). The joint likelihood based on data from s units is therefore the product of s such probability statements (one for each unit) of Eq. (7.1), i.e.,

$$L(\psi, p|y_1, y_2, \ldots, y_s) = \prod_{i=1}^{s} Pr(Y = y_i|\psi, p).$$

The basic strategy to developing models that allow for heterogeneity in p is to view p as a *random effect*. That is, we suppose that p may potentially take on a different value for each sample unit and that its potential values are governed by a probability distribution say $f(p|\theta)$, where θ are the parameters of this distribution, which are to be estimated. In other areas of applied statistics, it is common to choose one of various continuous distributions such as the beta distribution, or a normal distribution on the logit-transformation of p. Alternatively, discrete distributions in which two or more values of p are allowed may be constructed. Regardless of the form of $f(p|\theta)$, the view of p as a random variable enables standard probability calculus to be used to remove unit-specific detection probabilities (p_i) from the likelihood. Analysis can therefore focus on the likelihood as a function of ψ, and the parameters of the random effects distribution, θ. To formalize this notion, recall that the probability of obtaining y_i detections from K surveys, conditional on the unit being occupied, is based on a binomial distribution with parameter p. As p is now considered as a random variable, using the results of Chapter 3 we can therefore calculate the *expected* probability of obtaining y_i detections. That is, the average probability of observing y_i detections from K surveys, where the averaging is taken over possible values of p.

Let $\pi^{(c)}(y_i|\theta)$ denote this expected probability, that is:

$$\pi^{(c)}(y_i|\theta) = E_p\left[\binom{K}{y_i} p^{y_i}(1 - p)^{K-y_i}\right],$$

where the superscript '(c)' indicates the expected probability is conditional on occurrence of the species at a unit. To calculate $\pi^{(c)}(y_i|\theta)$, note that the probability of obtaining the y_i detections is simply some function of the random variable p (i.e., $g(p) = \binom{K}{y_i} p^{y_i}(1 - p)^{K-y_i}$). Hence from Section 3.1.2, when $f(p|\theta)$ is a discrete distribution, such as a finite mixture,

$$\pi^c(y_i|\theta) = \sum_p f(p|\theta) \binom{K}{y_i} p^{y_i}(1 - p)^{K-y_i},$$

with the obvious extension to a continuous distribution, replacing the summation with an integration (Royle, 2006). Finally, we have to marginalize over occupancy state to account for the fact that observed zeros arise from two possible sources (i.e., present but not detected, or not present). This gives rise to the unconditional probability of observing y_i detections from K surveys, $\pi(y_i|\boldsymbol{\theta}, \psi)$, of the form:

$$\pi(y_i|\boldsymbol{\theta}, \psi) = \psi \pi^{(c)}(y_i|\boldsymbol{\theta}) + (1 - \psi)I(y_i = 0). \tag{7.2}$$

It is these probabilities that are used to construct the likelihood of each observation. As above, the likelihood for the observed data from s units is then given by:

$$L(\psi, \boldsymbol{\theta}|y_1, y_2, \dots, y_s) = \prod_{i=1}^{s} \pi(y_i|\boldsymbol{\theta}, \psi).$$

This general model can also be expressed in terms of the underlying latent and observed random variables:

$$z_i \sim Bernoulli(\psi),$$

$$y_i|z_i \sim binomial(K, z_i p_i),$$

$$p_i \sim f(p|\boldsymbol{\theta}),$$

where z_i is the binary-valued random variable for presence/absence of the species, y_i is the number of detections at unit i from K surveys, and p_i is the probability of detection at unit i. Alternatively, we could specify the random variables in terms of the individual detection/nondetection surveys (h_{ij}) rather than the number of detections from K surveys (y_i):

$$z_i \sim Bernoulli(\psi),$$

$$h_{ij}|z_i \sim Bernoulli(z_i p_i),$$

$$p_i \sim f(p|\boldsymbol{\theta}).$$

The general form given by Eq. (7.2) holds regardless of the choice of mixture distribution $f(p|\boldsymbol{\theta})$. For any choice of $f(p|\boldsymbol{\theta})$, evaluation of the likelihood requires only that the expected probabilities for the $K + 1$ possible values of y_i (i.e., $0, 1, \dots, K$) be computed. We now consider some special cases of models that permit heterogeneity in p among units.

7.1.2 Finite Mixtures

We begin by considering the finite mixture distribution to model heterogeneous detection probabilities. Finite mixtures have been used in mark–recapture abundance estimation where detection probabilities are allowed to vary among individuals (Norris and Pollock, 1996; Pledger, 2000). Here it is assumed there are M possible values for p_i, and the probability that a unit has detection probability p_m is f_m. That is, $Pr(p_i = p_m) = f_m$. Note that as units must have one of the M detection probabilities, the sum of the f_m values must be 1.0, i.e., $\sum_{m=1}^{M} f_m = 1$. Because of this constraint, only $M - 1$ values of f_m are required, and the last can be obtained by subtraction. The parameters to be estimated include ψ, the detection probabilities $p_1, p_2, \ldots, p_M$, and the $M - 1$ probability masses $f_1, f_2, \ldots, f_{M-1}$. The logic behind the finite mixtures distribution is that the population actually consists of M groups of units, each with a different detection probability for the species, but group membership is unknown (e.g., at some units the species may be easy to detect, and difficult at others, but looking at the units prior to the surveying, we could not classify which units would have a high detection probability and which would be low). Using the results of the previous section, the expected probability of the observed detection frequency y_i at an occupied unit is

$$\pi^{(c)}(y_i | \boldsymbol{\theta}) = \sum_{m=1}^{M} f_m \binom{K}{y_i} p_m^{y_i} (1 - p_m)^{K - y_i} . \tag{7.3}$$

Here, the parameters ($\boldsymbol{\theta}$) associated with the distribution are the M possible values for p (sometimes referred to as the support points for the distribution), and the relative frequency for $M - 1$ of the possible values for p (i.e., f_m; with the last obtained by subtraction).

An equivalent approach to developing an expression for the probability of observing y_i detections considers the fact that the unit may have any of the M detection probabilities. This is easily incorporated by determining the probability for observing y_i detections assuming the detection probability is p_m (conditional upon occupancy state) and multiplying this component by the probability that the detection probability is p_m (i.e., by f_m). This is repeated for all the M possible values for p_m. This process is similar to that used in the previous chapter for developing the basic occupancy models. For example, suppose we want to consider a finite mixture model with $M = 2$, i.e., there are only two possible detection probabilities; p_1 and p_2 (say, high and low). Consider the case where four surveys were conducted at a unit and the species was detected once, hence $K = 4$ and $y_i = 1$. From Eq. (7.3), the expected conditional probability of observing one detections will be

$$\pi^{(c)}(1 | \boldsymbol{\theta}) = f_1 \binom{4}{1} p_1^1 (1 - p_1)^{4-1} + f_2 \binom{4}{1} p_2^1 (1 - p_2)^{4-1},$$

where $f_2 = 1 - f_1$. While this expression has been derived from considering the expected value of a function of a random variable, the expression could also be obtained simply by noting that there are two possible mechanisms that could have produced the observation of one detection, corresponding to binomial draws with probabilities either p_1 or p_2. Applying Eq. (7.2) (note that here $I(y_i = 0)$ evaluates to 0) the unconditional probability of observing one detection is therefore

$$\pi(1|\psi, \boldsymbol{\theta}) = \psi \sum_{m=1}^{2} f_m \binom{4}{1} p_m^1 (1 - p_m)^{4-1}. \tag{7.4}$$

The finite mixture model can also be developed in terms of probability statements for individual detection histories, e.g., $\mathbf{h}_i = 0010$. Under a finite mixture model with two possible detection probabilities (with no survey specificity so subscript relates to m, not survey j), the probability statement is:

$$Pr(\mathbf{h}_i = 0010|\psi, \boldsymbol{\theta}) = \psi \left(\begin{array}{c} f_1(1-p_1)(1-p_1)\,p_1(1-p_1) \\ + f_2(1-p_2)(1-p_2)\,p_2(1-p_2) \end{array} \right)$$
$$= \psi \left(f_1 p_1^1 (1-p_1)^{4-1} + f_2 p_2^1 (1-p_2)^{4-1} \right)$$
$$= \psi \sum_{m=1}^{2} f_m p_m^1 (1-p_m)^{4-1}.$$

Note the similarities with this probability statement and Eq. (7.4), both of which are for cases with one detection from four surveys of a unit. The only difference between the two is the binomial coefficient $\binom{4}{1}$ $(= 4)$, which is the number of different detection histories that have exactly one detection (i.e., 1000, 0100, 0010, and 0001). As the binomial coefficient is constant with respect to the parameters to be estimated, it's inclusion or omission has no effect on resulting parameter estimates.

A third way of developing the finite mixture occupancy model is in terms of a multi-state model (Chapter 5). For example, suppose there are three possible true occupancy states, where:

0 = species absent,
1 = species present with detection probability p_1,
2 = species present with detection probability p_2.

The probability of a unit being in each of these respective states is $(1 - \psi)$, ψf_1, and $\psi f_2 = \psi(1 - f_1)$. There are, however, only two possible observations that could be made during a survey '0' or '1'; an observation of '2' is impossible (i.e., cannot differentiate between the two different types of occupied units from the field observations). Our detection probability matrix for survey j could be defined as:

$$\mathbf{p}_j = \begin{bmatrix} 1 & 0 & 0 \\ 1-p_1 & p_1 & 0 \\ 1-p_2 & p_2 & 0 \end{bmatrix},$$

where rows relate to the possible true occupancy states and columns the observations. This could be simplified to:

$$\mathbf{p}_j = \begin{bmatrix} 1 & 0 \\ 1-p_1 & p_1 \\ 1-p_2 & p_2 \end{bmatrix}.$$

Using the matrix formulation, the conditional detection probability vector for $\mathbf{h}_i = 0010$ (for example) would be:

$$\mathbf{p}_{0010} = \mathbf{p}_1^{[\bullet 0]} \odot \mathbf{p}_2^{[\bullet 0]} \odot \mathbf{p}_3^{[\bullet 1]} \odot \mathbf{p}_4^{[\bullet 0]}$$

$$= \begin{bmatrix} 1 \\ 1-p_1 \\ 1-p_2 \end{bmatrix} \odot \begin{bmatrix} 1 \\ 1-p_1 \\ 1-p_2 \end{bmatrix} \odot \begin{bmatrix} 0 \\ p_1 \\ p_2 \end{bmatrix} \odot \begin{bmatrix} 1 \\ 1-p_1 \\ 1-p_2 \end{bmatrix}$$

$$= \begin{bmatrix} 0 \\ (1-p_1)(1-p_1)\,p_1(1-p_1) \\ (1-p_2)(1-p_2)\,p_2(1-p_2) \end{bmatrix}.$$

Then the probability statement would be determined as:

$$Pr(\mathbf{h}_i = 0010 | \psi, \boldsymbol{\theta}) = \boldsymbol{\phi}_0 \mathbf{p}_{0010}$$

$$= [1-\psi \quad \psi f_1 \quad \psi f_2] \begin{bmatrix} 0 \\ (1-p_1)(1-p_1)\,p_1(1-p_1) \\ (1-p_2)(1-p_2)\,p_2(1-p_2) \end{bmatrix}$$

$$= \psi \left(\begin{array}{l} f_1(1-p_1)(1-p_1)\,p_1(1-p_1) \\ +f_2(1-p_2)(1-p_2)\,p_2(1-p_2) \end{array} \right),$$

as above. Note that if a '2' was observable, i.e., in the field we could sometimes record detections of lots of individuals in our surveys (and only a few in other surveys; a '1'), then we could use the multi-state model to account for this detection heterogeneity.

Also note that in the case when $M = 1$, the finite mixture model reduces to the models given in Chapter 4, as we might expect.

Finite mixture models are implemented in Program PRESENCE (Hines, 2006) and Program MARK (White and Burnham, 1999) using maximum likelihood estimation.

The use of finite mixtures to approximate variation in detection probabilities extends back to the work of Carothers (1973) on heterogeneous capture probabilities in capture–recapture models. Finite mixture models were then used for capture–recapture inference by Norris and Pollock (1996), Pledger (2000), and many others. It is important to recognize that neither Carothers (1973) nor later users of this approach ever viewed mixture models as perfect descriptions of reality (i.e., they never thought that animal populations actually consisted of two or three groups of animals characterized by different detection probabilities). Rather, Carothers (1973) noted that it was the variation in detection probabilities (e.g., $CV(p_i)$) that was the key determinant of influence in capture–recapture models. Finite mixtures were simply viewed as one tractable means of approximating such variation. We share this view when utilizing this approach to occupancy modeling. In keeping with this view, we note that two-point mixtures are adequate for the vast majority of purposes.

7.1.3 Continuous Mixtures

A natural choice for f is the beta distribution, because this is the conjugate prior distribution for p of the binomial distribution. The beta distribution is defined by:

$$f(y|\alpha, \beta) = \frac{\Gamma(\alpha)\Gamma(\beta)}{\Gamma(\alpha + \beta)} y^{\alpha-1}(1 - y)^{\beta-1}$$

with 'sample size' parameters α and β, and where $\Gamma(\cdot)$ is a mathematical function called the gamma function. Conjugacy is the property that the posterior distribution is of the same parametric form as the prior distribution. The utility of this is not so important in the present context, but conjugate priors are often convenient in Bayesian analysis. Under a beta prior distribution for p, the expected conditional probability of y_i detections has a closed form; it is a zero-inflated beta-binomial with

$$\pi^{(c)}(y|\boldsymbol{\theta}) = \frac{\Gamma(K + 1)}{\Gamma(y + 1)\Gamma(K - y + 1)} \frac{\Gamma(\alpha + y)\Gamma(K + \beta - y)}{\Gamma(\alpha + \beta + K)} \frac{\Gamma(\alpha + \beta)}{\Gamma(\alpha)\Gamma(\beta)}.$$

This is commonly parameterized in terms of the mean $\mu = \alpha/(\alpha + \beta)$ and precision (i.e., the inverse of the variance) $\tau = \alpha + \beta$, or the mean μ and standard deviation $\sigma = \sqrt{\mu(1 - \mu)/(\tau + 1)}$. The beta mixture is appealing primarily because of its convenient form, although the likelihood must still be maximized numerically.

Another natural class of continuous models for describing variation in p is the logit-normal class of models (Coull and Agresti, 1999) in which $\text{logit}(p)$ is assumed to have a normal distribution with mean μ and standard deviation σ.

In this case, $\pi^{(c)}(y_i|\theta)$ does not have a closed form and requires computing the marginal likelihood by integration. However, the integration can be done numerically, and so the additional complexity here has no practical consequence as a computer program can do the calculations.

7.1.4   Abundance-Induced Heterogeneity Models

The mixture models considered previously are constructed by specifying a mixing distribution on the detection probability parameters p_i. In contrast, Royle and Nichols (2003) noted that heterogeneity in p_i can be induced by variation in abundance among units, and they developed a model of heterogeneity based on this consideration. That is, they placed the mixture distribution on abundance N and assumed a model for linking p_i to N. Specifically, let N_i be the abundance at unit i. Then, binomial sampling considerations yield that the probability of detection (i.e., of at least one individual) is $p(N_i, r) = 1 - (1 - r)^{N_i}$ where r is the individual detection probability, and the notation $p(N_i, r)$ is used simply to denote that the probability of detecting the species at unit i is now a function of both N_i and r. That is, the probability of detecting at least one individual of the species is 1 minus the probability of detecting none of the N_i individuals at the unit. However as the unit level abundance parameters N_i are unknown, by assuming they are random values from an appropriate discrete distribution, the expected probability of observing y_i detections can be calculated in a manner similar to above. Although here we do not have to condition upon occupancy status, as this is accounted for through the mixture distribution on N_i (i.e., a unit is occupied when $N_i > 0$ and unoccupied when $N_i = 0$). The expected unconditional probability of observing y_i detections at a unit can therefore be calculated as:

$$\pi(y_i|\eta, \theta) = \sum_{N_i=0}^{\infty} \left[Pr(N = N_i|\eta) \binom{K}{y_i} p(N_i, r)^{y_i} \left[1 - p(N_i, r)\right]^{K-y_i} \right]$$

$$(7.5)$$

where $Pr(N = N_i|\eta)$ is the probability of N_i individuals being present at the unit according to the mixture distribution for abundance (N) with parameters η. For example, suppose that unit-specific abundance has a Poisson distribution with mean λ. Then, the probability of observing y_i detections at unit i is:

$$\pi(y_i|\lambda, \theta) = \sum_{N_i=0}^{\infty} \left[\frac{e^{-\lambda}\lambda^{N_i}}{N_i!} \binom{K}{y_i} p(N_i, r)^{y_i} \left[1 - p(N_i, r)\right]^{K-y_i} \right].$$

Other discrete probability distributions can be considered for N_i, for example the negative binomial or other models that allow for more complex mean/variance relationships.

Expressed in terms of random variables, this model is:

$$N_i \sim g_N(\eta),$$

$$y_i | N_i \sim binomial(K, p_i),$$

$$p_i = 1 - (1 - r)^{N_i},$$

where $g_N(\eta)$ denotes the discrete-valued distribution for determining $Pr(N = N_i | \eta)$ (e.g., Poisson, negative-binomial, zero-inflated Poisson, etc.).

Note that evaluation of the likelihood under this model requires summations over all possible population sizes. Practical implementation of these models requires that these infinite summations be replaced by summations that stop at some large population size expected to exceed any that would be possible in practical situations, which is done by the implementations in the R package un marked (Fiske and Chandler, 2011) and Program PRESENCE (Hines, 2006).

Haines (2016) demonstrated that the infinite sum in Eq. (7.5) can be simplified to the sum of a finite number of terms for the cases of the Poisson and negative binomial abundance distributions, and their zero-inflated versions. Similar results likely hold for other abundance mixture distributions as well. The keys to the simplification are noting that: (1) the term $p(N_i, r)^{y_i}$ can be expressed as (using an algebraic rule known as the binomial expansion):

$$
\begin{aligned}
p(N_i, r)^{y_i} &= \left(1 - (1 - r)^{N_i}\right)^{y_i} \\
&= \sum_{k=0}^{y_i} \binom{y_i}{k} \left(-(1-r)^{N_i}\right)^k \\
&= \sum_{k=0}^{y_i} \binom{y_i}{k} (-1)^k (1-r)^{kN_i};
\end{aligned}
$$

and (2) that the sums of certain infinite series converge to specific quantities (e.g., Anton, 1988). For example, in the case of the Poisson abundance model, Haines (2016) showed that the probability of observing y_i detections can be expressed as:

$$\pi(y_i | \lambda, \theta) = exp(-\lambda) \binom{K}{y_i} \sum_{k=0}^{y_i} \binom{y_i}{k} (-1)^k exp\left(\lambda(1-r)^{K-y_i+k}\right).$$

However, while Haines (2016) reported that using the exact, simplified expression instead of halting the summation of an infinite number of terms after a sufficiently large value (as done currently in commonly available software) typically resulted in improved speed of the estimation routines, the maximum likelihood estimates were practically identical for the situations considered.

A number of extensions of occupancy models that allow for abundance-induced heterogeneity have been developed. Wenger and Freeman (2008) extended this approach to accommodate zero-inflated abundance distribution models (e.g., zero-inflated Poisson) allowing for more unoccupied units than specified by the standard abundance distributions. Rossman et al. (2016) extend the model to accommodate temporal dynamics of occupancy, and Yamaura et al. (2011) extend the Royle–Nichols models to community level studies (see Chapter 15).

One appeal of this abundance-based model is that in some cases it may be reasonable to view the expected value of N, $E(N)$, determined from the assumed abundance distribution $g_N(\eta)$, as density. For example, if individuals are assumed to be distributed in space according to a Poisson process, and the detection of individuals is independent, then $E(N) = \lambda$ is the expected number of individuals per sample unit (the product of the density of individuals and sample unit area). This interpretation should be considered only within the context of the underlying assumptions, that abundance is Poisson and that detection of individuals is independent. While such assumptions are not likely to be valid in most situations, some model extensions are possible. For example, one might consider alternative abundance distributions such as the negative binomial. However, such models can be difficult to fit, given data only on detection/nondetection. In general, there is no reason that N_i must be interpreted as abundance *per se*, nor that λ be interpreted as density (per sampling unit). Rather, N_i could be viewed as a random effect that yields variation in p_i, and thus we might view this model merely as an alternative mixing distribution that accommodates heterogeneity in detection probability. Indeed, it need not even be an integer. Royle (2006) noted a relationship between this model and a model that has a normally distributed random effect for detection using the complementary log–log (*cloglog*) link function. Specifically:

$$\begin{aligned} cloglog(p_i) &= ln(ln(1 - p_i)) \\ &= ln\big(ln\big((1 - r)^{N_i}\big)\big) \\ &= ln(ln(1 - r)) + ln(N_i). \end{aligned}$$

This yields a linear model using the *cloglog* link for p_i with intercept $ln(ln(1 - r))$ and random effect $u_i \equiv ln(N_i)$, which might be assumed normal with mean μ and variance σ^2.

The point of this is that while the Royle–Nichols model is constructed as a mixture over abundance states, it can be made to resemble a model that is a mixture on (the complement of) p_i, consistent with the construction of the other models that we have considered previously. That is, the source of heterogeneity in detection that is being attributed to variation in abundance in the development of the model, may actually be due to other sources or mechanisms. Therefore parameters associated with 'abundance', may not accurately reflect real abundance if there are other sources of heterogeneity in detection probability not accounted for in the model. As such, users are advised to exercise caution when making inferences about 'abundance' from this model, although it should certainly yield improved inferences about occupancy over the simpler approaches that do not account for extra heterogeneity (Chapter 4), irrespective of the actual source of that heterogeneity.

Note that under the RN model, occupancy is a derived parameter. For example, under the assumption of Poisson abundance, $\psi = Pr(N > 0) = 1 - e^{-\lambda}$. Also, even though the model is parameterized in terms of individual detection probability, one can extract a parameter that is analogous to the conditional (on occurrence) detection probability that one usually considers in occupancy modeling. The more familiar expected conditional detection probability is:

$$p_c = \sum_{N=1}^{\infty} \frac{Pr(y > 0|N, r)Pr(N|\eta)}{Pr(N > 0|\eta)}.$$

Finally, we note that the Royle and Nichols (2003) abundance-induced heterogeneity model provides interesting opportunities for integrated population modeling that entails combining data of different types. If a subset of units surveyed for occupancy included additional sampling permitting direct inference about abundance (e.g., double-observer, see Nichols et al., 2000b), then that additional information on individual detection probability and abundance would inform the parameters of the Royle and Nichols (2003) model, presumably leading to better estimation. Conroy et al. (2008) is one example where researchers have used such an integrated modeling approach, combining detection/nondetection data from an occupancy study with capture–recapture data collected at a subset of patches to enable direct estimation of patch-level abundance.

7.1.5 Evaluation of Model Fit

To evaluate model adequacy, we consider a typical deviance statistic based on the expected cell frequencies under the model in question:

$$D_g = 2 \sum_{k=0}^{K} n_k \left[ln(\pi_{sat}(k)) - ln\left(\pi_g(k|\hat{\theta})\right) \right]$$

where $\pi_{sat}(k)$ is the observed proportion of units with k detections and $\pi_g(k|\hat{\theta})$ are the model-based estimates obtained by plugging the MLEs of model parameters into Eq. (7.3). Asymptotically, and under the hypothesis that the model is correctly specified, we can expect this statistic to have a chi-square distribution with $K - Npar_g$ degrees-of-freedom where $Npar_g$ is the number of parameters in the heterogeneity model g.

In practice we often fail to achieve large samples, which renders the null distribution of this statistic invalid. In such cases we would probably rely on the usual strategies for assessing goodness-of-fit. That is, in small samples we might pool cells where the expected cell count is less than 5. Alternatively, we might estimate the small sample null distribution using common resampling strategies such as the parametric bootstrap (see Chapter 4). This would be preferred when the number of cell frequencies is small so that pooling cells reduces the number of cells to the point of yielding a test with low power. We address some important theoretical issues of assessing model fit in Section 7.6.

7.2 EXAMPLE: BREEDING BIRD POINT COUNT DATA

Here we demonstrate the application of these models for heterogeneous detection probabilities using avian survey data. These data originate from a study conducted to evaluate sources of variation in bird count data (Link et al., 1994) on North American Breeding Bird Survey (BBS) routes (e.g., see Robbins et al., 1986, for description of this survey). The specific data used here are from a single route (50 'stops' or sample locations at which number of detected birds is recorded over a three-minute period). Each sample location was sampled 11 times during an approximately one-month 'season'. The actual count data were reduced to observations of detection/nondetection. We consider data on four species: blue jay (*Cyanocitta cristata*), common yellow-throat (*Geothlypis trichas*), song sparrow (*Melospiza melodia*), and gray catbird (*Dumetella carolinensis*). The data for each species are given in Table 7.1.

No covariates thought to influence either detection or occurrence are available and so, in the analyses presented here, we have considered only the basic heterogeneity models described in the preceding section. That is, in addition to the constant-p model (described in Chapter 4, const), we consider heterogeneity models based on the logit-normal (LN), beta-binomial (BB), two-component finite mixture (FM2), and the Royle and Nichols Poisson abundance (RN) models. We note that the sample size is relatively small here ($s = 50$), which impairs the direct application of the deviance statistic described previously for assessing fit. Rather than undertaking a bootstrap characterization of the proper null distribution for each model, we will present deviance statistics merely as indices of

TABLE 7.1 Number of detections (out of $K = 11$ surveys) for each of four bird species at each of the 50 BBS stops. Species codes are: blue jay (JAY); common yellow-throat (CYT); song sparrow (SOSP); catbird (CATB). 'Naïve occupancy' is the observed proportion of occupied units

Species	Number of detections												Naïve occupancy
	0	1	2	3	4	5	6	7	8	9	10	11	
JAY	17	9	11	6	5	2	0	0	0	0	0	0	0.66
CYT	14	6	7	5	3	1	4	5	3	2	0	0	0.72
SOSP	24	5	1	5	1	5	3	1	3	1	0	1	0.52
CATB	31	6	4	5	2	0	2	0	0	0	0	0	0.38

TABLE 7.2 Parameter estimates and summary statistics from each occupancy model fit to the BBS survey data. Model designations are: constant-p model (const), logit-normal (LN), beta (BB), two-point finite mixture (FM2), and the Royle and Nichols (2003) abundance model (RN). $\hat{E}(p)$ and $\hat{\sigma}_p$ are the mean and standard deviation of the estimated heterogeneity distribution, $\hat{\psi}$ is the estimated proportion of units occupied, $-2l$ is twice the negative log-likelihood. *Npar* is the number of parameters for each model, and *DEV* is the model deviance, used as a relative measure of model fit for each species

Species	Model	$\hat{E}(p)$	$\hat{\sigma}_p$	$\hat{\psi}$	$-2l$	*Npar*	*DEV*
JAY	const	0.199	0	0.723	164.08	2	1.88
	LN	0.197	0.023	0.728	164.07	3	1.87
	BB	0.197	0.024	0.728	164.07	3	1.87
	FM2	0.191	0.046	0.753	164.01	4	1.81
	RN	0.178	0.085	0.806	165.45	2	3.25
CYT	const	0.385	0	0.723	250.55	2	41.22
	LN	0.341	0.218	0.809	218.26	3	8.93
	BB	0.330	0.221	0.835	218.26	3	8.93
	FM2	0.360	0.218	0.773	212.40	4	3.07
	RN	0.374	0.147	0.765	222.77	2	13.44
SOSP	const	0.419	0	0.521	215.08	2	37.89
	LN	0.377	0.237	0.578	188.43	3	11.24
	BB	0.365	0.243	0.597	188.18	3	10.99
	FM2	0.398	0.206	0.549	192.18	4	14.99
	RN	0.403	0.132	0.567	196.25	2	19.06
CATB	const	0.219	0	0.407	132.11	2	8.05
	LN	0.196	0.102	0.454	130.08	3	6.02
	BB	0.191	0.106	0.466	130.04	3	5.98
	FM2	0.206	0.095	0.433	129.99	4	5.92
	RN	0.209	0.075	0.428	130.33	2	6.27

relative fit. Results of fitting these various models to the avian point count data are presented in Table 7.2.

For the blue jay, we note that there does not appear to be much heterogeneity among units in detection probability because, under the logit-normal model $\hat{\sigma} = 0.023$, and none of the heterogeneity models yields more than a marginal improvement in log-likelihood. Consequently, we are inclined to favor the constant-p model for this species. Estimates of occupancy are similar across most of the heterogeneity models (except the RN model). Heterogeneity in detection probability is indicated for the remaining three species. For common yellow-throat, the two-component finite mixture appears favored, with $\hat{\psi} = 0.773$. For song sparrow, the beta or logit-normal models appear favored with $\hat{\psi} = 0.597$ under the beta model and $\hat{\psi} = 0.578$ under the logit-normal model. Moderate heterogeneity is indicated for the catbird data, but it is less clear which heterogeneity model is to be preferred. All heterogeneity models fit the data equally well and all have similar negative log-likelihoods. Strict adherence to the use of AIC as a model selection tool would lead us to choose the Royle and Nichols model because it wins out on parsimony grounds. Regardless, the four estimates of ψ range from 0.428 (RN) to 0.466 (Beta) and are about 10–15% higher than the estimate from the constant-p model.

Note that we might conduct a formal test of the hypothesis of no heterogeneity based on Self and Liang (1987) who demonstrated the asymptotic distribution of the usual likelihood ratio statistic to be a 50/50 mixture of a χ_0^2 and a χ_1^2 (thus the common advice to "halve the p-value" of the conventional test when used for testing that a variance component is zero). Such a strategy would yield a conclusion of no heterogeneity for the jay, significant heterogeneity for the common yellow-throat and song sparrow, and a more equivocal conclusion for the catbird.

7.3 MODELING COVARIATE EFFECTS ON DETECTION

In many problems, the possibility that detection varies temporally, or in relation to measurable covariates might be considered, perhaps in addition to heterogeneity among units. For such purposes, we require a detection history formulation of the likelihood. For example, let $\mathbf{h}_i = 0101$. The probability of observing this particular detection history at an occupied unit (hence ψ does not appear in the probability statement) is $Pr(\mathbf{h}_i = 0101|\text{unit is occupied}) = (1 - p_{i1})p_{i2}(1 - p_{i3})p_{i4}$. A convenient manner in which to parameterize heterogeneity in detection among units is to assume a linear logit-link function between p_{it} and covariates according to:

$$\text{logit}(p_{ij}) = \alpha + \beta x_{ij} \tag{7.6}$$

where x_{ij} is a covariate measured at unit i during survey j, and α and β are the parameters to be estimated (Section 3.4.1, where $\alpha = \beta_0$ and $\beta = \beta_1$, which is done to avoid confusion with the notation below). The extension of this basic model to accommodate heterogeneity among units is straightforward, and that is to replace α with α_i, where α_i is a unit-specific random effect endowed with a suitable distribution. For example, we might suppose that α_i has a normal distribution: $\alpha_i \sim \text{Normal}(\mu, \sigma^2)$, or there may be a small number of discrete values that α_i could take as under the finite mixture models described previously. Under this detection history formulation of the likelihood, parameter estimates may be obtained by integrating the likelihood contribution of each detection history over the specified mixing distribution and then zero-inflating the resulting marginal probabilities. For example, under a two-point finite mixture model where there are two values of α_i, say α_1 and α_2 with masses f_1 and $f_2 = 1 - f_1$, then the likelihood contribution for a unit with the detection history $\mathbf{h}_i = 0101$ is:

$$\pi(\mathbf{h}_i|\alpha_1, \alpha_2, f_1, \beta) = \psi \pi^{(c)}(\mathbf{h}_i|\alpha_1, \alpha_2, f_1, \beta) + I(\mathbf{h}_i = 0)(1 - \psi),$$

where

$$\pi^{(c)}(\mathbf{h}_i|\alpha_1, \alpha_2, f_1, \beta) = \sum_{m=1}^{2} f_m \left[(1 - p_{m,1}) p_{m,2} (1 - p_{m,3}) p_{m,4} \right],$$

and $p_{m,j}$ is obtained by substituting each α_m into Eq. (7.6).

Under the logit-normal model, the conditional (on occurrence) probability of obtaining capture history $\mathbf{h}_i$ is computed by evaluating the integral:

$$\pi^{(c)}\left(\mathbf{h}_i|\beta, \mu, \sigma^2\right) = \int_{-\infty}^{\infty} Pr(\mathbf{h}_i|\alpha, \beta) g\left(\alpha|\mu, \sigma^2\right) d\alpha,$$

for which the unconditional likelihood contribution is obtained by zero-inflating this probability:

$$\pi\left(\mathbf{h}_i|\beta, \mu, \sigma^2\right) = \psi \pi^{(c)}\left(\mathbf{h}_i|\beta, \mu, \sigma^2\right) + I(\mathbf{h}_i = 0)(1 - \psi).$$

The full model likelihood of observing the detection histories for all units is the product (over i) of each unit's likelihood contribution, i.e., $\prod_{i=1}^{s} \pi\left(\mathbf{h}_i|\beta, \mu, \sigma^2\right)$.

For the Royle and Nichols (2003) formulation of the model, covariates are modeled on the parameter r (individual detection probability) so that, for example, $\text{logit}(r_{ij}) = \alpha + \beta x_{ij}$, and this is substituted into the expression for net detection probability $p(N_i, r_{ij}) = 1 - (1 - r_{ij})^{N_i}$ and then used in constructing the likelihood of each detection history as before. As with finite mixture models, the likelihood of each detection history is averaged over all possible values of the detection probability parameter, i.e., values of N_i. Under this model, the intercept, α, remains constant because the heterogeneity is induced by mixing over the abundance distribution.

While not detailed above, covariate effects on occupancy probabilities can also be incorporated, as for the simpler models already discussed.

7.4 EXAMPLE: ANURAN CALLING SURVEY DATA

To illustrate the application of models with covariates on detection, we consider data from a study conducted to evaluate sources of variation in detection probability in North American Amphibian Monitoring Program (NAAMP) survey data. The particular data consist of observations of anuran detection/nondetection at 220 roadside 'stops' in Maryland, USA. Each stop is associated with potential breeding habitat of anurans. See Weir et al. (2005) for further details on this study and results. Between three and 14 visits were made to each of the 220 stops between early March and the end of July 2002. We focus here on data for the gray treefrog (*Hyla versicolor*).

We consider two covariates that are likely to influence detection probability. Most importantly, the breeding phenology of all North American anurans is strongly seasonal, and can be expected to vary even within the putative breeding season. In particular, while one might detect a given species over a relatively long period, there should be an increase in breeding activity subsequent to the onset of calling activity, with a distinctive period of peak calling behavior, followed by a gradual decline. Here we choose to model this seasonality in detection that derives from breeding behavior as a quadratic on (integer) sample day, defining March 1 to be $day = 1$. The only other covariate on detection probability considered here is air temperature at the time of survey, and we suppose that the response to temperature may also be quadratic.

Models containing these covariates, in addition to the null heterogeneity model without covariates, and also one containing no heterogeneity, were fitted to the gray treefrog data. The parameter estimates for each model, ordered by AIC, are presented in Table 7.3. First, we note that moderate heterogeneity is indicated (comparing the null model without heterogeneity to that with heterogeneity), with $\hat{\sigma} = 0.76$ under the best fitting model, which contains a quadratic response to temperature and a linear day effect.

The fitted quadratic temperature indicates a convex response with detection probability achieving a maximum at about 20.4°C, suggesting a preference for calling under warmer conditions for this species. Also, the fitted response to *day* indicates increasing detection probability as the season progresses, also consistent with the late onset of breeding in this species. Finally, the estimated occupancy rate of the species at stops along NAAMP routes in Maryland is $\hat{\psi} = exp(0.21) / \left[1 + exp(0.21)\right] = 0.55$.

7.5 ON THE IDENTIFIABILITY OF ψ

Link (2003) demonstrated that in closed population models for estimating abundance, abundance is not identifiable in the presence of heterogeneity in the sense that different mixture distributions may give rise to identically distributed data

TABLE 7.3 Parameter estimates for detection probability covariate models with heterogeneity fit to the gray treefrog data. The logit-normal model (intercept μ and standard deviation σ) with covariates was used to model heterogeneity. The estimated coefficients for each covariate are indicated in the column labeled with the corresponding covariate. β_0 is the logit-transform of ψ. No entry indicates that the particular covariate is absent from the model. $-2l$ is twice the negative log-likelihood, $Npar$ is the number of model parameters, and AIC is Akaike's information criterion

Model	$\hat{\mu}$	$log(\sigma)$	$\hat{\beta}_0$	$temp$	$temp^2$	day	day^2	$-2l$	$Npar$	AIC
$p\left(temp + temp^2 + day + het\right)$	−2.32	−0.28	0.21	0.73	−0.59	0.68		747.41	6	759.41
$p\left(temp + temp^2 + day + day^2 + het\right)$	−2.24	−0.26	0.25	0.68	−0.51	0.84	−0.24	745.83	7	759.83
$p\left(temp + day + day^2 + het\right)$	−2.47	−0.17	0.32	0.29		1.10	−0.48	754.61	6	766.61
$p\left(+day + day^2 + het\right)$	−2.50	−0.15	0.38			1.28	−0.47	757.44	5	767.44
$p\left(+day + het\right)$	−2.75	−0.19	0.28			1.00		764.78	4	772.78
$p\left(temp + temp^2 + het\right)$	−2.03	−0.45	0.10	1.23	−0.69			766.95	5	776.95
$p\left(temp + het\right)$	−2.50	−0.33	0.15	0.84				787.77	4	795.77
$p\left(het\right)$	−2.33	−0.45	0.26					842.81	3	848.81
$p\left(\cdot\right)$	−2.01		−0.09					845.04	2	849.04

(or nearly so), yet produce substantially different inferences about abundance. Although occupancy models with heterogeneous detection probabilities are simple to construct, as demonstrated previously, it is natural to consider whether Link's result applies to these classes of models. In closed population models for estimating abundance, the inability to observe the 'zero frequency' (number of uncaptured individuals) is the genesis of the non-identifiability problem reported by Link (2003). However, in occupancy models, the zero frequency is observed. Unfortunately, occupancy models introduce uncertainty about nondetection and absence of the species (by introduction of the occupancy parameter) that may largely mitigate any ability to partition the observed zeros into those due to nondetection and those due to non-occurrence of the species in the presence of heterogeneity.

Royle (2006) considers this problem in some detail and demonstrates that the same basic phenomenon should be of some concern in the context of occupancy models. It can be established empirically that, in certain instances, one cannot reasonably expect to distinguish between alternative mixture distributions from data. To demonstrate this, he chose g (e.g., logit-normal) and f (e.g., beta) to minimize twice the Kullback–Liebler distance (Burnham and Anderson, 1998, p. 37) between π_g and π_f:

$$KL_f = 2 \sum_{k=0}^{K} \pi_g(k) ln \left(\frac{\pi_g(k)}{\pi_f(k)} \right).$$

In some cases, KL_f is close to zero, yet ψ_f and ψ_g differ markedly (note that Royle, 2006, provides an example of a g and f where in fact $KL_f = 0$, but this is not generally the case). Link (2003) noted that s (number of sampled units) times KL_f is the non-centrality parameter that can be used to assess the power of a goodness-of-fit test of f against the g alternative. Consequently, there will be little power to distinguish between f and g, and hence to make the correct inference about ψ, when the KL distance is small.

For example, suppose that data are collected on $K = 5$ sampling occasions, that the logits of p_i have a normal distribution with $\mu = -2$ and $\sigma = 1$, and that $\psi = 0.75$. The closest (in the Kullback–Liebler sense) models in each of the other classes are given in Table 7.4.

The marginal cell probabilities are very similar across the five models, and the non-centrality parameter is small in all cases, indicating low power to choose among them, consistent with the results reported by Link (2003). For example, if one obtained a sample of size 200 units, the power to correctly reject the constant-p model is 0.274, but the heterogeneity models are all very similar. In the free software package R (Ihaka and Gentleman, 1996), this calculation is done by issuing the command:

TABLE 7.4 Multinomial cell probabilities and Kullback–Liebler distance between models compared to the specified logit-normal model for heterogeneity. The mixture models are logit-normal (LN; the 'true' model), beta (BB), two-point finite mixture (FM2), and the Poisson abundance model of Royle and Nichols (RN)

Model	ψ	Cell probabilities						$2 \times KL$
		$k = 0$	$k = 1$	$k = 2$	$k = 3$	$k = 4$	$k = 5$	
LN ($\mu = 2, \sigma = 1$)	0.750	0.630	0.222	0.097	0.037	0.011	0.002	0.000000
Const. $p = 0.23$	0.510	0.630	0.205	0.122	0.037	0.005	0.000	0.017000
BB ($\mu = 13, \tau = 5.87$)	0.910	0.630	0.223	0.096	0.037	0.011	0.002	0.000028
FM2 ($p = [0.15, 48], f = 0.87$)	0.620	0.630	0.222	0.097	0.036	0.012	0.002	0.000066
RN ($r = 0.15, \lambda = 0.84$)	0.570	0.629	0.213	0.111	0.037	0.008	0.001	0.004800

```
R>   1 - pchisq(qchisq(0.95,df=6-2),df=6-2,ncp=200*0.17)
```

These results are consistent with Link's (2003) conclusion that "there is virtually no power to distinguish the beta and logit-normal models, except with very large samples." The same can be said about the set of four heterogeneity models considered in Table 7.2. Importantly, the models imply very different values of ψ.

Consider this issue in the context of some of the data analyzed in Section 5.2. The results for the song sparrow suggested the presence of fairly extreme heterogeneity. Furthermore, the LN or BB models appeared to be favored and they provided nearly identical fits (as measured by deviance) to the data. If we suppose the LN is truth (with parameters given by those in Table 7.2), and carry out the calculation of the KL distance, we obtain the results given in Table 7.5.

Note that while the differences between the logit-normal and several of the mixture models are substantial (in particular, between the LN and constant and RN models), the LN and BB models are nearly indistinguishable. Their expected cell frequencies (the expected data) are nearly identical, and in reasonable sample sizes there will be no power to reject the BB model if the LN model is the correct model. However, this ambiguity between the LN and BB is mitigated to some extent by the fact that the estimated occupancy rates under the two models are not very different.

In this instance, having $K = 11$ and a moderately high detection probability partially mitigates the problem, either enunciating differences between the models (in terms of fit) or by minimizing differences between ψ under different models. Some might argue that this is a situation wherein statistical methods are not useful because most of the occupied units have been detected. However, we note that we are not interested in making inferences about the *apparent* level of occupancy, and we know that the naïve estimate (Table 7.1) of $\psi = 0.52$ is biased low. Consequently, the adjustment of ψ upward by about 15% indicated by the LN or BB models probably seems reasonable in light of the evidence resulting from investigation of a suite of what are generally regarded as being reasonable models.

Next, consider the catbird data for which a lower mean detection probability and moderate heterogeneity were indicated. For this species, the naïve estimate (Table 7.1) of occupancy is 0.38. All heterogeneity models are fairly similar in terms of fit and the KL distances between LN and const, BB, FM2, and RN models are (0.04179, 0.00012, 0.00094, 0.00434). In this case, the models are much more similar, and the estimated level of occupancy ranges from about 0.43 to 0.47. Pragmatically, we might feel comfortable thinking that truth is in the vicinity of 0.43–0.47. However, the basis for making a formal inference is muddled at best in light of the ambiguity among these heterogeneity models.

TABLE 7.5 Multinomial cell frequencies and Kullback–Liebler distance (KL) of closest heterogeneity models to the best-fitting logit-normal (LN) model to the SOSP data ($K = 11$ visits). Models are: constant-p (const), beta (BB), two-point finite mixture (FM2), Poisson abundance model (RN)

Model	ψ	Expected cell frequencies												KL
		0	1	2	3	4	5	6	7	8	9	10	11	
LN	0.578	0.480	0.074	0.074	0.069	0.062	0.056	0.049	0.042	0.036	0.028	0.020	0.010	0.000000
const	0.521	0.480	0.011	0.039	0.083	0.119	0.119	0.086	0.044	0.016	0.004	0.001	0.000	0.563191
BB	0.595	0.480	0.075	0.072	0.068	0.062	0.056	0.050	0.043	0.035	0.028	0.019	0.010	0.000176
FM2	0.535	0.480	0.054	0.088	0.087	0.061	0.040	0.039	0.048	0.049	0.035	0.015	0.003	0.040050
RN	0.568	0.439	0.031	0.071	0.101	0.104	0.087	0.066	0.047	0.030	0.016	0.006	0.001	0.146583

7.6 DISCUSSION

In this chapter, we extended occupancy models to allow for heterogeneity in detection probability among units. We considered several continuous and discrete mixture distributions for p_i. We also presented a model based on the consideration that heterogeneity in detection probability is due to variation in abundance. These models extend easily to the situation when covariates thought to influence detection are available. Similar extensions can be considered to other classes of occupancy models, for example to the multi-season models described by MacKenzie et al. (2003) (see Chapters 8 and 10).

We considered the problem of identifiability raised by Link (2003) who demonstrated that in closed population capture–recapture models for estimating abundance, abundance is not identifiable in the presence of heterogeneity in the sense that different mixture distributions may give rise to identically distributed data (or nearly so), yet produce substantially different inferences about abundance. In closed population models for estimating abundance, the inability to observe the 'zero frequency' (number of individuals never captured) is the genesis of the non-identifiability problem reported by Link (2003). However, in occupancy models, the zero frequency is observed. Unfortunately, occupancy models introduce uncertainty about nondetection and non-occurrence (by introduction of the occupancy parameter) that largely mitigates any ability to partition the observed zeros into those due to nondetection and those due to non-occurrence in the presence of heterogeneity. We believe that the problem raised by Link (2003) should be considered when attempting to estimate occupancy probabilities, because the same general phenomenon can occur in these models.

The effect of misspecification of heterogeneity models is related to the degree of heterogeneity and the mean detection probability. As Link (2003) noted, differences among mixtures are more pronounced as the mass of $g(p)$ is concentrated near zero (i.e., when most units have a very low detection probability). One might view low mean detection or high levels of heterogeneity as suggesting that the species of interest cannot be reliably, or effectively, sampled. This should be viewed as a biological sampling issue to consider in survey design prior to data collection (Chapter 11), and not a statistical issue to rectify after the fact by considering complex models of the detection process. The results described in Section 7.5 suggest that the latter may not be a viable option in some situations.

When faced with discrepant estimates of ψ, obtained from models that all appear to fit the data, it is not clear what conclusion (if any) can be drawn. Certainly if the heterogeneity models yield consistent results, we might feel comfortable with an estimate that is based on one or several of the competing models (e.g., use model averaging). However, when estimates are much more variable, model averaging may be less useful (although the standard error of the model-averaged estimate will be large, reflecting a high degree of uncertainty). The practical consequence of this issue is that monitoring programs that emphasize occupancy as a metric of population status must consider the possibility that the existence of heterogeneity may diminish the utility of such metrics, and they should take steps to minimize heterogeneity or to increase mean detection probability. For example, establishing rigorous sampling protocols, or identifying covariates that affect detection probability may reduce heterogeneity to the point of being unimportant in terms of selecting among possible mixture models. While this may be difficult in the context of estimating population size, in surveys for estimating occupancy it is often possible to measure a number of covariates about the unit being sampled, and the conditions under which sampling occurs, and sampling methods are more flexible in many cases. A key problem in dealing with heterogeneous capture probabilities in capture–recapture studies involving individual animals is that covariates cannot be measured on the animals that are never captured. In contrast, as noted above, occupancy studies do permit covariate measurement at all units, including those at which detections never occur. This provides a substantive advantage for occupancy studies and argues for careful consideration of covariates affecting detection probability.

For occupancy models that are based on detection/nondetection data, heterogeneity may be due to variation in abundance, and it is less clear how to deal with this issue in survey design. In sampling of populations wherein there is thought to exist considerable variation in abundance, in which case occupancy is a less useful summary of demographic state, demographic summaries that focus on abundance should be considered more explicitly by collecting data that are more informative than simple detection/nondetection (e.g., counts of individuals detected per unit). These counts could be modeled explicitly using N-mixture models (e.g., Royle, 2004b), or used to categorize observations in terms of relative abundance, e.g., none, some and lots of individuals, for use with multi-state occupancy models (Chapter 5). Alternatively, one might consider restricting attention to models of heterogeneity that are derived from models of variation in abundance, if that is thought to be the primary source of heterogeneity. While biological arguments to restrict the classes of models under consideration may be appealing, this may be less appealing when faced with a wide array of seemingly reasonable models for describing heterogeneity (e.g., due to the environment or habitat), and so *a priori* exclusion of such models is limiting. In Chapter 11 we consider the design of single-season occupancy studies. As noted above, some sources of heterogeneity can be controlled through the careful design of occupancy studies, and many of the suggestions made in that chapter are given to achieve this end.

Part III

Single-Species, Multiple-Season Occupancy Models

Chapter 8

Basic Presence/Absence Situation

In the preceding chapters we have considered the problem of estimating the probability of occupancy (or proportion of units occupied) in a single season. The methods we have detailed may provide some indication of the current patterns in occupancy within that season; a snapshot of the population at a single point in time. However as discussed in Chapter 1, despite the popularity of doing so, it is not always appropriate to attempt to infer process from an observed pattern. Often there are many processes that could result in the same pattern being observed at any given time (e.g., Pirsig, 1974; Romesburg, 1981; Nichols, 1991; Williams et al., 2002).

A much more reliable approach to understanding the processes occurring within a system is to observe how the system behaves over a longer timeframe. This should not be at all surprising. As an analogy, suppose that you are given a randomly selected photograph from a stack of photographs taken throughout a football game. You are then asked to comment on the current state of the game, and how the game has progressed up to that point. It would be possible to tell something about the current state of play, such as which team has the ball and possibly the score; however, it would be impossible to make further comment on how the game has progressed. Not until you are able to go through the entire stack of photographs (in order) would you be able to get some idea of how the game progressed. It is the same situation in ecological studies where processes of population dynamics can only be fully understood by observing the population at systematic points in time, noting how the patterns change and modeling these changes in terms of relevant rate parameters. As emphasized in Chapter 1, strong inferences arise when system behavior (e.g., estimated changes in rate parameters) is compared against predictions of *a priori* hypotheses, especially when system dynamics are generated by experimental manipulations within the context of experimental design. The models of this chapter were developed to provide the estimates needed for such investigations.

In this chapter we turn our attention to the problem of estimating occupancy over multiple seasons and, in particular, understanding the underlying population dynamics that may cause changes in the occupancy state of a unit. These dynamic parameters are of interest in many areas of ecology, including

Occupancy Estimation and Modeling. http://dx.doi.org/10.1016/B978-0-12-407197-1.00011-9

metapopulation studies where the processes of local extinction and colonization (often hypothesized to be functions of patch size and isolation from neighboring patches, respectively) produce an *incidence* function (e.g., Diamond, 1975; Hanski, 1994a, 1994b). However most of the methods used to study these parameters do not explicitly account for detection probability. Moilanen (2002) found false absences to be the greatest contributor of bias to the estimation of the incidence function parameters. In monitoring programs, often the rate of change in occupancy may be of as much or greater interest than the absolute level of occupancy at any point in time. Changes in the use of different habitats over time will also be of interest in many species-habitat studies. For example, are the same habitats used by a species in summer and winter, or what effect has a change in the habitat had on the species patterns of use?

We consider two general approaches for modeling changes in occupancy over time: (1) a model where underlying dynamics are implied but not explicitly accounted for (effectively combining several single-season models); and (2) explicitly modeling potential changes in the occupancy state of a unit over time with colonization and local extinction probabilities.

8.1 BASIC SAMPLING SCHEME

We assume a situation where s units are selected from an area of interest with the intent of establishing the presence or absence of a species, as in single-season studies, although now the assessment is for multiple points in time. Units may constitute naturally occurring sampling units such as discrete ponds or patches of vegetation, investigator-defined monitoring stations, or quadrats chosen from a predefined area of interest.

The timeframe of the study can now be considered at two scales. Firstly, at the larger scale, the study is conducted over multiple (T) seasons (e.g., years or breeding seasons, denoted by t). Each season is common to all units, with the occupancy state of units able to change between seasons, but not within seasons. Within each season, the smaller time scale, appropriate sampling methods are used to survey units K_t times (Fig. 8.1). Such a design is similar to Pollock's robust design (Pollock, 1982) used in mark–recapture studies where seasons represent the primary sampling periods and surveys within seasons represent secondary sampling periods. Effectively, the general design considered here is a sequence of single-season studies conducted at (usually) the same units for multiple seasons.

At each survey of a unit, the target species is detected (1) or not detected (0) and is never falsely detected when absent. The resulting sequence of detections and nondetections for unit i, conducted during season t, is denoted as the detection history $\mathbf{h}_{t,i}$. The complete detection history for unit i is denoted as $\mathbf{h}_i$,

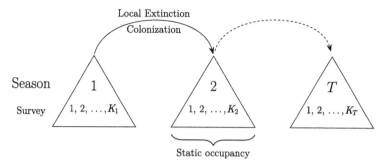

FIGURE 8.1 Graphical representation of the sampling situation for a multi-season occupancy study. Each triangle represents a season (t), with multiple (K_t) surveys within seasons. Occupancy status is static at units within seasons, but may change between seasons through the processes of colonization and local extinction.

and is the sequence of the T single season detection histories. For example the detection history $\mathbf{h}_i = 110\,000\,010$ represents a three-season study (with three surveys per season) where the target species was detected in the first and second surveys in season 1, was never detected in season 2, and detected only in the second survey in season 3. Similar to the single-season situation, due to imperfect detectability, we do not know whether the species was present but undetected in season 2 or was genuinely absent. That is, we do not know whether the species persisted at the unit for all three seasons (i.e., never went locally extinct), or went locally extinct and then (re)colonized the unit.

8.2 AN IMPLICIT DYNAMICS MODEL

One approach to modeling detection/nondetection data from multiple seasons is to effectively apply a single season model to the data collected in each of the T seasons. Under this approach, occupancy in one season is considered to be a random process in the sense that the occupancy status of a unit in the previous season has no effect on the probability of occupancy at the units in the current season. Regardless of the underlying processes of change in occupancy, only the resulting pattern or level of occupancy each season is modeled. Here, let ψ_t be the probability a unit is occupied in season t, and $p_{t,j}$ be the probability of detecting the species in the jth survey of a unit during season t (given the species was present at the unit in season t). Using the model-based approach of MacKenzie et al. (2002) (as detailed in Chapter 4), the observed data likelihood for season t would be

$$ODL_t\left(\psi_t, \mathbf{p}_t | \mathbf{h}_{t,1}, \mathbf{h}_{t,2}, \ldots, \mathbf{h}_{t,s}\right) = \prod_{i=1}^{s} Pr\left(\mathbf{h}_{t,i} | \psi_t, \mathbf{p}_t\right),$$

with the observed data likelihood evaluated for the full T seasons being the product of the seasonal likelihoods, i.e.,

$$ODL(\boldsymbol{\psi}, \mathbf{p}|\mathbf{h}_1, \mathbf{h}_2, \ldots, \mathbf{h}_s) = \prod_{t=1}^{T} ODL_t\left(\boldsymbol{\psi}_t, \mathbf{p}_t|\mathbf{h}_{t,1}, \mathbf{h}_{t,2}, \ldots, \mathbf{h}_{t,s}\right).$$

This same model can also be developed directly from the detection histories using the same techniques as in the previous chapters: taking a verbal description of the detection histories and translating them into a mathematical equation. Consider again the detection history $\mathbf{h}_i = 110\,000\,010$. A verbal description of these data would be:

In season 1: the unit was occupied with the species being detected in the first and second surveys, but not in the third.

In season 2: the unit was either occupied with the species not being detected in any of the 3 surveys, or the unit was unoccupied.

In season 3: the unit was occupied with the species being detected in the second survey, but not in the first or third surveys.

Translating these statements into mathematical equations using the defined model parameters we have:

Season 1: $\psi_1 p_{1,1} p_{1,2}\left(1 - p_{1,3}\right)$,

Season 2: $\psi_2\left(1 - p_{2,1}\right)\left(1 - p_{2,2}\right)\left(1 - p_{2,3}\right) + \left(1 - \psi_2\right)$,

Season 3: $\psi_3\left(1 - p_{3,1}\right) p_{3,2}\left(1 - p_{3,3}\right)$.

Therefore the probability of observing the entire detection history would be:

$$\begin{aligned}
Pr\left(\mathbf{h}_i = 110\,000\,010|\boldsymbol{\psi}, \mathbf{p}\right) = {} & \psi_1 p_{1,1} p_{1,2}\left(1 - p_{1,3}\right) \\
& \times \left[\psi_2\left(1 - p_{2,1}\right)\left(1 - p_{2,2}\right)\left(1 - p_{2,3}\right) + \left(1 - \psi_2\right)\right] \\
& \times \psi_3\left(1 - p_{3,1}\right) p_{3,2}\left(1 - p_{3,3}\right). \qquad (8.1)
\end{aligned}$$

This procedure can be used to obtain the probability statement for each of the s observed detection histories, and the observed data likelihood would be calculated as

$$ODL(\boldsymbol{\psi}, \mathbf{p}|\mathbf{h}_1, \mathbf{h}_2, \ldots, \mathbf{h}_s) = \prod_{i=1}^{s} Pr(\mathbf{h}_i|\boldsymbol{\psi}, \mathbf{p}).$$

Expressed in terms of the underlying random variables, the implicit dynamics model would be:

$$z_{t,i} \sim Bernoulli(\psi_t),$$

$$h_{t,ij}|z_{t,i} \sim Bernoulli\left(z_{t,i}\, p_{t,j}\right),$$

which could be used to construct the complete data likelihood.

As in Chapter 4, this model can be easily generalized so that the probabilities of occupancy and detection are functions of covariates, and to allow for missing observations. Models can also be considered where there is some structural relationship among probabilities in different seasons. For example, Field et al. (2005) modeled a systematic decline in occupancy over time by defining seasonal occupancy probabilities with a linear trend on the logit scale, i.e., $logit(\psi_t) = \beta_0 + \beta_1 t$.

Finally, we note that although the above modeling may appear to be relatively phenomenological, in the sense that vital rates (probabilities of local extinction and colonization) governing the dynamic process do not appear explicitly in this model, it actually makes fairly restrictive assumptions about these vital rates. In Section 10.4, we show that the implicit dynamics model is based on the assumption that the probability of the species not going locally extinct at a previously occupied unit is equal to the probability of colonization of a previously unoccupied unit. In the next section, we discuss a more general explicit model of occupancy dynamics, from which the above implicit dynamics model can be obtained as a special case.

8.3 MODELING DYNAMIC CHANGES EXPLICITLY

As noted in the previous section, the dynamic processes governing changes in the occupancy state variable are the colonization of an unoccupied unit by the species and the local extinction of the species at an occupied unit. In this section we consider models that directly incorporate these dynamic processes, as they are often of direct interest. They are somewhat analogous to the birth and death processes of the abundance state variable and as such supply information relevant to the long-term sustainability of a population. As the drivers of the system (with respect to occupancy), understanding how these dynamic processes are affected by changes of habitat or climatic conditions (for example) may be important for the successful management of ecological systems.

For the remainder of this chapter we consider the dynamic changes in occupancy as a first-order Markov process. That is, the probability of a unit being occupied in season t depends upon the occupancy state of the unit in the previous season, $t - 1$. In some situations, higher order Markov processes (e.g., occupancy probability at t depends upon state of occupancy at both $t - 1$ and $t - 2$) may be biologically reasonable to represent long-term 'memory' about the occupancy state of a unit. For example from mark–resight data, Hestbeck et al. (1991) modeled transition probabilities between different wintering grounds for Canada Geese (*Branta canadensis*) as a second-order Markov process to represent long term fidelity of individual birds to each region. Such an extension could be applied using the multi-season, multi-state model (Chapter 9) where the state of a unit is defined in terms of occupancy status in both the current and previous season (compared to a first-order Markov process model where, as developed below, states are defined in terms of the current season only). This is very similar to the approach used in the mark–recapture multi-event models (Pradel, 2005), although as in the mark–recapture setting, we would expect such modeling to be quite data hungry (Cole et al., 2014). Green et al. (2011) considered a second-order Markov model to accommodate the maturation period for wood frogs (*Lithobates sylvatica*), but inference methods were not fully developed.

Modeling changes in occupancy as a Markov process also accounts for a form of temporal autocorrelation. When observations on the same sampling unit are positively correlated, values close in time are more similar than those separated by longer periods (i.e., the sampling variance for a short time series will tend to be less than that of a longer time series). In the occupancy context this equates to the expectation that a unit that is occupied now may be more likely to be occupied again in the near future than one that is currently unoccupied. A Markov process adequately models this autocorrelation process.

Markovian changes in occupancy can also be considered as inducing a form of heterogeneity in occupancy probabilities where the probability of a unit being occupied in season t will be different for units that were occupied in the previous season, compared to the units that were unoccupied.

Formally, we define colonization (γ_t) and local extinction (ϵ_t) probabilities to be:

γ_t = the probability that an unoccupied unit in season t is occupied by the species in season $t + 1$; and

ϵ_t = the probability that a unit occupied in season t is unoccupied by the species in season $t + 1$.

These dynamic processes represent the probabilities of a unit transitioning between the occupied and unoccupied state between consecutive seasons (Fig. 8.2).

Below we detail three approaches to modeling multiple-season occupancy data that explicitly account for the processes of colonization and local extinction. First, we briefly discuss some historical approaches that were developed for situations where the species is (assumed to be) always detected when present at a unit (i.e., detection probability equals 1). We then focus on two methods that allow for the imperfect detection of the species; a 'conditional' and an 'unconditional' approach. The 'conditional' approach exploits the similarities between the type of data collected in the current context and capture–recapture data collected from individuals. This approach is 'conditional' in the sense that the detection history for a unit is only modeled from the season in which the

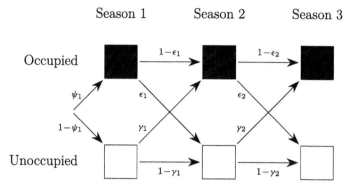

FIGURE 8.2   Representation of how the occupancy state of a unit may change between seasons in terms of the processes of first season occupancy (ψ_1), colonization (γ), and local extinction (ϵ). Filled boxes indicate the unit is occupied (species present) in that season while empty boxes indicate the unit is unoccupied (species absent).

species is first detected, i.e., the modeling conditions upon the first detection of the species. This is due to the fact that in capture–recapture, an individual is unknown to the researchers prior to the first time it is caught and marked. As a result, it is not generally possible to obtain seasonal estimates of occupancy, only estimates of the dynamic processes themselves. The second approach (upon which we largely focus) is an 'unconditional' approach, where the modeling makes full use of the detection histories. Here it is possible to estimate both occupancy and the dynamic parameters.

8.3.1   Modeling Dynamic Processes when Detection Probability is 1

In the late 1960s and throughout the 1970s, the ecological literature contained a number of studies of animals on islands in which the presence or absence of breeding populations (frequently of birds) was assessed over a number of different years (see Diamond and May, 1977, and papers cited therein). This work was motivated largely by the models of MacArthur and Wilson (1967), which suggested that species richness on islands reflected a dynamic equilibrium between rates of local extinction and colonization. Diamond and May (1977) recommended that such data be viewed as having resulted from a stationary Markov process defined by corresponding rates of extinction and colonization, a recommendation that had been anticipated by Simberloff (1969). Diamond and May (1977) focused on the implications of this model for (1) detection–nondetection data collected at varying time intervals and (2) computation of various turnover statistics.

Clark and Rosenzweig (1994) (also see Rosenzweig and Clark, 1994) considered the problem of estimating extinction and colonization rates for such a Markov model from detection–nondetection data. They provided maximum likelihood estimates assuming both a stationary process with rate parameters constant over time, and detection probabilities of 1. Erwin et al. (1998) expanded this general modeling approach, considering reduced-parameter non-Markovian models, as well as Markovian models permitting time-specific rates of local extinction and colonization. These models only provide reasonable estimates in the situation where presence and absence can be ascertained (Clark and Rosenzweig, 1994; Erwin et al., 1998) and are thus of limited usefulness.

8.3.2 Conditional Modeling of Dynamic Processes

Barbraud et al. (2003) considered models for colony unit dynamics, the same problem considered by Erwin et al. (1998), but wanted to relax the assumption of detection probabilities equal to 1. They considered the estimation problem by focusing on the analogy between occupancy dynamics of colony units and population dynamics of individual animals. The simplest form of colony detection history data for multiple seasons consists of 1's and 0's denoting detection or nondetection, respectively, at each study unit. These data are analogous to capture history data for individual animals in animal populations open to gains and losses. For example, the capture history 1 0 1 0 would indicate an animal caught in periods (analogous to 1 survey per season) 1 and 3, but not in periods 2 and 4. The usual approach to modeling such data uses parameters for survival from one sample period to the next and capture probability at each period (e.g., see Lebreton et al., 1992). Interior 0's (followed and preceded by one or more 1's; e.g., period 2) are usually viewed unambiguously as 'present but not captured' and thus modeled with the complement of capture probability. However, this analogy is not especially useful for occupancy studies in which an interior 0 can reflect either 'present but not detected' or 'absent, but followed by recolonization'.

Barbraud et al. (2003) recognized that there is a close analogy between occupancy studies and capture–recapture studies with temporary emigration (Kendall et al., 1997; Kendall, 1999; Williams et al., 2002). In the case of temporary emigration, an interior 0 can result from either 'presence without capture' or 'temporary emigration' of the animal. Unfortunately, the probability of being a temporary emigrant is confounded with capture probability in standard models for open populations (Burnham, 1993; Kendall et al., 1997). However, Kendall and Nichols (1995) and Kendall et al. (1997) recognized that the robust design (described above, also see Pollock, 1982) provides the information needed to estimate capture probability conditional on presence in the sampled area and thus allows separate estimates of this conditional capture probability and the probability of being a temporary emigrant.

Barbraud et al. (2003) viewed the Markovian temporary emigration model of Kendall et al. (1997) as potentially most useful for estimation in the context of occupancy dynamics. This temporary emigration model contains two parameters for the probability of being a temporary emigrant (i.e., the probability that an animal was not in the study population, but still part of the 'superpopulation') at any sampling period: (1) for animals that were not temporary emigrants the previous period, $t - 1$; and (2) for animals that were temporary emigrants the previous period. The temporary emigration parameter for animals that were not temporary emigrants the previous period (i.e., within the study population at $t - 1$, outside the study population at t) was viewed as local extinction probability in the context of occupancy studies (i.e., unit was occupied at $t - 1$, unoccupied at t). The complement of the second temporary emigration parameter, i.e., $1 - Pr$ (an animal outside the study population at $t - 1$, was also outside the study population at t), was viewed as a probability of colonization in occupancy studies. Under this analogy, the usual survival probability of open capture–recapture models was set equal to 1, as it reflected the probability that a unit always 'survived' (i.e., units will never 'die' in the sense that the species could always recolonize the unit a later time). Note, one situation where this might not be a reasonable assumption is where a change in the habitat or local environment of a unit renders it uninhabitable for the species, in which case the joint modeling of habitat and occupancy may be useful (see Chapter 13). To complete the analogy between the temporary emigration problem and occupancy dynamics, we note that the random temporary emigration model of Kendall et al. (1997) is equivalent to the implicit dynamics model for occupancy (Section 8.2).

The advantage of recognizing this analogy between the modeling of temporary emigration and occupancy dynamics involved software and computations. Software had been developed by Kendall and Hines (1999) and White and Burnham (1999) to obtain parameter estimates under the temporary emigration models of Kendall et al. (1997) using robust design data. Barbraud et al. (2003) thus used these programs (MARK: White and Burnham, 1999; RDSURVIV: Kendall and Hines, 1999) with data from two survey flights per year over the Camargue delta in southern France to estimate local rates of extinction and colonization for purple heron (*Ardea purpurea*) and gray heron (*Ardea cinerea*) breeding colonies at reed bed sites. Of particular interest, biologically, was the modeling of time-specific local colonization in one area of the Camargue as a function of local extinction probability in a neighboring disturbed area. This ultrastructural modeling dealt explicitly with spatial dependencies in occupancy and provided indirect inference about animal movement without using marked individuals (Barbraud et al., 2003).

We do not present the actual temporary emigration model here, as it is best viewed as a special case of the model of MacKenzie et al. (2003) which is presented in detail below. Readers especially interested in the details of the temporary emigration modeling approach for occupancy studies are directed to Barbraud et al. (2003). The primary difference between the two approaches (Barbraud et al., 2003; MacKenzie et al., 2003) is the conditional nature of the temporary emigration modeling. In capture–recapture studies of animal populations, most models condition on the release of individual animals at their periods of first capture (exceptions include the temporal symmetry models of Nichols et al., 1986, 2000a; Pradel, 1996; Williams et al., 2002). Stated differently, the 0's occurring in a capture history before an animal's first capture are not typically modeled. However, in occupancy studies in which potential units are identified at the beginning of the study, such conditioning is not needed, and initial 0's can be modeled. We thus refer to the temporary emigration modeling of Barbraud et al. (2003) as 'conditional', and contrast this with the 'unconditional' approach of MacKenzie et al. (2003). Both approaches should provide approximately un-biased estimates of the dynamic processes, but the approach of MacKenzie et al. (2003) should be more efficient and leads more readily to estimates of occupancy for each season of the study.

8.3.3   Unconditional Modeling of Dynamic Processes

MacKenzie et al. (2003) used the colonization and local extinction probabilities defined above (γ_t and ϵ_t) to extend the single-season model of MacKenzie et al. (2002). Once occupancy state (the probability of occupancy) is established in the first season (ψ_1), potential changes in the occupancy state of a unit between seasons are simply incorporated using the dynamic parameters. To construct their model, MacKenzie et al. (2003) used the now familiar approach of taking a verbal description of a detection history and translating it into a mathematical equation, giving the probability of observing the detection history. For example, consider again the detection history $\mathbf{h}_i = 110\,000\,010$ where the occupancy status of the unit in the second season is unknown. A verbal description of these data that incorporates the concepts of colonization and local extinction would be:

In season 1: the unit was occupied with the species being detected in the first and second surveys, but not in the third.

From the end of season 1 to the start of season 3 (immediately before surveying commenced): either the species did not go locally extinct between seasons 1 and 2, was not detected in any of the 3 surveys within season 2 and did not go locally extinct between seasons 2 and 3, *or* the species went locally extinct between seasons 1 and 2, then recolonized the unit between seasons 2 and 3.

In season 3: the species was detected in the second survey, but not in the first or third survey.

Translating these statements into mathematical equations using the model parameters defined earlier we have:

Season 1:
$$\psi_1 p_{1,1} p_{1,2}(1 - p_{1,3}),$$

From the end of season 1 to the start of season 3:
$$(1 - \epsilon_1)\left(\prod_{j=1}^{3}(1 - p_{2,j})\right)(1 - \epsilon_2) + \epsilon_1 \gamma_2,$$

Season 3:
$$(1 - p_{3,1}) p_{3,2}(1 - p_{3,3}).$$

The probability of observing the complete detection history would be,

$$Pr(\mathbf{h}_i = 110\,000\,010|\boldsymbol{\theta}) = \psi_1 p_{1,1} p_{1,2}(1 - p_{1,3})$$
$$\times \left[(1 - \epsilon_1)\left(\prod_{j=1}^{3}(1 - p_{2,j})\right)(1 - \epsilon_2) + \epsilon_1 \gamma_2\right]$$
$$\times (1 - p_{3,1}) p_{3,2}(1 - p_{3,3}), \tag{8.2}$$

where $\boldsymbol{\theta}$ is the set of parameters in the model.

Note the differences between Eqs. (8.2) and (8.1), where the dynamic processes are not explicitly modeled. Here the model incorporates a mechanistic process for how the occupancy state of a unit may change between seasons, whereas in the former model only the state of occupancy each season is considered.

Formally, the explicit dynamics model could be defined in terms of the underlying latent and observed random variables as:

$$z_{1,i} \sim Bernoulli(\psi_1),$$

$$(z_{t,i}|z_{t-1,i} = 0) \sim Bernoulli(\gamma_{t-1}) \text{ for } t = 2, \ldots, T,$$

$$(z_{t,i}|z_{t-1,i} = 1) \sim Bernoulli(1 - \epsilon_{t-1}) \text{ for } t = 2, \ldots, T,$$

$$h_{t,ij}|z_{t,i} \sim Bernoulli(z_{t,i} p_{t_j}).$$

Note that unlike for the implicit dynamics model, the Bernoulli distribution associated with the occupancy random variables $z_{t,i}$ for $t = 2 \ldots, T$, is different dependent upon the presence or absence of the species at the unit in the previous season. If the species was absent from unit i in the previous season ($z_{t-1,i} = 0$), the species is present in season t with probability γ_t (i.e., probability of the unit being colonized by the species), whereas, if the species was present at unit i in the previous season ($z_{t-1,i} = 1$), the species is present in season t with probability $1 - \epsilon_t$ (i.e., probability the species did not go locally extinct at the unit). In the implicit dynamics model, the probability of unit i being occupied in season t is always ψ_t, irrespective of the occupancy status of the unit in season $t - 1$. The conditional nature of the explicit dynamics model is what characterizes a first-order Markov process.

Generally, there may be a number of different possible pathways that could result in an observed detection history. MacKenzie et al. (2003) therefore suggest it might be most convenient to describe the model using matrix notation (see Appendix for a brief introduction to matrices). Let $\boldsymbol{\phi}_t$ be the 2×2 matrix denoting the probability of a unit transitioning between occupancy states from season t to $t + 1$. Rows of $\boldsymbol{\phi}_t$ represent the occupancy state of the unit in season t (state $0 =$ unoccupied; state $1 =$ occupied), and columns represent the occupancy state at $t + 1$, that is:

$$\boldsymbol{\phi}_t = \begin{bmatrix} 1 - \gamma_t & \gamma_t \\ \epsilon_t & 1 - \epsilon_t \end{bmatrix}.$$

Further, let $\boldsymbol{\phi}_0$ be the row vector:

$$\boldsymbol{\phi}_0 = \begin{bmatrix} 1 - \psi_1 & \psi_1 \end{bmatrix},$$

where ψ_1 is the probability the unit is occupied in the first season. This vector models whether a unit was occupied or unoccupied in the first season. Next, define $\mathbf{p}_{t,j}$ to be the detection probability matrix, defining the probability of each type of observation (or observed state) in survey j of season t, given the true occupancy state of a unit in season t (as used in Chapters 5 and 6). At this stage, there are only two possible outcomes of each survey, nondetection or detection of the species. Therefore:

$$\mathbf{p}_{t,j} = \begin{bmatrix} 1 & 0 \\ 1 - p_{t,j} & p_{t,j} \end{bmatrix}$$

where rows represent the true occupancy state of a unit in season t, and columns the observed state in survey j of season t. As we described for the single-season multi-state model, the probability of observing the detection history $\mathbf{h}_{t,i}$ for unit i in season t, conditional upon occupancy state, can be represented as the column vector $\mathbf{p}_{\mathbf{h},t}$. This is found by element-wise multiplication of the respective columns of $\mathbf{p}_{t,j}$, column 1 for nondetections and column 2 for detections, associated with the observation made during each survey within the season (see

Chapter 5 for details). For instance, $\mathbf{p}_{\mathbf{h},t}$ for the within-season detection history $\mathbf{h}_{t,i} = 101$ would be:

$$\mathbf{p}_{101,t} = \begin{bmatrix} 0 \\ p_{t,1} \end{bmatrix} \odot \begin{bmatrix} 1 \\ 1 - p_{t,2} \end{bmatrix} \odot \begin{bmatrix} 0 \\ p_{t,3} \end{bmatrix}$$

$$= \begin{bmatrix} 0 \\ p_{t,1} \left(1 - p_{t,2}\right) p_{t,3} \end{bmatrix}.$$

This expression indicates that the probability of observing this detection history is 0 if the unit was unoccupied (the first element; as the species could not be detected if it was absent from the unit), and $p_{t,1}\left(1 - p_{t,2}\right) p_{t,3}$ if the unit was occupied by the species (the second element). The first element of $\mathbf{p}_{\mathbf{h},t}$ will always be 0 whenever the species is detected at least once at the unit during season t. Using similar reasoning, if the species is never detected at a unit during season t ($\mathbf{h}_{t,i} = 000$), the first element will always be 1, as this is the only observable detection history for a unit that is unoccupied, that is:

$$\mathbf{p}_{000,t} = \begin{bmatrix} 1 \\ \prod_{j=1}^{3} \left(1 - p_{t,j}\right) \end{bmatrix}.$$

Using this matrix notation, the probability statement for an observed detection history, for all seasons, could be calculated as:

$$Pr\left(\mathbf{h}_i | \psi_1, \boldsymbol{\gamma}, \boldsymbol{\epsilon}, \mathbf{p}\right) = \boldsymbol{\phi}_0 \prod_{t=1}^{T-1} \left(D\left(\mathbf{p}_{\mathbf{h},t}\right) \boldsymbol{\phi}_t\right) \mathbf{p}_{\mathbf{h},T}, \tag{8.3}$$

where $D\left(\mathbf{p}_{\mathbf{h},t}\right)$ is a diagonal matrix with the elements of $\mathbf{p}_{\mathbf{h},t}$ along the main diagonal (top left to bottom right), zero otherwise. Diagonalizing the vector is required merely for the matrix algebra to work out correctly. Note that what Eq. (8.3) is doing is performing a series of matrix multiplications (see Appendix) that automatically sum together the various possible outcomes that could have resulted in the same observed detection history, rather than manually evaluating the different options as was done when developing Eq. (8.2).

Initially Eq. (8.3) may look somewhat confusing, but stepping through the various components, we see that it does have an intuitive interpretation. $\boldsymbol{\phi}_0$ establishes the probability that a unit is either unoccupied or occupied immediately prior to surveys commencing in season 1. The term $\left(D\left(\mathbf{p}_{\mathbf{h},t}\right) \boldsymbol{\phi}_t\right)$ calculates the probability of observing the particular sequence of detections and nondetections in season t (conditional upon occupancy state), and then the probability of the unit transitioning to the occupied or unoccupied state immediately before

TABLE 8.1 Examples of detection histories (h_i) and the associated probabilities of observing them ($Pr(h_i|\psi_1, \gamma, \epsilon, p)$) using the unconditional explicit dynamics model

h_i	$Pr(h_i|\psi_1, \gamma, \epsilon, p)$
11 10 01	$= \phi_0 D(p_{11,1}) \phi_1 D(p_{10,2}) \phi_2 p_{01,3}$

$$= \begin{bmatrix} 1-\psi_1 & \psi_1 \end{bmatrix} \begin{bmatrix} 0 & 0 \\ 0 & p_{1,1}p_{1,2} \end{bmatrix} \begin{bmatrix} 1-\gamma_1 & \gamma_1 \\ \epsilon_1 & 1-\epsilon_1 \end{bmatrix} \begin{bmatrix} 0 & 0 \\ 0 & p_{2,1}(1-p_{2,2}) \end{bmatrix} \begin{bmatrix} 1-\gamma_2 & \gamma_2 \\ \epsilon_2 & 1-\epsilon_2 \end{bmatrix} \begin{bmatrix} 0 \\ (1-p_{3,1})\,p_{3,2} \end{bmatrix}$$

$$= \psi_1 p_{1,1}p_{1,2}(1-\epsilon_1)\,p_{2,1}(1-p_{2,2})(1-\epsilon_2)(1-p_{3,1})\,p_{3,2}$$

| 00 10 00 | $= \phi_0 D(p_{00,1}) \phi_1 D(p_{10,2}) \phi_2 p_{00,3}$ |

$$= \begin{bmatrix} 1-\psi_1 & \psi_1 \end{bmatrix} \begin{bmatrix} 1 & 0 \\ 0 & \prod_{j=1}^{2}(1-p_{1,j}) \end{bmatrix} \begin{bmatrix} 1-\gamma_1 & \gamma_1 \\ \epsilon_1 & 1-\epsilon_1 \end{bmatrix} \begin{bmatrix} 0 & 0 \\ 0 & p_{2,1}(1-p_{2,2}) \end{bmatrix} \begin{bmatrix} 1-\gamma_2 & \gamma_2 \\ \epsilon_2 & 1-\epsilon_2 \end{bmatrix} \begin{bmatrix} 1 \\ \prod_{j=1}^{2}(1-p_{3,j}) \end{bmatrix}$$

$$= \left((1-\psi_1)\gamma_1 + \psi_1 \left(\prod_{j=1}^{2}(1-p_{1,j})\right)(1-\epsilon_1)\right) p_{2,1}(1-p_{2,2}) \left(\epsilon_2 + (1-\epsilon_2)\prod_{j=1}^{2}(1-p_{3,j})\right)$$

| 00 00 00 | $= \phi_0 D(p_{00,1}) \phi_1 D(p_{00,2}) \phi_2 p_{00,3}$ |

$$= \begin{bmatrix} 1-\psi_1 & \psi_1 \end{bmatrix} \begin{bmatrix} 1 & 0 \\ 0 & \prod_{j=1}^{2}(1-p_{1,j}) \end{bmatrix} \begin{bmatrix} 1-\gamma_1 & \gamma_1 \\ \epsilon_1 & 1-\epsilon_1 \end{bmatrix} \begin{bmatrix} 1 & 0 \\ 0 & \prod_{j=1}^{2}(1-p_{2,j}) \end{bmatrix} \begin{bmatrix} 1-\gamma_2 & \gamma_2 \\ \epsilon_2 & 1-\epsilon_2 \end{bmatrix} \begin{bmatrix} 1 \\ \prod_{j=1}^{2}(1-p_{3,j}) \end{bmatrix}$$

$$= (1-\psi_1)\left((1-\gamma_1)\left((1-\gamma_2)+\gamma_2\prod_{j=1}^{2}(1-p_{3,j})\right) + \gamma_1\left(\prod_{j=1}^{2}(1-p_{2,j})\right)\left(\epsilon_2+(1-\epsilon_2)\prod_{j=1}^{2}(1-p_{3,j})\right)\right)$$
$$+ \psi_1\left(\prod_{j=1}^{2}(1-p_{1,j})\right)\left(\epsilon_1\left((1-\gamma_2)+\gamma_2\prod_{j=1}^{2}(1-p_{3,j})\right) + (1-\epsilon_1)\left(\prod_{j=1}^{2}(1-p_{2,j})\right)\left(\epsilon_2+(1-\epsilon_2)\prod_{j=1}^{2}(1-p_{3,j})\right)\right)$$

surveying begins in season $t + 1$. This is done recursively from season 1 to immediately before the final season of surveying (season T), hence the product term $\prod_{t=1}^{T-1} \left(D\left(\mathbf{p}_{h,t}\right) \boldsymbol{\phi}_t \right)$. At this stage the equation has calculated the probability of observing the particular detection history up to the end of the second to last season of surveying, and the unit being in either the occupied or unoccupied state immediately prior to the surveying in season T. Therefore to complete the probability statement, the probability of observing the sequence of detections and nondetections in the final season (conditional upon occupancy state) is required, i.e., $\mathbf{p}_{h,T}$. This final term is not diagonalized (i.e., is just a 2×1 column vector) so the result of the series of matrix multiplications is just a single number, as the first term was a 1×2 row vector and all intervening matrices were of dimension 2×2 (see Appendix for more details on the aspect of matrix multiplication). Some examples of observed detection histories and their probability statements, according to the above model, are given in Table 8.1. We encourage readers to take the time to work through these examples to cement their understanding of the model. From the probability statements for each observed detection history, the model observed data likelihood can be calculated in the usual manner (assuming independence of detection histories), i.e.,

$$ODL(\psi_1, \boldsymbol{\gamma}, \boldsymbol{\epsilon}, \mathbf{p}|\mathbf{h}_1, \ldots, \mathbf{h}_s) = \prod_{i=1}^{s} Pr(\mathbf{h}_i|\psi_1, \boldsymbol{\gamma}, \boldsymbol{\epsilon}, \mathbf{p}).$$

Note that if $T = 1$, that is the study is only conducted for a single season, then the above equation reduces to $Pr(\mathbf{h}_i|\psi_1, \mathbf{p}) = \boldsymbol{\phi}_0 \mathbf{p}_{h,1}$. This is an equivalent formulation for calculating the probability of observing a detection history for the single season model of MacKenzie et al. (2002) (and described in Chapter 4), and this matrix form is the same as that presented for the single-season multi-state model in Chapter 5.

The complete data likelihood (CDL) for the unconditional explicit dynamics model can also be determined by considering the underlying latent and observed random variables, similar to the approach outlined in Section 4.4.1, but with additional complexity to account for changes in occupancy over time. Briefly, the key is to again assume that the latent variable for the occupancy status of a unit is known in each season. If so, the joint probability of observing the detection history data and presence/absence of the species in each season for unit i could be expressed as:

$$Pr(\mathbf{h}_i, \mathbf{z}_i|\psi_1, \boldsymbol{\gamma}, \boldsymbol{\epsilon}, \mathbf{p}) = \prod_{t=1}^{T} Pr\left(\mathbf{h}_{t,i}|\mathbf{p}_t, z_{t,i}\right)$$

$$\times \prod_{t \in \tau_0} Pr\left(z_{t,i} | \gamma_{t-1}, z_{t-1,i} = 0\right)$$

$$\times \prod_{t \in \tau_1} Pr\left(z_{t,i} | \epsilon_{t-1}, z_{t-1,i} = 1\right)$$

$$\times Pr\left(z_{1,i} | \psi_1\right),$$

where τ_0 is the set of seasons between 2 and T in which unit i was unoccupied by the species in the previous season, and τ_1 is the set of seasons between 2 and T in which unit i was occupied by the species in the previous season. Therefore, using a CDL approach, there are four independent components associated with the joint probability statement that determines the probability of:

1. first-season occupancy, $Pr\left(z_{1,i} | \psi_1\right)$;
2. extinctions, $\prod_{t \in \tau_1} Pr\left(z_{t,i} | \epsilon_{t-1}, z_{t-1,i} = 1\right)$;
3. colonizations, $\prod_{t \in \tau_0} Pr\left(z_{t,i} | \gamma_{t-1}, z_{t-1,i} = 0\right)$;
4. detection of the species in each survey given the species presence at the unit in each season, $\prod_{t=1}^{T} Pr\left(\mathbf{h}_{t,i} | \mathbf{p}_t, z_{t,i}\right)$.

Note that these terms are ordered 4–1 in the expression for the CDL above, following convention for the ordering of the conditional events. The relevant thing to note with the CDL approach is that unlike using the ODL, there is no summation of terms, as assuming the $z_{t,i}$ values are known removes the ambiguity associated with nondetection of the species during a season. However, a different set of estimation algorithms must be used to account for the fact that the $z_{t,i}$ values are actually unobserved (e.g., expectation-maximization algorithm or MCMC).

8.3.4   Missing Observations

Missing observations can be easily accounted for using this type of modeling approach, as in the single-season case. If the missing observations occur within season t, then the vector $\mathbf{p}_{h,t}$ is adjusted by removing the corresponding $p_{t,j}$ parameter(s). For example, if the history 11- is obtained at primary period t (where "-" indicates a missing observation), then:

$$\mathbf{p}_{11\text{-},t} = \begin{bmatrix} 0 \\ p_{t,1} p_{t,2} \end{bmatrix}.$$

This represents that fact that no information, on either detection or nondetection, has been collected about the parameter $p_{t,3}$ from the unit with this detection history.

Similarly, the model can be adjusted to allow for situations where a unit was not surveyed for an entire season. Consider the following detection history, where the unit was not surveyed at all in the second season, $\mathbf{h}_i = 10 - - 11$. Again, no information has been collected regarding either the detection or non-detection of the species, although here the occupancy state of the unit at season 2 is also unknown, hence all possibilities must be allowed for. This can be achieved by (effectively) omitting $\mathbf{p}_{\mathbf{h},2}$ entirely; i.e. the probability of this detection history is

$$Pr(\mathbf{h}_i = 10 - - 11 | \psi_1, \gamma, \epsilon, \mathbf{p}) = \phi_0 D(\mathbf{p}_{10,1}) \phi_1 \phi_2 \mathbf{p}_{11,3}.$$

By having the ability to accommodate missing observations, the unconditional model of MacKenzie et al. (2003) provides a great deal of flexibility in the way the data can be collected in the field, and still be analyzed using this technique. Not only can there be unequal sampling effort across units within seasons, but potentially, not all units have to be surveyed each season (within reason). However, it is important to note that even though no data were collected from this unit during the second season, the associated colonization and local extinction probabilities still appear in the probabilistic statement (within ϕ_2). As such, it is assumed that these probabilities are either the same, or are functions of the same covariates, at units that are and are not surveyed within that season. This assumption of the model must be carefully considered if a study design is proposed that intentionally avoids surveying all units each season.

8.3.5 Including Covariate Information

Thus far in this chapter, an implicit assumption has been made that all model parameters are constant across all units. Failure of this assumption results in heterogeneity in model parameters, which could result in inferences that are inaccurate. As already discussed in Chapters 4 and 7, one approach to dealing with potential heterogeneity is the inclusion of information on variables that may affect the value of one or more parameters, or covariates. Indeed, the relationship between the covariates and certain parameters of interest may often be the primary motivation for conducting the study (e.g., habitat variables in habitat modeling, or measures of isolation and patch size in metapopulation studies). As noted in Chapters 3 and 4, covariate information can be included in the model by use of an appropriate link function, e.g., the logit link (Chapter 3). The mechanics for doing so are identical to those presented in Section 4.4.8, hence we do not cover this material again here.

Similar to the single season case, occupancy, colonization and local extinction probabilities could all be functions of variables that have a single, constant value for the duration of the season (season-specific covariates). These may be variables that characterize units during each sampling season (e.g., habitat type, average value of a weather-related variable measured during the season, elevation or patch size) or variables that characterize the change in a quantity between seasons (e.g., changes in habitat composition). Detection probabilities can be functions of season-specific covariates, but also functions of variables that may change with each survey of a unit (e.g., rainfall in preceding 24 hours, air temperature or observer).

There is one pertinent point about potential covariates whose values may change from one season to the next and missing observations. As noted above, these methods can allow for situations where units may not be surveyed in some seasons, however in doing so, the occupancy-related parameters associated with that season are still included in the probability statement and model likelihood. When those parameters are being modeled as functions of covariates, the value of the respective covariates for each unit, including the unsurveyed ones, must be known to calculate the parameter value. There will be some classes of covariates for which this is problematic; covariates whose values are dynamic over time and can only be determined during a survey of that unit. In such a case investigators will have to develop some reasonable means to determine what the covariate values may have been at the unsurveyed units, or else recognize that they can not use the set of affected covariates in an analysis of the full data set (but possibly on a subset of the data with no missing values). Covariates that are unchanging over time, or the value of which can be determined independently of the detection/nondetection surveys, could still be used.

Having the ability to incorporate covariate information of these types provides a great deal of flexibility in the models that could be considered as reasonable descriptions of the processes that give rise to the data. Moreover, different hypotheses about the system can often be expressed as models that involve different sets of covariates for each parameter type. The strength of evidence for each hypothesis can then be determined by fitting the suite of models with the different sets of covariates and making a formal comparison of the models (e.g., by using the AIC model selection criterion). For example, in metapopulation studies, local extinction probabilities are frequently assumed to be decreasing functions of patch area (e.g., Moilanen, 1999). That is, the species is more likely to go extinct from small patches than large patches. This may be reasonable for some species, but perhaps not in every case. Furthermore in some situations, variation in the areas of the sampled patches may be insufficient to discern such an effect on local extinction probabilities. Therefore, a second hypothesis would be that local extinction probabilities are constant with respect to patch area. These competing hypotheses could be formulated as two models with different sets of covariates for ϵ_t. To represent the area hypothesis, a model could be fit to the data where 'patch area' is included as a covariate for extinction probability, and a second model without the 'patch area' covariate for ϵ_t (but identical in all other respects) could be fit to represent the second hypothesis. The level of support for each of the two models would then reflect the degree of support of each hypothesis. Note that one could also use a similar approach to determine the functional form of such a relationship (e.g., linear or quadratic), or even compare link functions. However, we caution that such comparisons should only be done on the basis of sound biological reasoning, not in the pursuit of a 'best' model.

8.3.6   Alternative Parameterizations

MacKenzie et al. (2003) noted that in some situations quantities other than the probability of occupancy in the first season, seasonal colonization and local extinction probabilities may be of interest. They suggested that these quantities could be derived from the estimated parameters, or the model could be reparameterized so that the quantities are estimated directly.

One immediate option is to parameterize the explicit dynamics model in terms of the probabilities of first-season occupancy, seasonal colonization and seasonal persistence; where persistence is defined as the probability of the unit being occupied by the species in successive seasons. That is:

$$\phi_t = Pr(\text{species present at unit in season } t + 1 \mid \text{species present at unit in season } t)$$
$$= 1 - \epsilon_t.$$

As the above indicates, persistence is the complement of local extinction probability, and distinguishing between the two is similar to distinguishing between survival and mortality probabilities of individuals. One advantage of parameterizing the explicit dynamics model in terms of persistence rather than local extinction is that all occupancy-related parameters are in terms of the probability of the species being present at a unit, whereas local extinction is in terms of the probability of the species being absent from a unit. We would point out, however, that persistence probabilities can be easily derived from estimates obtained using the original parameterization by simply substituting $\hat{\epsilon}_t$ into the above equation. Because persistence is the complement of local extinction, the standard error of $\hat{\epsilon}_t$ is also the standard error for $\hat{\phi}_t$. Furthermore, if covariates have been included in the modeling of local extinction and the effect sizes estimated, to interpret the effect of those covariates in terms of persistence probability, one just changes the sign of the estimated effect size. Once again, no adjustment to the standard errors of the effect sizes is necessary.

Seasonal estimates of occupancy are another such quantity. In some applications (e.g., monitoring), the processes of colonization and local extinction may not be of direct interest with the main focus of the study being how occupancy changes over time. The three probabilities are simply related by the recursive equation:

$$\psi_{t+1} = \psi_t(1 - \epsilon_t) + (1 - \psi_t)\gamma_t, \qquad (8.4)$$

i.e., units occupied next season are a combination of those units occupied this season where the species does not go locally extinct, $\psi_t(1 - \epsilon_t)$, and the units that are currently unoccupied that are colonized by the species before next season, $(1 - \psi_t)\gamma_t$. This is analogous to how the abundance of a species at a particular point in time is comprised of the survivors from the previous period, and new recruits. Eq. (8.4) can be rearranged to make either of the dynamic processes the subject, that is:

$$\gamma_t = \frac{\psi_{t+1} - \psi_t(1 - \epsilon_t)}{(1 - \psi_t)}$$

or

$$\epsilon_t = 1 - \frac{\psi_{t+1} - (1 - \psi_t)\gamma_t}{\psi_t}.$$

The same model as described above would be used, except rather than estimate the γ_t and ϵ_t parameters directly, one would directly estimate, for example, the seasonal occupancy and local extinction probabilities. The value for γ_t could then be derived using the above formula and used in the model to evaluate the likelihood.

There may be a temptation to use the recursive occupancy equation above in association with the implicit dynamic modeling approach described in Section 8.2 as a means of incorporating colonization and local extinction probabilities into a multi-season occupancy model. However, doing so does not yield the explicit dynamics approach described above. For example, consider the simple detection history $\mathbf{h}_i = 10\,01$. Using the unconditional approach of MacKenzie et al. (2003), the probability of observing this history would be (where θ denotes the set of parameters in the model):

$$Pr(\mathbf{h}_i = 10\,01|\theta) = \psi_1 p_{1,1}(1 - p_{1,2})(1 - \epsilon_1)(1 - p_{2,1})p_{2,2},$$

while using the implicit dynamics model from Section 8.2, the probability would be:

$$Pr(\mathbf{h}_i = 10\,01|\theta) = \psi_1 p_{1,1}(1 - p_{1,2})\psi_2(1 - p_{2,1})p_{2,2}.$$

Substituting the expression for ψ_2 given by Eq. (8.4) into the implicit dynamics model, does not give the equivalent of the unconditional explicit dynamics model. That is:

$$
\begin{aligned}
Pr(\mathbf{h}_i = 10\,01|\boldsymbol{\theta}) &= \psi_1 p_{1,1}\left(1 - p_{1,2}\right)\psi_2\left(1 - p_{2,1}\right)p_{2,2} \\
&= \psi_1 p_{1,1}\left(1 - p_{1,2}\right)\left[\psi_1(1 - \epsilon_1) + (1 - \psi_1)\gamma_1\right]\left(1 - p_{2,1}\right)p_{2,2} \\
&\neq \psi_1 p_{1,1}\left(1 - p_{1,2}\right)(1 - \epsilon_1)\left(1 - p_{2,1}\right)p_{2,2}.
\end{aligned}
$$

Another quantity suggested by MacKenzie et al. (2003) is the rate of change in occupancy. By analogy with population size (where the comparable measure is known as the *finite rate of change* or *growth rate*), they suggest it could be defined as:

$$
\lambda_t = \frac{\psi_{t+1}}{\psi_t}.
$$

Using the recursive occupancy equation as an intermediate step, the unconditional model could be reparameterized so that λ_t is estimated directly. However there are some practical problems that limit the usefulness of this parameterization. First, there are bounds on the allowable values of λ_t that vary with ψ_t. For example, suppose that currently the probability of occupancy is 0.5 (i.e., $\psi_t = 0.5$), then the maximal rate of change in occupancy, as defined above, must be 2, otherwise the probability of occupancy in the next season will exceed 1.0. However, if currently $\psi_t = 0.2$, then the maximal rate of change in occupancy would be 5. Second, you cannot have a constant, long-term, rate of change greater than 1 as eventually it will result in an estimate of $\psi_{t+1} > 1$. For instance, suppose the probability of occupancy in season 1 is 0.2. A long-term rate of change of 1.2 (i.e., occupancy probability increases by 20% each season) would suggest that in season 10 occupancy is greater that 1.0 (0.20, 0.24, 0.29, ..., 0.72, 0.86, 1.03).

An alternative definition for the rate of change in occupancy is to use odds ratios, that is:

$$
\lambda'_t = \frac{\psi_{t+1}/(1 - \psi_{t+1})}{\psi_t/(1 - \psi_t)}. \tag{8.5}
$$

While it may seem more complicated to interpret, it has the advantage of not suffering from the restrictions of the above definition. Also, the general concept is similar to that of using a logit or log-odds link function (recall from Chapter 3 that the odds ratio is the amount by which the odds of occupancy in season t is multiplied to get the odds of occupancy in season $t + 1$). Further, if λ'_t is constant across time (i.e., $\lambda'_1 = \lambda'_2 = \cdots = \lambda'_{T-1}$), then $ln(\lambda')$ will correspond to the trend parameter when modeling occupancy as a linear function of time on the logit scale:

$$logit(\psi_t) = \beta_0 + \beta_1 t$$
$$= \beta_0 + ln(\lambda') t.$$

In some applications, researchers express an interest in 'turnover' of the species, which could be defined in multiple ways. One definition would be the probability that a unit that is occupied has just become occupied. That is:

$$\tau_t = \frac{(1 - \psi_t) \gamma_t}{\psi_{t+1}}.$$

Another definition for 'turnover' would be the probability of a unit changing occupancy status between seasons:

$$\tau_t' = \psi_t \epsilon_t + (1 - \psi_t) \gamma_t.$$

Using either of these definitions it would be possible to reparameterize the unconditional explicit dynamics model, through a series of calculations, such that these alternative quantities of interest can be estimated directly. As noted above, however, instead of attempting to estimate these other parameters of interest directly, they could be derived by substituting the estimated values from the original parameterization into the respective series of equations, with the associated standard error determined by application of the delta method (Chapter 3). The main advantage of reparameterizing the model in terms of these alternative quantities is when there is a desire to directly model such quantities (e.g., consistency over time or as functions of covariates). When covariates are being incorporated, for some quantities a link function other than the logit-link may have to be used.

The choice of which parameterization may be most appropriate in a given situation depends on the goals of the study and scientific questions being addressed. If the main focus is on the underlying dynamic processes and factors that may affect them, then the original parameterization should be used. In many management scenarios, it is natural to focus upon occupancy estimates and changes in occupancy over time (e.g., trends in occupancy), suggesting that one of the alternative parameterizations may be more appropriate. However, we point out that while identifying whether the level of occupancy is increasing or decreasing over time has some utility, oftentimes a deeper understanding of the underlying dynamics will lead to a better understanding about how management actions may influence occupancy dynamics, and therefore which actions will obtain management goals most efficiently.

The results from fitting different parameterizations of the model to the same data are comparable, including the comparison of model selection metrics such as AIC. There is nothing inherently wrong with the comparison of multiple parameterizations of the model, however we suggest that choice of parameterization should be generally governed by the study objective rather than fitting models with all possible parameterizations and using model selection criteria to differentiate among them. We have often found the original parameterization to be the most numerically stable, particularly when a model contains a large number of covariates. As colonization and local extinction probabilities must take values in the 0–1 interval, there are constraints on allowable values for the occupancy probability. Enforcing these constraints when using a reparameterized version of the unconditional model can make the computer algorithms unstable.

8.3.7 Example: House Finch Expansion in North America

House finches (*Carpodacus mexicanus*) are native to western, but not eastern, North America. However, they were released in 1942 on Long Island, New York, and have exhibited an impressive westward expansion since that time. The magnitude of this expansion is such that it is obvious in the raw data of the North American Breeding Bird Survey (BBS; Robbins et al., 1986). Here we subject the BBS data to the probabilistic modeling of this chapter in an effort to draw formal inferences about this expansion. The BBS has been conducted annually since the mid-1960s by volunteer observers. The counts are conducted during the peak of the breeding season, usually during June. Observers follow a route along roads for ∼39.2 km, stopping every 0.8 km for 50 consecutive stops. At each stop, a point count is conducted for three minutes with observers counting all birds detected within a 400 m radius. There are now >4000 BBS routes throughout North America, so the geographic coverage is extensive.

The BBS protocol specifies that routes be run once each breeding season, so the data do not contain the temporal replication that we typically use for occupancy modeling. We thus take a different approach and view each of the 50 stops as a replicate count from the area covered by the route. This is far from ideal. For example, under the view that the area covered by each stop is a random selection from the area covered by all stops, we would ideally be sampling with replacement. Nevertheless, given the survey design and protocol of the BBS, we view our approach as not only reasonable, but better than most available approaches.

To investigate the westward expansion during the period 1976–2001, data from 694 BBS routes within 2600 km from the Long Island point of release were considered at 5-year intervals (i.e., 1976, 1981, 1986, . . .). We used a relatively phenomenological kind of modeling in which we focused on how these probabilities were related to distance from Long Island in each year. Thus, distance (*d*) and year (*year*) were covariates in our analysis. Distance was measured at

100 km increments, at the scale of 1000 km (i.e., $d = 0.1 = 100$ km). Year was included in the models as a categorical covariate, or factor, with the final year value for each parameter type treated as the standard or reference category (1996 for colonization and extinction as these are between period events, and 2001 for detection). The only other covariate used an *ad hoc* approach to dealing with relative abundance of birds that was possible because of the large number of stops per route. Specifically, we created a categorical variable for observed frequency of occurrence (f) indicating whether house finches were detected on > 10 stops in the route in any previous year or not (i.e., were locally highly abundant). This covariate was used to model detection probability, along with an interaction between year and distance, and is similar in intent to defining a 'trap response' covariate to allow the detection probability to be different (higher) for routes after they reached this observed frequency threshold. The logit-link function was used for all parameter types.

These covariates were used for detection probability in all models that were fit to the data, denoted as $p(year \times d + f)$. Only distance and year were used to model local rates of extinction and colonization, and only distance was used to model the initial occupancy level, $\psi_{76}(d)$. We assumed house finch were initially more common closer to the release point, hence did not consider a model where occupancy probability in 1976 was the same at all distances, i.e., $\psi_{76}(\cdot)$, to be biologically reasonable. Our prediction was that there would be an increase in rate of colonization with distance as time progressed, i.e., the effect of distance on colonization would increase over time, suggesting an interaction between the year and distance covariates. We had no real expectation about rate of extinction probabilities, other than they would be generally low. Therefore, we did not consider models that included an interaction between the distance and year covariate to guard against obtaining a spurious result with such a small number of events. We expected occupancy to increase with distance as time progressed, in much the same manner as colonization (although note that occupancy probabilities were derived using Eq. (8.4) for 1981 onward).

Twenty models were fit to the data using the R package RPresence and the top eight ranked models appear in Table 8.2. The model with lowest AIC received a model weight of 0.77, indicating a good degree of support, with the second-ranked model having an AIC model weight of 0.23. No other models considered were supported in comparison. There is very strong evidence of an interaction between the year and distance covariates for colonization (as included for both top models), with the effect of distance on colonization being different in different years. There is also very strong evidence that extinction probability changes with distance (as appears as a covariate in both top models), with some evidence of an additive year effect (as included in second-ranked model).

TABLE 8.2 Summary of model selection results for the house finch example. Factors affecting occupancy, colonization, and local extinction probabilities include distance (d) and year ($year$). Occupancy and detection probabilities were modeled as functions of these same factors and a categorical variable for observed frequency of occurrence (f); specifically $\psi_{76}(d)$ and $p(year \times d + f)$, respectively. Models are notated in terms of the factors included for colonization (γ) and extinction probabilities. Given are the relative differences in AIC values compared to the top ranked model (ΔAIC), AIC model weights (w), the numbers of parameters in the models ($Npar$), and twice the negative log-likelihood ($-2l$). Results are only presented for the top eight ranked models

Model	ΔAIC	w	$Npar$	$-2l$
$\gamma(year \times d)\epsilon(d)$	0.00	0.77	27	44,414.09
$\gamma(year \times d)\epsilon(year + d)$	2.41	0.23	31	44,408.50
$\gamma(year \times d)\epsilon(year)$	12.72	0.00	30	44,420.81
$\gamma(year \times d)\epsilon(\cdot)$	14.62	0.00	26	44,430.71
$\gamma(year + d)\epsilon(d)$	29.25	0.00	23	44,451.35
$\gamma(year + d)\epsilon(year + d)$	31.03	0.00	27	44,445.12
$\gamma(year + d)\epsilon(year)$	40.70	0.00	26	44,456.79
$\gamma(year + d)\epsilon(\cdot)$	47.44	0.00	22	44,471.54

TABLE 8.3 Estimated regression coefficients ($\hat{\beta}$) and associated standard errors ($SE(\hat{\beta})$) for the probability house finch were present at a Breeding Bird Survey route in 1976, on the logit-scale, i.e., $logit(\psi_{76,i})$. Given are the estimates from the two models ranked highest by AIC (Table 8.2). w is the AIC model weight for each model. Distance from Long Island (d) was measured in 1000 km units, to the nearest 100 km

Term	Model 1: $w = 0.77$		Model 2: $w = 0.23$	
	$\hat{\beta}$	$SE(\hat{\beta})$	$\hat{\beta}$	$SE(\hat{\beta})$
Intercept	-0.83	0.41	-0.81	0.41
d	-1.22	0.48	-1.17	0.53

The estimated regression coefficients for each parameter type, on the logit-scale, are given in Tables 8.3–8.6 from the two highest AIC-ranked models. There is generally very good agreement in the estimated effect sizes from the two models, especially considering the magnitude of the standard errors associated with the estimates. As predicted, the estimated effect of distance on occupancy in 1976 is negative (Table 8.3), indicating the probability of house finch being present at a BBS route was lower farther away from the release point. The interaction terms for the year and distance covariates on colonization ($year_y:d$; Table 8.4) indicate how the effect of distance on colonization is different in the respective years compared to the effect of distance in 1996 (the reference year in this case; d). To obtain the effect of distance in each year, the

TABLE 8.4 Estimated regression coefficients ($\hat{\beta}$) and associated standard errors (SE($\hat{\beta}$)) for the probability house finch colonize a Breeding Bird Survey route between survey periods t and $t + 1$, on the logit-scale, i.e., $logit(\gamma_{t,i})$. Given are the estimates from the two models ranked highest by AIC (Table 8.2). w is the AIC model weight for each model. Year effects $year_t$ are additive terms for the indicated years, distance from Long Island (d) was measured in 1000 km units, to the nearest 100 km, and $year_t{:}d$ are the interaction terms between year and distance

Term	Model 1: $w = 0.77$		Model 2: $w = 0.23$	
	$\hat{\beta}$	SE($\hat{\beta}$)	$\hat{\beta}$	SE($\hat{\beta}$)
Intercept	0.54	0.53	0.63	0.55
$year_{76}$	0.89	0.80	0.83	0.82
$year_{81}$	1.79	0.71	1.63	0.71
$year_{86}$	1.76	0.72	1.52	0.72
$year_{91}$	0.13	0.72	0.03	0.73
d	−0.74	0.34	−0.78	0.34
$year_{76}{:}d$	−7.44	3.15	−7.37	3.13
$year_{81}{:}d$	−3.50	0.85	−3.33	0.85
$year_{86}{:}d$	−1.38	0.49	−1.26	0.48
$year_{91}{:}d$	0.10	0.44	0.15	0.45

main effect of distance and the interaction terms must be added together. For example, from the top model:

$$d_{76} = d + year_{76}{:}d$$
$$= -0.74 - 7.44$$
$$= -8.18,$$

$$d_{81} = d + year_{81}{:}d$$
$$= -0.74 - 3.50$$
$$= -4.24,$$

$$d_{86} = d + year_{86}{:}d$$
$$= -0.74 - 1.38$$
$$= -2.12,$$

$$d_{91} = d + year_{91}{:}d$$
$$= -0.74 + 0.10$$
$$= -0.64,$$

and

$$d_{96} = d$$
$$= -0.74.$$

Clearly, these models suggest the effect of distance on colonization generally increased over time, as predicted. For extinction probability, the estimated effect of distance is positive from both models (Table 8.5) so local extinction probability was estimated to be higher farther away from Long Island, the location of the population 'source', which is biologically reasonable. The year effects in the second-ranked model suggests some temporal variation in the overall level of extinction, although this model is not well supported. The estimated regression coefficients for detection probability (Table 8.6) suggest that the effect of distance on detection also increased over time, similar to colonization probability, so house finch were more detectable at greater distances from Long Island as time progressed. This may have been due to local abundance increasing over time after house finch colonized an area. The estimated effect of f is positive, hence house finch were more detectable in a survey after being detected at > 10 stops on a BBS route in a previous year.

The model averaged estimates of occupancy, colonization and extinction probabilities in each year are presented in Fig. 8.3. The occupancy probabilities in 1976 were estimated directly in the modeling, while the values for subsequent years were derived from the estimated colonization and extinction probabilities in each year, and occupancy probabilities for the preceding year, using Eq. (8.4). No measure of uncertainty has been presented for clarity of

TABLE 8.5 Estimated regression coefficients ($\hat{\beta}$) and associated standard errors (SE($\hat{\beta}$)) for the probability that house finch went locally extinct from a Breeding Bird Survey route between survey periods t and $t + 1$, on the logit-scale, i.e., $logit(\epsilon_{t,i})$. Given are the estimates from the two models ranked highest by AIC (Table 8.2). w is the AIC model weight for each model. Year effects $year_t$ are additive terms for the indicated years, and distance from Long Island (d) was measured in 1000 km units

Term	Model 1: $w = 0.77$		Model 2: $w = 0.23$	
	$\hat{\beta}$	SE($\hat{\beta}$)	$\hat{\beta}$	SE($\hat{\beta}$)
Intercept	−3.39	0.26	−2.93	0.33
$year_{76}$			0.33	1.29
$year_{81}$			−0.13	0.72
$year_{86}$			−1.17	0.68
$year_{91}$			−0.60	0.37
d	1.17	0.25	1.00	0.26

TABLE 8.6 Estimated regression coefficients ($\hat{\beta}$) and associated standard errors (SE($\hat{\beta}$)) for the probability house finch were detected on a Breeding Bird Survey route at which they were present in survey period t, on the logit-scale, i.e., $logit(p_{t,ij})$. Given are the estimates from the two models ranked highest by AIC (Table 8.2). w is the AIC model weight for each model. Year effects $year_t$ are additive terms for the indicated years, distance from Long Island (d) was measured in 1000 km units, to the nearest 100 km, $year_t$:d are the interaction terms between year and distance, and f is the effect on detection of house finch being detected at > 10 stops in a previous year

Term	Model 1: $w = 0.77$		Model 2: $w = 0.23$	
	$\hat{\beta}$	SE($\hat{\beta}$)	$\hat{\beta}$	SE($\hat{\beta}$)
Intercept	−2.09	0.05	−2.09	0.05
$year_{76}$	−0.26	0.21	−0.26	0.21
$year_{81}$	0.29	0.10	0.29	0.10
$year_{86}$	0.62	0.08	0.62	0.08
$year_{91}$	0.66	0.06	0.66	0.06
$year_{96}$	0.18	0.06	0.18	0.06
d	−0.43	0.05	−0.43	0.05
$year_{76}$:d	−9.76	2.00	−9.86	2.01
$year_{81}$:d	−2.85	0.43	−2.86	0.43
$year_{86}$:d	−1.75	0.24	−1.76	0.24
$year_{91}$:d	−0.41	0.08	−0.41	0.08
$year_{96}$:d	0.08	0.07	0.08	0.07
f	0.94	0.03	0.94	0.03

the plots. The odds-ratio rate of change in occupancy in each year (λ'_t, Eq. (8.5); lower right) highlights at what distances were the fastest rates of change between years. Note that colonization, extinction, and rate of change are between season processes, thus there are $T - 1$ season-specific values, while there are T season-specific values for occupancy. This same information could also be presented in terms of maps. For example, Fig. 8.4 presents the model averaged occupancy probabilities estimated at different distances overlaid with a map of the eastern USA. Similar maps could be created for the other parameters if desired. Maps could also be produced in more complex situations involving additional covariates, e.g., habitat or elevation, although we stress, as noted previously, ideally one would not only present maps of estimates, but also some measure of uncertainty (e.g., standard error, confidence interval limits, or CV) to convey to the reader how reliable the estimates might be in different areas.

In this example, we have modeled expansion of house finch distribution as a function of distance from its release point at Long Island, allowing the relationship between distance and the model parameters to change over time. This enabled us to make inferences about the underlying processes associated with

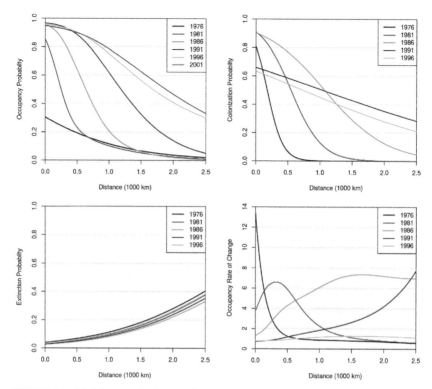

FIGURE 8.3 Model averaged estimates of occupancy (upper left), colonization (upper right), and extinction (lower left) probabilities for house finch at Breeding Bird Survey routes in the eastern USA, as a function of distance from the Long Island release point, in each year. The derived odds-ratio based rate of change in occupancy as a function of distance is also presented (lower right).

changes in the house finch distribution with respect to distance from Long Island, including rates of change as a function of distance. The underlying logic easily transfers to other covariates that might be of interest in other applications. For example, how is a species distribution changing over time relative to elevation, and at what elevations are the fastest rates of change in the species distribution? Such information would provide insights about how species are responding to changes in the environment.

We believe that the general topic of range expansion and contraction will become increasingly important in the future with the spread of invasive species and range changes induced by climate change. Monitoring programs designed to permit estimation of occupancy will be ideally suited to study these changes. We would like to extend these methods to more mechanistic models of range expansion. In particular, we would like to use an approach similar to that of Wikle (2003) to model colonization as a function of occupancy of nearby sample

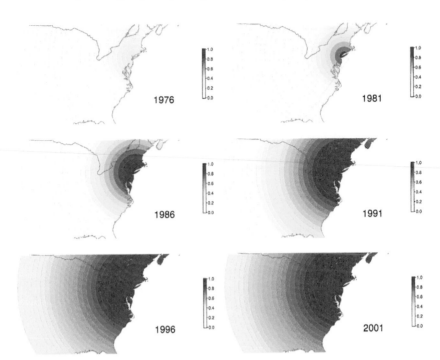

FIGURE 8.4 Maps of the model averaged occupancy probabilities in each year for house finch at Breeding Bird Survey routes in the eastern USA.

units, rather than simply as a time×distance interaction. Incorporating spatial correlation in the underlying occupancy dynamics would be one approach for doing so (e.g., Bled et al., 2011, 2013; Yackulic et al., 2012; Eaton et al., 2014; Chapter 10).

Before finishing with this example, we would like to point out that there were a number of practical challenges associated with its preparation that users should be aware of. These challenges were actually anticipated at the beginning of our modeling effort because of the nature of the invasion process that we sought to model. In early years following release, we knew that there were very few data for medium to large distances from the release site. Specifically, with very low occupancy at even intermediate distances, extinction events are expected to be rare, as extinction is conditional on species presence. Similarly, colonization simply did not occur at greater distances until later in the time series. The point is that we were trying to model rate parameters corresponding to rare events (hence little data), and this will always be a difficult task. With our treatment of year as a categorical variable, we were on the verge of asking too much of the data. We found that the likelihood function for many of the models considered exhibited multiple maxima, and the optimization routines used by the RPresence R package, and also used by Program PRESENCE, would sometimes converge to different maxima depending on the starting values used for the regression coefficients. Our solution was to use five random sets of starting values for each model, and use the results from the set that converged to the highest likelihood value. In our experience the issue of multiple maxima is more problematic for more complex models, and we advise practitioners to check for the possibility of multiple maxima with their data. We did not check how sensitive the optimization routines used by other software were to multiple maxima with these data, but this is a general issue for any application of maximum likelihood estimation so we expect all software to exhibit some degree of sensitivity to it. Another practical issue we encountered related to how we coded the *year* categorical covariate, in particular, which year was treated as the standard or reference category. Initially the first category (1976) was used, but we found that in doing so, the software had numerical problems obtaining the variance–covariance matrix (from which standard errors are obtained) for the estimated regression coefficients. Recoding the *year* covariate such that the final year was used as the reference category removed this issue. This is likely due to the fact that at the beginning of the time series for these data, there were many fewer detections than later in the time series. We therefore recommend that when using a categorical covariate, you should choose your reference category to be one that has a fair number of detections associated with it, and not choose one with relatively few detections.

The files used for this example can be downloaded from http://www.proteus.co.nz.

8.4 VIOLATIONS OF MODEL ASSUMPTIONS

The assumption of no unmodeled heterogeneity in any of the parameters (occupancy, colonization, extinction or detection probabilities) is one of several assumptions for the multiple season models presented in the chapter. Additional assumptions include: (1) occupancy state at each unit does not change over surveys within a season; that is, consistent with Pollock's robust design, units are 'closed' to changes in occupancy within seasons or primary periods, (2) detection of species and detection histories at each location are independent, and (3) the target species are never falsely detected (i.e., species are identified correctly).

If these assumptions are not met, some or all estimators may be biased, and inferences about factors that influence both occupancy and occupancy dynamics may be erroneous. Even within the capture–recapture arena, there has been little investigation of effects of heterogeneity on robust design estimators; rather, it is believed that these estimators behave in a manner similar to those for separate closed and open capture–recapture models (Williams et al., 2002).

Assumption violations for single-season (closed) models were presented in Section 4.4.9. In this section we briefly review anticipated impacts of occupancy closure violations and possible solutions and then focus primarily on the impacts of assumption violations on rate parameters estimated between seasons (i.e., during the open periods). We caution readers that there have been few formal investigations of assumption violations within the occupancy context, and that information in this section is based mostly on the analogy with capture–recapture population models that may not always have parallels to occupancy studies.

As mentioned in previous chapters, the closed occupancy state assumption within seasons can be relaxed, provided changes in occupancy are random (sensu Kendall et al., 1997; Kendall, 1999). The species of interest is viewed as having some non-negligible probability of being present in the unit at the time of any survey, which is unaffected by whether the species was present at the previous survey (that would be non-random or Markovian changes within a season). When there are random changes within each season, the occupancy estimator is approximately unbiased, but interpreted as the probability units are *used* by the target species, and detection probability is the probability the species is present at the time of the survey and detected at occupied or used units. Therefore, changes in 'occupancy' should be interpreted as changes in 'use', and colonization and local extinction probabilities are the underlying dynamic parameters governing changes in use.

Nonrandom movement of a species in and out of sample units likely causes bias in occupancy estimators; nevertheless, if movement is always either only in or only out of the unit(s) (i.e., immigration or emigration only) then Kendall (1999) describes ways in which surveys can be combined to likely eliminate bias in occupancy estimators. As noted in Section 4.4.9, Kendall's (1999) recommendations involve pooling survey data into two surveys per season and then using models with survey-specific detection probabilities. Specifically, for the case of only emigration, the first survey is retained for each unit, and the last $K - 1$ surveys are combined into a second 'survey'. In the case of only immigration, the first $K - 1$ surveys are combined and treated as the initial survey, and survey K becomes the second survey. Under this approach, approximately unbiased estimates can likely be obtained for either occupancy at the beginning of each season, for emigration-only situations, or occupancy at the end of each season for immigration-only situations (see Section 4.4.9 and Kendall, 1999, for details). Kendall (1999) also mentions that this pooling approach is valid within the robust design context, yielding unbiased estimates of survival rate between primary periods. In the context of multiple season occupancy models, we would anticipate that similar pooling to accommodate emigration- or immigration-only movement within seasons would yield approximately unbiased estimates of extinction and colonization probabilities. Bias in occupancy estimates will likely remain if analyses are conducted using more than two surveys per season or models with constant detection probability (Section 4.4.9; Kendall, 1999). Kendall's (1999) work suggests that unmodeled heterogeneity or permanent trap response (see below) in detection probabilities will cause bias in occupancy and vital rate estimators.

Another option to deal with the closure assumption is to restrict the data to include surveys between times when the availability of the species is uninterrupted (i.e., during periods of closure) as demonstrated by MacKenzie et al. (2003) with tiger salamanders in Minnesota. Here detection/nondetection information was only included during a time period where the life history of the species dictated that individuals would be confined to the pond (eggs, larvae and early metamorphs). Time periods when adults may be migrating to ponds or when metamorphs may be transitioning to a terrestrial life phase were not included in the analysis. Again, investigators should use their knowledge about the phenology of the target species and design their studies to try to minimize violations in the closure assumption.

The impact of unmodeled variation in occupancy, colonization, and extinction probability among units is virtually unexplored, and more thorough simulation studies are still needed. Effects of heterogeneous survival rates have been investigated for open population capture–recapture estimators (Nichols et al., 1982; Pollock and Raveling, 1982; Pollock et al., 1990). However, the analogy between extinction and the complement of survival is not sufficiently close that we are comfortable in drawing inferences about effects on extinction estimators based on inferences about survival estimators. Recall, for example, to use capture–recapture models to estimate parameters of occupancy dynamics, Barbraud et al. (2003) equated temporary emigration parameters with colonization and extinction. We are aware of no investigation of the effects of heterogeneity on the temporary emigration estimators presented by Kendall et al. (1997), so we conclude that the effects of heterogeneous rates of extinction and colonization are a topic of future investigation.

There are some extensions of the multi-state occupancy models discussed so far that may be more appropriate when some of the above model assumptions are suspected to be violated. These are discussed in the next two chapters. We would also point out that, to the best of our knowledge, robust methods for assessing model fit, and thereby identifying evidence of assumption violations, are yet to be developed for multi-season models.

8.5   DISCUSSION

We believe that this chapter may be the most important in the book. Although most previous occupancy studies have focused on single-season patterns (the topic of Part II), the objectives of most of these previous investigations involved dynamic processes (see review of Chapter 2). Because of the difficulties inherent in attempts to infer process from observation of pattern (see Chapters 1 and 2), we believe that studies of units extending over multiple seasons are likely to provide the strongest inferences about occupancy dynamics and the processes that produce these dynamics. We thus believe that the models of this chapter should see a great deal of use and that future work should focus on extensions and elaborations of these approaches.

The implicit dynamics modeling approach of Section 8.2 essentially involved multiple applications of the single-season models of Chapter 4 to species detection data from a sequence of seasons. However, if occupancy dynamics are best viewed as a Markov process (i.e., non-random changes in occupancy over time), as will be reasonable in many situations, then the explicit dynamics models of Section 8.3 should provide better descriptions of the data. In cases for which they are appropriate, the implicit dynamics models of Section 8.2 yield time-specific estimates of occupancy, as well as estimates of rate of change or 'trend' in occupancy. Trend estimates are the focus of many current animal monitoring programs, and are justified as providing a basis for prioritization of conservation efforts. Species and areas in which rapid reductions in occupancy are occurring make prime targets for conservation efforts. However, estimation of trends does not provide much information about the causes of observed declines or, more importantly, about the kinds of management actions that are likely to reverse them. It is possible to model rate of change in occupancy as functions of environmental or management covariates, and such modeling can prove useful. Although, we believe that direct modeling of the processes governing change is likely to be even more useful.

The models of Section 8.3 explicitly incorporate parameters for the vital rates responsible for changes in occupancy, rates of local extinction and colonization. Covariate modeling can be used to investigate effects of environmental variables and management actions on these rate parameters. We believe that the models of this chapter deserve much more attention than they have thus far received, as they permit direct investigation of such topics as metapopulation dynamics, range dynamics, and the relationship between occupancy dynamics and habitat change. These models also provide alternative means of investigating population dynamics that do not require detailed studies of marked individuals. For example, Barbraud et al. (2003) were able to draw inferences about bird movement (shifting of colony units) by modeling the vital rates in one location as a function of vital rates in a neighboring location. These inferences were indirect, and thus not as strong as those based on observed movements of marked animals (e.g., Nichols, 1996; Kendall and Nichols, 2004). However, such indirect inferences can be obtained for areas too large to permit comprehensive capture–recapture studies, thus providing a useful complement to more detailed intensive investigations. Finally, we discussed some of the consequences of model assumption violations.

The models presented in this chapter can be used not only as a basis for parameter estimation and data analysis, but also for prediction of species distributions into the future or under alternative scenarios for the underlying dynamic processes. Such prediction is based on the transition probability matrix, ϕ_t, and we revisit this topic during the next two chapters.

In the following chapters we discuss a number of useful extensions and applications of these dynamic occupancy models, including multi-state models (Chapter 9), false-positive detections, spatial correlation, investigation of the fundamental properties of the stochastic processes governing changes in occupancy, heterogeneous detection probabilities (Chapter 10), and study design considerations (Chapter 12). In Chapter 13 we present an approach that enables joint modeling of both habitat and occupancy dynamics, allowing for some form of dependence between the two sets of processes. The multi-species cooccurrence dynamics model outlined in Chapter 14 should prove very useful, for example for dealing with competition between native species of conservation concern and related species experiencing range expansions (Olson et al., 2005; Yackulic et al., 2014). Finally, we believe that similar models of occupancy dynamics hold great promise for investigations of such topics as multi-species community dynamics (Chapter 15).

Chapter 9

More than Two Occupancy States

In Chapter 5 we introduced the idea that occupancy state of a unit may be more than just a dichotomous-valued variable, and that it could be defined with a greater number of categories, for example: (1) species absent; (2) species present without breeding; and (3) species present with breeding. Imperfect detection means that the observed state from field surveys will not always reflect the true occupancy state, which needs to be accounted for when making inferences about the underlying pattern of occurrence, and factors that may affect that pattern. We detailed methods in Chapter 5 that have been developed for investigating landscape-wide patterns in these multiple occupancy states for a single-season, accounting for the potential ambiguity of the true occupancy state of a unit by utilizing the information about the detection process from repeat surveys conducted within the season.

Here, the concepts of Chapters 5 and 8 are combined to enable investigation of occupancy dynamics for situations with more than two occupancy states over multiple seasons. We refer to this as a multi-season multi-state occupancy model. The general framework presented in this chapter is very useful, and as will be demonstrated in later chapters, provides the structure for a number of interesting extensions. We believe that multi-state occupancy models dramatically increase our ability to address interesting questions about ecological processes associated with concepts such as source–sink dynamics, disease spread, and changes in the (relative) abundance distribution of species.

For example, often in source–sink applications each type of location is considered as fixed, with some areas being inherently productive and others inherently unproductive, but the status is assumed to be constant over time. Such a view is reasonable for situations where habitat is relatively static and is the primary determinant of reproductive success at a location. However, if there is temporal variation in habitat quality with respect to reproductive output, then locations may transition between 'source' and 'sink' states/habitat over time. For low density species that exhibit high fidelity to sites of previous breeding, certain phenotypic 'costs of reproduction' at the individual level (reduced probability of successful reproduction in years following successful reproduction) might yield, at the landscape level, predictable patterns in the species occurrence and

Occupancy Estimation and Modeling. http://dx.doi.org/10.1016/B978-0-12-407197-1.00012-0

reproductive dynamics. For instance, units where the species successfully reproduced in one year would have a high probability of no reproduction occurring in the following year, while units where the species was present but did not reproduce might have a higher probability of containing successful reproduction in the following year. The combination of temporal variation in habitat quality and the existence of reproductive costs could produce oscillating patterns in state-transition probabilities.

Alternatively, in many avian species an individual's reproductive success in one breeding season is believed to serve as a predictor of reproductive success the next season, i.e., some individuals are considered to be inherently better breeders. Individuals may have a higher fidelity to breeding sites where they have successfully reproduced compared to those sites with no, or failed, reproduction (e.g., Greenwood and Harvey, 1982; Johnson et al., 1992). Even in the absence of substantial habitat variation over time, such behavior would tend to induce a relationship between reproductive state and species occurrence. Units with successful reproduction in one year would tend to be occupied the next, whereas units without successful reproduction would be less likely to be occupied the next year.

9.1   BASIC SAMPLING SCHEME

The same basic sampling scheme described in the previous chapter is used here, with suitably defined sampling units being surveyed for the species repeatedly within a season, and units being resurveyed in subsequent seasons at systematic points in time. The prime difference being that, here, the outcome of the surveys may be more than just detection/nondetection of the target species. As in Chapter 5, we begin by assuming a natural hierarchy to the true states such that the observed state must be $\leq$ the true state of the unit. For example, using the above scenario, if the species is present without breeding (true state $= 1$), that precludes the possibility of observing breeding at the unit (observed state $= 2$), but the species could be detected without breeding (observed state $= 1$) or the species could be undetected in a survey (observed state $= 0$). This creates a situation where there is potential one-way misclassification of units from the observed data; the true state cannot be lower than the observed state, but it may be higher. In the present context of investigating occupancy dynamics, this potential ambiguity creates uncertainty about the true transition in occupancy states and can therefore lead to misleading inferences. Conducting repeat surveys at a unit, thereby having multiple opportunities to observe the true state of the unit, provides information on the detection and (mis)classification probabilities, enabling this ambiguity to be accounted for. Note that the maximum observed state at a unit within a season equates to the minimum possible true

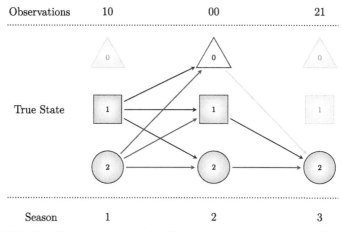

FIGURE 9.1 Graphical representation of ambiguity in true occupancy state each season and changes in state between seasons. Non-gray symbols indicate the possible true states for a unit each season given the detection observations in that season, and arrows indicate the possible state transitions. Arrow color indicates the true state of the unit at the beginning of the interval.

occupancy state for the unit, as the true state is assumed fixed within a season (i.e., closure).

The detection history for a unit therefore consists of a sequence of observed states, grouped by season. For example, if unit i is surveyed twice per season over three seasons, the resulting detection history ($\mathbf{h}_i$) for that unit could be denoted as:

$$\mathbf{h}_i = 10\,00\,21.$$

This history would be interpreted as:

> *In season 1 the species was detected and state 1 observed in the first survey, and the species was not detected in the second survey (so true state may be either 1 or 2), the species was undetected in both surveys in season 2 (so true state may be 0, 1, or 2), then detected in both surveys in season 3, with evidence of state 2 being observed in the first survey, and state 1 observed in the second (so true state must be 2).*

Note ambiguity exists in some seasons for the true occupancy state of the unit, therefore ambiguity also exists in the change in occupancy states between seasons (Fig. 9.1).

9.2 DEFINING AN EXPLICIT DYNAMICS MODEL

Multi-season multi-state occupancy data could be analyzed using an extension of the implicit dynamics models defined in the previous chapter, which would amount to a series of multi-state, single-season analyses with some possible constraints on parameters across seasons. Such an approach might be reasonable when interest is primarily in the patterns of occurrence in each season, or when units are not consistently surveyed each season. Here we focus in the approach detailed by MacKenzie et al. (2009) that explicitly considers the dynamic processes of change in occupancy state between seasons.

Within any given season, conditional upon the true state of a unit, detection probabilities can be defined as in Chapter 5. That is, let $p_{t,j}^{[m,l]}$ be the probability of observing the evidence of occupancy state l in survey j of season t, given the true occupancy state is m. We will denote the matrix of detection probabilities for survey j of season t as $\mathbf{p}_{t,j}$, where each row relates to the possible true states ($m = 0$, 1, or 2) and columns relate to observed states ($l = 0$, 1, or 2):

$$\mathbf{p}_{t,j} = \begin{bmatrix} 1 & 0 & 0 \\ 1 - p_{t,j}^{[1,1]} & p_{t,j}^{[1,1]} & 0 \\ 1 - p_{t,j}^{[2,1]} - p_{t,j}^{[2,2]} & p_{t,j}^{[2,1]} & p_{t,j}^{[2,2]} \end{bmatrix}.$$

Further, as defined previously, let $\mathbf{p}_{t,j}^{[m\bullet]}$ denote the row of $\mathbf{p}_{t,j}$ for true state m, and $\mathbf{p}_{t,j}^{[\bullet l]}$ denote the column of $\mathbf{p}_{t,j}$ for observed state l. Here $\mathbf{p}_{t,j}$ is defined using a multinomial parameterization, but we note other parameterizations are possible including a conditional binomial parameterization that was described in Chapter 5, and we provide further details of that parameterization in this chapter.

The probability of a unit being in each occupancy state in season 1 can be defined with the row vector $\boldsymbol{\phi}_0$, as in the single-season case, e.g.,

$$\boldsymbol{\phi}_0 = \begin{bmatrix} 1 - \varphi_1^{[1]} - \varphi_1^{[2]} & \varphi_1^{[1]} & \varphi_1^{[2]} \end{bmatrix}.$$

To model changes in the occupancy state between seasons, let $\varphi_t^{[m,n]}$ be that probability of a unit transitioning from state $m \to n$ between t and $t + 1$. This can be expressed as the transition probability matrix, $\boldsymbol{\phi}_t$:

$$\boldsymbol{\phi}_t = \begin{bmatrix} \varphi_t^{[0,0]} & \varphi_t^{[0,1]} & \varphi_t^{[0,2]} \\ \varphi_t^{[1,0]} & \varphi_t^{[1,1]} & \varphi_t^{[1,2]} \\ \varphi_t^{[2,0]} & \varphi_t^{[2,1]} & \varphi_t^{[2,2]} \end{bmatrix},$$

with the constraint that each row must sum to 1.0 (Fig. 9.2). Note that the row of $\boldsymbol{\phi}_t$ corresponding to a unit being in state m in season t (which we denote as

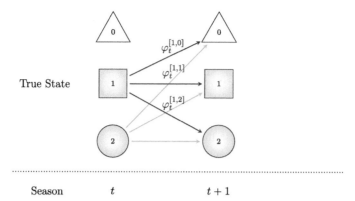

FIGURE 9.2 Illustration of possible state transitions between seasons t and $t + 1$. Arrow color indicates the true state of the unit in season t (state m), and the probabilities associated with the transitions from state m must sum to 1.0. The defined transition probabilities where $m = 1$ are given as an example.

$\phi_t^{[m\bullet]}$) gives the probability of such a unit being in the possible occupancy states in season $t + 1$.

Formally, let $z_{t,i}$ denote the random variable for the occupancy state of unit i in season t, and $h_{t,ij}$ denote the random variable for the outcome of survey j of unit i in season t. Then:

$$z_{1,i} \sim Categorical\left(\phi_0\right),$$

$$\left(z_{t,i}|z_{t-1,i} = m\right) \sim Categorical\left(\phi_{t-1}^{[m\bullet]}\right) \text{ for } t > 1,$$

$$\left(h_{t,ij}|z_{t,i} = m\right) \sim Categorical\left(\mathbf{p}_{t,j}^{[m\bullet]}\right).$$

As each row of ϕ_t must sum to 1.0, one term for each row will be obtained by subtraction of the remaining terms in the row. The choice of which term to obtain by subtraction is arbitrary, but different options may be more appropriate depending upon the questions of interest. For example, setting the terms in the first column to be obtained by subtraction means that all of the estimated parameters relate to the probability of being in an occupied state (i.e., in state 1 or 2) in season $t + 1$. That is:

$$\phi_t = \begin{bmatrix} 1 - \varphi_t^{[0,1]} - \varphi_t^{[0,2]} & \varphi_t^{[0,1]} & \varphi_t^{[0,2]} \\ 1 - \varphi_t^{[1,1]} - \varphi_t^{[1,2]} & \varphi_t^{[1,1]} & \varphi_t^{[1,2]} \\ 1 - \varphi_t^{[2,1]} - \varphi_t^{[2,2]} & \varphi_t^{[2,1]} & \varphi_t^{[2,2]} \end{bmatrix}.$$

Alternatively, the elements on the main diagonal could be obtained by subtraction so that the parameters being estimated are the ones that relate to the probabilities of a change in state:

$$
\phi_t = \begin{bmatrix} 1 - \varphi_t^{[0,1]} - \varphi_t^{[0,2]} & \varphi_t^{[0,1]} & \varphi_t^{[0,2]} \\ \varphi_t^{[1,0]} & 1 - \varphi_t^{[1,0]} - \varphi_t^{[1,2]} & \varphi_t^{[1,2]} \\ \varphi_t^{[2,0]} & \varphi_t^{[2,1]} & 1 - \varphi_t^{[2,0]} - \varphi_t^{[2,1]} \end{bmatrix}.
$$

The parameterizations used above for ϕ_0 and ϕ_t are referred to as multinomial parameterizations because the probabilities are estimated independently, except for the constraint that each row must sum to 1.0. As for the multi-state single-season model (Chapter 5), in many cases a conditional binomial parameterization may be more practical where the cell probabilities are modeled as a series of binary outcomes. For example, if state 2 represents occupied with breeding, state 1 occupied with no breeding, and state 0 unoccupied, then ϕ_t could be defined as:

$$
\phi_t = \begin{bmatrix} 1 - \psi_{t+1}^{[0]} & \psi_{t+1}^{[0]}\left(1 - R_{t+1}^{[0]}\right) & \psi_{t+1}^{[0]} R_{t+1}^{[0]} \\ 1 - \psi_{t+1}^{[1]} & \psi_{t+1}^{[1]}\left(1 - R_{t+1}^{[1]}\right) & \psi_{t+1}^{[1]} R_{t+1}^{[1]} \\ 1 - \psi_{t+1}^{[2]} & \psi_{t+1}^{[2]}\left(1 - R_{t+1}^{[2]}\right) & \psi_{t+1}^{[2]} R_{t+1}^{[2]} \end{bmatrix}
$$

where $\psi_{t+1}^{[m]}$ is the probability of a unit being occupied in season $t + 1$ given the unit was in state m in season t, and $R_{t+1}^{[m]}$ is the probability of reproduction occurring at an occupied unit in season $t + 1$ given the unit was in state m in season t. A similar parameterization could be used for ϕ_0, as in Chapter 5. Obviously the above binomial parameterization is not unique and other parameterizations could be used depending on the biological questions of interest, although we have found this parameterization to be useful in many circumstances.

When there are only two possible states, then the explicit dynamics multi-season multi-state model described above simplifies to the regular explicit dynamics multi-season model of Chapter 8. For example, ϕ_t reduces to:

$$
\begin{bmatrix} 1 - \varphi_t^{[0,1]} & \varphi_t^{[0,1]} \\ \varphi_t^{[1,0]} & 1 - \varphi_t^{[1,0]} \end{bmatrix}
$$

or

$$
\begin{bmatrix} 1 - \psi_{t+1}^{[0]} & \psi_{t+1}^{[0]} \\ 1 - \psi_{t+1}^{[1]} & \psi_{t+1}^{[1]} \end{bmatrix}
$$

where $\varphi_t^{[0,1]}$ or $\psi_{t+1}^{[0]}$ is the probability of the unit being occupied in season $t + 1$ given it was unoccupied in season t, i.e., the probability of the unit being colonized by the target species, γ_t (Chapter 8). Similarly, $\varphi_t^{[1,0]}$ or $1 - \psi_{t+1}^{[1]}$ is the probability of the unit being unoccupied in season $t + 1$ given it was occupied in season t. That is, the probability of the species going locally extinct at the unit, ϵ_t. This highlights that the multi-season multi-state model is a natural extension of the multi-season model of MacKenzie et al. (2003) (Chapter 8).

Finally, we note that $\boldsymbol{\phi}_0$ and $\boldsymbol{\phi}_t$ can be used as the basis for an occupancy-based population model, as they are equivalent to the initial population vector and stage-based population projection matrix in abundance-based population models (e.g., Caswell, 2001). Using this framework predictions can be made about the future occupancy state of the species either at the landscape- or unit-level. Ideally the transition probabilities for the underlying dynamics will be estimated from real data, but these could always be specified by other means, or 'what if' scenarios could be considered to determine how changes in these parameters are likely to affect the long-term status of the species. If the transition probabilities are related to habitat or other covariates, the impact of future changes in these covariates on the population can also be assessed. A closely related issue is assessing the sensitivity of the species occupancy status to the dynamic parameters. Martin et al. (2009b) have considered this in the simpler case of two occupancy states, and Green et al. (2011) and Miller (2012) for the multi-state situation. We return to this topic in Chapter 10.

9.3 MODELING DATA AND PARAMETER ESTIMATION

As we have noted throughout the book, there are two general approaches to estimation that could be used with this model, either the observed data likelihood or complete data likelihood (Chapter 3). Here we briefly outline these approaches as they pertain to the multi-season multi-state model.

The observed data likelihood is constructed by considering the detection history for each surveyed unit and determining the unconditional probability of observing the data. This is calculated by summing the probabilities for all possible outcomes that could have resulted in the observed detection history at a unit. For example, consider the detection history $\mathbf{h}_i = 10\ 21\ 00$, representing a unit that was surveyed for three seasons, with two surveys per season.

In season 1: the unit may have been in state 1 with state 1 observed in the first survey and the species undetected in the second survey, *or* the unit was actually in state 2 with state 1 observed in the first survey and nondetection in the second survey.

In season 2: the unit must have been in state 2, with state 2 being observed in the first survey and state 1 observed in the second survey. Therefore, the possible state transitions between seasons 1 and 2 are between true states $1 \rightarrow 2$ *or* $2 \rightarrow 2$.

In season 3: the unit may have been in any of the three possible states as the species was not detected in either survey (so possible state transitions are $2 \to 0$, *or* $2 \to 1$, *or* $2 \to 2$).

To obtain the probability of observing this detection history we again translate this verbal description of the history using the model parameters (for the more general formulation) defined above:

$$Pr(\mathbf{h}_i = 10\ 21\ 00)$$

$$= \left(\begin{array}{l} \varphi_1^{[1]} p_{1,1}^{[1,1]} \left(1 - p_{1,2}^{[1,1]}\right) \varphi_1^{[1,2]} \\ + \varphi_1^{[2]} p_{1,1}^{[2,1]} \left(1 - p_{1,2}^{[2,1]} - p_{1,2}^{[2,2]}\right) \left(1 - \varphi_1^{[2,0]} - \varphi_1^{[2,1]}\right) \end{array} \right)$$

$$\times\ p_{2,1}^{[2,2]} p_{2,2}^{[2,1]}$$

$$\times \left(\begin{array}{l} \varphi_2^{[2,0]} \\ + \varphi_2^{[2,1]} \left(1 - p_{3,1}^{[1,1]}\right) \left(1 - p_{3,2}^{[1,1]}\right) \\ + \left(1 - \varphi_2^{[2,0]} - \varphi_2^{[2,1]}\right) \left(1 - p_{3,1}^{[2,1]} - p_{3,1}^{[2,2]}\right) \left(1 - p_{3,2}^{[2,1]} - p_{3,2}^{[2,2]}\right) \end{array} \right).$$

Note the terms in the first set of parentheses include the probabilities of the unit being in either state 1 or 2 in season 1, then transitioning to state 2 in season 2. The parameterization used here allows the transition probabilities associated with changing states to be estimated, and probabilities of remaining in the same state between seasons are obtained by subtraction (e.g., $\varphi_t^{[2,2]} = 1 - \varphi_t^{[0,2]} - \varphi_t^{[1,2]}$). The middle term represents the probability of the detection in season 2, given the unit was in state 2. The final term is the probability of transitioning from state 2 to any of the 3 states between seasons 2 and 3, and the respective probabilities of nondetection.

The probability of observing other detection histories can be calculated in a similar manner, although as with the two-state multi-season model described in the previous chapter, these probability statements can be expressed succinctly using the matrix formulation given in Eq. (8.3). That is:

$$Pr(\mathbf{h}_i | \boldsymbol{\theta}) = \boldsymbol{\phi}_0 \prod_{t=1}^{T-1} D\left(\mathbf{p}_{\mathbf{h},t}\right) \boldsymbol{\phi}_t \mathbf{p}_{\mathbf{h},T},$$

where $\boldsymbol{\theta}$ is the set of all parameters in the model, and the dimensions of the vectors and matrices equal the number of defined states. The vector $\mathbf{p}_{\mathbf{h},t}$ is formed in the same manner as for the multi-state single-season model (Chapter 5) and two-state multi-season model (Chapter 8). $D\left(\mathbf{p}_{\mathbf{h},t}\right)$ is a diagonal matrix with the elements of $\mathbf{p}_{\mathbf{h},t}$ on the main diagonal.

The observed data likelihood for the s sampled units is therefore:

$$ODL(\theta|\mathbf{h}_1, \ldots, \mathbf{h}_s) = \prod_{i=1}^{s} Pr(\mathbf{h}_i|\theta).$$

Recall that the true occupancy state of the unit at each time period $(z_{t,i})$ is considered a latent variable, and that in using the observed data likelihood approach, this is accounted for by including terms for all possible combinations of values that $z_{t,i}$ could have been, given the observed data (technically known as integrating out the latent variable; Chapter 3). An alternative approach to estimation is to use the complete data likelihood which assumes the values for $z_{t,i}$ are actually known. This greatly simplifies the construction of the likelihood function as the specific transitions between states are presumed to be known exactly, so there is no need for the summation expressed via matrix multiplication above. Instead, the appropriate terms would just be multiplied together. However $z_{t,i}$ is generally unknown, so the estimation techniques used with the complete data likelihood must account for this by either imputing and updating values for $z_{t,i}$ (e.g., with Markov chain Monte Carlo) or replace $z_{t,i}$ with its expected value (e.g., the expectation-maximization (EM) algorithm). We will not provide further details on the estimation here, except to note that typically MCMC is employed with the complete data likelihood approach for occupancy models.

One advantage of using the complete data likelihood approach with MCMC is that the true state of each unit is predicted each season (with different values being predicted when there is ambiguity as the Markov chain progresses), thus relevant summaries of the system, such as the number of units in each state each season, can be calculated relatively easily. Similar approaches have been used for multi-state capture–recapture data (e.g., Dupuis, 1995), dichotomous (e.g., Royle and Kéry, 2007) and multi-state (e.g., MacKenzie et al., 2009) dynamic occupancy applications, and estimating species richness and accumulation (e.g., Dorazio et al., 2006).

Again we stress that the observed and complete data likelihood approaches are based upon the same underlying model structure and should yield similar inferences, although more complex models can be easier to implement using the complete data likelihood approach because of the estimation tools available.

9.4 COVARIATES AND 'MISSING' OBSERVATIONS

As with the modeling described elsewhere in this book, incorporating predictor variables or covariates into the analysis for various probabilities can be easily accomplished through the use of link functions (Chapter 3). There is a range of possible link functions available, although we note the logit and multinomial-logit link functions may often be preferred, as these correspond to logistic and multinomial-logistic regression. The logit link is appropriate when the model is parameterized in terms of a series of binary outcomes, and the multinomial-logit link is appropriate for the multinomial outcomes case.

Similarly, accounting for missing observations or unequal sampling effort is done in exactly the same way as for the other models in this book. Essentially the survey occasions that correspond to the missing observations at a unit are skipped, with no detection probability (or nondetection probability) parameters appearing in the probability statement for the corresponding survey occasion. Provided that the cause for the missing observation is independent of the parameters and quantities of interest in the model, it is not necessary to explicitly account for the missing observations within the analysis. An example where such a dependency exists (and hence should be avoided) is if budget cuts necessitate a reduction in the number of sampling units being monitored, and investigators respond by discontinuing monitoring of those units they think are unoccupied by the species (i.e., knowledge about the likely occupancy state of units is used to determine which ones will become 'missing' observations).

9.5 MODEL ASSUMPTIONS

The main assumptions of the multi-season multi-state model are essentially the same as for the simpler occupancy models. Namely:

1. occupancy state is static within a season;
2. there is no unmodeled heterogeneity;
3. observations are independent.

We are not aware of any study that has investigated the consequences of violations of these assumptions, but we would expect similar consequences as for assumption violations in the simpler models. However, we would note that many of the extensions that have been developed for the regular occupancy models were done so to address violations of the assumptions of the initial occupancy model developed by MacKenzie et al. (2002) (Chapter 6). We believe that in time, many of the same extensions could be incorporated into the multi-state models to broaden the range of situations where these models could be applied.

9.6 EXAMPLES

We now provide two examples of multi-season multi-state occupancy modeling. The first example is from MacKenzie et al. (2009), examining changes in the relative abundance of green frogs (*Rana clamitans*) in Maryland, USA, using Bayesian methods of inference. The second example examines the effect of broad-scale environmental covariates on the distribution and reproductive success of California spotted owls (*Strix occidentalis occidentalis*) in Sierra Nevada, California, USA (MacKenzie et al., 2012). Files associated with these analyses are available for download from http://www.proteus.co.nz.

9.6.1 Maryland Green Frogs

The North American Amphibian Monitoring Program (NAAMP; see Weir and Mossman, 2005, for sampling protocol details) was an amphibian monitoring program consisting of randomly selected routes (each $\approx$24 km long) along secondary and smaller roads, predominately in the eastern US, that ran from 1997 to 2015. Each route included 10 listening stations spaced at least 0.8 km apart. Some routes had assigned equidistant (0.8 km) listening stations, whereas others had listening stations within hearing distance of wetland habitat. The protocol specified that surveys were to begin at least 30 minutes after sunset with observers proceeding along a route stopping at each listening station. A five-minute audial survey was conducted at each station with observers recording the calling intensity or chorus index for each species: '0' for species that were not detected, '1' if individual calls could have been counted (low call intensity), '2' if individual calls could be distinguished but some overlapping calls (moderate call intensity), and '3' if a full chorus was present for a species, with constant, overlapping calls (high call intensity). Each station was surveyed three times annually with one survey early, middle and late in the breeding season.

MacKenzie et al. (2009) analyzed the chorus index data for green frogs collected along routes in Maryland, USA, during 2001–2005, using each station as a sampling unit. Ideally, the sampling units would be independently selected, hence the fact that stations were actually situated along selected routes means there was an element of clustering to the sampling which should be taken into account for a full analysis. MacKenzie et al. (2009) did not do so in their example analysis. Alternatively, the route could have been considered as the sampling unit, although data from other states would have been needed to increase the sample size.

Each station was assumed to be capable of generating a maximum chorus index of either 0, 1, 2, or 3 during a season, which represented the true occupancy state of a station and was considered a measure of relative abundance. However, on any given evening when a station was surveyed, the level of green frog activity recorded by an observer may have been less than or equal to the true maximal value. For example, suppose local-abundance was sufficient to generate a maximum chorus index of 2 during the breeding season, but the level of calling activity at the time of a survey may have resulted in either a 0, 1, or 2 being recorded. The number of stations surveyed each year was 240, 280, 250,

270, and 250 for 2001–2005, respectively. MacKenzie et al. (2009) expected within-season variation in detection probabilities because of the frog's breeding phenology (a relative late breeder).

MacKenzie et al. (2009) used the complete data likelihood approach described above to analyze the data, by defining the model in terms of the latent and observed random variables ($z_{t,i}$ and $h_{t,ij}$), implemented using the WinBUGS software (a precursor to OpenBUGS, and an alternative to JAGS) for an analysis with Bayesian methods. They used a multinomial parameterization of the initial occupancy state probability vector (ϕ_0) and transition probability matrices (ϕ_t). No constraints were placed on detection probabilities, allowing them to be year-, state-, and survey-specific, although they were the same across all stations otherwise. Simpler detection probability models could be considered if desired.

Flat prior distributions were defined for all probabilities in the model, i.e., standard uniform distributions (equivalent to a beta(1, 1) distribution) for probabilities associated with random variables that had binary outcomes, and Dirichlet prior distributions with all parameters equal to 1 for probabilities associated with random variables with greater than two outcomes. The Dirichlet distribution is a multivariate extension of the beta distribution. Two chains of 50,000 iterations were used to approximate the posterior distributions of the model parameters after discarding an initial 10,000 iterations of each chain as the burn-in period. MacKenzie et al. (2009) reported the chains demonstrated good mixing and rapidly achieved convergence.

The intent of the analysis performed by MacKenzie et al. (2009) was not directed at estimation of the probabilities associated with the model, but to make inferences based on the latent occupancy random variables $z_{t,i}$, namely, how many of the 280 stations surveyed at some stage during this period were in each occupancy state in each year. The underlying logic for inferences of this type is that it might reflect the information required by a manager from a monitoring program (e.g., number of occupied units within a certain area). Note that this is a case of a finite population of units discussed in Chapter 6.

Using the complete data likelihood approach with MCMC for the estimation algorithm, the unobserved value of $z_{t,i}$ is imputed for each surveyed unit in each year. Hence, determining the number of units that are of occupancy state 0, 1, 2, and 3 in each year is simply a matter of counting the number of $z_{t,i}$ with the respective value. A value for each $z_{t,i}$ is imputed for each iteration of the MCMC procedure. At units where a definitive observation has been made during a season (i.e., where an observation has been made that confirms the true occupancy state of a unit; in this case recording a 3, or full chorus), the value for $z_{t,i}$ will be the same for all MCMC iterations, as the observations are unambiguous about the true state in that season. At units where only ambiguous

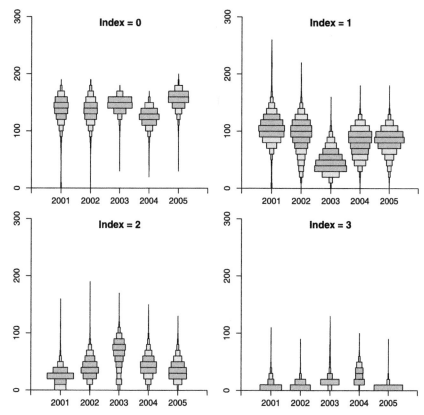

FIGURE 9.3 Posterior distribution for the number of North American Amphibian Monitoring Program (NAAMP) stations in Maryland, USA (from the 280 surveyed), with populations of green frogs capable of generating chorus indices of 0, 1, 2, and 3. Darker shading indicates bars that are within the central 50% of the posterior distribution. The widths of the bars indicate relative frequencies that are comparable between years within each chorus index, but not between indices. *Source: MacKenzie et al. (2009).*

observations were made during a season, the imputed value for $z_{t,i}$ will be different in different iterations, although the maximum observed state for a unit within a season will define the minimum value of $z_{t,i}$. As the imputed $z_{t,i}$ value will change for some units in some years, the number of units in each occupancy state will also change. The imputed value for $z_{t,i}$ can change with each iteration reflecting uncertainty in the true state of a unit due to imperfect detection and is accounted for using the complete data likelihood approach with MCMC.

The posterior distributions for the number of NAAMP stations in Maryland (out of a maximum of 280) estimated to be capable of generating chorus indices of 0, 1, 2, or 3 from 2001 to 2005 are summarized in Fig. 9.3. There is little indication of a strong trend in the number of stations with each chorus level

TABLE 9.1 Probability of not detecting green frogs during call surveys of North American Amphibian Monitoring Program (NAAMP) monitoring stations in Maryland from 2001 to 2005, for stations capable of generating maximal chorus indices of 1, 2, or 3. Given are the mean of the posterior distribution and limits of a central 95% credible interval

Year	Survey	Maximal chorus index		
		1	2	3
2001	1	0.94 (0.87–0.99)	0.81 (0.17–0.98)	0.68 (0.30–0.94)
	2	0.86 (0.73–0.96)	0.55 (0.11–0.83)	0.51 (0.16–0.82)
	3	0.47 (0.20–0.73)	0.24 (0.02–0.63)	0.18 (0.01–0.53)
2002	1	0.99 (0.95–1.00)	0.94 (0.79–0.99)	0.71 (0.32–0.95)
	2	0.74 (0.35–0.89)	0.57 (0.18–0.86)	0.21 (0.01–0.66)
	3	0.68 (0.36–0.95)	0.19 (0.01–0.63)	0.32 (0.05–0.71)
2003	1	0.96 (0.87–1.00)	0.94 (0.78–0.99)	0.83 (0.55–0.97)
	2	0.63 (0.35–0.91)	0.61 (0.06–0.88)	0.49 (0.16–0.87)
	3	0.32 (0.07–0.62)	0.27 (0.04–0.54)	0.19 (0.01–0.50)
2004	1	0.97 (0.90–1.00)	0.81 (0.49–0.96)	0.82 (0.53–0.97)
	2	0.78 (0.57–0.98)	0.27 (0.01–0.61)	0.29 (0.05–0.55)
	3	0.45 (0.05–0.73)	0.51 (0.08–0.91)	0.10 (0.00–0.42)
2005	1	0.99 (0.94–1.00)	0.93 (0.78–0.99)	0.47 (0.02–0.92)
	2	0.68 (0.34–0.91)	0.50 (0.19–0.82)	0.30 (0.01–0.77)
	3	0.49 (0.21–0.77)	0.24 (0.01–0.57)	0.26 (0.01–0.72)

suggesting that perhaps the system is relatively stable. The number of stations capable of generating each chorus index decreases as the chorus index value increases, i.e., more stations are estimated to generate a chorus index of 0 (or green frogs are essentially absent), with fewest stations capable of generating a chorus index of size 3. However, these results are characterized by a relatively large degree of uncertainty about the estimated number of stations in each state. This is primarily due to only three surveys being conducted at each station per year, and the probability of nondetection is fairly high (Table 9.1), particularly in the first survey of each year (reflective of the relatively late breeding of green frogs). Better precision on the estimated number of stations capable of generating each chorus index for green frogs would have been obtained with a greater number of surveys per year, particularly if they had been conducted later in the breeding season. Finally, note that the probability of nondetection tends to be lower at stations that have a higher potential maximal chorus index (i.e., true occupancy state), which would be expected if the chorus index is a reasonable indication of relative abundance, and if there is abundance-induced heterogeneity in the probability of detection of green frogs (at the species level) among stations. This has been easily accounted for in this framework.

9.6.2 California Spotted Owls

MacKenzie et al. (2012) used the multi-season multi-state occupancy model to investigate the population dynamics of California spotted owls from a landscape perspective, namely owl distribution and the level of reproduction within that distribution. They analyzed data collected from 66 potential nesting territories at the Eldorado study site in central Sierra Nevada (California, USA), for eight years from 1997 to 2004. This was a continuation of the data set considered by Nichols et al. (2007a), and in Chapter 5, where the single-season multi-state model was used to account for imperfect detection and misclassification in estimates of reproductive success. MacKenzie et al. (2012) defined the same three occupancy states: unoccupied potential territory (0); occupied territory without successful reproduction (1); and occupied territory with successful reproduction. Surveys were conducted using the same protocols that were recounted in Chapter 5. Briefly, initial call surveys were performed in a potential nest territory, and, if an adult owl responded, it would be offered a live mouse as a prey item. If the owl took the mouse, the researchers would then follow the adult in an attempt to locate its nest. If the nest was located, researchers would look for owlets that were capable of leaving the nest, which was considered evidence of successful reproduction. Therefore, a survey could result in the nondetection of owls (0), detection of owls but not of owlets capable of leaving the nest (1), and detection of owlets capable of leaving the nest (2).

A primary focus of MacKenzie et al. (2012) was whether climatic and environmental variables explained annual variation in the underlying dynamic processes responsible for changes in owl distributions, particularly the probability of successful reproduction. They considered three such variables: Southern Oscillation Index (SOI), level of precipitation during the April incubation period ($INCP$), and energy expenditure during the incubation period ($INCE$). See MacKenzie et al. (2012) for further details. Using the conditional binomial parameterization of the transition probabilities, combinations of these variables were considered for the dynamic occupancy probabilities ($\psi_t^{[m]}$: probability owls present in year t given territory in state m in year $t-1$) and successful reproduction probabilities ($R_t^{[m]}$: probability successful reproduction at an occupied territory in year t given territory in state m in year $t-1$), along with other factors such as state-dependent probabilities (denoted as *State*). The broad-scale variables were all predicted to have negative relationships with these probabilities. No covariates were considered for the $R_t^{[0]}$ probabilities, which were made time invariant in all models as it was considered very rare for a territory to transition from being unoccupied by owls in one year, to being occupied and having successful reproduction the following year (MacKenzie et al., 2009). It will typically take a number of years for a pair to establish themselves in a territory before breeding.

Following Nichols et al. (2007a) and MacKenzie et al. (2009), the conditional-binomial parameterization was also used for detection probabilities, where the probability of detecting reproduction during a survey early in the season (given owls were detected, δ) was assumed to be different than surveys conducted later in the season. The early season probability was expected to be very close to zero and was modeled as constant across all years, while the late season probability was modeled as year-specific. The probability of detecting owls (p) was allowed to vary by true occupancy state and year.

MacKenzie et al. (2012) used AIC_C model selection (AIC corrected for small sample bias; Chapter 3), with an arbitrarily selected effective sample size of 1500 (they suggested it could reasonably be argued to range from 528 to 2227, so picked a middling value). A two-step model selection approach was used where model selection was first performed on the dynamic occupancy probabilities while maintaining a general structure on the dynamic reproduction probabilities, allowing them to vary by both state in the previous year and survey year (i.e., *State* × *Year* interaction). The top-ranked dynamic occupancy structure was then retained while performing model selection for the reproduction probabilities. We note that other strategies could have been used. The logit-link function was used to model all covariate relationships with probabilities, and the modeling was performed in Program PRESENCE by maximum likelihood methods.

The six models considered by MacKenzie et al. (2012) for dynamic occupancy probabilities, and ranked according to AIC_C, are given in Table 9.2. In the top-ranked model, the dynamic occupancy probabilities depended upon the state of the territory in the previous year, but there was no other annual variation. The remaining models all included some form of annual variation, either with year-specific fixed effects, or through a covariate whose value varied annually, with the second- to fourth-ranked models all having non-negligible AIC_C weight. However, the log-likelihood values for these models ($-2l$) are very similar to that of the top-ranked model, and they only differ in structure from the top model by the addition of one or two extra parameters. Hence, the inclusion of these extra covariates is explaining little additional variation in the data, and the model is being ranked high due to the other factors in the model, not those particular covariates. Anderson (2008) refers to such covariates as "pretending variables". Use of model-averaging would be one approach to dealing with this

TABLE 9.2 Model selection summary of the six models for the dynamic occupancy component of the multi-season multi-state models fit to the California spotted owl data. Models were compared using AIC_C with an effective sample size of 1500. Given are the relative difference in AIC_C (ΔAIC_C), AIC_C model weight w, total number of parameters in the model (*Npar*), and twice the negative log-likelihood value ($-2l$)

Model	ΔAIC_C	w	*Npar*	$-2l$
ψ (*State*)	0.00	0.48	45	2826.26
ψ (*State + SOI*)	1.53	0.23	46	2825.66
ψ (*State + Trend*)	1.98	0.18	46	2826.11
ψ (*State + SOI + SOI*^{2})	2.98	0.11	47	2824.98
ψ (*State + Year*)	10.35	0.00	51	2823.79
ψ (*State × Year*)	18.55	0.00	63	2806.04

Source: MacKenzie et al. (2012).

TABLE 9.3 Estimated dynamic occupancy probabilities for California spotted owl data from the top-ranked model

State (m) in year $t - 1$	$\hat{\psi}^{[m]}_{1998-2004}$
Unoccupied	0.16 (0.10–0.25)
Occupied w/o reproduction	0.85 (0.78–0.90)
Occupied w/ reproduction	0.94 (0.85–0.97)

issue, or alternatively there would be some justification to ignore these models. If the extra variables are explaining little additional variation in the data, their estimated effect sizes will be near zero, presumably, hence the estimated probabilities will be similar to those from other models, so little would be achieved by performed model-averaging. MacKenzie et al. (2012) took this latter approach, but we point out that this may not always be appropriate.

MacKenzie et al. (2012) reported that the estimated probability of California spotted owls occupying a potential territory in 1997 was 0.89 (95% CI: 0.76–0.95). The estimated dynamic occupancy probabilities are given in Table 9.3, and clearly demonstrate a large effect of previous occupancy state. If a territory was unoccupied in the previous year, the probability of the territory being occupied in the current year is relatively low, but if the territory was occupied in the previous year, there was a very high probability of the territory also being occupied in the current year (irrespective of reproductive success in the previous year). This should not be surprising for a long-lived species that exhibits high site fidelity. Note that the absence of substantial annual variation in the estimated transition probabilities suggests a certain degree of stability in the distribution of the spotted owls.

TABLE 9.4 Model selection summary of the four top models for the dynamic reproduction component of the multi-season multi-state models fit to the California spotted owl data. The $R(State)$ model is presented for comparison of a model with no annual variation. Models were compared using AIC_C with an effective sample size of 1500. Given are the relative difference in AIC_C (ΔAIC_C), AIC_C model weight w, total number of parameters in the model ($Npar$), and twice the negative log-likelihood value ($-2l$). Model weights do not sum to 1 in table because of the model results that have not been presented

Model	ΔAIC_C	w	$Npar$	$-2l$
$R(State \times Year)$	0.00	0.84	45	2826.26
$R(State \times (SOI + SOI^2))$	5.55	0.05	37	2848.73
$R(State + SOI + SOI^2)$	6.64	0.03	35	2854.03
$R(State + Year)$	7.02	0.03	39	2845.99
$R(State)$	18.92	0.00	33	2870.50

Source: MacKenzie et al. (2012).

In comparison, the models that included annual variation on the reproduction probability, either with fixed year effects or a covariate, were ranked much higher, on the basis of AIC_C, than the model that included no annual variation (Table 9.4). The $R(State)$ model was ranked 16th out of the 22 reproduction models considered by MacKenzie et al. (2012). The top-ranked $R(State \times Year)$ model was the most general model considered, suggesting that there is strong evidence of annual variation in the probabilities of successful reproduction at an occupied territory, and the nature of that variation was different depending on whether there was successful reproduction at the territory in the previous year (Fig. 9.4). Attempting to explain this variation with covariates, or a simpler year effect structure (where the nature of the annual variation was the same for territories; i.e., an additive year effect $R(State + Year)$), was partially successful. While these structures did not rank higher than the more complex model, in terms of AIC_C, they were a clear improvement over the model with no annual variation. Models with SOI included as a linear or quadratic function appeared to be the best predictor of annual variation (Fig. 9.5). MacKenzie et al. (2012) interpreted this result to indicate that SOI appeared to explain some of the annual variation in the reproductive probabilities, but there was still substantial additional annual variation.

Overall, the results of MacKenzie et al. (2012) suggested that California spotted owl distribution within the Eldorado study site was relatively stable during this period. However, the territories that successfully produced young were much more dynamic. This inference is in keeping with what might be expected for California spotted owls at an individual level; annual survival will exhibit

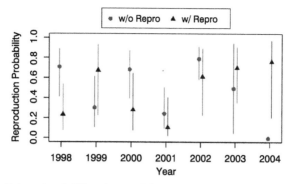

FIGURE 9.4 Estimated probability of successful reproduction at an occupied territory in the current year is a function of the state of the territory in previous year. Estimates are from the top-ranked $R(State \times Year)$ model for the California spotted owl example.

less annual variation than reproductive success because the latter will be more limited by climatic condition, local availability of food resources, etc.

9.7 DISCUSSION

In this chapter we described the multi-season multi-state occupancy model, a natural combination of the regular multi-season occupancy model described in Chapter 8 and the single-season multi-state model from Chapter 5. We believe this framework can be used to address a wide range of interesting and relevant landscape-level ecological questions about species dynamics, in situations where individuals of the species are not uniquely identifiable and the true state of the unit can not be discerned exactly. The flexible multi-state occupancy framework can be used as the basis of constructing models to examine questions of joint habitat-occupancy dynamics (Chapter 13) and species co-occurrence (Chapter 14). The key is defining the possible occupancy states for units in terms of the questions of interest (e.g., which species are present at the unit) and using a suitable parameterization for the various matrices. Identification of possible applications of this approach is facilitated by recognizing that the principle aim of the general set of problems addressed in this book is, ultimately, not about finding the focal species of interest, but about sampling landscape units and collecting relevant information to characterize those landscape units given the objectives of the study. Which target species are present at the units is one set of characteristics that we are attempting to measure, albeit imperfectly, while habitat type, elevation, distance from major urban areas, etc., are other characteristics we could measure. Potentially, the true state of a unit could be defined based upon any such characteristic.

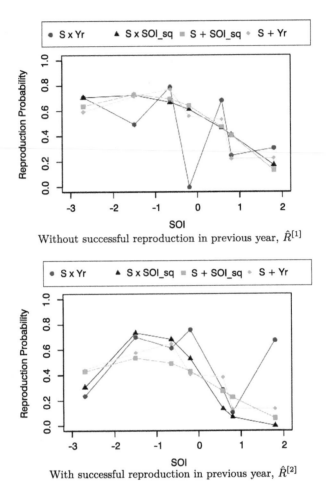

FIGURE 9.5 Estimated probability of successful reproduction at an occupied territory in the current year, for territories that were occupied without (top panel) and with (bottom panel) successful reproduction in the previous year. Estimates are from the four highest-ranked models for the California spotted owl example and are plotted against the Southern Oscillation Index for the current year (SOI). Only point estimates are presented for clarity.

Although we view the primary use of these models as a means of asking interesting ecological questions, we note that even in the absence of interesting ecology, the use of multi-state occupancy models can be viewed as a means of dealing with heterogeneous probabilities of detection, occupancy, and transition parameters. The basic dynamic occupancy models of the previous chapter assume constancy of parameters for all units, or at least constant relationships between parameters and sampling unit covariates. Multi-state models can be viewed as an approach for dealing with heterogeneity among occupied units by admitting the possibility of different parameter values depending on occupancy state (e.g., the Maryland green frog example). For example, one would expect the probability of extinction to be different if there are lots of individuals at a unit than if there are fewer individuals present. The basic multi-season occupancy model does not allow for this variation unless a covariate is used that represents the number, or relative number, of individuals present at each time period. However, one of the main tenets of this book is that number, even in a relative sense, will not be known accurately, i.e., there is uncertainty about the covariate value that an analyst would like to include in a model. The multi-season multi-state model would be one approach for dealing with this issue by defining states in terms of relative abundance (e.g., none, some and lots of individuals), which would explicitly account for the fact that true relative abundance (the covariate in this context) is not observed perfectly. When making inferences about extinction probabilities, the transition probabilities of interest would be $\varphi_t^{[1,0]}$ and $\varphi_t^{[2,0]}$. When the occupancy states are defined in terms of relative abundance, one might predict *a priori* that $\varphi_t^{[2,0]} < \varphi_t^{[1,0]}$. Indeed, one could evaluate how much evidence there is that extinction probabilities, for example, depend upon the relative abundance of the species by comparing a model in which these parameters are estimated independently and a second model where these parameters are constrained to have the same values.

As with any estimation problem, the reliability of the estimated parameters depends upon the amount of information contained in the data about those parameters. Obtaining reliable estimates for parameters associated with infrequent transitions will require a larger sample size compared to transitions that occur more frequently. Therefore, as the number of possible occupancy states increases, the required sample sizes will also increase. Prior to embarking upon a study that utilizes these methods, we recommend exploring/evaluating the level of sampling required to provide reliable inference for a given study and its associated defined states. As with the regular occupancy models we suggest that when detection and classification probabilities are lower, increasing the number of repeat surveys may lead to greater improvements in parameter estimates than increasing the number of units sampled. Computer simulation studies are recommended to gain realistic expectations about the reliability of inferences that could be made given available resources for data collection.

Chapter 10

Further Topics

Chapters 8 and 9 introduced the underlying frameworks for modeling species occurrence dynamics in the cases where 'occupancy' may be defined as a dichotomous or > two-state variable, respectively. In this chapter we outline some extensions, potential uses, and practical insights for these models. We believe that investigating the dynamic processes of species occurrence with multi-season models, provides much greater insight about the ecology of species than investigating whether the pattern of occurrence has changed over time. That is, the explicit-dynamics multi-season model is an attempt to move beyond describing *how* species occurrence has changed, and instead to explore *why* it has changed. A better understanding of the *why* should lead to improved knowledge of how the species functions within the landscape, and may result in more targeted management actions in a management or conservation context. The topics of this chapter seek to provide practitioners with a further set of tools for exploring occupancy dynamics.

10.1 FALSE POSITIVE DETECTIONS

As outlined in Section 6.2, false positive detections (species mis-recorded as present when absent) can result in biased estimates of the probability the target species occupies a unit in single-season applications (see also McClintock et al., 2010b; Miller et al., 2015). Clearly then, false positive detections, or more generally misclassification errors, could result in biased estimates of the underlying dynamic parameters as well.

Standard multi-season occupancy models (Barbraud et al., 2003; MacKenzie et al., 2003) described in this book can be viewed as hidden Markov process models; 'hidden' because occupancy state of a unit with no detections is not known with certainty. In the case of false positives, the uncertainty is extended to some units that do have detections, as well. As described previously, the multi-season model of MacKenzie et al. (2003) uses a similar sampling design to Pollock's robust design that is often used in capture–recapture studies (Pollock, 1982). Multiple survey occasions within each season permit inference about detection probability, and thus unit-level occupancy within seasons. Changes in occupancy of units across seasons are modeled using two basic parameters, ϵ_t, the probability that a unit occupied in season t is not occupied in season $t + 1$, and γ_t, the probability that a unit not occupied in season t is occupied in season $t + 1$; local extinction and colonization probabilities, respectively. Thus, changes in occupancy of a unit are modeled with state-specific parameters determined by the state of the unit (occupied or not) at the beginning of the focal interval (season t). Species occurrence dynamics in the basic multi-season models are parameterized with the probability a unit is occupied in the first season, ψ_1, and season-specific probabilities of extinction, ϵ_t, and colonization, γ_t. The outcome of the within-season surveys is modeled with detection probabilities, p_{tj}, that may be season- and survey-specific.

Extension of single-season false positive models to multiple seasons requires that we model occupancy dynamics between seasons in the same manner as MacKenzie et al. (2003) using extinction and colonization parameters, but that the modeling of within-season data follows an approach that accommodates false positives. Multi-season models can be based on any of the three study designs described for single-season false positive models. The only published multi-season models that accommodate false positives (Miller et al., 2013, as of late 2016) are based on the unit confirmation design of Miller et al. (2011). Our description will thus follow this same approach, but it can be readily extended to the calibration and observation confirmation designs as well. In a subsequent section, we will describe a very general approach, following Miller et al. (2013), which utilizes the multi-season multi-state occupancy model described in the previous chapter that is a form of a hidden Markov model. The unit confirmation design requires a set of observations that are unambiguous, such that detections at a unit indicate occupancy, with no uncertainty. These unambiguous observations can either be obtained as part of a single survey approach or by implementation of an additional special survey. Here we describe multi-season models that use each of these approaches for obtaining unambiguous observations.

10.1.1 Confirmation Designs

Two Detection Types

Under this design, surveys at a unit can yield any of three different observations. We define an observation of '0' to indicate a survey at which the focal species was not detected at the unit. A '1' indicates one or more detections of the focal species during a survey, all of which are of the ambiguous type. A '2' indicates that at least one unambiguous observation of the focal species was obtained during the survey. Thus the detection history, $\mathbf{h}_i = 000\ 100$ indicates that the focal species was not detected during any of three surveys of unit i in season 1, but was detected with an ambiguous observation(s) during survey 1 of season 2.

Models for these data use the detection and classification probability notation described for single season models with two detection methods (Section 6.2). Specifically, p_{10} is the probability of a false positive detection, and is 0 for unambiguous observations and typically > 0 for ambiguous observations. At an occupied unit, detection via an unambiguous observation is denoted as p_{11}, and b is the probability that an observation at an occupied unit is unambiguous. Thus at occupied units, each certain detection is modeled as bp_{11} and each uncertain detection as $(1-b)p_{11}$ (see Table 6.4).

Thus, the probability statement for this detection history $\mathbf{h}_i = 000\ 100$ would be as follows:

$$
\begin{aligned}
Pr(\mathbf{h}_i = 000\ 100|\theta) =\ & (1-\psi_1)(1-p_{10})^3(1-\gamma_1)\,p_{10}(1-p_{10})^2 \\
& +(1-\psi_1)(1-p_{10})^3\,\gamma_1\big[(1-b)p_{11}\big](1-p_{11})^2 \\
& +\psi_1(1-p_{11})^3\,\epsilon_1 p_{10}(1-p_{10})^2 \\
& +\psi_1(1-p_{11})^3(1-\epsilon_1)\big[(1-b)p_{11}\big](1-p_{11})^2,
\end{aligned}
$$

where θ denotes the set of parameters in the model. Note that to simplify the notation, we have omitted season and survey indices from the detection-related parameters. The four lines of the above expression correspond to the following probabilities for the unit:

1. unoccupied in both seasons, with no false-positive detections in season 1, one false-positive detection, and two false-positive nondetections in season 2;
2. unoccupied in season 1 and then colonized and thus occupied in season 2, with no false-positive detections in season 1, one uncertain detection, $(1-b)p_{11}$, and two nondetections in season 2;
3. occupied in season 1 followed by local extinction and thus unoccupied in season 2, with no detections in season 1, one false-positive detection and two false-positive nondetections in season 2;
4. occupied in season 1 and not going locally extinct and thus occupied in season 2, with no detections in season 1, one uncertain detection, and two nondetections in season 2.

Consider another unit with detection history $\mathbf{h}_i = 201\ 101$ and its associated probability:

$$
\begin{aligned}
Pr(\mathbf{h}_i = 201\ 101|\theta) \\
=\ & \psi_1(bp_{11})(1-p_{11})\big[(1-b)p_{11}\big]\epsilon_1(p_{10})^2(1-p_{10}) \\
& +\psi_1(bp_{11})(1-p_{11})\big[(1-b)p_{11}\big](1-\epsilon_1)\big[(1-b)p_{11}\big]^2(1-p_{11}).
\end{aligned}
$$

The certain detection (2) in season 1 reduces the uncertainty associated with this history, as there are only two possibilities:

1. occupied in season 1 followed by local extinction and thus unoccupied in season 2, one certain detection, bp_{11}, one nondetection, and one uncertain detection in season 1, with two false-positive detections and one false-positive nondetection in season 2;
2. occupied in season 1 and not going locally extinct and thus occupied in season 2, one certain detection, one nondetection, and one uncertain detection in season 1, with two uncertain detections and one nondetection in season 2.

In both of these possibilities the unit is occupied in season 1, which is known because of the certain detection in season 1.

Two Detection Methods

This alternative unit confirmation design is based on different survey methods. One method yields unambiguous observations and the other ambiguous observations. Detection history data are again separated to distinguish observations within a season from those corresponding to a different season. In addition, a '/' is used to distinguish the method providing ambiguous observations from that providing unambiguous observations (ambiguous detection method first). For example, $\mathbf{h}_i = 00/0 \; 01/0$ specifies a unit at which there were no observations of the focal species during season 1 for either method. In season 2, there was an observation corresponding to the ambiguous method in survey 2 of season 2.

The probability structure underlying the sampling process for this model is given in Table 6.5. The uncertain method permits detections at occupied units with probability p_{11}, and false positive detections at unoccupied units with probability p_{10}. When using the certain method, false positive detections at unoccupied units cannot occur. The probability of detection at an occupied unit using the certain method is denoted r_{11}.

Using the above notation, the probability associated with detection history $\mathbf{h}_i = 00/0 \; 01/0$ can be written as:

$$
\begin{aligned}
Pr(\mathbf{h}_i = 00/0 \; & 01/0|\boldsymbol{\theta}) \\
= & (1 - \psi_1)(1 - p_{10})^2(1 - \gamma_1)(1 - p_{10})\,p_{10} \\
& + (1 - \psi_1)(1 - p_{10})^2\,\gamma_1(1 - p_{11})\,p_{11}(1 - r_{11}) \\
& + \psi_1(1 - p_{11})^2(1 - r_{11})\,\epsilon_1(1 - p_{10})\,p_{10} \\
& + \psi_1(1 - p_{11})^2(1 - r_{11})(1 - \epsilon_1)(1 - p_{11})\,p_{11}(1 - r_{11}).
\end{aligned}
$$

Again, there are four possibilities for the unit:

1. unoccupied in season 1 and not colonized, thus unoccupied in season 2, no false-positive detections by the uncertain method in season 1, with one false-positive nondetection and one false-positive detection by the uncertain method in season 2;
2. unoccupied in season 1 and then colonized, thus occupied in season 2, no false-positive detections by the uncertain method in season 1, with one non-detection and one detection by the uncertain method, and a nondetection by the certain method, in season 2;
3. occupied in season 1 followed by local extinction, thus unoccupied in season 2, no detections by the uncertain method and no detections by the certain method in season 1, with one false-positive nondetection and one false-positive detection by the uncertain method in season 2;
4. occupied in season 1 not going locally extinct, thus occupied in season 2, no detections by the uncertain method and no detections by the certain method in season 1, with one nondetection and one detection by the un-certain method, and a nondetection by the certain method, in season 2.

Note that the certain method occupancy probabilities, r_{11}, appear only when the unit is occupied, as the certain method always yields detection probabilities of 0 for unoccupied units.

Now consider detection history $\mathbf{h}_i = 10/1\ 10/0$. The associated probability is:

$$
\begin{aligned}
Pr(\mathbf{h}_i = 10/1\ 10/0|\boldsymbol{\theta}) = {} & \psi_1 p_{11}(1 - p_{11}) r_{11} \epsilon_1 p_{10}(1 - p_{10}) \\
& + \psi_1 p_{11}(1 - p_{11}) r_{11}(1 - \epsilon_1) p_{11}(1 - p_{11})(1 - r_{11}).
\end{aligned}
$$

Because of the detection by the certain method in season 1, the unit must have been occupied in season 1, hence the only ambiguity is associated with the potential for false-positive detections in season 2, therefore there are only two possibilities for the species at the unit:

1. occupied in season 1 followed by local extinction, thus unoccupied in season 2, one detection and one nondetection by the uncertain method and one detection by the certain method in season 1, with one false-positive detection and one false-positive nondetection by the uncertain method in season 2;
2. occupied in season 1 and not going locally extinct, thus occupied in season 2, one detection and one nondetection by the uncertain method and one detection by the certain method in season 1, with one detection and one nondetection by the uncertain method, and one nondetection by the certain method in season 2.

Once again the certain method detection probability, r_{11}, only enters the probability structure when the unit is occupied.

As noted by Miller et al. (2013) any of the above parameters can vary among units, seasons, and sampling occasions within seasons. Covariate relationships can be modeled generally using, for example, linear-logistic relationships: $logit(\theta_{it}) = \mathbf{X}_{it}\mathbf{B}$, where θ_{it} is a parameter for unit i and season t, β is a column vector of the regression equation coefficients (i.e., effect sizes; to be estimated), and $\mathbf{X}_{it}$ is a row vector of covariate values for unit i in season t.

Example Analysis

Miller et al. (2013) illustrated use of their multi-season site confirmation models for false positives using hunter survey reports of gray wolf (*Canis lupus*) detections in Montana. The original publication should be consulted for study details, as we simply provide a summary here. Gray wolves are an endangered species, and the size of, and changes in, its range are important criteria used to evaluate their conservation status. After years of spending substantial funds on radio-telemetry studies of wolves, the state of Montana sought an approach for drawing inferences about range that would be less expensive. They developed an approach based on occupancy modeling of data from reports obtained via telephone surveys of hunters, developed initially for inferences about harvest of deer (*Odocoileus virginianus; O. hemionus*), elk (*Cervus elaphus*), and moose (*Alces alces*). Initial analysts recognized the likelihood of false positives, with hunters possibly mistaking coyotes for wolves or providing unreliable reports for other reasons. Rich et al. (2013) analyzed these hunter report data and dealt with the false positive issue in an *ad hoc* manner, by requiring at least three independent hunter reports of 2–25 wolves as constituting observations of 'occupancy' (i.e., disregarding geographic cells within which wolf observations by only one or two hunters were reported). They believed that this approach to false positives produced reasonable inferences about range, but they recognized that a more formal approach would be preferable

Miller et al. (2013) developed the general approach described here and used it to estimate gray wolf range dynamics for the years 2007–2010 using Montana hunter survey data and radio telemetry data. The hunter survey was viewed as the uncertain method, as false positive hunter reports were believed to be possible. Radio-telemetry data on wolf packs throughout the state were viewed as the certain method, as telemetry locations documented pack presence in survey cells. The area of inference was northern Montana, which was subdivided into 600 km^2 grid cells, and the focal state variable was the proportion of these cells occupied during the five-week rifle hunting season in each year of the study. The sample of telemetered wolves and their packs was believed to be representative, hopefully satisfying the assumption of no correlation between detection probabilities associated with the two methods. For example, if radio-collared wolves were only present in areas having higher or lower detection, and/or false positive, probabilities for the uncertain method (the hunter surveys), then biased estimates would be expected to result. Each of the five weeks of the hunting season was treated as a sampling occasion for the uncertain method. Hunters were asked where and when they observed wolves, providing the cell-specific detection history data for this method. Telemetry data were treated as a single occasion for the certain method, with data typically coming immediately before and/or after the hunting season. A detection was recorded for the certain method in a cell if the centroid of telemetry locations for a pack fell within that cell.

An important assumption associated with the use of telemetry data for making inference about species distributions is that animals with a tracking device go to the same units as animals without a tracking device. Otherwise, the probability of detecting the species in a unit where the species is present, but there are no animals with a tracking device, is zero for this method. Therefore, the resulting species distribution is restricted to those members of the species with a tracking device, not for the species in general. This is most likely an issue when relatively few tracking devices are deployed for a territorial species, or other situations where animals with and without trackers do not mix freely. In this case, radio-collars were attached to wolves throughout northern Montana, and Miller et al. (2013) felt this assumption was reasonable.

Grid cells were stratified by proportion of good wolf habitat in them, leading to strata of high, medium and low quality habitat. Hunter effort also varied across northern Montana, so hunter-days were used as a covariate for cell-specific detection probabilities. Multiple models were fit to the data, including those with and without the hunter effort covariate on detection and an effect of habitat quality on local occupancy, extinction and colonization. In addition to the multi-season false positive model framework, models assuming no false positives and no false positives or negatives (nondetections) were fit and included as part of their analysis.

As our purpose here is to illustrate the model itself, we do not present any details of the model selection process for this particular example, but will instead provide only summary comments. Model selection results based on the set of models that included false positives provided evidence that detection probability (p_{11}) increased with hunter effort and was somewhat lower in low quality habitat than elsewhere. False positive probabilities (p_{10}) also increased with hunter effort and were > 0.5 in areas with the highest hunter effort. Such large occasion-specific probabilities of false positives produced a probability of about 0.97 that at least one false positive observation would occur during the five sampling occasions at an unoccupied cell. Detection probabilities for the certain method (telemetry, r_{11}) were greater in higher quality habitat, and all detection probabilities increased with time (years).

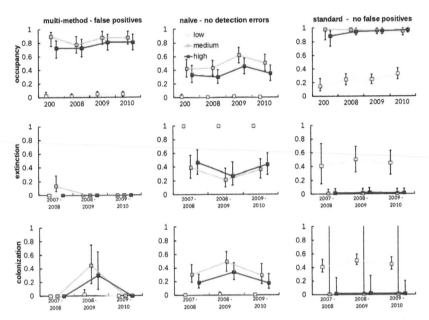

FIGURE 10.1 Estimates of occupancy, extinction, and colonization for gray wolves in northern Montana from 2007 to 2010. Cells were divided into low, medium, and high habitat quality. Parameters were estimated using a naïve approach where false positives and false negatives were assumed not to occur, using a standard multi-season occupancy model where false positives were assumed not to occur, and using the multi-season occupancy model that allows for false positives. *(Source: Miller et al., 2013)*

With respect to gray wolf occupancy dynamics, the best supported model included additive habitat and time effects on initial occupancy, and both local extinction and colonization probabilities. Occupancy for all years was substantially greater in cells of high and medium habitat (Fig. 10.1). Comparison of these occupancy estimates based on the multi-season models allowing for false positives with estimates based on models that did not include all of these components of the detection process produced predictable results. Naïve modeling that assumed wolves were detected in all occupied cells, with no false detections in unoccupied cells, provided occupancy estimates that were substantially lower than those of the preferred model (Fig. 10.1). Modeling that included nondetections but did not deal with false positives produced estimates that were larger than those that included the possibility of false positives (Fig. 10.1). In summary, the analysis provided strong evidence of the presence of both nondetections and false positives in the hunter survey data. Inferences based on the preferred site confirmation model that dealt with these two sampling issues were substantially different from inferences based on models that ignored one or both sampling issues.

10.1.2   More Observation and Occupancy States: General Approach

Miller et al. (2013) (also see Miller et al., 2011) noted that the probability statement for any observed detection history for a multi-season unit confirmation
model can be calculated using the same equation that we have already encountered in Chapters 8 and 9. Namely:

$$Pr\left(\mathbf{h}_i | \boldsymbol{\theta}\right) = \boldsymbol{\phi}_0 \prod_{t=1}^{T-1} D\left(\mathbf{p}_{\mathbf{h},t}\right) \boldsymbol{\phi}_t \mathbf{p}_{\mathbf{h},T},$$

where $\boldsymbol{\theta}$ denotes the set of all parameters in the model. $\boldsymbol{\phi}_0$ and $\boldsymbol{\phi}_t$ are the initial
state probability vector and transition probability matrix, respectively, which can
either be defined in terms of either two occupancy states (Chapter 8) or $>$ two
occupancy states (Chapter 9). Therefore, generalization of false-positive models to situations with greater than two true occupancy states is, conceptually,
straightforward. The key issue identified by Miller et al. (2013) is the appropriate formulation of $\mathbf{p}_{\mathbf{h},t}$, the conditional detection probability vector (probability
of observing the detection history during season t conditional upon each of the
possible true states), which, in turn, depends on defining a suitable detection
probability structure for the possible observations that could be made using the
employed methods. That is, as we noted at the start of this section, the only
difference between multi-season models that do, and do not, account for false
positive detections is how the within-season data are modeled.

Details of how $\mathbf{p}_{\mathbf{h},t}$ can be constructed for a false-positive model are given
in Section 6.2, which we briefly revisit here. Under a unit confirmation design
with three possible observations (i.e., nondetection, ambiguous detection and
unambiguous detection of the species), the detection probability matrix, $\mathbf{p}_{t,j}$,
defines the probability of the possible observation (with one column per possible
observation) given the true state of the unit (one row per possible true state) in
survey j of season t. That is:

$$\mathbf{p}_{t,j} = \begin{bmatrix} 1 - p_{10} & p_{10} & 0 \\ 1 - p_{11} & (1-b)\,p_{11} & bp_{11} \end{bmatrix}.$$

$\mathbf{p}_{\mathbf{h},t}$ is then formed by the element-wise multiplication of the columns of $\mathbf{p}_{t,j}$ that
correspond to each of the survey outcomes within season t (see Chapter 5). In
some applications the degree of certainty about observations may vary, and it
may be useful to incorporate such variation in the modeling. For example, identification of tiger and leopard pugmarks to species is based largely on pugmark
size. Very large pugmarks are likely to belong to tigers, and very small pugmarks are more likely to belong to leopards, whereas pugmarks of intermediate

size are characterized by the greatest degree of uncertainty. In such a case, false positive probabilities could be modeled with different parameters for discrete size classes of pugmarks. Incorporating this into the modeling simply requires defining $\mathbf{p}_{t,j}$ with additional columns to represent the different observations (nondetection, small pugmarks, intermediate pugmarks, and large pugmarks), and parameterizing $\mathbf{p}_{t,j}$ in a manner that reasonably describes the uncertainty associated with each type of observation. To accommodate additional occupancy states, $\mathbf{p}_{t,j}$ should be defined with additional rows to represent the additional true occupancy states and suitable probability structure to model the uncertainty in the observations, as they relate to the true states.

Unit confirmation designs with multiple methods also generalize easily, theoretically, either in terms of more observation types, more occupancy states, or more methods. In Section 6.2, and above, we considered the case with two occupancy states, two types of observations, and an uncertain and a certain method of detection. Two detection probability matrices were defined, one for each method:

$$\mathbf{p}_{1,t,j} = \begin{bmatrix} 1 - p_{10} & p_{10} \\ 1 - p_{11} & p_{11} \end{bmatrix}$$

and

$$\mathbf{p}_{2,t,j'} = \begin{bmatrix} 1 & 0 \\ 1 - r_{11} & r_{11} \end{bmatrix},$$

for the uncertain and certain detection methods, respectively. Once again, incorporation of additional observation types for either or both methods simply requires the respective detection probability matrix to be defined with additional columns (note that the number of columns does not have to equal for all methods), and with additional rows for a greater number of true occupancy states (the number of rows does have to be equal for all methods). If there is a greater number of methods, with different levels of certainty associated with the possible observations, then a detection probability matrix could be defined for each method, and $\mathbf{p}_{h,t}$ calculated as for the single-season case, but where the within-season detection history $\mathbf{h}_{t,i}$ now consists of $>$ two methods.

For example, suppose there are three detection methods, A, B, and C, each with different types of uncertainty structure. Method A may yield three observation types: nondetection, ambiguous and unambiguous (as in the two detection types confirmation design, above), method B yields two observations, but there is uncertainty about the true occupancy state from each (e.g., $\mathbf{p}_{1,t,j}$), and method C yields two observations, but one observation confirms species presence (e.g., $\mathbf{p}_{2,t,j}$). Denote the detection probability matrices for each method as

$\mathbf{p}_{A,t,j}$, $\mathbf{p}_{B,t,j}$, and $\mathbf{p}_{C,t,j}$, respectively. Consider the within-season detection history $\mathbf{h}_{t,i} = 20/01/0$, representing that two surveys were conducted at the unit for methods A and B, and one survey with method C. Using the matrix notation established in Chapters 5, 8, and 9, the conditional detection probability vector $\mathbf{p}_{\mathbf{h},t}$ for this portion of the detection history observed in season t would be calculated as:

$$\mathbf{p}_{\mathbf{h}=20/01/0,t} = \mathbf{p}_{A,t,1}^{[\bullet 2]} \odot \mathbf{p}_{A,t,2}^{[\bullet 0]} \odot \mathbf{p}_{B,t,1}^{[\bullet 0]} \odot \mathbf{p}_{B,t,2}^{[\bullet 1]} \odot \mathbf{p}_{C,t,1}^{[\bullet 0]},$$

where $\mathbf{p}_{X,t,j}^{[\bullet l]}$ indicates the column of $\mathbf{p}_{X,t,j}$, corresponding to observation l in the jth survey using method X. Note that where methods have the same uncertainty structure, an alternative to defining them as separate methods in the above approach, would be to consider them as a single method type (with a common uncertainty structure), and use 'field method' as a covariate for the relevant parameters.

A similar calculation would be performed for all within-season detection histories using either of the unit confirmation approaches. Once $Pr(\mathbf{h}_i|\boldsymbol{\theta})$ has been determined for all units, the observed data likelihood would be calculated as:

$$ODL(\boldsymbol{\theta}|\mathbf{h}_1,\ldots,\mathbf{h}_s) = \prod_{i=1}^{s} Pr(\mathbf{h}_i|\boldsymbol{\theta}),$$

which could be used to obtain parameter estimates in the usual manner.

Alternatively, the false-positive multi-season model could be defined in terms of the underlying random variables, and a complete data likelihood approach could be used. The matrices $\boldsymbol{\phi}_0$, $\boldsymbol{\phi}_t$, and $\mathbf{p}_{X,t,j}$ define the probability structure for the latent and observed random variables, and the methods outlined elsewhere (e.g., Chapter 9) could be applied.

This general approach for the false-positive multi-season models is a generalized case of the multi-season multi-state occupancy model (MacKenzie et al., 2009; Chapter 9) that allows for two-way state misclassification. This highlights the usefulness of the multi-state framework, not only for parameter estimation and prediction, but also as the basis for conceptualizing the ecological and sampling problem at hand. As outlined by Miller et al. (2013), the generalized framework is similar in concept to the multi-event approach to dealing with state uncertainty in multi-state capture–recapture models (Pradel, 2005), and both are a type of hidden Markov model.

While the above development would suggest a great deal of potential flexibility in how detection probability matrices could be defined, and parameterized, to accommodate false-positive detections and other forms of state misclassification, we highlight there will be limitations on the structures that could be implemented for the type of sampling situations we have considered, without auxiliary data. There may be insufficient information in the data to enable all parameters to be estimated for some detection probability matrix structures. That is, not all parameters may be uniquely identifiable without additional sources of information about some of those parameters or constraints being placed on those values. For example, in the single-season case Royle and Link (2006) defined a detection probability matrix structure to accommodate potential misclassification of unoccupied units as occupied, however, not all detection-related parameters were estimable unless the probability of a false-positive detection was constrained to be less than the probability of detecting the species when it was truly present (Royle and Link, 2006; Section 6.2). In developing new parameterizations and extensions of these general misclassification models, we recommend practitioners ensure that all parameters can be estimated.

10.1.3 Discussion

At the time of writing, multi-season modeling has been implemented only for the unit confirmation design (Miller et al., 2013, and above), but calibration and observation confirmation design approaches (Chambert et al., 2015a) could be used as well. Extension to these design approaches would simply require using suitably defined detection probability matrices (as done in Section 6.2 for these designs) to describe the within-season observation process for the calculation of $p_{h,t}$, as detailed above. Once $p_{h,t}$ is determined, construction of a multi-season (multi-state) model is the same as for the unit confirmation designs, and designs that assume no false-positive detections.

In this section we have detailed how the logic behind the single-season false-positive models (e.g., Royle and Link, 2006; Miller et al., 2011; Chambert et al., 2015a), can be extended to the multi-season case by modifying standard multi-season models to incorporate a more complicated within-season detection process, which may include false positives (Miller et al., 2013). Such modeling provides the framework for investigating species distribution dynamics while acknowledging that for some species, and survey methods, there is the potential for both false-negative (present but not detected) and false-positive (absent and detected) results. Survey methods where these modeling approaches may be useful include, but are not limited to, auditory surveys of some avian, anuran and bat species (where researchers have demonstrated non-negligible false-positive detection rates, e.g., Simons et al., 2007; McClintock et al., 2010a; Clement et al., 2014), sign surveys and sightings by the general public. These methods may be particularly useful for analyzing the data from 'citizen science' programs where species sightings are reported by members of the general public, who will undoubtedly vary in their ability to accurately identify species.

10.2   AUTOCORRELATED WITHIN-SEASON DETECTIONS

In Section 6.4 we described the single-season model developed by Hines et al. (2010) to account for autocorrelated repeat surveys, where the probability of detecting the target species depends on the outcome of the previous survey. Such correlation may arise with temporal or spatial repeat surveys of a unit within a season, particularly when 'surveys' are close in time or space. In that section, we described the model in terms of an intermediate (between occupancy and detection) availability process, which could be described as a first-order Markov process. That is, the probability of the species being *available* for detection in survey j, depends upon whether it was available for detection in survey $j - 1$. However, there are no direct observations of the availability process, and only two levels of sampling (units and repeat surveys of units). The following parameters were defined to describe this non-random availability process for the single-season case:

$\theta_1 = Pr$(species available in first survey);

$\theta_j^{[0]} = Pr$(species available in survey j | species not available in survey $j - 1$)
for $j = 2, \ldots, K$;

$\theta_j^{[1]} = Pr$(species available in survey j | species available in survey $j - 1$) for
$j = 2, \ldots, K$.

Note the superscripts on $\theta_j^{[0]}$ and $\theta_j^{[1]}$ indicate the availability status of the species in the previous survey occasion.

Hines et al. (2014) extended this modeling to the multi-season situation where changes in occupancy are modeled as a first-order Markov process (e.g., Chapter 8; MacKenzie et al., 2003), occupancy state is considered static within a season, but availability is modeled as a first-order Markov process between surveys (given unit occupied in that season). In each survey the species may be detected (or not) given it was available for detection. Essentially, the only difference between the multi-season modeling approach of MacKenzie et al. (2003) and Hines et al. (2014) is the modeling of the within-season detection process, in the same manner that false positive detections were accounted for in the previous section. Thus, the modeling approach of Hines et al. (2014) is a combination of the MacKenzie et al. (2003) model to describe occupancy dynamics, and the Hines et al. (2010) model to describe the within-season availability and detection process.

For example, suppose the detection history $\mathbf{h}_i = 111\,011$ was observed at unit i. Under the multi-season model of MacKenzie et al. (2003) the conditional detection probability vectors for each season, which describe the within-season detection process given the true occupancy state of unit, would be:

$$\mathbf{p}_{111,1} = \begin{bmatrix} 0 \\ p_{1,1}p_{1,2}p_{1,3} \end{bmatrix},$$

and

$$\mathbf{p}_{011,2} = \begin{bmatrix} 0 \\ (1 - p_{2,1})\, p_{2,2}p_{2,3} \end{bmatrix},$$

where $p_{t,j}$ denotes the probability of detecting the species in survey j of season t, given the species is *present at the unit*. Under the Hines et al. (2014) model to account for correlated surveys, the conditional detection vector for season 1 becomes:

$$\mathbf{p}_{111,1} = \begin{bmatrix} 0 \\ \theta_{1,1}p_{1,1}\theta_{1,2}^{[1]}p_{1,2}\theta_{1,3}^{[1]}p_{1,3} \end{bmatrix},$$

where the subscript on the θ parameters indicates season t and survey j, respectively, and $p_{t,j}$ denotes the probability of detecting the species in survey j of season t, given the species is *present at the unit and available for detection in survey j*. Note that 'p' has a slightly different interpretation under each of the modeling approaches. The species was detected at the unit in each survey of season 1, hence the species must have been available for each survey. However, when the species was not detected in a survey, the species may have been available but not detected, or not available, which the modeling must take into account. For example, the conditional detection vector for season 2 under the correlated detections model would be:

$$\mathbf{p}_{011,2} = \begin{bmatrix} 0 \\ \{\theta_{2,1}(1 - p_{2,1})\,\theta_{2,2}^{[1]} + (1 - \theta_{2,1})\,\theta_{2,2}^{[0]}\}\, p_{2,2}\theta_{2,3}^{[1]}p_{2,3} \end{bmatrix},$$

where the term in curly brackets describes the ambiguity about the availability of the species because of the nondetection of the species in the first survey of season 2. Under both models, the probability statement for the detection history could be expressed as:

$$Pr(\mathbf{h}_i = 111\,011|\boldsymbol{\theta}) = \boldsymbol{\phi}_0 D(\mathbf{p}_{111,1})\,\boldsymbol{\phi}_1 \mathbf{p}_{011,2},$$

where $\boldsymbol{\theta}$ denotes the set of all model parameters, and $\boldsymbol{\phi}_0$ and $\boldsymbol{\phi}_1$ are defined as previously.

Clement et al. (2016) extended this multi-season model with correlated detections to incorporate heterogeneous detection probabilities using a finite mixture model approach (e.g., see Chapter 7, and also later this chapter). Applications of Hines et al. (2014) and Clement et al. (2016) were based on the North American Breeding Bird Survey (e.g., Robbins et al., 1986; Peterjohn and Sauer, 1993), consisting of avian point counts at 50 roadside stops along each route. In both of these analyses, model selection provided evidence of the need for the correlated detection model when analyzing such data by comparing models with and without the within-season correlation structure.

These multi-season approaches for dealing with correlated within-season survey replicates are incorporated into program PRESENCE (Hines, 2006). The continuous Poisson process model of Guillera-Arroita et al. (2011) could likely be substituted for the within-season component of multi-season models as well, providing another approach to dealing with correlated surveys for multi-season modeling.

10.3   SPATIAL CORRELATION IN DYNAMICS

In Section 6.6 we detailed methods that could be, and have been, used to incorporate spatial correlation into patterns of species occurrence each season (e.g., spatially explicit covariates, autologistic, conditional autoregressive, and restricted spatial regression models). The intent is to acknowledge that species will tend to aggregate at some spatial scale due to a variety of ecological and biophysical factors; therefore if a species is present at a sampling unit, it is arguably more likely to be also present at a neighboring unit than one randomly selected from the population of units of interest. This is particularly the case when units are defined in a contiguous manner. It would therefore stand to reason to also expect that spatial correlation will be present in the dynamic processes of species occurrence. That is, colonization and extinction probabilities, for example, will be more similar at nearby units than units that are farther apart. In this section we provide an overview of approaches that could be used to achieve this. We will avoid the mathematical detail of many of the approaches, focusing on the conceptual issues and making reference to the relevant literature.

However before doing so, it is worth highlighting that, as in the single-season situation for patterns of occurrence, while spatial correlation in the dynamic processes is a near-certainty in many ecological applications, it is possible to make reliable population-level inferences about occurrence dynamics without explicitly accounting for that spatial correlation, provided a suitable probabilistic sampling scheme has been used. Incorporating spatial correlation may provide a more biologically-realistic representation of the dynamics at the unit-scale, but the additional complexity of the modeling may not result in substantially different conclusions at the population-scale. For example, suppose that the presence of an anuran species at ponds within a wildlife park is of interest, and that managers are primarily interested in occupancy, and changes in occupancy, at the park scale as that is the applicable scale for management reporting and decision making. In other words, the state variable used in the management decision making process is the estimated proportion of occupied ponds in the park. Decisions are not based on the status of individual ponds, or based upon maps of which ponds are thought to be occupied, which does not mean such information would not be useful to managers, but simply that such information is not of prime relevance to the decision making process. That is, given the management objectives, it is not relevant which particular ponds were occupied, just what overall proportion of ponds in the park was occupied. If a random sample of ponds from the park was surveyed, estimates and predictions from models that do not account for spatial correlation will give unbiased estimates at the population-level. As in the single-season case, this is because the modeling requires that the observations (i.e., detection histories) are independent, which arises from the selection of which units will be surveyed, not the underlying biological processes (Section 4.4.9). Accounting for spatial correlation can be more important when non-random sampling has been used, or when inference is required for individual units.

One approach for incorporating spatial correlation in the transition probabilities governing occupancy dynamics is with covariates or predictor variables. These may be based on environmental variables such as vegetation type, elevation or annual rainfall, or some distance-related measure, e.g., distance from nearest urban area. Variables may also be defined to create a smooth surface through the use of bivariate splines on the Cartesian coordinates of the units, for example. Spatial correlation that is due to, or well approximated by, such covariate relationships will be accounted for by their inclusion. As covariates are regarded as fixed values in the modeling, it is of no material consequence if their values are spatially correlated or not. How to incorporate covariates into the modeling of species occurrence has been a central theme of this book and covered elsewhere.

Autologistic models could also be used to account for spatial correlation in the occupancy state transition probabilities. For example, the probability of unit i being colonized by the focal species between seasons t and $t + 1$ could be defined as:

$$logit(\gamma_{t,i}) = \beta_0 + \beta_1 g_{t,i},$$

where $g_{t,i}$ is a function of the number of occupied neighboring units (Section 6.6). When data analyses are being performed using Bayesian methods, then $g_{t,i}$ can be calculated based on the imputed values of the latent occupancy variables, $z_{t,i}$ (e.g., Bled et al., 2011, 2013). As $g_{t,i}$ is based on the occupancy of units at the beginning of the transition period, it could also be approximated from the $\hat{\psi}_{t,i}$ values, which enables such analyses to be conducted within a maximum likelihood framework (e.g., Yackulic et al., 2012, 2014; Eaton et al., 2014). That is, $\hat{g}_{t,i}$ could be approximated as:

$$\hat{g}_{t,i} = \frac{\sum_{l \in n(i)} \hat{\psi}_{t,l}}{|n(i)|},$$

where $n(i)$ is the set of units defined to be the 'neighbors' of unit i and $|n(i)|$ is the number of neighbors of unit i. Note that where $n(i)$ is defined to be all units in the population, including unit i, then the autologistic model becomes a form of 'density-dependent' model, where 'density' is in terms of occupancy rather than abundance. The $\hat{\psi}_{t,l}$ term in the above expression could also be replaced by the occupancy probabilities conditional upon the observed detection history for that unit, i.e., $\hat{\psi}_{condl,t,l}$ (Chapter 4). This would provide a closer approximation of the latent occupancy variables than use of $\hat{\psi}_{t,i}$. An advantage of the autologistic approach is its close correspondence to metapopulation ideas about influences of neighboring patches on occupancy dynamics. Conditional autoregressive (CAR) models have also been used to account for spatial correlation in dynamic parameters (e.g., Bled et al., 2013), which have a similar intent to autologistic models, but are based on a spatially correlated random effect rather than an occupancy random variable (Section 6.6).

Broms et al. (2016) used restricted spatial regression (RSR; Johnson et al., 2013; Section 6.6) to incorporate spatial correlation into the dynamic processes. An advantage of RSR is that it focuses on any residual spatial correlation after accounting for spatial correlation that has been explained by covariates included in the model. Other approaches do not, hence inferences about covariates that have been included with an autologistic model, for example, may be unreliable (Johnson et al., 2013) where the covariate itself is spatially correlated. As such, we believe RSR to be a more generally applicable approach to simultaneously incorporate covariate information and spatial correlation.

Modeling structures typically used in metapopulation studies can be considered as another approach for incorporating spatial correlation into occupancy dynamics, as distance to nearest occupied patch is often included in a measure of patch connectivity, which is a determinant of colonization probability (e.g., Hanski, 1994a; Moilanen, 2002). Therefore, patches that tend to be near other occupied patches will have higher colonization probabilities than more isolated patches, or patches farther away from an occupied patch. Multi-season occupancy models have been parameterized using metapopulation theory (e.g., Risk et al., 2011; Eaton et al., 2014; Chandler et al., 2015) to incorporate both a spatially explicit structure and imperfect detection of populations in patches. Sutherland et al. (2014) have also extended the methods to incorporate demographic data on the abundance of adults and juveniles.

Most applications of multi-season occupancy models that incorporate spatial correlation in the state transition probabilities have used the complete data likelihood approach with Bayesian analytic methods (i.e., MCMC). Primarily, this has been a matter of convenience, as with the complete data likelihood approach using MCMC, as implemented in popular software such as OpenBUGS and JAGS, the underlying model can be defined in terms of the latent and observed random variables, without having to integrate the unobserved values of the latent variables out of the model likelihood (as with the observed data likelihood). This greatly simplifies model specification, particularly for customized applications of the methods. However, it is possible to also fit models with spatial correlation using the observed data likelihood and maximum likelihood methods (e.g., Yackulic et al., 2012, 2014; Eaton et al., 2014). This illustrates that there is a variety of analytic tools already available to the practitioner for applying this class of model, and the number, and availability, of such tools will only improve over time.

10.4   INVESTIGATING OCCUPANCY DYNAMICS

The multi-season models described in Chapters 8 and 9 can be used to address interesting hypotheses about species biology and ecology through occupancy dynamics; specifically hypotheses about the mechanistic processes underlying changes in occupancy and hypotheses that the population is at some form of equilibrium with respect to occupancy dynamics. Such hypotheses differ fundamentally from those dealing with covariate relationships discussed earlier in this chapter, focusing instead on constraints that could be made upon the dynamic probabilities themselves. Obviously the two approaches could be combined to assess not only the factors that may affect occupancy, for instance, but also whether changes in occupancy appear to be Markovian or random in nature. Our basic philosophy for investigating competing hypotheses, as mentioned throughout this book, is simple: articulate the different hypotheses through different model structures, fit the associated models to the observed data, and formally compare them to determine the strength of evidence for each model and hence each biological hypothesis.

10.4.1   Markovian, Random, and No Changes in Occupancy

The colonization (γ_t) and local extinction (ϵ_t) probabilities defined in the unconditional explicit dynamics model (MacKenzie et al., 2003; Chapter 8) allow Markovian changes in occupancy, i.e., the probability that a unit is occupied this season (season $t + 1$) depends upon the unit's occupancy state last season (season t). If a unit was occupied last season, the probability the unit is occupied

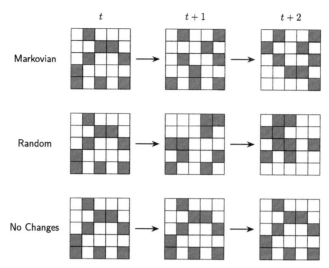

FIGURE 10.2 Illustration of Markovian, random, and no changes in occupancy of units over time. Under Markovian dynamics, changes in occupancy would typically be expected to be more gradual than under random dynamics.

this season is $(1 - \epsilon_t) = \phi_t$ (i.e., species did not go locally extinct, or persisted at the unit), whereas if a unit was unoccupied last season, the probability of occupancy this season is γ_t (i.e., species colonized the unit). In many applications it may be reasonable to expect that the changes in occupancy could be Markovian in nature, particularly when a 'unit' is defined in terms of the behavioral nature of the species at certain times of the year. For instance in a study of northern spotted owls (*Strix occidentalis caurina*), Olson et al. (2005) defined a 'unit' as a potential breeding territory. As many bird species will return to the same breeding or nesting area each year, one would expect that potential breeding territories that were occupied in year 1 would be more likely to be occupied in year 2, than those unoccupied in year 1. When the sampling season is defined to coincide with the owl's breeding season, and each season is separated by a single year, modeling changes in occupancy as a Markov process seems reasonable. However there are two other types of changes in occupancy that may be of biological interest: (1) random changes in occupancy; and (2) no changes in occupancy (or closure of the system; Fig. 10.2).

If changes in the occupancy state of units over time is random, the probability of occupancy in season $t + 1$ does not depend upon the occupancy state of the unit in season t. That is, all units have the same probability of species presence (given any covariate effects) in season $t + 1$, irrespective of whether the species was present or absent at the unit in the previous season. Random changes in occupancy may suggest that the species as a whole has a low degree of unit fidelity (not to be confused with, although likely to be a function of, the fidelity of individuals to the unit), as the species tends to occur randomly at units within the area over time, rather than being more likely to persist at the same units. It may also be that changes in occupancy appear to be random when the length of time between seasons is sufficiently large that several Markovian changes can occur (e.g., if the potential spotted owl territories were only surveyed every five years rather than annually).

Changes in occupancy are random when the colonization and persistence, the complement of local extinction, probabilities are equal. Mathematically, changes are random when $\gamma_t = \phi_t = (1 - \epsilon_t)$, or equivalently, $\epsilon_t = (1 - \gamma_t)$, for $t = 1, 2, \ldots, T - 1$, i.e., the probability of a unit becoming occupied is the same as the probability a unit stays occupied. Note the effect this constraint has on Eq. (8.4) for recursively calculating occupancy probabilities:

$$\psi_{t+1} = \psi_t(1 - \epsilon_t) + (1 - \psi_t)\gamma_t$$
$$= \psi_t\gamma_t + (1 - \psi_t)\gamma_t$$
$$= \gamma_t = \phi_t = (1 - \epsilon_t).$$

When changes are random, the colonization and persistence probabilities are the probability a unit is occupied. Furthermore, this constraint affects the explicit dynamics model. Consider the example given in Eq. (8.2) with the above constraint to define a random process:

$$Pr(\mathbf{h}_i = 110\,000\,010|\boldsymbol{\theta}) = \psi_1 p_{1,1} p_{1,2}(1 - p_{1,3})$$
$$\times \left[(1 - \epsilon_1)\prod_{j=1}^{3}(1 - p_{2,j})(1 - \epsilon_2) + \epsilon_1\gamma_2\right]$$
$$\times (1 - p_{3,1})\,p_{3,2}(1 - p_{3,3})$$
$$= \psi_1 p_{1,1} p_{1,2}(1 - p_{1,3})$$
$$\times \left[\psi_2 \prod_{j=1}^{3}(1 - p_{2,j})\psi_3 + (1 - \psi_2)\psi_3\right]$$
$$\times (1 - p_{3,1})\,p_{3,2}(1 - p_{3,3})$$
$$= \psi_1 p_{1,1} p_{1,2}(1 - p_{1,3})$$
$$\times \left[\psi_2 \prod_{j=1}^{3}(1 - p_{2,j}) + (1 - \psi_2)\right]$$
$$\times \psi_3(1 - p_{3,1})\,p_{3,2}(1 - p_{3,3}).$$

This probability statement now has the same structure as the probability statement derived for this history using the implicit dynamics model of Section 8.2 (Eq. (8.1)), which demonstrates that imposing the constraint of random changes in occupancy on the explicit model of Section 8.3, leads to the implicit dynamics model of Section 8.2. That is, even though the implicit dynamics model could be used to describe the resulting *patterns* in occupancy caused by any underlying mechanistic process (random, first-order Markovian or otherwise), the model makes the de-facto assumption that any changes are random in nature. One would conclude changes to be random if the strength of evidence for such a model (or group of models) was sufficiently greater than the evidence for similar models where changes are Markovian.

Under the assumption of random changes, the transition probability matrix, ϕ_t, can be expressed as:

$$\phi_t = \begin{bmatrix} (1 - \gamma_t) & \gamma_t \\ (1 - \gamma_t) & \gamma_t \end{bmatrix}.$$

The important aspect, here, is that the two rows of ϕ_t are identical, hence the probability of a unit being in each occupancy state in season $t + 1$ (represented by the columns) is the same for each state the unit may have been in during season t (represented by the rows); occupancy state in the previous season has no effect on the probability of occupancy in the current season.

Rather than Markovian or random changes in occupancy, in some situations it may be reasonable to consider the system as static, or closed to changes in occupancy over the duration of the study. That is, the occupancy status of units does not change between seasons. While perhaps a biological unreality over a longer timeframe, when data have been collected for only a few seasons, this assumption of no changes in occupancy may be reasonable depending upon the target species and study design. Such a model could be considered by enforcing the constraint $\gamma_t = \epsilon_t = 0$ for $t = 1, 2, \ldots, T - 1$ (i.e., there is zero probability that units change occupancy status). In effect, the unconditional explicit dynamics model reduces to the MacKenzie et al. (2002) single season model detailed in Chapter 4 when these constraints are imposed. For example, consider again the probability statement for the above detection history under this set of constraints:

$$
\begin{aligned}
Pr(\mathbf{h}_i = 110\,000\,010) &= \psi_1 p_{1,1} p_{1,2}\left(1 - p_{1,3}\right) \\
&\times \left[(1 - \epsilon_1) \prod_{j=1}^{3} \left(1 - p_{2,j}\right)(1 - \epsilon_2) + \epsilon_1 \gamma_2 \right] \\
&\times \left(1 - p_{3,1}\right) p_{3,2}\left(1 - p_{3,3}\right) \\
&= \psi_1 p_{1,1} p_{1,2}\left(1 - p_{1,3}\right) \\
&\times \left[\prod_{j=1}^{3} \left(1 - p_{2,j}\right) \right] \left(1 - p_{3,1}\right) p_{3,2}\left(1 - p_{3,3}\right).
\end{aligned}
$$

Given the assumption that the occupancy state of units does not change between seasons, when the species was not detected during the second season, it must have been present but undetected because the species was detected at the unit during the other seasons.

While the three types of changes have been detailed above in terms of the basic two-state multi-season model, the same logic can be applied to multi-state multi-season models. For example, in the case with three occupancy states using the multinomial parameterization of the model (Chapter 9), $\varphi_t^{[m,n]}$ is the probability of a unit transitioning from state m in season t, to state n in season $t+1$, where m and $n = 0$, 1, or 2. Whenever these probabilities are unconstrained such that they are dependent on the state of the unit at t, then changes are of a Markovian nature, and the transition probability matrix could be expressed as:

$$\Phi_t = \begin{bmatrix} \varphi_t^{[0,0]} & \varphi_t^{[0,1]} & \varphi_t^{[0,2]} \\ \varphi_t^{[1,0]} & \varphi_t^{[1,1]} & \varphi_t^{[1,2]} \\ \varphi_t^{[2,0]} & \varphi_t^{[2,1]} & \varphi_t^{[2,2]} \end{bmatrix},$$

where each row must sum to 1.0. Defining the transition probabilities such that they no longer depend on occupancy state in season t, i.e., $\varphi_t^{[\cdot,n]} = \varphi_t^{[0,n]} = \varphi_t^{[1,n]} = \varphi_t^{[2,n]}$ for all n, results in a multi-state model with random changes in occupancy state. Hence, each row of Φ_t will be identical:

$$\Phi_t = \begin{bmatrix} \varphi_t^{[\cdot,0]} & \varphi_t^{[\cdot,1]} & \varphi_t^{[\cdot,2]} \\ \varphi_t^{[\cdot,0]} & \varphi_t^{[\cdot,1]} & \varphi_t^{[\cdot,2]} \\ \varphi_t^{[\cdot,0]} & \varphi_t^{[\cdot,1]} & \varphi_t^{[\cdot,2]} \end{bmatrix}.$$

Finally, when there are no changes in occupancy state, then $\varphi_t^{[m,n]} = 1$ when $m = n$, and zero otherwise. With this constraint, the transition probability matrix becomes the identity matrix:

$$\Phi_t = \begin{bmatrix} 1 & 0 & 0 \\ 0 & 1 & 0 \\ 0 & 0 & 1 \end{bmatrix}.$$

10.4.2 Equilibrium

An assumption required by many of the published techniques for analyzing single-season occupancy data is that the metapopulation is stationary, or in a state of equilibrium (e.g., Hanski, 1994a), i.e., the net number of units that are colonized each season is equal, on average, to the net number of units where the species goes locally extinct. Given the changeable nature of most ecosystems around the world, and the outside pressures they are frequently subjected to, the validity of this assumption must be questioned. Yet, it is an assumption that goes untested in many metapopulation studies. However using the unconditional, explicit dynamics model of MacKenzie et al. (2003), it is a biological assumption that can be assessed. The concept of equilibrium also exists in multi-state occupancy applications, where the multi-season multi-state model of MacKenzie et al. (2009) can be used.

The equilibrium assumption could be defined in at least two ways. First it may be expressed in terms of the overall probability of the unit being in each occupancy state each season (unconditional on unit state in the previous season), where the probability is constant for the duration of the study. For example, in a two-state occupancy system (Chapter 8), $\psi_t = \psi$ for $t = 1, 2, \ldots, T$ (or equivalently $\lambda_t = \lambda' = 1$). Secondly, equilibrium could also be defined in terms of the state-specific transition probabilities being constant across time, which results in a stationary Markov process. For a two-state application, this means $\gamma_t = \gamma$ and $\epsilon_t = \epsilon$ for $t = 1, 2, \ldots, T - 1$. When these probabilities are constant, the population will attain an equilibrium level of occupancy, which can be calculated as:

$$\psi_{Eq} = \frac{\gamma}{\gamma + \epsilon}.$$

These definitions of equilibrium are not equivalent, as in the former case the overall level of occupancy may be constant even though the transition probabilities vary over time, but balance each other each season. Such compensating vital rates may arise through resource limitations (a local extinction frees resources enabling colonization), or from movement of individuals of the species (if all individuals of the target species move from an occupied unit, to an unoccupied unit, that results in both a local extinction and colonization). In the latter definition of equilibrium, occupancy may not be constant either: (1) because the population is still heading toward the new equilibrium level, i.e., is experiencing transient dynamics (see Yackulic et al., 2012); or (2) as an artifact of how units were sampled from the population (see Chapter 12). Regardless of exactly how equilibrium is defined, it is a biological hypothesis that can be represented by applying appropriate constraints to the parameters of the multi-season models.

Units could be in a state of equilibrium with respect to occupancy, whilst either Markovian, random or no changes in occupancy are occurring between seasons. Each combination represents a different set of biological mechanisms that may be of interest in different circumstances.

Finally, we would note that the above definitions of equilibrium are based on the common concept that time-invariance equates to a 'stable' system. This is a somewhat restrictive interpretation of stability, as one can also imagine situations where a cyclical population could be stable, with regular oscillations about a long-term mean value. We have not explored this concept with dynamic occupancy modeling ourselves, but are simply noting that there may also be other ways in which 'equilibrium' could be defined.

10.4.3 Example: Northern Spotted Owl

To illustrate the above approach for investigating occupancy dynamics using the unconditional explicit dynamics model (Chapter 8), we now revisit the example considered by MacKenzie et al. (2003) of the northern spotted owl (*Strix occidentalis caurina*) in Alan Franklin's Willow Creek study area in northern California. As part of the example we also illustrate the idea of formally comparing assumptions commonly used in investigations of metapopulations, specifically that the population is at equilibrium in terms of occupancy dynamics. We also illustrate how one could compare models to make inference about whether changes in occupancy are best described as Markovian, random, or static. As with other examples, different hypotheses about the occupancy state of the biological system and the associated vital rates are expressed by a suite of candidate models. We use AIC model selection procedures to rank the models within the candidate set. Files associated with this analysis can be downloaded from http://www.proteus.co.nz.

In northern California, monitoring of potential spotted owl habitat commenced even before the species was listed as threatened by the United States Fish and Wildlife Service in 1990 (U.S. Fish and Wildlife Service, 1990). The data considered by MacKenzie et al. (2003) consisted of 55 potential breeding territories ($s = 55$ units) surveyed each year between 1997 and 2000 ($T = 5$ breeding seasons). Each unit was surveyed up to eight times per season to determine whether the unit was occupied by a pair of breeding owls. Surveys were not conducted simultaneously across units, resulting in some missing observations ($K_{max} = 8$ surveys, $K_{average} = 5.3$ surveys); however, when conducted, surveys followed a well established protocol that was consistent across years (Franklin et al., 1996). For simplicity, we assume that detection probabilities were constant for all surveys within seasons, but the modeling could be extended to allow detection probabilities to vary in time or in accordance with measured covariates. In addition, MacKenzie et al. (2003) found strong evidence that detection probabilities were year specific, $p(year)$, thus we adopt the same detection probability structure for all models in this analysis. We do not include a term for detection probability in our model notation below, as it is the same for all models.

Our primary goal is to illustrate how practitioners could investigate changes in occupancy and equilibrium issues with models presented in Chapter 8. Specifically, we explore whether our set of potential breeding units was in a state of equilibrium in terms of occupancy dynamics for breeding pairs over the five-year period, and whether the changes were better represented by random or Markovian processes.

We represent these hypotheses with six competing models, using both the original parameterization of MacKenzie et al. (2003) that estimates only first-season occupancy directly, and reparameterizations that allow direct estimation of occupancy probabilities in each season (Section 8.3.6). Which parameterization was used is identified in this example by the notation $\psi_{init}()$ and $\psi_{seas}()$, respectively, where the terms in parentheses indicate what factors were included as covariates, which, generally, may include time effects or unit-specific covariates. However, no unit-specific covariates were used here. Constraints that were placed on the model parameters are indicated in square brackets.

First, we considered a model where the occupancy status of units did not change, (i.e., by constraining both colonization and extinction probabilities equal to zero for all years), denoted as $\psi_{init}(\cdot)[\gamma_t = \epsilon_t = 0]$. This model suggests that owl territories were well established and the breeding pair occupancy state was static for all five breeding seasons. Next, we explored possible random changes in occupancy, where the probability that a breeding pair occupied a unit did not depend on whether the unit was occupied in the previous breeding season. We would expect this type of occupancy change if owl pairs had low unit fidelity (e.g., if breeding pairs chose units randomly each year). This is probably not the case for spotted owls, but we are able to formally assess this expectation by setting $\epsilon_t = (1 - \gamma_t)$. Further, the population may be at a point of equilibrium or not. The model denoted as $\psi_{seas}(year)[\epsilon_t = 1 - \gamma_t]$ suggests changes in occupancy are random, with the model parameterized such that the seasonal occupancy probabilities are estimated directly. Recall that under an assumption of random changes in occupancy $\psi_{t+1} = \gamma_t$, hence the constraint $\epsilon_t = (1 - \gamma_t)$ is equivalent to $\epsilon_t = (1 - \psi_{t+1})$. This model allows the probability of occupancy to vary among years (i.e., population may not be in equilibrium according to the definition of time-constancy of occupancy). To fit a model that assumes the population is at equilibrium with random changes in occupancy, the probability of occupancy in each year was constrained to be equal (i.e., the first definition of equilibrium provided above), which we denote as $\psi_{seas}(\cdot)[\epsilon_t = 1 - \gamma_t]$. Again, this model was parameterized in terms of seasonal occupancy. In fact, this model also represents the second equilibrium definition, i.e., a stationary Markov process, through the relationship between ψ_{t+1} and γ_t under the assumption of random changes in occupancy, noted above. Previous capture–recapture studies have shown spotted owls to be a territorial, with high unit fidelity and little migration among adult breeders, thus *a priori* we would expect Markovian changes in occupancy (Franklin et al., 1996). To allow Markovian changes in occupancy, we remove the constraint $\epsilon_t = (1 - \gamma_t)$, such that both colonization and local extinction probabilities are separately estimated. The models $\psi_{init}(\cdot)\,\gamma(\cdot)\,\epsilon(\cdot)$ and $\psi_{init}(\cdot)\,\gamma(year)\,\epsilon(year)$ represent the equilibrium and non-equilibrium hypotheses with Markovian changes in occupancy, respectively, using the parameterization where only first-season (1997) occupancy was estimated directly. The equilibrium definition used here is in terms of time-invariant vital rates, i.e., a stationary Markov process. Occupancy probabilities in years 1998–2001 can be calculated based upon the initial value and the colonization and extinction probabilities using the recursive equation (Eq. (8.4)). This is the initial parameterization used by MacKenzie et al. (2003). Finally, we considered one other model; $\psi_{seas}(\cdot)\,\gamma(year)$. This model presumes Markovian changes in occupancy (as colonization and extinction probabilities are unconstrained), but was parameterized such that seasonal occupancy and colonization probabilities were estimated directly, with extinction probabilities now being derived parameters. That is:

$$\epsilon_t = 1 - \frac{\psi_{t+1} - (1 - \psi_t)\,\gamma_t}{\psi_t}.$$

However, the notation '$\psi_{seas}(\cdot)$' indicates that while the model includes a separate occupancy probability for each year, they have been constrained to be constant over time, which again corresponds to the first equilibrium definition. Therefore, our model set contained two models that assumed the population was at equilibrium, with Markovian changes in occupancy, but they differed in how equilibrium was defined. We note that the model $\psi_{init}(\cdot)\,\gamma(\cdot)\,\epsilon(\cdot)$ represents a stationary Markov process, which is sometimes assumed in metapopulation studies that use so-called 'incidence functions' for inference about extinction and colonization (e.g., Hanski, 1994a).

Table 10.1 presents the six candidate models ranked according to AIC, with parameter estimates given in Table 10.2. Consistent with our *a priori* expectation, changes in occupancy are best represented by a Markov process. Models with random or no changes in occupancy have essentially no support (ΔAIC values are much greater than 10). Additionally, the hypothesis that these 55 units are in some form of equilibrium state appears to be well supported. The top two ranked models both represent some form of equilibrium situation, and have a combined model weight of $>95\%$. Note that from the top-ranked model, with time-invariant occupancy, years with higher or lower colonization probabilities are matched by higher or lower local extinction probabilities. This highlights that overall occupancy may remain constant over seasons, but have compensating season-specific dynamic probabilities. For the top-ranked model, this relationship is enforced due to the constraints made on the model structure, however note that a similar pattern in the estimated colonization and extinction

TABLE 10.1 Summary of model selection procedure results for northern spotted owl example. ΔAIC is the relative difference in AIC values compared with the top ranked model; w is the AIC model weight; $Npar$ is the number of parameters in the model; $-2l$ is twice the negative log-likelihood value. The type of changes in occupancy represented by the model (Changes; M = Markovian, R = Random, N = None), and equilibrium assumption (Eq.; Y_1 = time-constant occupancy, Y_2 = time-constant dynamic probabilities, N = none) are indicated. For all models, detection probability varied among years, but not within years, $p(year)$. Note that under random changes in occupancy, $\epsilon_t = (1 - \gamma_t)$ is equivalent to $\epsilon_t = (1 - \psi_{t+1})$

Model	ΔAIC	w	$Npar$	$-2l$	Changes	Eq.
$\psi_{seas}(\cdot)\gamma(year)$	0.00	0.86	10	1329.34	M	Y_1
$\psi_{init}(\cdot)\gamma(\cdot)\epsilon(\cdot)$	4.18	0.11	8	1337.52	M	Y_2
$\psi_{init}(\cdot)\gamma(year)\epsilon(year)$	6.30	0.04	14	1327.64	M	N
$\psi_{seas}(\cdot)[\epsilon_t = 1 - \gamma_t]$	92.46	0.00	6	1429.80	R	Y_1, Y_2
$\psi_{seas}(year)[\epsilon_t = 1 - \gamma_t]$	99.98	0.00	10	1429.32	R	N
$\psi_{init}(\cdot)[\gamma_t = \epsilon_t = 0]$	205.22	0.00	6	1542.56	N	Y_1

probabilities is apparent for the third-ranked model, where these dynamic parameters were estimated independently. The second-ranked model, which has substantially less support than the top-ranked model, represents the assumption of a stationary Markov process, with time-invariant colonization and extinction probabilities. As noted above, in this case equilibrium occupancy can be calculated as:

$$\psi_{Eq} = \frac{\gamma}{\gamma + \epsilon}$$
$$= \frac{0.18}{0.18 + 0.14}$$
$$= 0.56,$$

which is the value the probability of occupancy will converge to over time while colonization and extinction probabilities remain at that level. From Table 10.2, the derived occupancy probabilities for 1998–2001 under the second-ranked model were trending toward 0.56, as would be expected. While the models that assumed random changes in occupancy are not well supported (ranked 4 and 5 in Table 10.1) in comparison to the models with Markovian changes, the estimated probabilities of occupancy resulting from the 'random' models are similar to those from the corresponding 'Markovian' models, for this data set (Table 10.2). This suggests that 'random' models could be used to provide reasonable inferences about *patterns* of species occurrence over time, even though they do not adequately model the underlying dynamic processes. The final model in Table 10.1 assumes no changes in occupancy during the five-year period; breeding

TABLE 10.2 Estimated probabilities of occupancy ($\hat{\psi}$), colonization ($\hat{\gamma}$), and local-extinction ($\hat{\epsilon}$) from the six models fit to the northern spotted owl data from 1997–2001 (Table 10.1). Bold values indicate values that are constrained by the defined model structure, and underlined values are derived parameter estimates (not estimated directly). Standard errors are not presented

Model	$\hat{\psi}_{97}$	$\hat{\psi}_{98}$	$\hat{\psi}_{99}$	$\hat{\psi}_{00}$	$\hat{\psi}_{01}$	$\hat{\gamma}_{97}$	$\hat{\gamma}_{98}$	$\hat{\gamma}_{99}$	$\hat{\gamma}_{00}$	$\hat{\epsilon}_{97}$	$\hat{\epsilon}_{98}$	$\hat{\epsilon}_{99}$	$\hat{\epsilon}_{00}$
$\psi_{seas}(\cdot)\,\gamma(year)$	0.59	**0.59**	**0.59**	**0.59**	**0.59**	0.12	0.13	0.37	0.15	0.08	0.09	0.26	0.10
$\psi_{init}(\cdot)\,\gamma(\cdot)\,\epsilon(\cdot)$	0.62	0.60	0.59	0.58	0.57	0.18	**0.18**	**0.18**	**0.18**	0.14	0.14	**0.14**	**0.14**
$\psi_{init}(\cdot)\,\gamma(year)\,\epsilon(year)$	0.63	0.61	0.56	0.60	0.57	0.11	0.07	0.39	0.12	0.09	0.13	0.24	0.12
$\psi_{seas}(\cdot)\,[\epsilon_t = 1 - \gamma_t]$	0.59	**0.59**	**0.59**	**0.59**	**0.59**	**0.59**	**0.59**	**0.59**	**0.59**	0.41	0.41	0.41	0.41
$\psi_{seas}(year)\,[\epsilon_t = 1 - \gamma_t]$	0.62	0.60	0.57	0.60	0.57	0.60	0.57	0.60	0.57	0.40	0.43	0.40	0.43
$\psi_{init}(\cdot)\,[\gamma_t = \epsilon_t = 0]$	0.84	**0.84**	**0.84**	**0.84**	**0.84**	**0.00**	**0.00**	**0.00**	**0.00**	0.00	0.00	0.00	0.00

territories that were occupied in 1997 were still occupied in 2001, and territories that were not occupied in 1997, were never occupied through until 2001. Essentially this model assumes the territories (units) were closed to changes in occupancy for the duration of the five years of surveying, which is the exact same assumption as the single-season occupancy model of MacKenzie et al. (2002) (Chapter 4). In this case, the static occupancy model was not well supported, but this general approach provides a framework to 'test' for unit closure in single-season studies if there are sufficient repeat surveys to break a nominal season into smaller periods that include some replicate surveys (e.g., early season and late season).

The above results suggest that this population of spotted owl territories was in a period of stability during 1997–2001, which is consistent with the conclusions drawn from a demographic study conducted in the same region over a similar period (Anthony et al., 2006).

10.4.4   Further Insights

There is a number of further insights into the modeling of occupancy dynamics with these methods that we now offer. Primarily these are in terms of the two-state situation, with which we have had the greatest experience, but some of the issues will also generalize to multi-state applications as well.

Inferring Population Trajectory from Dynamic Parameters

We have often encountered researchers who attempt to infer the trajectory of a population, in terms of increasing or decreasing occupancy probabilities, based upon estimated probabilities of the occupancy dynamics. For example, from the second model of Table 10.2, colonization probability was estimated to be greater than extinction probability, therefore some would infer that occupancy of spotted owl territories must have been increasing. However, looking at the estimated occupancy probabilities we see that quite the opposite appears to be true; occupancy is estimated to be declining over time. The reason for this result, that some may find contradictory, is that the dynamic probabilities are conditional upon the occupancy state of the unit. For example, in the two-state case, colonization events only happen at unoccupied units, and extinction events only happen at occupied units. Therefore, inference about whether the level of occupancy is increasing or decreasing also requires that the current level of occupancy is taken into account. To illustrate, from model $\psi_{init}(\cdot)\,\gamma(\cdot)\,\epsilon(\cdot)$ for the spotted owl example, $\hat{\psi}_{97} = 0.62$, $\hat{\gamma}_{97} = 0.18$, and $\hat{\epsilon}_{97} = 0.14$. The estimated net probability of colonization (e.g., fraction of all units that are colonized) will therefore be:

$$\left(1 - \hat{\psi}_{97}\right)\hat{\gamma}_{97} = 0.38 \times 0.18 = 0.07,$$

where the estimated net probability of local-extinction will be:

$$\hat{\psi}_{97}\hat{\epsilon}_{97} = 0.62 \times 0.14 = 0.09.$$

Hence, as the net probability of extinction is larger, the effect on overall occupancy will be a decrease.

This holds true irrespective of whether covariates have been included to explain temporal or spatial variation in some, or all, of the occupancy-related probabilities. When covariates have been included, the net result on occupancy may be different in different times or places, e.g., the species distribution may be expanding in some areas, but contracting in others.

Time-Invariant Dynamic Parameters Can Induce an Apparent Trend

A consequence of assuming a stationary Markov process (time-invariant transition probabilities, e.g., colonization and extinction) during data analysis is that the estimated seasonal occupancy probabilities will converge toward the equilibrium value suggested by the time-invariant dynamic parameter estimates. The rate of convergence will depend on how different the estimated initial level of occupancy is from the derived equilibrium value. This convergence may be interpreted as an apparent trend, especially over a shorter time series where surveying halts before occupancy stabilizes at the equilibrium level, which may lead to erroneous inferences, particularly if the assumption of a stationary Markov process is unreasonable. Again, this holds true irrespective of whether covariates have been included in the estimating model to explain static spatial variation in the occupancy and transition probabilities; the equilibrium level that estimated occupancy converges to will just be different in different areas.

This is not a failing of the modeling approaches; the models are simply behaving according to the structure defined by the practitioner. We highlight this point here, however, as we have seen examples where practitioners may specify relatively complex models for first-season occupancy, when using the original parameterization of the MacKenzie et al. (2003) multi-season model, but maintain a fairly simple structure on colonization and local extinction probabilities, particularly, time-invariance. They have done so claiming that they are primarily interested in occupancy rather than the underlying dynamics, hence have used simple model structures on the transition probabilities to reduce the number of parameters to estimate, and reduce the number of possible models that could be fit to the data. While we applaud the desire to focus analyses on the key questions of interest, practitioners must be aware of how the various model parameters interrelate and any unintended consequences of decisions around which models they choose to fit to their data. For example, it is difficult to explain how covariates could influence occupancy, yet not influence the dynamic rate parameters that determine all changes in occupancy.

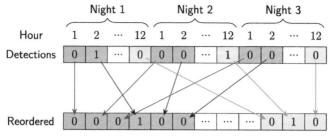

FIGURE 10.3 Illustration of reordering detection data from chronological order, into the required order for an analysis. Each color indicates a different 'season' as defined by the study objective.

Whenever there is no temporal variation in the transition probabilities that govern occupancy dynamics, a stationary Markov process results, and estimated occupancy probabilities will always converge to the equilibrium values over time. We recommend that temporal variation should always be considered in the modeling of the transition probabilities (e.g., models representing this assumption be included in the set of models fit to data, along with models that assume time-invariant dynamic parameters, if desired), allowing for the possibility of a non-stationary Markov process (i.e., the population is not in, or converging to, an equilibrium state).

10.4.5   Chronological Order of Surveys

While multi-season models are typically applied to the data in the same chronological order in which they were originally collected, there is no reason why that has to be the case. For example, suppose you wish to investigate how bat distribution generally changes within a night, as they emerge, and then return to their roosts or hibernacula. To do so, 50 acoustic detection devices could be set up throughout the study area, with all devices run simultaneously for a three-night period. As interest is in within-night changes in distribution, one approach would be to divide the nightly surveys into twelve one-hour periods, and model the changes in bat distribution between one-hour periods as a first-order Markov process. As interest is in the *general* distribution of bats, the detection data collected each night could be treated as repeat surveys for the respective one-hour intervals. Therefore, the ordering of the data would be changed to format it for analysis (Fig. 10.3).

10.4.6   Discussion

In this section we have outlined some ways in which the multi-season models detailed in Chapters 8 and 9 could be used to investigate certain aspects of the underlying dynamic processes of species occurrence. We believe that exploring the dynamics of species occurrence will ultimately provide much greater insight about the population, which will be much more useful to scientists and managers in their efforts to learn about, and manage, ecological systems.

10.5 SENSITIVITY OF OCCUPANCY TO DYNAMIC PROCESSES

In abundance-based population modeling, the sensitivity of population growth rate, λ, to underlying vital rates (e.g., individual survival and fecundity) is an important topic in studies of evolutionary and population ecology (e.g., Caswell, 2001). Identification of which vital rates λ is most sensitive to, under certain conditions, can provide important insights about the population's dynamics. Such insights might be used to identify which vital rates might be targeted by managers to manipulate population dynamics according to management objectives, make predictions about the future state of the system, or to identify which parameters future sampling should target to reduce uncertainty about population dynamics (e.g., Tuljapurkar, 1990; Caswell, 2001; Martin et al., 2009b; Miller, 2012).

The same thinking has been applied to occupancy-based state variables, for both the regular dichotomous occupancy situation (Martin et al., 2009b), and the multi-state occupancy situation (Green et al., 2011; Miller, 2012). Here we present some key results of these investigations, and direct readers to these papers for further details, particularly the general treatment provided by Miller (2012).

10.5.1 Two-State Situation

The sensitivity of a population parameter ω to a vital rate θ represents the amount of change in ω induced by a small change in θ, i.e., the rate of change in ω with respect to θ. Mathematically, this can be calculated as the partial derivative of ω with respect to θ, $\partial \omega / \partial \theta$.

Martin et al. (2009b) considered a regular two-state occupancy situation, focusing on equilibrium occupancy, ψ_{Eq}, as the population parameter, and its sensitivity to colonization and extinction probabilities. Recall that for a stationary Markov process:

$$\psi_{Eq} = \frac{\gamma}{\gamma + \epsilon}.$$

Therefore, taking the partial derivative with respect to each dynamic probability (the vital rates), the sensitivities (s_θ) are:

$$s_\gamma = \frac{\partial \psi_{Eq}}{\partial \gamma} = \frac{\epsilon}{(\gamma + \epsilon)^2},$$

and

$$s_\epsilon = \frac{\partial \psi_{Eq}}{\partial \epsilon} = -\frac{\gamma}{(\gamma + \epsilon)^2}.$$

The sign of s_θ simply indicates whether ψ_{Eq} will increase (if $s_\theta > 0$) or decrease (if $s_\theta < 0$) as θ increases, hence Martin et al. (2009b) focus on the magnitude of the sensitivity, by taking the absolute value, $|s_\theta|$. They concluded that:

$$|s_\epsilon| > |s_\gamma|; \text{ if } \gamma > \epsilon, \text{ or } \psi_{Eq} > 0.5,$$

$$|s_\epsilon| < |s_\gamma|; \text{ if } \gamma < \epsilon, \text{ or } \psi_{Eq} < 0.5,$$

and

$$|s_\epsilon| = |s_\gamma|; \text{ if } \gamma = \epsilon, \text{ or } \psi_{Eq} = 0.5.$$

That is, a small change in extinction probability will have a greater effect on equilibrium occupancy than the same-sized change in colonization probability when the species is present at $>$ 50% of units, and when the species is present at $<$ 50% of units, a small change in colonization probability will have a greater effect on equilibrium occupancy than the same-sized change in extinction probability. When the species is present at exactly 50% of units, then a small change in colonization or extinction probabilities produces the same change in equilibrium occupancy.

The reason for this is similar to that discussed in the previous section with respect to inferring the population trajectory from dynamic parameter estimates. The processes of colonization and local extinction operate on unoccupied and occupied units, respectively. Therefore, when occupancy is higher there are more units in the population whose occupancy state in the next season will be governed by the local extinction probability, hence a small change in that value will result in a greater effect on occupancy than the same change in colonization probability. When occupancy is low, there are more unoccupied units that could potentially be colonized than occupied units where the species could go locally extinct. As such, small changes in the colonization probability will potentially affect a greater number of units, resulting in a potentially larger effect on overall occupancy.

The practical application of this result, for example, is that when conservation efforts are focused on a rare species, i.e., low occupancy, management actions that can change colonization probability by an amount x, will have a larger effect on the species occupancy than changing extinction probability by x amount.

Martin et al. (2009b) also presented results for elasticities (sensitivities scaled to be proportional changes) and variance-stabilized sensitivities (VSS;

Link and Doherty, 2002) of the dynamic parameters. They showed that the magnitude of the elasticities for colonization and extinction are equal ($|e_\gamma| = |e_\epsilon|$), and the results using VSS were consistent with the regular sensitivities.

10.5.2 General Situation

Miller (2012) detailed sensitivities calculations for the more general multi-state occupancy situation (see also Green et al., 2011), for both the equilibrium and non-equilibrium cases. Not only did Miller (2012) derive sensitivities for the transition probabilities of population dynamics, but also for "lower-level" parameters, such as regression coefficients when incorporating covariates, through application of the chain rule for partial derivatives, and for incorporating environmental variation.

We do not provide further details of these methods here, referring the interested reader to Miller (2012), although note that these results are applicable to a wide range of ecologically relevant problems, and predict they will receive greater use in the future.

10.6 MODELING HETEROGENEOUS DETECTION PROBABILITIES

The same methods that have been used to account for heterogeneity in detection probabilities for single-season occupancy models (e.g., finite-mixtures, abundance-induced heterogeneity, and random effects; Chapter 7), could also be applied to the multi-season case. This requires additional model structure for the within-season detection process, with the between-season model structure for the occupancy dynamics being, largely, unchanged.

The use of finite-mixtures to account for detection heterogeneity has been incorporated into the software packages Program PRESENCE (Hines, 2006) and Program MARK (White and Burnham, 1999). There are some additional considerations involved with the use of finite-mixture models in the multi-season situation compared to the single-season case. Finite-mixtures involve defining that the occupied units consist of a fixed number of groups, each with a different detection probability. From the available data, group membership is unknown, hence the analysis must account for that fact that a unit may belong to any of the defined groups each season. One consideration is whether group membership is independent each season, or whether a unit will tend to be in the same group, e.g., the group with lower detection probability, in consecutive seasons. In other words, is group membership randomly 'assigned' each season or does it follow a first-order Markov process. Random group membership has been implemented in PRESENCE and MARK, which involves group membership probabilities (for occupied units) being estimated for each season, each group with a potentially different detection probability. Which group a unit may have belonged to in the previous season, has no effect on which group that unit may belong to in the current season. Alternatively, a finite-mixture approach could be developed where group membership in a season does depend on group membership in the previous season. For example, if a unit belonged to the 'low detection' group in one season, it is more likely to be in the 'low detection' group in the next season.

This can be accomplished using the multi-season multi-state occupancy model, particularly the conditional-binomial parameterization, where the dynamic occupancy parameters (for a finite-mixture with two groups; $\psi_t^{[0]}$, $\psi_t^{[1]}$, and $\psi_t^{[2]}$; Chapter 9) are the colonization and persistence (the complement of local extinction) probabilities. That is, $\psi_t^{[0]}$ equates to colonization, and $\psi_t^{[1]}$ and $\psi_t^{[2]}$ are both persistence probabilities (species present in consecutive seasons), for each detection group. The dynamic 'reproduction' probabilities ($R_t^{[0]}$, $R_t^{[1]}$, and $R_t^{[2]}$) now become the probabilities of group membership, conditional upon the unit's group in the previous year. A key consideration is the structure of the detection matrix. Presuming there are only two possible observations, nondetection and detection, the detection probability matrix may therefore be defined as:

$$
\mathbf{p}_{t,j} = \begin{bmatrix} 1 & 0 \\ 1 - p_{1,t,j} & p_{1,t,j} \\ 1 - p_{2,t,j} & p_{2,t,j} \end{bmatrix},
$$

where $p_{1,t,j}$ and $p_{2,t,j}$ are the detection probabilities for each group in survey j of season t. Recall that each column of $\mathbf{p}$ is associated with a type of observation, in this case the first column relates to nondetection and the second column to detection. This is the same structure as that used in the derivation of the finite-mixture model as a multi-state model in Chapter 7. We have not actually attempted to fit this model to data ourselves, and it may be that some constraints on parameters are required to ensure identifiability, but the model is certainly conceptually possible.

Another approach to the modeling of detection heterogeneity in multiple season data would be to extend the abundance distribution modeling of Royle and Nichols (2003) (see Chapter 7). Under this model, the probability of detecting the species at unit i, in survey j of season t ($p_{t,ij}^{[N_{t,i}]}$ is the probability of detecting at least one individual at the unit if there are $N_{t,i}$ individuals at the unit in season t) is modeled as a function of the detection probability of individual animals ($r_{t,ij}$):

$$
p_{t,ij}^{[N_{t,i}]} = 1 - \left(1 - r_{t,ij}\right)^{N_{t,i}}.
$$

The N_{ti} could be modeled as independent values each season (i.e., $N_{t+1,i}$ does not depend on $N_{t,i}$), where abundance at each unit is assumed to be a random value from an appropriate distribution (e.g., Poisson or negative binomial distribution), which may be different in each season. This would be a non-Markovian approach. A Markovian dynamics model could be constructed by defining $N_{t+1,i}$ in terms of the number of 'survivors' from season t at the unit and number of 'recruits' (Rossman et al., 2016). That is:

$$N_{t+1,i} = \varphi_t N_{t,i} + G_{t,i},$$

where φ_t is the probability of an individual 'surviving' at unit i between seasons t and $t + 1$, and $G_{t,i}$ is the number of 'recruits' or gains to the local population of individuals between seasons t and $t + 1$, which could be assumed to be a random value from an appropriate discrete distribution (e.g., Poisson; Rossman et al., 2016). Local extinction of the species would therefore be the probability of all individuals dying or dispersing from the unit and no new individuals arriving at the unit:

$$\epsilon_{t,i} = (1 - \varphi_t)^{N_{t,i}} \times Pr\left(G_{t,i} = 0\right).$$

The probability of the species colonizing an unoccupied unit i between seasons t and $t + 1$, hence $N_{t,i} = 0$, would be:

$$\gamma_{t,i} = Pr\left(G_{t,i} > 0\right).$$

In this approach, colonization and local extinction probabilities are derived parameters and the modeling would have to be performed in terms of changes in local abundance because of the defined model for detection probabilities. As the $N_{t,i}$ values are unknown, they would have to be integrated out of the probability statement, as in the single-season model of Royle and Nichols (2003), by accounting for all possible combinations of $N_{t,i}$ and $N_{t+1,i}$ that are consistent with the data. Alternatively, the complete data likelihood approach could be used which would avoid complex integration terms, particularly with MCMC estimation methods (e.g., Rossman et al., 2016).

However, we would advise caution in the interpretation of the resulting estimates of the abundance-related parameters for two reasons. First, the demographic parameters associated with individual 'survival' and 'recruitment' include an among-unit movement aspect. If all individuals move from one unit to another, that would be modeled as 100% mortality at one unit and recruitment at another unit, although there has been no actual change in the number of individuals within the area of interest. Second, all of the information about 'abundance' and parameters associated with changes in 'abundance' is from among-unit variation in detection probabilities, and changes in that variation over time, from detection/nondetection data, that may be due to sources other than abundance. Therefore, as with the Royle and Nichols (2003) model (Chapter 7), we would suggest this above approach may be a suitable method for improving inferences about occupancy-level dynamic processes by accounting for detection heterogeneity, but will typically provide weaker inference about abundance-level dynamic processes, particularly where individuals are more numerous or likely to move among units.

A third approach would be a 'random effects' model where each parameter for a specific unit is a random value from some defined distribution, similar to the approach outlined in Chapter 7. In some circumstances it may be reasonable to consider that some random values for a unit may be correlated. For instance units with a high detection probability (due to abundance say) may have a low extinction probability. As such the 'random effects' could be modeled as a random draw from a multivariate distribution with some covariance or correlation structure. Such a model could be difficult to implement using a maximum likelihood approach, but could be very easily implemented using MCMC algorithms, including software such as OpenBUGS or JAGS.

10.7 DISCUSSION

In this chapter we have covered a number of important topics and extensions that could be used with multi-season, or dynamic, occupancy models. These topics and extensions may not be of universal interest, but we have attempted to highlight key issues and insights that are not covered elsewhere in the book, which we have encountered since 2002 in working with researchers attempting to apply these models. As noted as the beginning of this chapter, we believe that multi-season models that attempt to investigate the underlying dynamic processes of species occurrence, can be used to address a much richer set of questions about species ecology, and we predict such models will gain greater use in the future.

Part IV

Study Design

Chapter 11

Design of Single-Season Occupancy Studies

In Chapters 1 and 3 we briefly described some of the basic principles and reasons for carefully designing a study. A poorly designed study may not yield the type and quality of information required to achieve study objectives. In a worst-case scenario, a poorly designed study may yield no useful information and may have only succeeded in wasting precious time, effort, and money. There is no statistical magic-wand for data that have been poorly collected that will improve inferential power. Statisticians may occasionally pull a veritable rabbit out of the hat by developing new methods for such data sets, but there is no substitute for a well-designed (and executed) study. Generally the principle of 'Garbage In, Garbage Out' stands.

In this chapter we focus on the issues associated with designing a single-season study for estimating occupancy for a target species, based upon the analysis methods covered in Chapters 4–7. Most of the research into study de-sign issues has been conducted for the simpler dichotomous occupancy case (e.g., Chapter 4), but many of the topics covered in this chapter are also relevant to multi-state occupancy models and other more complex approaches.

In particular we consider issues related to how to define a sampling unit, how to select a unit, how to define a season, how to incorporate repeated surveys, and how to allocate effort. This is not an instruction book for designing *your* study, but a playbook with different ideas and possible approaches that may be appropriate in some situations and not in others. We draw from our own and others' practical experiences, computer simulation studies, and careful thought and intuition.

First, we argue very strongly that, for most species, it is important that data be collected in such a manner that the probability of detecting the species in a survey can be estimated. Generally we suggest that this requires repeated sur-veys be conducted at, at least, a subset of units. At units where the species is detected, the repeated surveys provide the necessary information about the chances of detecting the species at an occupied unit in any single survey, i.e., detection probability. Without repeated surveys it is impossible to reliably dis-entangle false and genuine absences, unless there is some other form of auxiliary

Occupancy Estimation and Modeling. http://dx.doi.org/10.1016/B978-0-12-407197-1.00015-6

439

information. It is important to realize that to perform repeated surveys, units do not have to be visited more than once. There are a number of practical options for conducting repeated surveys in the field, some of which will be covered later in this chapter.

We stress that generally studies need to be designed on a case-by-case basis. In most situations, there are unique aspects that must be addressed. Often the best studies arise when biologists, statisticians, and other relevant experts work in unison as a team; each person examining a proposed study design from his/her own perspective, looking for potential problems either with the practicalities of the design or with the adequacy of the information to be collected. We strongly suggest that input should always be sought from people with the relevant expertise.

We also stress the overarching importance of a clear objective when designing a study (Chapter 1). Subtle differences in the study objective may have a major impact on how a study should be conducted in the field. For example, suppose that there are two possible objectives for an occupancy study within a national park: (1) to compare occupancy for two specific habitat types, and (2) to obtain an overall estimate of occupancy for the entire national park. For both objectives, one potential design would be to randomly select units from across the national park and to collect the relevant data including habitat type at each unit. To address Objective 1, one could use the techniques detailed in Chapter 4 and determine whether there was evidence that occupancy varied by habitat type by using habitat type as a covariate (although note there may be more than just the two habitat types of interest represented in the data). A second design, tailored to address Objective 1, would be to first identify all areas within the park comprising the two habitat types of interest, and randomly select units solely from within those two habitats. Again the methods of Chapter 4 could be applied to determine whether there was evidence that occupancy differed for the two habitat types. The latter design is likely to be much more efficient for addressing Objective 1, but would not be appropriate for Objective 2 as the sample is not drawn from the entire park. Obviously these are not the only two possible study designs that could be used, and with other designs it may be possible to address both objectives (e.g., by stratifying the entire park into three habitat types, the two of direct interest and 'others', then randomly select units within each stratum). However, by attempting to adequately address both objectives simultaneously, a greater level of field effort may be required (i.e., surveying more units) than would be needed to achieve a single objective.

Readers may find it useful to review the main assumptions of these models and the consequences of their violation (Sections 4.4.1 and 4.4.9), as some of the key study design considerations relate to collecting the data in a manner such that those assumptions are met.

It is important to appreciate that a lot of the design issues covered in this chapter are interrelated, and decisions made about one aspect of the design will likely have consequences on other aspects of the design. The overriding consideration is to have a clear objective in mind and clear understanding of exactly what 'occupancy' means in a biological sense so the study is designed in a manner that is most conducive to achieving those goals.

11.1   DEFINING THE POPULATION OF INTEREST

As in any statistical endeavor, the first golden rule is to clearly identify your population of interest. Importantly, the *population* in this case is not the collection of individuals of the target species existing on, or in, the landscape, but the portions of the landscape that your study or monitoring program is attempting to draw inferences about. This is a fundamental conceptual issue that is often under-appreciated by those whose fieldwork has tended to be designed to maximize the chances of encountering their target species. In occupancy modeling, it is not useful to have every unit occupied by only focusing on those places where the species is likely to occur. To learn where a species is likely to be, we also need to understand where it's unlikely to be; it is actually desirable to not detect the target species in some places. Remember, a species range is not defined by where it is, but also by where it is not!

The presence or absence of the species at a sampling unit is just a characteristic of that unit, much like any other characteristic such as aspect, elevation, vegetation type, etc. From a design perspective, the intent is to sample the landscape and then measure the occupancy status of the target species at the selected locations, a task similar to that of measuring the average elevation for the area of interest. The intent should not be to maximize the number of detections of the species on the landscape, which would be analogous to targeting areas with higher elevations in the above example when trying to determine average elevation. That would clearly be undesirable.

Oftentimes, identifying the population of interest involves outlining the area(s) of interest on a map, thus delineating the scope of inference provided by a sample of units within that area. Key considerations that might restrict the area of inference include:

- study objective (e.g., prime objective may be to compare two specific vegetation types);
- accessibility due to land ownership or remoteness;
- areas of 'non-habitat' (e.g., waterbodies for a terrestrial species, or urban centers).

Once the population has been clearly defined, other design aspects follow naturally.

11.2   DEFINING A SAMPLING UNIT

We use *unit* simply as a generic term to represent the patches or sections of the landscape that are available to be sampled. These may be naturally occurring (e.g., ponds or habitat remnants) or defined arbitrarily (e.g., five-hectare blocks within a forest). However at the spatial scale of the unit, the intent is to establish the occupancy state of the unit as a categorical measure. That is, for a given unit, the desired outcome of the surveying (briefly ignoring the issue of detection probability) is to observe its occupancy status which could be one of a discrete number of values, e.g., species presence or absence in a dichotomous case, or categories of relative abundance in a multi-state setting. Only by combining the data from a number of units do we obtain a measure of occupancy that could take values between 0 and 1. Clearly, the first consideration with respect to defining a unit should be the spatial scale at which an observed category value is meaningful. For example, suppose a landscape contains a number of remnant forest stands and a management agency wants to monitor for a rare species within these stands using the occupancy state variable. Does determining whether the species is present within the stand (yes/no) provide the manager with sufficient information to make informed decisions, or is information required at a finer level of resolution (i.e., the fraction of each stand that is occupied)? In the former case, it may be appropriate to consider the individual stands as 'units', while in the latter case each stand would have to contain multiple units; hence a unit would be defined at a smaller spatial scale. Which option would be appropriate in any given situation clearly depends upon a number of additional factors such as management objectives (i.e., what level of information do managers require), and the size and number of the stands within the landscape.

A second point with respect to defining a unit is realizing that measures of occupancy are scale dependent, particularly for arbitrarily defined units within a contiguous habitat. In a given situation, a larger unit is likely to have a higher probability of occupancy (i.e., contains at least one individual of the target species) than a smaller unit. For example, consider Fig. 11.1 which represents a one-hectare (100 m × 100 m) area of interest with the black cells representing the location of the species, and where units have been defined at two different scales: (A) 50 m × 50 m; or (B) 20 m × 20 m. Sampling and detectability issues aside, the true proportion of units occupied is 1.0 if the first definition is used (i.e., the species is present within all four units) and 0.36 using the latter definition (or even 0.09 if a unit was defined to be 10 m × 10 m). It is therefore important to appreciate that a measure of occupancy is intrinsically linked to how units have been defined; if units are defined differently in different times, places or studies, any difference in occupancy is likely to be influenced by the design and may not reflect a biologically relevant difference. Generally we suggest that a unit should be large enough to have a reasonable probability of the

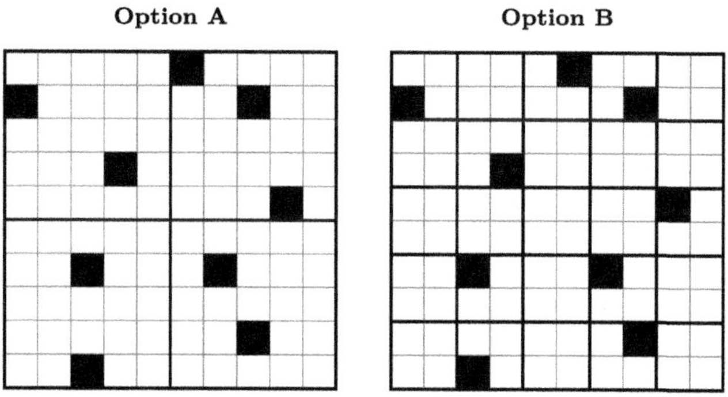

FIGURE 11.1 Graphical representation of a 1-hectare (100 m × 100 m) area with the location of the target species indicated by the black cells. 'Units' have been defined at two different scales. (A) 50 m × 50 m; or (B) 20 m × 20 m.

species being there (we suggest that the probability should be 0.2–0.8 in most situations); but small enough that the measure of occupancy is meaningful and that the unit can be surveyed with a reasonable level of effort.

Occupancy may change as the size of a unit is reduced because the species is only present within a portion of an occupied unit. That is, there is not spatial closure of units. Analogous to temporal closure, spatial closure is the property that if a unit is occupied then all sub-units are occupied (i.e., the species is either everywhere or nowhere within a unit). As the size of the defined unit reduces, the areas where the species is absent are more readily identified, hence the proportion of units that is *un*occupied increases and occupancy decreases. This phenomenon continues to the point where the size of a unit is small enough such that spatial closure is obtained; further sub-division of an occupied unit yields only smaller occupied units. For example in Fig. 11.1, at the scale of the individual 10 m × 10 m cells the species is present in 9 out of 100 units, so occupancy is 0.09, and all units are either occupied or not. If a unit was defined at the scale of 5 m × 5 m by subdividing each cell into four smaller units, 36 out of 400 units will be occupied so occupancy is still 0.09. Clearly, one does not know *a priori* whether units will be completely occupied or not, but it is an important consideration that is less likely to be true when units are defined at a larger (relative to the species ecology) spatial scale.

With arbitrarily defined units, the size of the unit is analogous to the pixel size on a television screen or monitor. Larger units will lead to a courser, grainier picture with smaller units providing a higher definition image. Obviously, the ideal pixel size will depend on how big the screen is, the desired image quality and whether you will be viewing it up close or from the other side of a stadium! This analogy is particularly apt if one of the desired outputs is to create species distribution maps. We stress that we are not suggesting that spatial closure is a

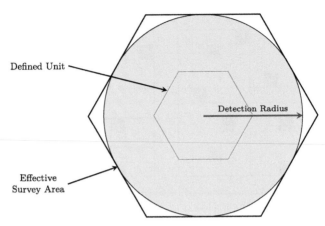

FIGURE 11.2 Example where employed field method (e.g., audial survey) is surveying an area greater than the nominally defined sampling unit.

requirement for these types of studies, but simply noting that it has an impact on how 'occupancy' should be interpreted.

The proposed field methods for detecting the species may also have an influence on how the sampling units should be defined. For example, if audial surveys are employed, what is the effective detection radius? Is it likely that the species could be recorded as detected within the defined unit, but the call actually came from outside of the unit? This is more likely to happen when the nominally defined unit is small relative to the detection radius, and would suggest sample units should be defined as something larger (Fig. 11.2), or detections should be restricted to those that appear to come from within the unit (if possible). A similar notion holds if a bait or lure is being used to attract individuals to a detection device (e.g., a camera trap); are individuals being attracted into a unit where they normally would not go? The converse can also hold true if only a small fraction of a unit is actually surveyed (e.g., Fig. 11.3), then the effective size of a unit may be smaller than its nominal size. When the entire unit is not being surveyed within a season (at least those parts of the units where the species has some chance of being detected), then it is assumed that observations obtained from the surveyed areas are representative of the unit in general, which may not always be a reasonable assumption. This is particularly relevant if the field protocols specify that only certain types of features or micro-habitats will be surveyed (e.g., road-side transect surveys within a grid cell, camera-traps placed along an animal trail).

When surveys will only be conducted within certain features or habitats, a further consideration is whether the population of interest, and therefore the sampling units, should be defined in terms of those features or habitat. For ex-

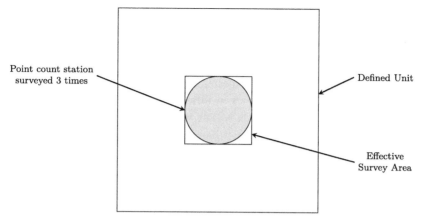

FIGURE 11.3 Example where only a fraction of a defined sample unit is being effectively surveyed; a point count station with 100 m detection radius is surveyed three times within a 600 m × 600 m sample unit.

ample if sample units are initially defined as relatively large cells, but only roads and accessible trails will be surveyed, a detection of the species along the road within a surveyed cell is sometimes presumed to indicate that the entire cell is occupied or used by the species. Whether the species detection is truly indicative of species presence in some broader area relies on assumptions that are not testable using the occupancy data themselves. That is, without any 'off-road' survey data, extrapolation of 'on-road' detections to 'off-road' areas relies on assumptions made by the researchers about how individuals of the focal species move through the landscape. This may often be a reasonable approach, but when the road network (for example) is sparse, hence the fraction of the cell that could actually be surveyed is small, the appropriateness of such assumptions may be more questionable. This is particularly true when the coverage of a cell by the features or habitats to be surveyed is uneven (e.g., roads may pass through the center of some cells, but in other cells, roads are only near the edge of the cell).

An alternative, and arguably more honest, approach would be to define the statistical population of interest as limited to the road network and its immediate environs, as areas further off-road have no chance of being surveyed. A buffer area around the road network could also be defined, if desired, to incorporate some sense of the species occupying an area that is greater than just the road (although researchers are again making assumptions about species movement patterns when doing so). Sample units might then be defined in terms of road segments of a suitable length rather than larger grid cells. In each case, the practical aspects of the fieldwork are likely to be similar, the main difference is an acknowledgment of the limits of what can be reliably determined from the data.

Note that this general approach is also pertinent to situations where the population of interest represents a more linear network, such as stream-based studies or studies of riparian corridors.

When the purpose of the occupancy study is to identify covariates that are important for the presence of the species, or to create maps of species distribution based upon covariate values at different places, we suggest that it is advisable to define sample units at a scale where the covariate values are relatively homogeneous. For example, if the relationship between occupancy and a certain habitat is of interest, rather than define units such that they include a variety of different habitat types (e.g., with larger cells or long transects that traverse multiple habitat types), it would be preferable to define units so that they are either one habitat type or another. When units are a mix of habitat types (for example), the proportion of each habitat type could be used a covariate in analysis, but it may be more difficult to tease out the effects of specific habitats, or perhaps the combination of certain habitats is more relevant at that scale.

A tangential issue related to covariates and sampling unit definition is that the predictor variables may be defined at a different spatial scale than a sampling unit. For example, researchers may define sample units as 1 ha cells, which fairly represents the area they can reliably survey to determine species occupancy, but define potential covariates as the proportional cover of a key habitat type within a 100 m, 500 m, and 1000 m radius of the centroid of each cell. Each variable is just a different characteristic of a cell, quantifying the surrounding habitat at different spatial scales. It is perfectly acceptable that the buffer zones used to calculate the variable values for some cells may overlap, because there is no requirement in the analysis for the value of a covariate to be independent for each cell. However, the sets of variables calculated at the different spatial scales are likely to be correlated, so should not be included in the same model due to the usual concerns about multicollinearity. This allows investigation of which spatial scale appears to be the best predictor of species occurrence.

Home-range or territory size has been suggested as the basis for defining the size of a sample unit, particularly in contiguous habitat (e.g., Karanth and Nichols, 2010; O'Connell and Bailey, 2011; Jathanna et al., 2015). Often authors do so from the perspective of trying to equate occupancy with abundance, or attempting to create sample units that are independent in the sense of not being used by the same individual animals. While well intentioned, we believe such arguments are sometimes misguided. If there are discernible features on the landscape that individuals, or groups of individuals, of the target species would tend to use to establish a territory (e.g., drainages), and those features are used to define sample units, there may be reasonable agreement between occupancy and abundance (at least proportionally). This is because the boundaries of the individual's territory and sample unit should correspond well. However, this is less likely when sample units are defined more arbitrarily (e.g., with a grid), as such correspondence is unlikely. An individual has no knowledge of where the grid lines have been superimposed over the landscape, and therefore a single individual may occupy multiple cells. While occupancy-based measures have been used as a surrogate for abundance or abundance-based measures (e.g., Linden et al., 2017), when inferences are truly desired about abundance, then a quite different set of methods should be used than those considered in this book. Practitioners have to accept that if they chose to use an occupancy-based measure, there are limitations on what can be reliably inferred about abundance. When individuals span multiple sample units, which is one argument for defining units at the scale of an individual's home range, this is often viewed as a lack of independence (nearby units are more likely to have a similar occupancy status), but from a species-level perspective it is immaterial whether neighboring units are occupied by the same or different individuals. Irrespectively, the species is present in those units. This simply induces a form of spatial correlation in occupancy which, in itself, is not necessarily problematic (Section 6.6). In conclusion, we suggest there is no clear theoretical advantage to defining sample units at the approximate size of a home range or territory, and believe it is perfectly acceptable to define units at a scale that may be much smaller, or larger. The prime consideration relates to the objective of the study or monitoring program, in particular, the desired biological interpretation of 'occupancy' that the data collection program is attempting to encapsulate.

A final point with respect to defining a suitable sample unit is to consider whether the potential unit itself may change over time. For example, in some wetland systems an area may be one pool in a wet year, or three smaller pools in a dry year. Or in a study of habitat patches, what is currently one patch may be multiple patches if further fragmentation is likely in the near future. This consideration is particularly relevant if the study or monitoring program is likely to continue for a number of time periods; we mention this here as some single-season studies may be repeated at some point in the future. While there is nothing particularly wrong with defining a unit as something that may change, it could make comparisons at different time points more difficult to interpret. An alternative is to use a grid-based approach (not necessarily a regular grid if there are large areas of non-habitat for the species) where the first assessment is whether the cell contains suitable habitat in that year, and if so field surveys are conducted. The amount of suitable habitat in nearby cells could be used as a covariate reflecting connectedness, if desired.

11.3   UNIT SELECTION

In earlier chapters we briefly mentioned various probabilistic sampling schemes that one might use to select a sample of units from the population or area of interest (e.g., simple random sampling; stratified random sampling; etc.). The importance of having a probabilistic sampling scheme is to be able to generalize the results from the analysis of data collected from the specific study units, to the wider population of interest. If there is no intent to generalize the results beyond the actual study units sampled, a probabilistic sampling scheme is not required (i.e., the study units represent the entire population of interest). However, if no probabilistic sampling scheme is used (i.e., units are selected purely in a haphazard manner or because of convenience), there is no statistical basis for generalizing the results to beyond the specific study units.

Implicitly, the methods detailed in this book assume that the study units represent a simple random sample from the wider population. If a stratified random sample is used, then one could analyze the data for each stratum separately (or within a single analysis with covariates defined to represent each stratum), and then combine the estimates for each stratum using standard results (e.g., Cochran, 1977; Thompson, 2002; Gould et al., 2012). If a systematic sample is used (e.g., by sampling at regularly-spaced cells on a grid) then, as in any systematic sample, it is assumed the occupancy status of each unit is random relative to the sampled cells. Random placement of the initial location, which determines the other sample locations, helps to justify this assumption.

For other sampling schemes there may be additional considerations that may limit the inferences that can be made from the results, or modifications of the analysis methods may be required. Selecting clusters of units is sometimes desirable from a logistical perspective. For example, one can select a vertex of a grid then and survey the four surrounding cells, or place units along a transect. In such cases the surveyed units have not been selected independently of each other, hence the effective sample size may be smaller than the number of units surveyed. If the occurrence of the species is spatially random at the scale of the defined units there may be little penalty, but in the presence of spatial correlation the effective sample size can be drastically reduced. The methods detailed in Section 6.6 to incorporate spatial correlation should be considered to analyze such data, or the multi-scale model (Section 6.3) might also be useful such that unit-level occupancy is made to be conditional upon cluster-level occupancy. However, from a study design perspective the relative benefits of a logistically convenient sampling scheme that may require a more complex analysis with a reduced effective sample size, should be compared to a scenario that is less logistically convenient. It may be possible to achieve a similar level of precision in the estimates by actually going to fewer units that are independently selected, resulting in less overall field work. A simulation study to assess the relative tradeoffs among potential designs is generally recommended.

Where random sampling occurs at multiple levels, e.g., first a random sample of forest stands within a region (primary-level units), and then a random sample of plots from within the selected stands (secondary-level units), with occupancy data being collected at the secondary-level units, the exact method of analysis depends upon the level at which inference is desired. No additional structure to the modeling is required if the results are not generalized beyond the selected primary-level units (e.g., inference is limited to the plots within the selected stands), as this sample represents the entire population of interest. However, if inference is to be generalized beyond the initial sample of primary-level units (e.g., to all stands within the region), then the method of analysis should reflect that random sampling has occurred at multiple levels. We believe the easiest way to accomplish this would be through the inclusion of a cluster random effect term in a Bayesian analysis. The multi-scale occupancy model (Section 6.3) might be also considered to analyze such data to allow for some localized spatial correlation.

If an adaptive sampling scheme (Thompson and Seber, 1996; Thompson, 2002) is used (i.e., neighboring units are included in the sample once the species is detected at a focal unit), then an important consideration is that detection probability not only affects the ability to observe occupancy, but also influences which set of units is eventually included in the sample. If the target species is present and detected at a unit, then the neighboring units are subsequently surveyed, however if the species is present but undetected then the neighboring units are not surveyed. The methods in this book would need to be extended to account for the probability of a unit being included in the sample, which is a function of detection probability (Thompson and Seber, 1996; Thompson, 2002). Using a model that accounts for spatial correlation may also be required with such a sampling scheme (Pacifici et al., 2016).

While there are some probabilistic sampling schemes that are not directly compatible with the methods discussed in this book at present, we do not necessarily discourage the use of such schemes (although note that 'convenience sampling' is not a probabilistic sampling scheme). We are confident that in the future, techniques will be developed to estimate and model occupancy data collected using these schemes. In addition, it is frequently possible to obtain model-based occupancy estimates for small spatial units using the methods described in this book and then use these estimates in replication-based estimation approaches (e.g., Skalski, 1994). Although this two-step approach may not be optimal, it provides a reasonable approach until more inclusive modeling is completed. However, we urge people to think very carefully about why they wish to use a more complicated scheme if similar results could be obtained using a much simpler design.

A final comment with respect to selecting units is that generally we advise against only selecting units based upon knowledge of their likely occupancy state (e.g., units that were known to be occupied by the species in the recent past, or based upon casual observations), when the population of interest consists of units about which such knowledge is and is not available. Unless this group of units actually represents the population of interest, estimates of occupancy for the entire population may be biased. For example, suppose occupancy is to be estimated within a stream system for a particular salamander species. Within this system, there are locations where the salamanders have been reported as present, based on sightings in the past by members of the public and local herpetologists. If these locations were selected as study units, then the estimated level of occupancy is likely to be higher than for a random sample of study units from throughout the stream system. Alternatively, if interest lies both in locations where the salamanders have been reported and in all other places within the stream system, then these regions could be treated as separate strata within the population (the stream system), and a random sample of units selected from each stratum. We return to this point in Chapter 12 where we consider design issues for multiple-season occupancy studies, but note here that preferential selection of units that are occupied could lead to apparent trends in occupancy even for a population that is currently stable. However, one situation where this type of design may be appropriate, is when the fraction of units that are still occupied now is of direct interest (e.g., the persistence probability for the species over the intervening time period). Now the population of interest consists only of units that were known to have been occupied during the past (e.g., based on museum records, field notes of naturalists and explorers, etc.) and 'occupancy' may be interpreted as a measure of persistence accordingly (e.g., Karanth et al., 2010). Asking questions about whether current 'occupancy' (of previously occupied units) is lower in human-developed units than less developed units (for example) may be worthwhile study objectives.

The key consideration is that all units within the population of interest must have a non-zero probability of being selected for surveying. Units that have no chance of being surveyed (e.g., because they are considered too difficult to access) are outside the scope of inference for the study. Extending the results to such units has no statistical support and is an act of faith rather than science. It is important to keep in mind that any deviation from a statistical 'ideal' in the name of convenience or expediency will have consequences on the type of analysis required (i.e., additional complexity) or the quality of the inferences that can be made (i.e., limitation of scope, additional caveats, or untestable assumptions).

11.4   DEFINING A 'SEASON'

In Section 4.4.9 we considered how violation of the closure assumption influenced estimates of occupancy. Provided changes in occupancy within a season are random (i.e., the probability of the species being present within a unit at one point in time, does not depend upon whether it was present at an earlier point in time), then occupancy estimates are unbiased if they are interpreted in terms of units that are 'used' (places at which the species is sometimes present during the season), rather than 'occupied' (places at which the species is *always* present during the season). For other kinds of changes in occupancy within a season, e.g., Markovian changes, units becoming permanently occupied (immigration), or units becoming permanently unoccupied (emigration) during the sampling season, the correlated detection (Section 6.4) or staggered entry (Section 6.5) models should be used. The key consideration is that a season should be defined such that it is reasonable to assume that the occupancy status of a unit is static over that time period. Therefore, times of the year when there are major changes in the species distribution (e.g., during periods of migration or dispersal) should ideally be avoided. If those changes are actually of interest, then a multi-season model is suggested.

It is also vitally important to consider the study objectives when defining a season. When changes in occupancy do occur at random over some timeframe, then at a smaller timescale it is likely that units are effectively closed with respect to occupancy. For example, over a five-day period, a wide-ranging carnivore may be physically present at random within a given unit as it roams throughout its territory, but on any single day the carnivore would be either present or absent at the unit. If the objective of the study was to investigate what habitats are used by the carnivore (perhaps to identify habitats to prioritize for conservation) then it may be appropriate to define a week (or perhaps a much longer period) as a season, but if the objective was to provide more of a snapshot of the carnivore population at a given point in time (say occupancy is being used as a surrogate for population size), then a season should be defined as a much shorter time interval. Careful consideration must be given to exactly how 'occupancy' is to be interpreted, and the timeframe associated with that interpretation. Even if detection is considered perfect, this notion of a season is often implicitly made to extrapolate the survey outcome beyond the actual survey period, e.g., when a species is detected in a five-minute survey, that is interpreted as indicative of its presence for the next two weeks.

Definition of a season can also depend upon how a unit is defined, particularly when unit size was arbitrarily chosen (e.g., grid cells of a certain size). Depending upon the species, the closure assumption may be more reasonable for a large unit than a small unit. Thus in some instances it may be possible to satisfy the closure assumption by increasing the size of a unit, provided that the relevant information (according to the study objectives) will still be collected. Recall that the closure assumption is at the species level, and not associated with individual animals (although there may be little distinction for low density species).

We stress that the estimate of occupancy applies to the entire population of units, and consideration of the closure assumption not only applies at the timeframe during which surveys are conducted at a single unit, but to the timeframe during which all units within the population are surveyed. If a week is thought to be a reasonable definition of a season at the unit level, but it takes a month to complete the surveying for the entire population, the season should be considered as a month-long period if the occupancy estimate is to be applied to all the units surveyed within that month, i.e., it is assumed that there are no major changes in the types of places being used by the species over that month.

While a season will often be a single time period, this may not always be the case for some objectives. For example, in some cases it may be reasonable to assume that the species distribution is very similar in March of each year, so March is defined as the season with a single survey conducted in each year for multiple years. Alternatively, for some species with very slow changes in occupancy or distribution it may be reasonable to define a season as spanning a number of years, e.g., the distribution of the species is considered static for multiple years.

Essentially, the objective of the field surveys is to provide a snapshot of species occurrence at that point in time. How a season is defined is analogous to the shutter speed of a camera; depending on the subject and type of photo desired, the shutter speed is adjusted accordingly (e.g., fast for an action shot, or slow for a time-lapse image). The same is true for season definition; how the researchers want to interpret 'occupancy' is a major determinant of how 'season' should be defined.

11.5 CONDUCTING REPEAT SURVEYS

11.5.1 General Considerations

Once a meaningful definition of a season has been determined, there are a number of practical options for how repeated surveys could be conducted at a unit. These include:

1. visit the unit multiple times and conduct a single survey on each visit;
2. conduct multiple surveys within a single visit, where surveys are separated by sufficient time so they can be considered independent (i.e., the probability of detecting the species in a subsequent survey does not depend upon whether the species was detected in a previous survey);

3. have multiple surveys conducted simultaneously by independent surveyors, or by using independent sampling methods, during a single visit;
4. within a larger unit, conduct surveys at multiple smaller plots (e.g., conduct multiple transect surveys within a five-hectare block).

There are various advantages and disadvantages to each approach, but the choice of which approach is most appropriate for a given situation really depends upon the nature of the biological question, the timings of the surveys in relation to the biology of the species in question, and the factors that may affect the probability of detecting the species.

It is worthwhile to try to ensure (as much as possible) that each survey is independent. That is, the probability of detecting the species in one survey should not depend upon the outcome (detection or nondetection) of another survey. In options 1 and 2 above, a lack of independence may occur, for example, if once the species is detected at a nest or den (say), then the probability of redetection may be higher because the observer knows where to look for the species or its sign in a subsequent survey. However, this could be easily accounted for during the design of the study either by having a different observer visit the unit each time or by using a 'removal' design, where surveys halt once the species is first detected. The effect could also be easily modeled with a survey-specific covariate that allows the probability of detection to be different after the first detection of the species (e.g., the survey-specific covariate $= 1$ for all surveys after the species is first detected, and $= 0$ otherwise).

When conducting multiple surveys simultaneously, as in options 3 and 4, a lack of independence may also be introduced if the number of opportunities for detecting the species is limited, such that it would not be possible for each of the surveys to detect the species. For example, suppose two observers are surveying a unit for blue jay (*Cyanocitta cristata*) nests. They conduct a ten-minute search of the unit, each searching one half of the unit. Now suppose that units have been defined such that there will only be either a single nest or no nest within the unit. Given that there is a nest at the unit, as each observer only surveys one half of the unit, it will be impossible for them to both find the nest (it cannot be in both halves), yet with no prior knowledge of the nest's location each observer has some non-zero probability of finding the nest (the probability that the nest is located within either half, multiplied by the probability of each observer finding the nest if it is in his/her half). That is, the *a priori* probabilities of observer 1 and observer 2 finding the nest are p_1 and p_2, respectively, which are both greater than zero. Assuming independence of the surveys, the probability of both observers finding the nest will be $p_1 p_2$, which again will be greater than zero. Yet because of the practical situation we know this probability to be zero, therefore the surveys could not be considered independent. If there

Unit	Design A Day 1	Day 2	Day 3	Unit	Design B Day 1	Day 2	Day 3
1	X X X			1		X	X
2			X X X	2	X	X X	X
3		X X X		3	X	X	X
4		X X X		4	X	X	X
5			X X X	5		X	X X
6	X X X			6	X	X	X
7		X X X		7	X	X	X X
8			X X X	8		X X	X
9	X X X			9	X		X X
p	0.5	0.3	0.8	p	0.5	0.3	0.8

FIGURE 11.4 Example of study design-induced heterogeneity when detection probability varies at some time scale. In Design A, all three surveys of each unit are only conducted on a single day, hence the detection probability is different for units surveyed on different days. In Design B, each unit is surveyed on different days thereby removing the heterogeneity.

are limited opportunities for detecting the species, then again this can be accommodated either through design (each observer randomly searches the entire unit for 20 minutes, or each tosses a coin to determine which of the two halves he/she will survey, such that both observers may survey the same or different halves), or via modeling of the data.

Determining which of the four options above may be appropriate involves consideration of how the biology of the species may affect a surveyor's ability to detect it, and the factors that may affect the detection probabilities (e.g., animal activity patterns; observer experience; weather conditions). The choice of whether to conduct repeated surveys during a single or multiple visits largely rests with the degree of variability in detection probabilities, and the timescale over which it varies. For example, suppose detection probability is thought to be relatively constant over a certain timeframe (e.g. a week). If all units can be surveyed within a week it is probably more efficient (logistically) to visit each unit once and conduct the repeated surveys within that single visit, than to visit units multiple times and conduct a single survey each time.

However, if the detection probability varied daily (say), and units were only visited once during a week-long study (i.e., a timescale during which the detection probability may have varied), a form of heterogeneity in detection probabilities may be introduced. The detection probability associated with each unit may be different as units were surveyed on different days (Fig. 11.4). If a moderate number of units (e.g., more than 20) is visited each day, then it may be possible to estimate a daily detection probability for each group of units. Alternatively, when the mechanisms that drive changes in detection probability are thought to be known (e.g., cloud cover), these could be measured as covariates and then included in the occupancy models of Chapters 4 and 7. However, in this case our resulting inference could be heavily dependent upon the appropriateness of that assumption about the mechanisms underlying variation in detection probability. Where practical, we suggest a more prudent course would be to conduct multiple visits, so that each unit can be surveyed under a range of conditions.

Consideration must also be given to whether the species (or sign of the species) is always available for detection during a season. If the species may be temporarily absent from a unit during a visit, then regardless of how many surveys are conducted during that visit the species will never be detected there even though it uses the unit at some point during the season. If 'use' is how occupancy is to be interpreted, then repeat visits will be required such that the species will be available for detection at some point during the season.

There are a number of other ways in which characteristics of the study design may unintentionally introduce heterogeneity into the detection probabilities. For example, MacKenzie et al. (2004b) discuss design aspects of the Mahoenui giant weta pilot study, the data for which were used as an example in Chapter 4. Two potential sources of heterogeneity they discuss are due to use of multiple observers and different survey times. In many situations, observers will vary in their ability to detect the target species caused by such factors as differences in experience, eyesight and hearing. As a result, the detection probabilities among units will be heterogeneous if the same observers always survey the same units. Again, if each observer surveys a moderate number of units then perhaps the effects could be estimated. MacKenzie et al. (2004b) suggest a simple design-based solution that enables better estimation of potential observer effects; that of rotating observers among the units so that no unit is ever surveyed by only a single observer. In a similar manner, heterogeneity could be introduced by always surveying the same units at the same time of day (e.g., due to the observers always surveying units in the same sequence). Many species exhibit different behavioral patterns at different times of the day that may result in their being more or less detectable at certain times (e.g., reptiles sunning themselves early in the day; morning song of breeding birds; temperature-related inactivity of butterflies early in the day; Casula and Nichols, 2003). If surveying for the species is conducted over a period of the day where detection probabilities change appreciably, and different units are being surveyed during that period, heterogeneity may be the result. While a 'time of day' covariate could be used to account for such variation, there is the potential for such an effect to be confounded with other unit-specific factors. MacKenzie et al. (2004b) again suggest a simple design-based solution that would provide more robust information about any potential time of day effect, which is to change the order in which units are surveyed each day.

In summary, we suggest that when considering the exact manner in which repeat surveys are to be conducted, careful thought must be given to the timescale and likely factors that may influence the probability of detecting the target species and changes in its availability during a season. If it is not possible to conduct the required number of repeated surveys at a moderate number of units within the timescale over which detection probability is believed to relatively constant, a prudent course of action would be to visit the units repeatedly, with a single survey of each unit per visit. One should also design a study to avoid inducing heterogeneity in detection probabilities with factors that could (partially) be controlled (e.g., conducting all repeat surveys in a single day when only a small number of units is surveyed per day), and to aim for independent surveys.

11.5.2 Special Note on Using Spatial Replication

Kendall and White (2009) examined the situation of having a limited number of opportunities for detecting the species in some detail, primarily from the perspective of using spatial replicate surveys (e.g., option 4 above: within a selected unit, a number of sub-units is surveyed where each provides an opportunity to detect the species if it is present in the unit). They considered two general approaches to how sub-units within a unit should be selected for surveying: sample without replacement (i.e., each sub-unit is only surveyed once at most) or sample with replacement (i.e., potentially, a sub-unit may be surveyed multiple times). Their findings were similar to the blue jay example above; if there is a limited number of detections that could be made, and that number is the same across all units, sampling without replacement will result in biased estimates of occupancy, but sampling with replacement (i.e., randomly selecting which half of the unit each observer surveys) does not. An example for why this is the case is as follows. Suppose that within a unit, there are ten sub-units that could be surveyed, of which only five contain evidence of the species. Also suppose that probability of detection, given presence of the species in an occupied subunit, is equal to 1. When sampling without replacement, the probability of selecting a sub-unit that contains evidence of the species, which is detection probability in this case, will be 5/10. However for the second selected sub-unit, which cannot be the sub-unit that was selected first, the probability of detection will be 5/9 if the first sub-unit did not contain evidence of the species, or 4/9 if it did. That is, the probability of detection in each survey (of different sub-units in this case) depends on the outcome of the previous surveys, which is a violation of the independence assumption. Fig. 11.5 illustrates this idea more fully. In contrast, when sampling with replacement, the probability of detection in each survey will be 5/10, as for each survey any of the ten sub-units may potentially be selected. Note that when there is a greater number of sub-units from which to select, then the effects of sampling without replacement will be less pronounced. For example, with 100 sub-units of which 50 contain evidence of

Survey   1   2   3

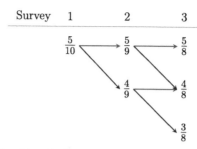

FIGURE 11.5 Probability of detection in each survey in the case of sampling ten sub-units without replacement, of which five contain evidence of the species. Red arrows indicate how the probability of detection changes in successive surveys when the species was not present in the sub-unit and the blue arrows indicate the change in detection probability when the species was present in the sub-unit. Here the probability of detection in a survey depends on the outcome of previous surveys, which is a violation of the independence assumption.

the species, the probability of detection will be 50/100 in the first survey, then either 50/99 or 49/99 depending on whether the first sub-unit did not contain evidence of the species, or it did, respectively. The distinction between 50/99 vs. 49/99 is much less than 5/9 vs. 4/9. Finally, note that the type of problem illustrated in Fig. 11.5 is most likely to occur when species are restricted to specific spatial subunits (e.g., as with plants, bird nests, etc.). When individual animals move about the sample unit, this problem becomes much less likely, as the animal movement itself provides the same independence as sampling with replacement.

Somewhat confusingly, Guillera-Arroita (2011) considered a very similar situation, but with contradictory results. Instead of assuming the number of sub-units containing evidence of the species was the same for all units, as Kendall and White (2009) did, Guillera-Arroita (2011) assumed that the number of sub-units with evidence of the species could vary according to a binomial distribution with a constant probability across units. Guillera-Arroita (2011) found via computer simulation that sampling with replacement resulted in a larger bias in estimated occupancy (using the basic occupancy model from Chapter 4) than sampling without replacement; the exact opposite of Kendall and White (2009). Importantly, within the simulations Guillera-Arroita (2011) allowed occupied units to have zero sub-units that contained evidence of the species. However, logically, if a unit is occupied, then at least one sub-unit must contain evidence of the focal species. We have conducted some very limited simulations using the methods of Guillera-Arroita (2011), but enforcing that occupied units must have at least one sub-unit with evidence of the species (which may or may not be selected for surveying). When this constraint is enforced, our findings were more similar to those of Kendall and White (2009): that sampling with replacement was preferable.

The fundamental issue here is that the basic occupancy model of MacKenzie et al. (2002) (and similar methods) was not developed for the case where the maximum number of opportunities for detections at a unit is limited, and gets depleted with each successive detection. Ideally, these models should be extended to better represent the realities of the sampling (e.g., that the number of detections at a unit, given presence, follows a hypergeometric distribution rather than a binomial distribution). In the mean time, from a practitioners perspective, when the number of detection opportunities is limited our best advice is to sample with replacement if possible.

11.6 ALLOCATION OF EFFORT, NUMBER OF SITES VS. NUMBER OF SURVEYS

Once the investigator has some idea of how repeat surveys should be conducted, given the biology of the species and the practicalities of attempting to detect it at a unit, the next obvious questions are "How many repeated surveys are required?" and "How many units should be surveyed?". Ultimately, increasing the number of units surveyed will improve the precision of the occupancy estimate. However, this should not be to the detriment of conducting repeated surveys, otherwise the variance component related to the uncertainty due to the species imperfect detection (see Chapter 4) will increase, and potentially negate any beneficial effect of surveying more units. In effect, this means a delicate balancing act is required to appropriately allocate resources between the number of units to survey and the number of repeated surveys per unit.

In this section we shall begin by considering a general design where all units are surveyed an equal number of times, before continuing to consider other designs. While in practice this general and simple design may not always be the most efficient design, it is useful to consider such a design to establish some general principles. That said, this general design may often be more robust than other more specialized designs. We also note that the results are specific to the basic occupancy model of Chapter 4, but many of the principles are likely to hold for more complex models.

To determine how resources should be allocated, three pieces of information are required:

1. the level of acceptable precision for an occupancy estimate;
2. initial guesstimates of the probability of occupancy and detection, or a range of possible values;
3. an indication of the maximum number of surveys (where a survey denotes a single search of a single unit) that could be conducted.

The level of acceptable precision or uncertainty should be clearly stated. Often such a statement will specify that the standard error (say) for the estimated occupancy is less than some desired value, or it may be expressed in terms of an acceptable confidence interval width. Without an *a priori* inclination of the level of acceptable uncertainty for the study, it is difficult to judge whether the resources that will be consumed by the fieldwork will yield sufficient information to be useful for the intended purpose. Commonly used metrics for specifying the level of acceptable uncertainty are the variance, $Var\left(\hat{\psi}\right)$, standard error, $\sqrt{Var\left(\hat{\psi}\right)}$, and coefficient of variation, $\sqrt{Var\left(\hat{\psi}\right)}/\hat{\psi}$.

As in all study design situations, initial estimates or guesses of the key population parameters are required to determine the number of samples needed to achieve the study's objective. This is because the measures of uncertainty depend on these values. For the basic dichotomous occupancy model, this relates to the occupancy and detection probabilities. Often a range of potential values should be suggested, so that the sensitivity of the required sample size to the values that are estimated can be examined.

Finally, it is important to have some idea of the maximum number of surveys that could be conducted (remembering that in some cases it may be appropriate to conduct multiple surveys within a single visit to a unit). This is determined by the amount of resources that are available and the total sampling effort that could be used. There is little point in designing a survey if there are insufficient resources to implement it (although it would highlight that to achieve the objective, more resources may be needed!). A good study design must have all the specified statistical properties, but still be practical to implement in the field.

Once all this information has been obtained, a useful starting point is to consider the (unrealistic) situation where species are always detected at units without error, i.e., $p = 1$. This represents a best-possible case and establishes the minimum number of units that would be required to obtain an occupancy estimate with the desired level of uncertainty (Chapter 4). If it is not possible to survey the minimum number of units within a season, then clearly the study's objectives will need to be reconsidered.

For example, suppose that it is thought that the probability of a unit being occupied by the target species within an area is 0.8 ($\psi = 0.8$), and suppose that we would like the standard error of the estimate to be 0.05 (i.e., an approximate 95% confidence interval would be ±0.1). Therefore assuming $p = 1$, the number of units that should be surveyed can be determined from the regular variance formula for an estimated binomial probability:

$$0.05 = \sqrt{Var\left(\hat{\psi}\right)}$$
$$= \sqrt{\frac{\psi(1-\psi)}{s}}$$
$$= \sqrt{\frac{0.8 \times 0.2}{s}}$$

which gives

$$s = \frac{0.16}{0.05^2}$$
$$= 64.$$

Hence, based upon the above information, a minimum of 64 units must be surveyed to achieve the desired level of precision. If it is not feasible to sample 64 units, then it may be necessary to reconsider the study's objective. If 64 units is an acceptable number, then the next issue is how many repeated surveys and additional units should be surveyed (given that $p < 1$) to achieve a standard error of 0.05.

Below we consider three general sampling schemes that are compatible with the model described in Chapter 4: (1) a *standard design* where s units are each surveyed K times; (2) a *double sampling design* where s_K units are surveyed K times and s_1 units surveyed once; and (3) a *removal design* where s units are surveyed up to a maximum of K_{max} times, but surveying halts at a unit once the species is first detected. We draw heavily on the work of MacKenzie and Royle (2005) who made a detailed examination of the important issue of allocating effort between the number of units and the number of surveys per unit in a single season, particularly when the total available effort may be limited. They did so under the assumption that one aspect of the study objective was based upon the level of uncertainty in an occupancy estimate, i.e., the variance of $\hat{\psi}$; $Var\left(\hat{\psi}\right)$. MacKenzie and Royle (2005) found that for each of the three sampling schemes listed above, there was an optimal number of repeated surveys that should be conducted regardless of whether the objective was to design a study to: (i) achieve a desired level of precision for minimal total survey effort; or (ii) minimize the variance of the occupancy estimator for a given total number of surveys. We note that these definitions of 'optimal' are not unique, and that other options will be feasible in some situations (e.g., Guillera-Arroita et al., 2010), although typically the suggested level of survey effort is similar. Aside from MacKenzie and Royle (2005), other useful papers on the topic of effort allocation in occupancy studies include Field et al. (2005), Bailey et al. (2007), and Guillera-Arroita et al. (2010).

TABLE 11.1 Optimum number of surveys to conduct at each unit (K) for a standard design where all units are surveyed an equal number of times with no consideration of survey costs, for selected values of occupancy (ψ) and detection probabilities (p)

p	ψ								
	0.1	0.2	0.3	0.4	0.5	0.6	0.7	0.8	0.9
0.1	14	15	16	17	18	20	23	26	34
0.2	7	7	8	8	9	10	11	13	16
0.3	5	5	5	5	6	6	7	8	10
0.4	3	4	4	4	4	5	5	6	7
0.5	3	3	3	3	3	3	4	4	5
0.6	2	2	2	2	3	3	3	3	4
0.7	2	2	2	2	2	2	2	3	3
0.8	2	2	2	2	2	2	2	2	2
0.9	2	2	2	2	2	2	2	2	2

11.6.1   Standard Design

No Consideration of Cost

Under a standard design where detection probability is assumed constant, $Var\left(\hat{\psi}\right)$ can be expressed as

$$Var\left(\hat{\psi}\right) = \frac{\psi}{s}\left[(1 - \psi) + \frac{1 - p^*}{p^* - Kp\,(1 - p)^{K-1}}\right] \qquad (11.1)$$

where $p^* = 1 - (1 - p)^K$ is the probability of detecting the species at least once during K surveys of an occupied unit (Chapter 4). Further, the total number of surveys (TS) will be

$$TS = s \times K. \qquad (11.2)$$

Table 11.1 (constructed from the results of MacKenzie and Royle, 2005) indicates the number of surveys that should be conducted at each unit (without consideration of survey costs). For given values of occupancy (ψ) and detection probabilities (p), these values for K will result in the most efficient standard design possible (given all model assumptions are met). There are a number of notable features about the values in Table 11.1 to which we wish to draw attention. The first is that the optimal course of action whenever there is a chance that the species may go undetected at an occupied unit is to always conduct two or more surveys at all the units. Second, when detection probability is moderately high (greater than 0.5) conducting only two or three surveys per unit results in a reasonably efficient design, but many more surveys may be required where detection probability is lower. That is, as detection probability decreases, the optimal number of surveys increases. Further, the optimal number of surveys also increases as the occupancy probability increases, the reason for which may not be immediately obvious. However, consider a species that is rare on the landscape (in terms of occupancy). A randomly selected unit will, therefore, have a low probability of occupancy; hence, expending a large amount of effort trying to detect a species that is unlikely to be present will be an inefficient use of resources available to the study. Whereas for a common species, it is more efficient to expend those resources to confirm species presence than being relatively uncertain about occupancy status and moving on to another unit where the species is also likely to be present (and having greater uncertainty about the presence of the species at that unit as well). For example, consider the optimal number of surveys in Table 11.1 for case where $p = 0.3$, and assume a total of 1000 surveys can be conducted. If $\psi = 0.1$, then 5 surveys per unit should be conducted so we can survey a total of 200 units, but if $\psi = 0.9$ then 10 surveys should be conducted per unit, at 100 units. From these results, MacKenzie and Royle (2005) suggest a crucial general principle for the design of occupancy studies: for a rare species one should survey more units less intensively (i.e. with fewer repeated surveys), while for a common species, one should survey fewer units more intensively (i.e., with more repeated surveys). Field et al. (2005) came to a similar conclusion.

In terms of designing a particular study, the required value for K can be obtained based upon the values of ψ and p that are assumed for the species. Then the number of units to survey to obtain a desired level of precision, or for a fixed total level of survey effort, can be simply calculated from Eqs. (11.3) or (11.4) respectively.

$$s = \frac{\psi}{Var(\hat{\psi})}\left[(1 - \psi) + \frac{1 - p^*}{p^* - Kp(1 - p)^{K-1}}\right], \qquad (11.3)$$

$$s = \frac{TS}{K}. \qquad (11.4)$$

Oftentimes one should consider a range of values for ψ and p if there is uncertainty about what values may be reasonable, and consider the resulting impact on the suggested design because of that uncertainty.

For example, previously the desired standard error was 0.05, $= \sqrt{Var(\hat{\psi})}$, the probability of occupancy was assumed to be about 0.8 ($\psi \approx 0.8$), and now suppose that the probability of detecting the species in a single survey of an occupied unit is thought to be about 0.4 ($p \approx 0.4$). From Table 11.1 we find that the optimal number of repeated surveys per unit in this situation is six.

This gives the expected probability of detecting the species at least once (p^*) as approximately $p^* = 1 - (1 - 0.4)^6 = 0.953$. To achieve a standard error of 0.05, the number of units to survey is:

$$
\begin{aligned}
s &= \frac{0.8}{0.05^2} \left[0.2 + \frac{0.047}{0.953 - 6 \times 0.4 \times 0.6^5} \right] \\
&= \frac{0.8}{0.05^2} [0.2 + 0.061] \\
&= 83.5.
\end{aligned}
$$

Therefore, approximately 84 units should be surveyed (with $504 = 84 \times 6$ total surveys) to give the desired level of precision (note that the 84 units required is greater than the 64 units required if the detection probability is 1.0).

However, suppose that there are only enough available resources to conduct 300 total surveys. What is the best standard error that could be achieved? Even though the criterion for the design has changed, the optimal number of repeated surveys per unit is still six, hence there are only enough resources to survey 50 units. Substituting these values into Eq. (11.1) and taking the square-root gives a standard error of 0.065. A decision should then be made whether a standard error that is 30% higher than what was originally desired, can be tolerated, or if the objective of the study should be reassessed. As noted earlier, it would be wise to check how sensitive these results are to the assumed values of ψ and p, before reconsidering the study's objective. In addition, there may be other designs where units are surveyed an unequal number of times, that may yield a standard error close to the desired level for a reasonable level of effort. We will discuss this later in the chapter.

Naturally, in some instances it may not be practical to conduct the optimal number of surveys at each unit (e.g., Table 11.1 suggests that 20 is the optimal number of surveys when $\psi = 0.6$ and $p = 0.1$). Eqs. (11.3) and (11.4) can be used to explore how much worse a sub-optimal design would perform, enabling an assessment as to whether the study's objectives can be practically achieved. In such situations (and sometimes more generally) the field methods and protocols being used to detect the species should also be reassessed to see if alternative methods could be used to improve detection probabilities.

Including Survey Cost

So far we have implicitly assumed that the 'cost' (as measured in some meaningful way) of conducting each survey is constant, or at least that cost is immaterial to the design of a study. However, the study's budget will be one of the greatest limiting factors in many situations, and the cost associated with each survey may vary. For example, if it is appropriate to conduct multiple surveys within a single unit visit, most of the cost will be associated with traveling to the unit and conducting the first survey. The cost of conducting subsequent surveys during the same visit may be relatively small. Another example is where the surveying of additional units becomes more costly as the number of units (s) increases, which could be due to increased travel cost, or the need for hiring of additional staff, etc.

The general techniques used by MacKenzie and Royle (2005) for finding the optimal allocation of resources can be easily extended to account for costs by defining a cost function. A study can then be designed in terms of minimizing cost to obtain a desired level of precision, or by seeking the best design to minimize the variance of the occupancy estimator for a fixed cost. Note Eq. (11.2), which calculates the total number of surveys in a standard design, could be considered as a simple cost function, where the cost for each survey is 1. The cost function itself could take a great variety of forms. For instance, if the cost of subsequent surveys is different than the cost of the first survey, then the cost function could be of the form:

$$C = c_0 + s[c_1 + c_2(K - 1)],$$

where c_0 is a fixed overhead cost; c_1 is the cost of conducting the first survey of a unit; and c_2 is the cost of conducting subsequent surveys. Alternatively, in a situation where the cost of surveying additional units continues to increase, the cost function may take the form:

$$C = c_0 + \sum_{i=1}^{s} K[c_1 + c_2(i - 1)],$$

where c_0 is the overhead cost; c_1 is the cost of conducting each survey at the first unit; and c_2 is the extra cost per survey for each additional unit (i.e., the cost per survey at unit 2 is $c_1 + c_2$, at unit 3 the cost per survey is $c_1 + 2c_2$, etc.).

In the first case, where the cost of an initial survey may be different than that of a subsequent survey of a unit, MacKenzie and Royle (2005) used analytic techniques to show that similar results to those presented above hold. That is, regardless of whether the study is being designed to achieve a desired level of precision with minimal cost or to minimize uncertainty for a fixed cost, the optimal value for K is the same. This is similar to the situation where cost was not considered; however, the optimal value for K does depend upon the relative cost of an initial and subsequent survey (c_1 and c_2, respectively) and on ψ and p, although K is relatively stable when the relative cost is in the range $0.5c_2 < c_1 < 2c_2$. Table 11.2 presents the optimal value of K in the situation when $c_1 = 10c_2$. Note that the same general trends occur (K decreases as p

TABLE 11.2 Optimum number of surveys to conduct at each unit for a standard design where all units are surveyed an equal number of times and the relative cost of conducting initial (c_1) to subsequent (c_2) surveys is $c_1 = 10c_2$, for selected values of occupancy (ψ) and detection probabilities (p)

p					ψ				
	0.1	0.2	0.3	0.4	0.5	0.6	0.7	0.8	0.9
0.1	18	19	20	21	22	24	26	30	37
0.2	10	10	11	11	12	13	14	15	18
0.3	7	7	7	8	8	9	9	10	12
0.4	5	5	6	6	6	6	7	8	9
0.5	4	4	4	5	5	5	5	6	7
0.6	3	3	4	4	4	4	4	5	5
0.7	3	3	3	3	3	3	3	4	4
0.8	2	2	2	2	2	3	3	3	3
0.9	2	2	2	2	2	2	2	2	2

increases, but increases as ψ increases), although the optimal value for K tends to be larger by one or two surveys than when cost was not considered. This result is intuitive as, if it is relatively cheap to conduct subsequent surveys, then it would be reasonable that the best use of resources is to reduce uncertainty related to the potential presence of the species at a unit rather than incur the higher costs associated with sampling a new unit.

Returning to our example, above we found that 504 was the minimum number of surveys required to obtain a standard error of 0.05, with $s = 84$ and $K = 6$, and that if there were only enough available resources to conduct 300 surveys the best design of $s = 50$ and $K = 6$ would give a standard error of 0.065 (given the assumed values of ψ and p). Now suppose that for the species in question it would be appropriate to conduct multiple surveys within a single unit visit, hence the cost of subsequent surveys ($20) is much less than the cost of the first survey ($200). Let us assume that the overhead cost is $3000, hence the cost function takes the form:

$$C = \$3,000 + s[\$200 + \$20(K - 1)].$$

The issue here is, what design would minimize the total cost while achieving a standard error of 0.05? The relative cost of an initial to subsequent survey is $c_1 = 10c_2$, hence from Table 11.2 the optimal value for K is eight. By setting $K = 8$ and $Var(\hat{\psi}) = 0.05^2$ in Eq. (11.1), solving for s suggests that $s \approx 70$ should give the best design to minimize the total cost at $26,800 (given $\psi \approx 0.8$ and $p \approx 0.4$). Compare this to the design where cost was not considered, where $s = 84$ and $K = 6$. Although 11% more surveys would be conducted (560 here vs. 504), the design that accounts for the cost of surveying would be $1400 cheaper (assuming the same cost function is used).

For more complicated cost functions the optimal choice for K may no longer be constant with respect to s or total cost. The techniques used by MacKenzie and Royle (2005) can be easily replicated for the new cost function using the numerical routines available in spreadsheet software such as Microsoft Excel. Most spreadsheet packages have a numerical routine that can be used to find a set of values that provide a solution to a particular optimization problem. Such routines can be used to find the best allocation of resources given some design criteria. The general procedure would be:

1. Specify values for ψ, p, s, and K in a series of cells.
2. Enter a formula to calculate $Var\left(\hat{\psi}\right)$ (or the appropriate uncertainty measure being used) based upon these cell entries.
3. Enter a formula corresponding to the study's cost function.
4. Run the numerical optimization routine where the values in the corresponding cells for s and K are to be changed. The exact procedure used here depends upon the design criteria. If $Var\left(\hat{\psi}\right)$ is to be minimized for a fixed (maximum) cost, then the routine should be run such that the cell corresponding to $Var\left(\hat{\psi}\right)$ is minimized subject to the constraint that the value in the cost function cell equals (or is less than) the available budget. If the study is designed to achieve a specific $Var\left(\hat{\psi}\right)$ while minimizing cost, then the cell corresponding to the cost function should be set as the target with a constraint imposed upon the $Var\left(\hat{\psi}\right)$ cell.
5. Numerical routines can sometimes be sensitive to starting values. Step 4 should be repeated with different starting values for s and K to verify that the best combination has been found.
6. The resulting values for s and K are likely to be non-integer, therefore the values will need to be rounded and adjusted to ensure the constraints are still met (e.g., above the numerical routine suggests the values of $s = 71.6$ and $K = 7.6$).

11.6.2   Double Sampling Design

As an alternative to surveying all units an equal number of times, repeated surveys could be conducted at a subset of units with all other units surveyed only once (MacKenzie et al., 2002, 2003, 2004b; MacKenzie, 2005a). When a unit is only surveyed once, then the probability of occupancy and detection are confounded unless there is additional information. The intent behind the double sampling design is that the additional information required to separate occupancy from detectability at those units only surveyed once, comes from the units surveyed repeatedly. Detection probabilities can be estimated (and modeled) from the data collected at units where repeated surveys were conducted, and that information is then applied to the units only surveyed once. Data collected in this manner are completely compatible with the model detailed in Chapter 4, as occasions when surveys were not conducted could be considered as missing values (Section 4.4.7). Using such an approach, the entire estimation procedure can be conducted within a single framework. However, using a double sampling scheme does require the assumption that detection probabilities (or the model used to describe them) is the same at the units surveyed repeatedly and those surveyed once.

One of the motivations for the suggested use of such a design is efficiency (MacKenzie et al., 2002, 2003, 2004b; MacKenzie, 2005a). Intuitively it might seem reasonable that at some point sufficient repeated surveys have been conducted across all units to allow precise estimation and modeling of detection probabilities. Conducting additional repeated surveys of units could, therefore, be inefficient in that the resources used on repeat surveys could be better utilized by surveying additional units, increasing the spatial replication of the study. While this may sound reasonable in theory, MacKenzie and Royle (2005) showed that this idea does not generally hold in practice.

They computed the fraction of the total survey effort that should be used to survey units only once, which minimized the asymptotic variance for $\hat{\psi}$ under such a design (assuming p is constant). Only when $p \geqslant 0.8$ was there a suggestion that a substantial fraction of the total surveys available should be devoted to surveying a number of units once, in most other cases the entire survey effort should be used to survey units repeatedly (i.e., $\hat{\psi}$ had a smaller variance under a standard design than a double sampling design; Table 11.3). Some entries in Table 11.3 may look unusual, e.g., when $\psi = 0.4$ and $p = 0.6$, 12% of the total survey effort should be used to survey units only once, while the entries in nearby cells are very small. This result (and others) is simply because splitting the survey effort in that manner gives an average number of surveys per unit that is very close to the *non-integer* optimal value for K under the standard design. As the numerical search used only considered optimal *integer* values for K with the optimal design, the variance in 'unusual' situations tended to be smaller. However, even in the situations where it is suggested that double sampling may be feasible, the reduction in the standard error tended to be minor unless ψ was small and p large (Table 11.4). When MacKenzie and Royle (2005) also considered survey costs, only if the cost of an initial survey was less than the cost of a subsequent survey (a situation that is difficult to conceive of in practice) did the double sampling scheme consistently perform better than an optimal standard design.

The reason why a double sampling scheme does not (generally) provide more precise estimates of ψ than an optimal standard design, is that the com-

TABLE 11.3 Optimal fraction of total survey effort that should be devoted to surveying sites only once using a double sampling design, for selected values of occupancy (ψ) and detection probabilities (p)

p	ψ								
	0.1	0.2	0.3	0.4	0.5	0.6	0.7	0.8	0.9
0.1	0	0	0	0	0	0	0	0	0
0.2	0	0	0	0	0	0	0	0	0
0.3	0	0	0	0	0	0	0	0	0
0.4	0	3	0	0	0	0	0	0	0
0.5	6	1	0	0	0	0	0	0	0
0.6	0	0	0	12	4	0	0	0	0
0.7	9	5	0	0	0	0	0	0	0
0.8	33	30	26	21	14	5	0	0	0
0.9	56	54	51	48	44	39	31	17	0

TABLE 11.4 Percent reduction in standard error of the occupancy estimator ($\hat{\psi}$) using an optimal double sampling design compared to an optimal standard design, for selected values of occupancy (ψ) and detection probabilities (p)

p	ψ								
	0.1	0.2	0.3	0.4	0.5	0.6	0.7	0.8	0.9
0.1	0	0	0	0	0	0	0	0	0
0.2	0	0	0	0	0	0	0	0	0
0.3	0	0	0	0	0	0	0	0	0
0.4	0	0	0	0	0	0	0	0	0
0.5	0	0	0	0	0	0	0	0	0
0.6	0	0	0	0	0	0	0	0	0
0.7	0	0	0	0	0	0	0	0	0
0.8	4	3	2	2	1	0	0	0	0
0.9	12	11	10	9	7	5	3	1	0

ponent of $Var\left(\hat{\psi}\right)$ related to the uncertainty in an estimate of p, is generally much smaller than the component associated with the imperfect detection of the species, see Eq. (4.10). That is, even if p is estimated precisely, the uncertainty related to being unable to confirm a species is absent from a unit may still be relatively large.

While the results of MacKenzie and Royle (2005) indicate that a double sampling scheme may not perform as well as an optimal standard design, there may be some situations where such a design could still be useful. For example, if some potential units are very remote and difficult to access, it may not be possible to implement a standard design, as resurveying the remote units may be prohibitively expensive. If it can be assumed that occupancy and detection probabilities are similar at the remote and more accessible units, then it may be reasonable to use a double sampling-type design.

11.6.3   Removal Sampling Design

Another type of design that has been used involves surveying units up to a maximum of K_{max} times in a season, but once the species is initially detected, then no further surveys of that unit are conducted. This is the type of design considered by Azuma et al. (1990) when they developed their occupancy estimation procedure for the monitoring of northern spotted owls in the Pacific northwest of the United States. Here we refer to such a design as a 'removal' design as units are removed from the pool of units being actively surveyed once the species is detected (also because of the analogy with removal experiments used in mark–recapture studies, e.g., Otis et al., 1978; Williams et al., 2002).

In addition to enabling the estimation of detection probabilities, the repeated surveying of units reduces the chance of the species being declared as falsely absent from a unit. Once the species presence has been confirmed then additional surveys only provide further information about detection probabilities, not about occupancy, which is often the main focus of the type of studies considered here. This is the logic behind using a removal design, that the main piece of information required is confirmation that the target species is present at a unit. The number of surveys required until the species is first detected provides the relevant information allowing detection probabilities to be estimated.

MacKenzie and Royle (2005) showed that an optimal removal design can provide more precise estimates of $\hat{\psi}$ than an optimal standard design with the same total number of surveys. Effectively, the resources not used to resurvey units, following the first detection of the species, can be used to sample additional units, increasing the spatial replication without increasing the total level of survey effort.

Using a removal design with constant detection probability, the variance of the occupancy estimator is:

$$Var\left(\hat{\psi}\right) = \frac{\psi}{s}\left[(1 - \psi) + \frac{p^*(1 - p^*)}{(p^*)^2 - K_{max}^2 p^2(1 - p)^{K_{max}-1}}\right]. \tag{11.5}$$

As for a standard design, MacKenzie and Royle (2005) found that, regardless of whether a study was being designed to minimize total survey effort to obtain a desired level of precision, or to minimize uncertainty in the occupancy estimate for a fixed level of effort, the optimal value for K_{max} (now the max-

TABLE 11.5 Optimal maximum number of surveys to conduct at each unit for a removal design (K_{max}) where all units are surveyed until the species is first detected, for selected values of occupancy (ψ) and detection probabilities (p)

p	ψ								
	0.1	0.2	0.3	0.4	0.5	0.6	0.7	0.8	0.9
0.1	23	24	25	26	28	31	34	39	49
0.2	11	11	12	13	13	15	16	19	23
0.3	7	7	7	8	8	9	10	12	14
0.4	5	5	5	6	6	6	7	8	10
0.5	4	4	4	4	4	5	5	6	8
0.6	3	3	3	3	3	4	4	5	6
0.7	2	2	2	3	3	3	3	4	5
0.8	2	2	2	2	2	2	3	3	4
0.9	2	2	2	2	2	2	2	2	3

imum number of surveys to be conducted) is consistent for given values of ψ and p (Table 11.5). They also found that incorporating differential costs for initial and subsequent surveys with a removal design has a similar effect to that for the standard design (e.g., increasing K_{max} by one or two surveys if $c_1 \geqslant 5c_2$). Note that under a removal design, the total number of surveys required will not be a fixed number because now there is an element of chance involved with the number of surveys that will be required before detecting the species for the first time at an occupied unit, but the expected number of surveys required can be calculated.

To give an indication of the relative efficiency of an optimal removal design to the optimal standard design, MacKenzie and Royle (2005) compared the expected standard errors of the estimators under the two designs with the same (expected) total number of surveys (Table 11.6). Values of the ratio less than 1.0 indicate situations where the optimal standard design is more efficient in terms of obtaining a smaller standard error, which only occurs when the level of occupancy is less than 0.3. Therefore, generally an optimal removal design will be more efficient than an optimal standard design, although one must be prepared to conduct a greater maximum number of surveys to fully realize the gain in efficiency. For example, if $\psi = 0.8$ and $p = 0.3$ the standard error of an optimal standard design with 8 repeat surveys per unit (Table 11.1) will be 42% greater than that of an optimal removal design, but units may have to be surveyed up to a maximum of 12 times under the removal design (Table 11.5).

Removal studies are likely to be most useful in situations where, once a species is detected at a unit, it is more (or less) likely to be detected again in

TABLE 11.6 Ratio of standard errors for optimal standard and removal designs. Values greater than 1 (in bold) indicate situations where an optimal removal design has a smaller standard error than the optimal standard design

p	ψ								
	0.1	0.2	0.3	0.4	0.5	0.6	0.7	0.8	0.9
0.1	0.90	0.94	0.98	**1.04**	**1.10**	**1.18**	**1.30**	**1.46**	**1.74**
0.2	0.91	0.94	0.99	**1.04**	**1.10**	**1.18**	**1.28**	**1.44**	**1.71**
0.3	0.92	0.95	0.99	**1.04**	**1.10**	**1.17**	**1.27**	**1.42**	**1.68**
0.4	0.93	0.96	0.99	**1.03**	**1.09**	**1.17**	**1.26**	**1.40**	**1.64**
0.5	0.93	0.96	1.00	**1.04**	**1.08**	**1.16**	**1.24**	**1.37**	**1.60**
0.6	0.94	0.97	**1.01**	**1.06**	**1.09**	**1.15**	**1.22**	**1.35**	**1.55**
0.7	0.95	0.96	0.97	**1.01**	**1.07**	**1.13**	**1.22**	**1.31**	**1.48**
0.8	1.00	**1.02**	**1.04**	**1.07**	**1.09**	**1.11**	**1.15**	**1.25**	**1.45**
0.9	**1.02**	**1.05**	**1.07**	**1.10**	**1.13**	**1.17**	**1.20**	**1.24**	**1.31**

subsequent surveys. For example, once a surveyor locates the species at its den or nest within the unit, the surveyor may simply return to the den or nest in future surveys of the unit resulting in a higher detection probability (or lower detection probability if the surveyor disturbed the species causing it to vacate the den or nest). By examining the model likelihood it is easy to see why the removal design can be used without a loss of efficiency compared to using a covariate to model the change in detection probability. Consider again the likelihood equation for a standard design with constant detection probability given in Eq. (4.7):

$$ODL(\psi, p | \mathbf{h}_1, \mathbf{h}_2, \ldots, \mathbf{h}_s)$$
$$= \left[\psi^{s_D} p^{\sum_{j=1}^{K_{max}} s_j} (1 - p)^{K_{max} s_D - \sum_{j=1}^{K_{max}} s_j} \right] \left[\psi (1 - p)^{K_{max}} + (1 - \psi) \right]^{s - s_D}.$$

This could be factored and written in two components:

$$ODL(\psi, p | \mathbf{h}_1, \mathbf{h}_2, \ldots, \mathbf{h}_s)$$
$$= \left\{ \left[\psi^{s_D} p^{s_D} (1 - p)^{\left(\sum_{i=1}^{s_D} t_i \right) - s_D} \right] \left[\psi (1 - p)^{K_{max}} + (1 - \psi) \right]^{s - s_D} \right\}$$
$$\times \left\{ p^{\left(\sum_{j=1}^{K_{max}} s_j \right) - s_D} (1 - p)^{(K_{max} + 1) s_D - \left(\sum_{j=1}^{K_{max}} s_j \right) - \sum_{i=1}^{s_D} t_i} \right\}.$$

The first component models the data up to and including the survey where the species was detected for the first time (t_i), and is the likelihood for data collected under the removal design with constant detection probability (Chapter 4). The second component models the detection/nondetection of the species in surveys conducted after the first detection and does not involve ψ. When the detection probability is the same before and after the first detection of the species, the second component does contribute some information toward estimating occupancy as it helps reduce uncertainty associated with having to estimate detection probability from the data, i.e., the third component of $Var\left(\hat{\psi}_{MLE}\right)$ in Eq. (4.10). However, when the detection probability is different after first detection of the species, then the second component will contribute no information toward estimation of the occupancy probability (this is also true of closed capture–recapture models in which individuals exhibit a behavioral response to capture; Pollock, 1974; Otis et al., 1978). Therefore additional modeling of the detection probabilities after first detection of the species may not greatly improve the estimation of ψ, hence in such situations, it may be advantageous to simply halt the surveys once the species is first detected at a unit.

However, a disadvantage of the removal design is that there is less flexibility in how the resulting data can be analyzed. For example, when data are collected according to a standard design, models can be fit where detection probability is either survey specific or constant for all surveys. However, for data collected from a removal design, models with fully survey-specific detection probabilities cannot be fit (recall from Chapter 4 that the parameters are not identifiable when p is survey-specific under a removal design). However, an equality constraint for even two detection probabilities (e.g., $p_{K-1} = p_K$) will permit estimation under a removal design. If such constraints are not reasonable (i.e., if it is not likely that detection probabilities for multiple surveys will be constant), then it may not be appropriate to use a removal design because the model cannot reflect the reality of the situation, and the resulting estimates of occupancy may be biased. If the changes in detectability are thought to occur in relation to some covariate, then the data from both types of designs should be able to fit appropriate models, although data from a standard design may have more success at differentiating between the effect of covariates that vary in space versus those that vary in time. In an effort to combine efficiency and robustness, MacKenzie and Royle (2005) suggest a hybrid design where a standard design protocol is conducted at some units and a removal design protocol at the remainder. To the best of our knowledge however, the properties of such a design have not been investigated to date.

11.6.4 More Units vs. More Surveys

A final point highlighted by MacKenzie and Royle (2005) is the realization that increasing the number of units surveyed at the expense of decreasing the number of repeat surveys may not result in a better design (in terms of precision of the occupancy estimate). They offer the example where if it is assumed $\psi \approx 0.4$ and $p \approx 0.3$, the asymptotic standard error of the occupancy estimator for a standard design where 200 units are each surveyed twice is 0.11. However, by surveying only 80 units five times the standard error is reduced to 0.07. Allocating the same total number of surveys in a more efficient manner resulted in a 36% reduction of the standard error. To achieve the same gain in precision with only two surveys of each unit, 500 units would need to be surveyed and total survey effort increased by 250%!

We believe the same holds true more generally where, for example, the objective of the study is to identify important habitat preferences for the species. Without a sufficient number of repeat surveys, the probability of a 'false absence' of an occupied unit, $= (1 - p)^K$, may be sufficiently large that it is difficult to identify any important factors associated with occupancy. In our experience, a key design consideration often seems to be reducing the probability of a false absence to an acceptable level to make robust inferences about occupancy patterns in the population. We suggest that inference is best when the probability of a false absence is in the range 0.05–0.15. Any smaller and one may be expending too much effort on repeat surveys that could be better utilized by surveying more units; any larger and one may not be expending enough effort on repeat surveys resulting in a greater level of uncertainty about the occupancy status of units where the species was never detected.

These results suggest that a key benefit of conducting the optimal number of repeat surveys within a season is not only so that it allows occupancy and detection probabilities to be disentangled during the estimation, but the data themselves will be more robust and inferences about occupancy will be less sensitive to what factors are included in the detection component of the occupancy models fit to the data. It is also worth noting that bias introduced by effects of unmodeled detection heterogeneity will also be less when using an optimal design.

11.6.5 Finite Population

All of the above design advice is for estimating the underlying probabilities of occupancy, or for estimating the proportion of units occupied, when the number of units surveyed represents a relatively small proportion (e.g., < approximately 20%) of the total population of interest. This distinction is discussed in greater detail in Section 6.1. When the population of units is small such that it is likely a substantial fraction of the population will be surveyed, then some of the above results may need to be reassessed. To the best of our knowledge, no one has investigated and published study design considerations for this situation, although one could use the results given in Section 6.1, particularly Eq. (6.4), to determine required sample sizes with a finite population.

11.7   DISCUSSION

Designing a good study is often as much an art as a science. The 'science' plays a role in giving guidance on what may be an optimal design, given a list of assumptions for specific situations, and also in providing knowledge about the biology of the species and the study objectives. The 'art' comes in taking the information provided by the science and turning it into a practical design that can be used in the field so that the required data can be collected to meet the study's objectives. The art can be viewed as subjective ways of taking account of additional considerations and constraints where the science can provide little insight. This may involve generalizing results, lateral thinking and even the occasional leap-of-faith. The science should be used to take as much of the guesswork out of the design of studies as possible, but sometimes intuition (both statistical and biological) has an important role to play, as there may well be a number of unknown factors on which the current science can shed little light. In such cases however, small pilot studies can often be helpful to collect additional information, or at least trial the proposed design.

In this chapter, there has been a lot of discussion related to optimal designs. These designs are optimal in the sense of providing the most precise estimate of the occupancy probability, given certain constraints on resources and relatively simple models (i.e., occupancy and detection probabilities are constant). However, an optimal design may not be a robust design. What if there is heterogeneity in occupancy or detection probabilities among units? What if detection probabilities are affected by some factor that has not been considered? Generally we believe the standard design will be the most robust, as it will provide the greatest degree of flexibility in modeling the data. Whether an optimal standard design should be used depends very much on the application; what may be optimal in one set of circumstances may be sub-optimal in another. As such, the values given in Tables 11.1, 11.2, and 11.5 for the 'optimal' number of repeat surveys, should only be considered as a bare minimum. Generally we would suggest that at least 3 surveys per unit be conducted, but even then there may be insufficient data to identify important sources of variation such as heterogeneity in detection probabilities (either with covariates or the mixture models of Chapter 7), in which case a greater number of surveys (5+) may be needed. Other general principles that may be useful in the design of single-season occupancy studies are:

1. A removal design is likely to more efficient than a standard design if detection probability is relatively constant, but may provide less flexibility for modeling.
2. Sampling more units with fewer surveys for a rare species vs. fewer units with more surveys for a common species with a similar detection probability should hold as a general strategy.
3. For all designs, there is likely to be some optimal value for K that gives the most efficient design, for given values of ψ and p. However, if ψ and p are thought to vary among units (in some known manner that can be modeled with a covariate or used as a basis for stratification), then the choice of K may also vary among units.

To combine robustness and efficiency for a fixed level of resources, a combination of study designs could also be used. For example, half of the total survey effort may be used to construct a standard design, while the other half of the survey effort is used to construct a removal design. The standard design provides additional flexibility for modeling, while the removal design increases spatial replication for the same level of effort. Given that the same physical sampling methods are used in each, the combined data could be analyzed within a single framework.

The study design recommendations discussed in this chapter have focused on minimizing properties of occupancy estimates, e.g., variance, coefficient of variation, confidence interval width. Sometimes study design will focus on a key question, such as "Is occupancy really greater in this habitat as opposed to that habitat?". When designs are focused on specific questions, then our optimization criterion will likely be power of a test for differences, or likelihood that model selection metrics will clearly favor the better approximating model. Most of the design principles discussed above will still apply, but the design optimization will now depend on these different criteria. Design analyses for such question-driven studies can be conducted using GENPRES (Bailey et al., 2007) and similar approaches.

In occupancy studies requiring use of false positive models, design tradeoffs may extend to a third dimension. In addition to considering the number of units and the number of surveys at each unit, the issue arises of how to distribute effort in the case of surveys that yield uncertain (false positives possible) versus certain (no false positives possible) detections. In some cases, such as when both types of detections occur during the same survey efforts (Miller et al. 2011; Section 6.2), the probability that a detection is certain is viewed as the outcome of a stochastic process and may not be under investigator control. However, for other models effort can be allocated to survey methods that do and do not admit false positives (Miller et al. 2011; Section 6.2), with the latter surveys often being more expensive. Effort allocated to methods yielding certain detections becomes a potentially important design issue. Clement (2016) has begun to consider designs for false-positive models, and we expect this to become an important area of future development.

As was noted at the beginning of this chapter, it is important to consider how the assumptions of modeling techniques match with the biology of the species and reality of collecting the data from the field. For example, the basic models of MacKenzie et al. (2002) assume no unmodeled heterogeneity in detection probability among units, yet it can be very easy to accidentally introduce such heterogeneity when designing a study (e.g., through observer effects). The purpose of this chapter, as mentioned earlier, was not to provide a recipe for designing your specific study, but to give some guidance on the various aspects of study design that need to be considered along with some general principles.

A good study design can make or break a project and thus deserves careful consideration. Studies should usually be designed on a case-by-case basis, as the details of each design (goals, species, environment, etc.) will often be different. Once a satisfactory design has been developed, simulation and pilot studies can often be useful tools for assessing whether the design will provide the type and quality of information required to meet the studies' objectives. Computer simulations can be particularly useful when sample sizes are going to be small, as some of the results presented above are based on large-sample approximations, hence may not hold in such situations (Bailey et al., 2007; Guillera-Arroita et al., 2010). Pilot studies can also be invaluable for trialling field protocols and resolving 'teething troubles' prior to implementing a full-scale design. Often there may be pressures from various quarters to begin data collection as soon as possible, but without a sound study design in place, the resulting data may be inappropriate or insufficient, and all that has been achieved is the wasting of precious economic resources.

Chapter 12

Multiple-Season Study Design

In the previous chapter we examined some of the issues related to the design of a single-season occupancy study. In particular we emphasized the importance of considering how the biology of the species corresponded to the model assumptions and timing of the surveys, and also presented results on the 'optimal' allocation of resources. We believe the issues considered in Chapter 11 (e.g., unit selection, defining a season, and number of repeat surveys) are equally pertinent to the design of multiple-season occupancy studies, as a multiple-season study is simply a sequence of single-season studies.

Very little has been published on design issues related to multi-season occupancy studies, some exceptions being Field et al. (2005), MacKenzie (2005b), and Bailey et al. (2007). Field et al. (2005) considered a situation where trend in occupancy (specifically a decline) over a three season time period is of interest, subject to budgetary constraints. Based upon a hypothesis testing objective, they assessed the effect that the number of repeated surveys has on the power of the test to detect a decline (assuming that increasing the number of repeat surveys decreases the number of units that can be sampled, given a fixed budget). Field et al. (2005, p. 476) recommended that "2 to 3 visits to each unit would perform adequately for most species (and) fewer than the optimal number of visits resulted in a harsher penalty than making more (visits)." MacKenzie (2005b) noted that Field et al. (2005) assumed detection probability was equal in each of the three seasons, an assumption which the analysis of empirical data sets has shown to be unlikely (e.g., MacKenzie et al., 2003; Bailey et al., 2004; MacKenzie et al., 2005; Olson et al., 2005). As such, a greater number of repeat surveys (visits) per season should be used, a suggestion more in line with the recommendations of MacKenzie and Royle (2005) (as reported in Chapter 11). Field et al. (2005) found that a general strategy of sampling more units, less intensively, for rare (i.e., low occupancy) species, and fewer units, more intensively, for common species was optimal for detecting a decline in occupancy. Given the similarities of this result (and others) with those of MacKenzie and Royle (2005) in the single-season case, we believe that many of the recommendations given in Chapter 11 are equally applicable to multi-season occupancy studies.

In particular we believe the previous recommendations on the 'optimal' number of repeat surveys per season holds for longer-term studies. Some prac-

Occupancy Estimation and Modeling. http://dx.doi.org/10.1016/B978-0-12-407197-1.00016-8

titioners have found that multi-season occupancy models can be applied to data sets that only have one survey per season, but a number of restrictive assumptions have to be made about temporal variation in detection and the dynamic parameters, with at least one of these sets of parameters having to be time invariant or functions of identified covariates. That is, it is not possible to fit models with season-specific estimates of detection and the dynamic parameters. However, as discussed in Chapter 11 on single-season design, a main benefit of the repeat surveys is that not only does it enable detection and occupancy probabilities to be estimated separately, but the repeat surveys provide more reliable data about the presence/absence of the target species by reducing the probability of a false absence, compared to only a single survey per season. Thus higher quality data are obtained about the true occupancy state of units, which is more beneficial than having data from a greater number of units, that are less reliable about the true presence/absence of the species (due to fewer, or no, repeat surveys). The same holds in the present context, where a benefit of repeat surveys (aside from enabling detection probabilities to be estimated for each season) is that they reduce the probability of observing false extinction and false colonization events, thereby having a better quality data set. That is, we strongly advocate for quality before quantity.

In this chapter we highlight additional issues that must be considered whilst designing a multiple-season occupancy study.

12.1 TIME INTERVAL BETWEEN SEASONS

It is important to contemplate the length of time that should separate two seasons. While generally the type of practical situations we have experience with are those where seasons are clearly defined by the species biology (i.e., breeding seasons), an arbitrary length of time could be used if that is consistent with the objectives of the study. The main considerations are the timescale at which the processes of colonization and local extinctions operate and how that relates to the objectives of the study. It is important to note that the estimated parameters relate to the length of time between two seasons and that attempts to rescale these estimated quantities onto other timescales may be misleading. For example, suppose a population is surveyed during a single week each year. Colonization and local extinction probabilities therefore relate to potential changes in the occupancy status of units over a 12-month period. Now suppose that managers decide they want estimates that correspond to six-month time intervals (e.g., estimates for summer and winter) without collecting data at this time scale. It is mathematically possible to compute average six-month estimates of colonization and extinction probabilities that assume constant instantaneous rates over the 12-month interval, although such calculations are not straightforward, as one must account for colonization and extinction events at the finer temporal scale. For example, the species may go locally extinct from a unit between two seasons 12 months apart, but at a six-month time scale the species may have gone locally extinct during the first six months and then failed to colonize the unit in the second six month period, or it did not go locally extinct in the first six months but did go locally extinct during the second period. That is:

$$\epsilon_{12} = \epsilon_6(1 - \gamma_6) + (1 - \epsilon_6)\epsilon_6,$$

where ϵ_t and γ_t denote the respective dynamic probability during an interval of duration t. Similarly, 12-month colonization could be expressed in terms of six-month dynamic processes as:

$$\gamma_{12} = \gamma_6(1 - \epsilon_6) + (1 - \gamma_6)\gamma_6.$$

By applying some algebra to these two equations we could solve for ϵ_6 and γ_6, and thus obtain estimates at this finer timescale. In fact, we could use the right-hand sides of these two expressions to redefine colonization and extinction probabilities in the multi-season model, and estimate them directly.

However, to solve these equations, constancy of the dynamic processes at the finer scale must be assumed (i.e., one cannot estimate different parameter values for each six-month period), hence it is not possible to draw inferences about potential seasonal variation in local extinction and colonization probabilities. Such inferences depend on the collection of additional data about when the colonization or local extinction event occurred (or indeed whether multiple such events may have occurred). Relatedly, in the metapopulation literature, some authors have postulated a "rescue effect" (e.g., Brown and Kodric-Brown, 1977; Hanski, 1991, 1994a, 1997; Moilanen, 1999, 2002), where between two survey periods the species may go locally extinct at a unit or patch and the patch is then recolonized by members of the species from a nearby patch. While we do not argue whether a rescue effect may be a biological reality, we hold the view that without additional information it cannot be reliably estimated. Present attempts to estimate parameters associated with the rescue effect rely on restrictive assumptions about the number of extinction and colonization events that can occur between surveying periods (e.g., Clark and Rosenzweig, 1994).

Our primary recommendation is to tailor the timing of sampling to the temporal resolution at which questions are being addressed. In addition, it may be possible in some situations to use ancillary data (e.g., on marked individuals) to address very specific mechanistic questions (e.g., regarding rescue effects) about dynamic processes (MacKenzie and Nichols, 2004). Furthermore, we suggest that a consistent time interval between seasons be used. That is, we generally recommend against designs where there is substantial variation in the time between seasons (e.g., seven months, five months, nine months, etc.). Although interval-specific estimates of rate parameters can be readily obtained under time-specific models, the resulting estimates do not lend themselves to straightforward comparisons. While we can think of a number of *ad hoc* approaches that could be used to allow for inconsistent time intervals in collected data (e.g., by reparameterizing the dynamic parameters as above, including a *time interval* covariate, or inserting missing observations into the detection histories to represent missing seasons), parameter estimates may be misleading and difficult to interpret. We also note that in such situations, fitting models with time-invariant dynamic parameters are unlikely to be biologically reasonable, i.e., for many species the probability of colonization over a five-month interval is unlikely to be the same as over a nine-month interval. However, there may be some cases where it is desirable to have unequal time periods between sampling seasons, for example if there is interest in how the distribution of a species changes between early and late summer each year, then the interval from early to late summer in the same year will obviously be different from the late summer to early summer the following year. Essentially, the nature of the questions being asked should dictate the appropriate time intervals between periods of data collection.

12.2   SAME VS. DIFFERENT UNITS EACH SEASON

The explicit dynamics models detailed in Section 8.3 rely on information collected from the same units each season to estimate colonization and local extinction probabilities, and hence rates of change in occupancy. Using the model, not all units would have to be surveyed each season because of the ability to incorporate missing values, but generally the expectation would be that a reasonable fraction of the units would be surveyed every season. However, the implicit dynamics model of Section 8.2 could be used equally well in situations where the same or different units are surveyed each year, because there is no necessary linkage of specific units in multiple seasons, and therefore no explicit modeling of changes in the occupancy status of units between seasons.

In many of the studies with which we are most familiar, most units are surveyed consistently each season (i.e., a longitudinal study), however, this is not the only design that could be used. Another type of design that is occasionally used in ecological studies is the rotating panel design (Urquhart and Kincaid, 1999). In a rotating panel design, rather than attempting to survey units that are randomly selected over the entire area of interest each season, units are selected from within a smaller sub-area, and each season a different sub-area is selected (e.g., surveying different watersheds within a forest or national park each year). Sub-areas are surveyed again in subsequent seasons, but with a consistent rotation period that is > one season (e.g., surveying is conducted annually across a national park, but each watershed is only surveyed every third year). Clearly, an advantage of such a design is that each season the logistical costs may be lower, as less time will probably be required to travel among units. Another justification that is sometimes given for such a design is that the spatial coverage of a study can be increased, i.e., if in any given season there are only enough resources to survey 20 units, using a three-year rotating panel design, the total sample size (viewed over the entire three years) can be increased to 60.

We have used simulation studies to compare the relative benefits of using a rotating panel-type design versus a design where all units are surveyed over multiple seasons (which was briefly reported in MacKenzie, 2005b). For all of the scenarios considered, the simulation study compared designs that had the same number of units surveyed per season, but differed in terms of which units were surveyed in consecutive seasons (e.g., the same 20 units surveyed each season, 20 units surveyed per season with a five-season rotation period, hence 100 units surveyed in all, or ten units surveyed each season and an additional ten units surveyed per season on a five-season rotation period, hence 60 units surveyed in all). One of the main results from our simulation study is that in terms of estimating the trend in occupancy (using either an implicit or explicit dynamics model), the overall design of the study is relatively unimportant. The precision of the trend estimate is largely determined by the number of units surveyed each season (provided sufficient repeat surveys have been conducted within each season). This means that using a rotating panel design with a total of 100 units surveyed over five seasons (i.e., 20 different units per season) does not provide any more information with regard to a *seasonal* trend in occupancy than a design where the same 20 units are surveyed over the same period. That is, using a rotating panel design does not increase the effective sample size for estimating a seasonal trend.

This result assumes that all units that could be surveyed in a particular season have the same occupancy, colonization and local extinction probabilities, i.e., units that are selected to be surveyed in two consecutive seasons do not come from two different sub-populations with different occupancy-related parameters. Potentially, with a rotating panel design where units are selected from different sub-areas each season, this assumption could be violated. Resulting inferences from a rotating panel design could be inaccurate, and possibly misleading; the reasons for an observed change in occupancy may be due to: (1) a change in occupancy across the landscape over time; (2) the fact that different units within the landscape have been sampled; or (3) a combination of both. It will not be possible to make strong statements about a trend in occupancy because it could be reasonable to argue that the observed change was a result of surveying different locations characterized by a different average level of occupancy, rather than temporal changes in occupancy within locations (trend). While additional structure could be built into a model to reflect any potential effect of the sub-areas, the inclusion of additional parameters to be estimated will increase the level of uncertainty in the trend parameter (i.e., increase its standard error), further hampering one's ability to make strong inference about trend. We believe that for either the processes underlying occupancy dynamics or trends in occupancy, a far more practical approach is to survey the same units each season thereby removing any confounding spatial variation.

Furthermore, a rotating panel design will not provide the necessary information required to make seasonal estimates of colonization and local extinction probabilities. Management actions may depend on the magnitudes of these vital rates, and virtually all management actions attempt to modify one or both of these rates. In some situations, the appropriate management action to address a downward trend in occupancy, for example, could be different if colonization probability was high compared to if it was zero. Hence, to understand the mechanisms of change in the population, and make appropriate decisions when necessary, rotating panel designs may be of limited utility.

A rotating panel design may be more useful when the temporal scale of interest for changes in occupancy matches the rotation period. For example, instead of being interested in annual changes the objective is to estimate changes at a five-year interval (i.e., with a five-year rotation period, there are only three time points for each unit after ten years). In this case, the rotating panel design could be considered as a type of longitudinal study, but where the surveying of units has to be staggered across multiple years. In this case we suggest that units are surveyed in the same order each rotation, in accordance with our above recommendation to maintain a consistent time interval between sampling seasons for each unit.

Our general recommendation is that when interest is in change, or the underlying factors driving change, a longitudinal study (i.e., repeat surveys of the same units) should be used with the interval between sampling seasons matching the time-scale of interest for change.

12.3 MORE UNITS VS. MORE SEASONS

In some situations it may be possible to consider a tradeoff between number of units surveyed per season, and the number of seasons for which the study is conducted. For example, given a fixed total budget, is it better to monitor a population with 100 units over ten years, or 200 units over five years. Of course, the ten-year study provides information about a period of time for which no formal inference is possible based on the five-year study. Thus, treatment of this deci-

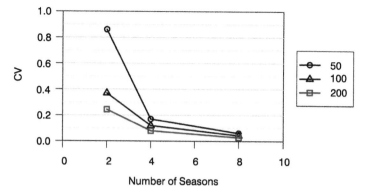

FIGURE 12.1 Coefficient of variation (CV) for an estimated 0.2 seasonal trend in occupancy on the logit scale (approximated via simulation) from an implicit dynamics model, given that 50, 100, or 200 units are surveyed for two, four, or eight seasons. Solid lines are for ease of interpretation only. Within seasons, three surveys were conducted at all units with a detection probability of 0.5.

sion as a design tradeoff is predicated on an assumption of similar dynamics over the entire ten-year period of interest. Nevertheless, 'trend' is frequently viewed as a parameter characterizing a relatively long time period, hence such design considerations may be reasonable in some situations. MacKenzie (2005b) gave evidence that such a tradeoff is possible by considering the coefficient of variation (CV) of the estimated trend in occupancy on the logistic scale (and using an implicit dynamics model for simplicity). Fig. 12.1 portrays the approximate CV for an increasing trend of 0.2 units on the logit scale per season. That is:

$$logit(\psi_t) = \beta_0 + \beta_1 t,$$

where $\beta_1 = 0.2$, which equates to the odds of occupancy increasing $e^{0.2} = 1.22$ each season (Chapter 3). Solid lines indicate the approximate CV (obtained via simulation) for the trend estimate assuming designs where 50, 100, or 200 units are surveyed each season. Clearly a tradeoff exists, as a similar level of precision can be achieved by surveying more units over fewer seasons vs. surveying fewer units over a longer time period (e.g., 200 units for four seasons vs. 50 units for eight seasons). An implication of this result is that if precise information about a trend in occupancy is required within a short timeframe, more sampling effort will be needed each season than if a longer timeframe is available. An alternative interpretation of this result is that if available funding only permits a small number of units to be surveyed each season, then managers and stakeholders must understand that a longer timeframe will be required to provide decisive information about the system. Hence managers should be prepared to make a long-term commitment to the program. Finally we caution readers that the results presented in Fig. 12.1 are illustrative only, and pertain to a specific situation. While we believe this tradeoff to exist generally, the magnitude of the tradeoff should be assessed on a case-by-case basis.

That said, we believe that a linear trend will often be an insufficient descriptor of temporal changes in a population. Trend estimates may even be misleading in some cases, e.g., for species whose population trajectory tends to be cyclical in nature. We also believe that 'trend detection' should not be a prime consideration when designing monitoring programs, as we do not view that as a useful management objective (Chapter 1). Hence, while it is useful to be aware that tradeoffs such as those described above exist, our view is that often there will be other descriptors of occupancy dynamics that investigators should be focusing on given the objectives of their study or management situation.

12.4 MORE ON UNIT SELECTION

The final issue we address here with respect to the design of multiple-season occupancy studies is unit selection. To generalize to the entire area of interest, we have so far assumed that units are selected in some probabilistic manner (i.e., the probability of a unit being selected from the population of units can be defined). If units are selected in a non-probabilistic manner then any generalization of the occupancy-related parameters to units beyond those that were surveyed, must be made with caution. This point is particularly relevant if units are included in the sample due to prior knowledge about their expected occupancy state, which may create bias in parameter estimates. For example, we have seen some studies where units are selected because the species has been known to historically inhabit those units. Depending upon the timeframe between that historic knowledge and the present study, it may be reasonable to expect that a sample of historic units will exhibit a higher proportion of units occupied than a random sample of units from the area of interest. While estimates of colonization and local extinction probabilities should remain unbiased, as they are conditional upon the occupancy state of units in the previous season, any estimate of a change or trend in occupancy could be biased because of the recursive relationship between occupancy, colonization and local extinction. For example, consider Fig. 12.2. Suppose the system was in a state of equilibrium, hence the probability of a unit being occupied was constant each season, i.e., there is no trend in occupancy. If the units are selected in such a manner that the proportion of units occupied in the sample is much higher than the level in the overall population, then an apparent downward trend in occupancy could be observed at the sampled units even though no trend actually exists. A similar result can occur when units are added to those being surveyed based upon knowledge

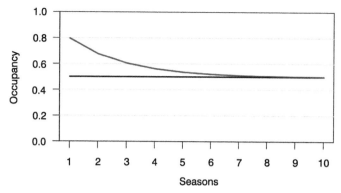

FIGURE 12.2 Apparent trend in occupancy caused by having an 'unrepresentative' sample of units with a higher occupancy probability than the population of interest. Solid red line shows the change in occupancy induced by selection of an unrepresentative sample. Solid black line shows the true level of occupancy in the population, which is constant over time. The same constant rates of local colonization and extinction apply to both red and black trajectories.

of the likely occupancy state (e.g., monitoring begins at a new unit because of casual reports of the species being present there).

While the surveying of historic units alone could create apparent trends in occupancy, it is one approach for targeting the surveying to provide better estimates of local extinction probabilities (as these are conditional upon units being occupied, and a sample of historic units may have a higher fraction of occupied units). To provide accurate trend estimates about the population in general, while utilizing the historic unit information, one could balance this information with data from a second sample of units of unknown historic status. Effectively the population is now stratified into units of known and unknown historic occupancy status, hence a form of stratified random sample could be used to select units. Of course an assessment should be made of whether a more complicated stratified random sample is likely to achieve the objective of the study more efficiently than a simple random sample. As with other design issues, there is no single approach that is likely to be optimal in all situations. Thus, we recommend that units should be selected for surveys in a manner that is consistent with study objectives.

As noted in the previous chapter, another issue related to unit selection that is especially relevant for multi-season studies, is consideration of potential system dynamics. For example, suppose a species may exhibit strong habitat preferences, and there are areas that would currently be considered as non-habitat for the species, but could potentially become habitat at some point in the future and hence be colonized by the species. To fully appreciate the system dynamics and changes in the distribution of the species, the potential sampling frame should include areas that are currently considered non-habitat. We are not advocating that such areas be intensively surveyed for the target species if it is known, or at least accepted, that the species will not be present at units in these areas, but simply that some survey effort be assigned to units in those areas with periodic checks of the habitat status. Once it is considered 'habitat', then more intensive surveys might be initiated at those units to detect the species once it colonizes those units. Similarly, in studies of species range changes, some survey effort should be devoted to units that are considered to be outside of the species current range so that changes of interest can be observed. Solely focusing on areas that are considered to be inside the species current range may give a misleading impression of changes in the species distribution, as the sampling frame has not been defined to allow for incursions into new areas to be observed.

12.5 DISCUSSION

In this short chapter we have briefly touched on a number of important issues associated with occupancy-based studies conducted over longer timeframes. Primarily we have focused on the simpler situation where occupancy is a binary measure, but most of the above issues are equally relevant to the multi-season multi-state occupancy case (e.g., time between seasons completely determines interpretation of the temporal aspect of the transition probabilities).

While we have tried to highlight some of the general principles we believe practitioners need to consider when designing a multiple season study, it is difficult to give more specific advice. When faced with a myriad of options during the design phase, our best advice is to continuously return to the objective to guide you through the options and to avoid the temptation of interesting 'add ons' that may take valuable, and likely limited, resources away from pursuing that objective.

As with single season studies, pilot studies and computer simulations are particularly valuable to assess alternative field methods, and to gain realistic expectations about the limits on what inferences could be drawn from the resulting data. Taking the time to refine methods and gain such understandings before embarking on the full data collection phase will enable resources to be used more efficiently, and may avoid future disappointments.

Part V

Advanced Topics

Chapter 13

Integrated Modeling of Habitat and Occupancy Dynamics

13.1 INTRODUCTION

Throughout this book we have noted that many species exhibit preferences for certain habitats such that the probability of the target species occupying a unit is a function of the unit's habitat. This idea holds equally for the underlying dynamic processes: that colonization and extinction probabilities, for example, are different in different habitats. However, often the habitat on the landscape is not static, but changes through time via processes such as vegetation succession, human activities, and environmental variation (e.g., the availability and duration of water at vernal pools depends on seasonal rainfall). When habitat is dynamic, it will frequently be useful from both scientific and conservation perspectives to partition species occurrence dynamics into components associated with habitat versus other factors. The theoretical literature on metapopulation dynamics contains multiple pleas to study dynamics of species occurrence and habitat simultaneously, focusing on habitat changes occurring as a result of ecological succession (Ellner and Fussmann, 2003), natural disturbance frequency (Amarasekare and Possingham, 2001), and human activities (Lande, 1987, 1988; Thomas, 1994).

Many researchers will often tend to restrict their surveys for a species of interest to only those units that are regarded to contain suitable habitat, i.e., units at which species occurrence is possible. In the face of dynamic habitat, solely focusing on species occurrence is likely to be inadequate as a sole descriptor of system state, as similar occurrence proportions may have very different meanings with respect to system well-being, depending on the number of units with suitable habitat. For example, consider the situation where the amount of suitable habitat within an area of interest is declining over time. Occurrence of the species in those suitable patches that remain may be high (most suitable units occupied), yet because the number of suitable units is decreasing, the overall well-being of the species, and level of occupancy in all patches, is declining (e.g., Lande, 1987). The type of management actions that might be effective to halt such a decline in this situation could be quite different from those required by an alternative situation, where the proportion of suitable patches remains

Occupancy Estimation and Modeling. http://dx.doi.org/10.1016/B978-0-12-407197-1.00018-1

constant, but the proportion of suitable patches occupied is declining. In these cases, characterizing the system by two state variables, the proportion of units at which habitat is suitable and the proportion of suitable units at which the species is present, would be more informative.

MacKenzie et al. (2011) developed models that explicitly accounted for both habitat and species occurrence dynamics, while also accounting for the potential imperfect detection of the species (although these models can also be used in the case where detection is presumed to be perfect, e.g., Martin et al., 2010). Here we utilize the multi-state, multi-season occupancy model framework in Chapter 9 to describe the modeling approach. We focus on the parameterizations used to set up the model structure, and avoid the specifics of how to construct the model likelihoods from the observed data. Those details are described in Chapter 9.

13.2 BASIC SAMPLING SITUATION

We envisage a similar situation to other multi-season applications, where a collection of sampling units is surveyed at systematic points in time and the habitat state of a unit is recorded, as is the presence or absence of the target species. When there is the possibility of false absences, repeat detection/nondetection surveys could be conducted within a season such that detection probability can be estimated and appropriately accounted for. Within a season, the habitat state and occupancy state of the species is assumed to be static or unchanging (i.e., the closure assumption). Initially the modeling will be described in terms of two possible habitat states (A and B) and two occupancy states, but we note that the modeling can be extended to accommodate additional states if required (e.g., Miller et al., 2012a). It is assumed that the habitat type at each unit is known without error and that each may have a nonzero probability of being occupied by the species, although one habitat could be defined as 'unsuitable' and constrained to always be unoccupied. We shall return to this situation later in the chapter.

Detection histories are therefore comprised of information on both the observed habitat state in each season and the outcome of the surveys for the species. There are multiple ways in which this information could be presented, but here we follow MacKenzie et al. (2011) where a detection history has a habitat component and a detection component. For example, suppose a unit was surveyed in three seasons, was habitat A in season 1 and habitat B in seasons 2 and 3. This could be represented as $\mathbf{H}_i = $ A B B. The detection component would be the same as for the regular occupancy models, e.g., $\mathbf{h}_i = $ 01 00 11.

Rather than a two-part detection history, an alternative would be to use a coding where unique values are used to denote different combinations of habitat type and detection/nondetection in each survey. For example, let the outcome of each survey of a unit be coded as:

- $0 =$ habitat type A and species not detected
- $1 =$ habitat type A and species detected
- $2 =$ habitat type B and species not detected
- $3 =$ habitat type B and species detected

The above example could therefore be recorded as $\mathbf{h}_i = 01\ 22\ 33$.

13.3 MODEL DEVELOPMENT AND ESTIMATION

The parameters associated with this model are defined in Table 13.1, where the π and η parameters are associated with the habitat component of the modeling; ψ, γ, and ϵ determine the occupancy component; and p denotes detection probability. Note that the parameters, as defined, allow a very general case where the species occurrence dynamics depend on the current habitat state at a unit, and the habitat dynamics can also depend on the current occupancy state of the species. This latter possibility might occur through habitat modification by the species itself when present at a unit (e.g., through grazing of vegetation; McNaughton, 1983; McNaughton et al., 1997), or indicate preferential selection of locations by the species within a defined habitat type that exhibit different dynamics (e.g., vernal pools selected by breeding amphibians may be less likely to dry as quickly as those not selected).

Using the multi-state framework, the state of the unit in season t can be defined with a combination of habitat and occupancy states. For example, with two habitat and two occupancy states there are four possible states:

- $0 =$ habitat type A and species absent
- $1 =$ habitat type A and species present
- $2 =$ habitat type B and species absent
- $3 =$ habitat type B and species present

The integrated dynamics model therefore describes how the occupancy state of a sampling unit changes between seasons. Using the parameters defined in Table 13.1, the initial state probability vector $\boldsymbol{\phi}_0$ would be:

$$\boldsymbol{\phi}_0 = \left[\ \pi^{[A]}\left(1-\psi^{[A]}\right)\quad \pi^{[A]}\psi^{[A]}\quad \left(1-\pi^{[A]}\right)\left(1-\psi^{[B]}\right)\quad \left(1-\pi^{[A]}\right)\psi^{[B]}\ \right].$$

For example, the first element of $\boldsymbol{\phi}_0$ is the probability of a unit being of habitat type A ($\pi^{[A]}$), and the species being absent given habitat type is A ($1-\psi^{[A]}$) in the first season.

The transition probability matrix, $\boldsymbol{\phi}_t$, is therefore populated with the various combinations of defined parameters to model the change in the occupancy state

TABLE 13.1 Definition of parameters used in the integrated modeling of habitat and species occurrence

Component	Parameter	Description
Habitat	$\pi^{[H]}$	Probability a unit is of habitat type H in first season
	$\eta_t^{[X_t,H_t,H_{t+1}]}$	Probability the habitat changes from state H_t in season t to state H_{t+1} in season $t+1$, given the species was either present ($X=1$) or absent ($X=0$) from the unit in season t
Species occurrence	$\psi^{[H]}$	Probability the species is present at a unit of habitat type H in first season
	$\gamma_t^{[H_t,H_{t+1}]}$	Probability species colonizes a unit between seasons t and $t+1$ given the habitat has transitioned from state H_t in season t to state H_{t+1} in season $t+1$
	$\epsilon_t^{[H_t,H_{t+1}]}$	Probability species goes locally extinct from a unit between seasons t and $t+1$, given the habitat has transitioned from state H_t in season t to state H_{t+1} in season $t+1$
Detection	$p_{t,j}^{[H_t]}$	Probability of detecting the species in survey j of season t, given the habitat state in season t is H

of a unit between seasons t and $t + 1$. It could be defined as:

$$
\phi_t = \begin{bmatrix}
\eta_t^{[0,AA]}\left(1-\gamma_t^{[AA]}\right) & \eta_t^{[0,AA]}\gamma_t^{[AA]} & \left(1-\eta_t^{[0,AA]}\right)\left(1-\gamma_t^{[AB]}\right) & \left(1-\eta_t^{[0,AA]}\right)\gamma_t^{[AB]} \\
\eta_t^{[1,AA]}\epsilon_t^{[AA]} & \eta_t^{[1,AA]}\left(1-\epsilon_t^{[AA]}\right) & \left(1-\eta_t^{[1,AA]}\right)\epsilon_t^{[AB]} & \left(1-\eta_t^{[1,AA]}\right)\left(1-\epsilon_t^{[AB]}\right) \\
\eta_t^{[0,BA]}\left(1-\gamma_t^{[BA]}\right) & \eta_t^{[0,BA]}\gamma_t^{[BA]} & \left(1-\eta_t^{[0,BA]}\right)\left(1-\gamma_t^{[BB]}\right) & \left(1-\eta_t^{[0,BA]}\right)\gamma_t^{[BB]} \\
\eta_t^{[1,BA]}\epsilon_t^{[BA]} & \eta_t^{[1,BA]}\left(1-\epsilon_t^{[BA]}\right) & \left(1-\eta_t^{[1,BA]}\right)\epsilon_t^{[BB]} & \left(1-\eta_t^{[1,BA]}\right)\left(1-\epsilon_t^{[BB]}\right)
\end{bmatrix}.
$$

Consider the element in the top left of ϕ_t, $\eta_t^{[0,AA]}\left(1-\gamma_t^{[AA]}\right)$, which denotes the probability of a unit being in occupancy state 0 (using the 0, 1, 2, 3 coding) in season t and remaining in state 0 in season $t + 1$, i.e., the probability a unit was of habitat type A and unoccupied in both seasons. The term $\eta_t^{[0,AA]}$ denotes the probability of the habitat being of type A in both seasons, given the species was absent in season t, and the term $1 - \gamma_t^{[AA]}$ denotes the probability of the species not colonizing the unit given the habitat was type A in both seasons. Essentially, the transition probabilities are defined as a two-step process where first the habitat transitions to either state A or B (where the probability of the transition is potentially different depending on which of the four states the unit was in during season t), then conditional upon the habitat transition, the pres-

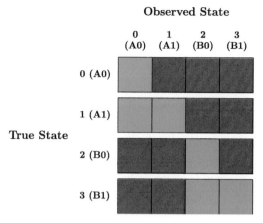

FIGURE 13.1 Illustration of the presumed relationship between observed and true states for the integrated model of habitat and species occurrence dynamics. States are denoted by the 0, 1, 2, 3 coding, and also in terms of habitat type and species presence/absence (or detection/nondetection), e.g., A0 indicates habitat type A and species absent/not detected for true/observed state, respectively. Green cells indicate possible combinations and red cells impossible combinations.

ence/absence of the species potentially changes at a unit (as determined through the probabilities of colonization and extinction).

To relate the possible outcomes from survey j in season t to the true occupancy states (Fig. 13.1), the detection probability matrix $\mathbf{p}_{t,j}$ (Chapters 5 and 9) needs to be defined as:

$$\mathbf{p}_{t,j} = \begin{bmatrix} 1 & 0 & 0 & 0 \\ 1-p_{t,j}^{[A]} & p_{t,j}^{[A]} & 0 & 0 \\ 0 & 0 & 1 & 0 \\ 0 & 0 & 1-p_{t,j}^{[B]} & p_{t,j}^{[B]} \end{bmatrix}.$$

Note that the structure of $\mathbf{p}_{t,j}$ indicates that the habitat state must be observed with certainty. For example, it is not possible to have an observation of habitat A (first two columns) when the true state is habitat B (last two rows; the four elements in the lower left corner are all 0), and vice versa. However, when the species is not detected in a survey (columns 1 and 3), there are two nonzero elements in each column indicating that the species was not observed either because the species was absent, or present and undetected. When the species is detected (columns 2 and 4), such observations are only possible if the species is present at the unit.

Once cast into the multi-state framework, parameter estimation by either the observed or complete data likelihood approaches proceeds as detailed in

Chapter 9. Instead of repeating such details, here we shall focus on how this integrated modeling approach could be used to address biologically interesting questions.

Finally, in the above development, changes in both habitat state and species occurrence have been assumed as stochastic, but in some situations certain types of transitions could be considered deterministic. This can be easily accomplished within this framework by setting the respective probability for the deterministic transition to be 1.0 and the probability of other transitions to 0.0. Such a constraint may be applied to all sampling units or only to a selection of sampling units as required (e.g., the probability of reverting from *mature forest* to *clear cut* could be set equal to 1.0 if the timber at a sampling unit is harvested between seasons).

13.3.1   Missing Observations

As with the other modeling described in this book, it is possible to account for missing observations and unequal sampling effort. A slight distinction here, however, is that we can foresee two possible options depending on whether the habitat state is known or not during a season. When the habitat state is known (e.g., the unit was surveyed at least once in the corresponding season, or habitat state is determined from a data source independent of the occupancy surveys), then accounting for missing survey outcomes proceeds exactly as for the standard occupancy models where the detection probabilities for the respective surveys and sampling units are effectively set to zero. In this case, instead of four possible types observations that could result from the survey (in terms of habitat and detection of the species) there are only two (i.e., the habitat state), although obviously there remain four possible true states for the unit. Therefore, the detection probability matrix for the corresponding survey becomes a 4×2 matrix:

$$\mathbf{p}_{t,j} = \begin{bmatrix} 1 & 0 \\ 1 & 0 \\ 0 & 1 \\ 0 & 1 \end{bmatrix}.$$

When habitat state is unknown but a detection survey was conducted, then essentially one must integrate over the possible habitat states in each season. Again, the number of possible observations from the survey reduces to two, although in this case the possible observations are the detection/nondetection of the species (i.e., no information about the habitat state). The detection probability matrix would therefore be:

$$
\mathbf{p}_{t,j} =
\begin{bmatrix}
1 & 0 \\
1 - p_{t,j}^{[A]} & p_{t,j}^{[A]} \\
1 & 0 \\
1 - p_{t,j}^{[B]} & p_{t,j}^{[B]}
\end{bmatrix}.
$$

In the case where both habitat state and species detection are missing (e.g., the unit was never surveyed during a season), there is only one 'observation' possible; see nothing. Now $\mathbf{p}$ becomes a column vector, or 4×1 matrix, of 1's. That is:

$$
\mathbf{p}_{t,j} =
\begin{bmatrix}
1 \\
1 \\
1 \\
1
\end{bmatrix}.
$$

13.3.2 Covariates

Covariate information or predictor variables can be easily included in these models by applying an appropriate link function (e.g., logit link) to the respective probabilities. The mechanics of doing so are the same as described elsewhere in the book (e.g., Chapter 3). In addition to the usual type of covariates that might be considered in similar modeling to explain variation in occupancy-related and detection probabilities, we note that broader-scale environmental covariates may be particularly relevant to explain variation in the probabilities associated with habitat change. For example, in a wetland system where ponds are defined as sample units and habitat state is the size of the pond (e.g., large or small), suitable predictors for habitat change could include the amount of rainfall between seasons or annual snowpack. This flexible framework should enable identification of what factors are important for the separate but interrelated processes of habitat and occupancy dynamics.

13.4 BIOLOGICAL QUESTIONS OF INTEREST

The integrated model of habitat and occupancy dynamics was motivated by situations in which the dynamics of the two processes are linked, to some degree. This enables a number of biologically interesting questions about these interrelationships to be explored with this modeling approach, in addition to the types of questions that could be addressed with occupancy-only dynamics models (e.g., Section 10.4), or modeling of the habitat only.

13.4.1 Effect of Habitat Change on Occupancy Dynamics

In Table 13.1, the colonization and extinction probabilities were defined in terms of the habitat state at both the beginning and end of the period during which change can occur, that is, between seasons t and $t + 1$ (e.g., $\gamma_t^{[H_t, H_{t+1}]}$). Therefore with two habitat states, there are potentially four different types of colonization and extinction probabilities depending on the transition of the habitat states, i.e., A→A, A→B, B→A, and B→B. Note that each of these types of probabilities may also vary temporally, or as a function of covariates. Allowing these probabilities to be different for each specific type of habitat transition is very general, and while it would be appropriate in some situations, such flexibility may not be required in others. Constraints could be placed on these probabilities reflecting different hypotheses about how the habitat state affects the occupancy dynamics. If it was thought that only the habitat state at the start of the period affected colonization probabilities, for example, then we could impose the constraints:

$$\gamma_t^{[A,A]} = \gamma_t^{[A,B]} = \gamma_t^{[A,\cdot]}$$

and

$$\gamma_t^{[B,A]} = \gamma_t^{[B,B]} = \gamma_t^{[B,\cdot]}.$$

Similarly, if the habitat state at the end of the period was thought to be the important determinant or colonization then the following constraints could be made:

$$\gamma_t^{[A,A]} = \gamma_t^{[B,A]} = \gamma_t^{[\cdot,A]}$$

and

$$\gamma_t^{[A,B]} = \gamma_t^{[B,B]} = \gamma_t^{[\cdot,B]}.$$

Obviously, if it was hypothesized that habitat state was unimportant for colonization, then all four types of probabilities could be set equal. That is:

$$\gamma_t^{[A,A]} = \gamma_t^{[A,B]} = \gamma_t^{[B,A]} = \gamma_t^{[B,B]} = \gamma_t.$$

The same types of constraints could be applied to the extinction probabilities if desired. Applying each of these sets of constraints, at least the sets related to the hypotheses of interest, results in a different model that can be fit to the data and formally compared using the analyst's preferred method.

Another approach to evaluating these same questions would be to use the occupancy-only dynamics model (Chapter 8), where a set of covariates is defined in terms of the habitat state at the beginning and end of each period. Advantages of the more complex integrated approach include: (1) the ability to allow for missing habitat state values; and (2) the modeling of habitat dynamics as functions of environmental covariates, management actions, and even species occupancy state (as detailed in the next section), permitting prediction of future habitat states.

13.4.2   Effect of Species Presence on Habitat Dynamics

The habitat transition probability, $\eta_t^{[X_t, H_t, H_{t+1}]}$, is defined in Table 13.1 enabling different transition probabilities if the species is present at the unit in time t ($X_t = 1$) or absent from the unit ($X_t = 0$). This is a general case and the constraint $\eta_t^{[1, H_t, H_{t+1}]} = \eta_t^{[0, H_t, H_{t+1}]}$ could be made and different models fit to the data to evaluate the evidence for habitat dynamics being different when the species is present at the beginning of the period of change.

Biologically, the ability to allow habitat transition probabilities to be different depending on whether the species is present or absent from the unit in season t allows one to consider how a species may impact a landscape. For example, herbivore grazing can promote plant growth and nutrient cycling in ways that increase habitat quality (McNaughton, 1983; McNaughton et al., 1997), whereas overgrazing and seed dispersal of noxious plants by animals may promote habitat deterioration. Alternatively, apparent focal species effects may indicate preferential selection by the species of units with different dynamics (e.g., an amphibian species may occupy vernal pools that are less likely to dry during a summer).

13.5   SYSTEM SUMMARIES

While the motivation for this joint modeling was to provide greater flexibility and a better understanding of system dynamics, often the probability of a unit being in each possible state within a given season may be of interest as a summary of the overall system state (Miller et al., 2012a). As noted in Chapter 10, let φ_t be the row vector containing the probability of being in each state in season t. For season 1:

$$\varphi_1 = \phi_0,$$

and in subsequent seasons, φ_t can be derived recursively as:

$$\varphi_t = \varphi_{t-1}\phi_{t-1}.$$

Note the similarities between this expression and that typically used in abundance based age- or stage-structured population modeling (e.g., Caswell, 2001; Williams et al., 2002) to project population sizes forward in time based on initial population size in each age-class or stage (i.e., ϕ_0) and the transition matrix (i.e., ϕ_{t-1}) which contains demographic parameters such as fecundity and survival. As in abundance-based population modeling, this recursive equation can be used to predict the probability of a unit being in each state into the future beyond the survey period. This is particularly useful to predict both how the habitat may be likely to change in the future, and also the distribution of the species, taking into account any relationships in the dynamics of the two processes. One can also consider 'what if' scenarios by modifying the values of some dynamic parameters and examining the resulting change in the predictions.

Relevant summaries of φ_t can also be calculated. For example, system-wide occupancy in season t can be determined by summing the first and third elements of φ_t. Other summaries of possible interest could include habitat-specific occupancies. In fact ratios of summed habitat-specific occupancies (e.g., ratio of sum across sample units of element 1, to similar sum of element 3, of φ_t) provides one measure of the relative importance of the two habitats to the focal species in the system.

13.6 MODEL EXTENSIONS

So far in this chapter we have assumed a situation with two habitat states and two occupancy states, however as noted by MacKenzie et al. (2011), the model could be generalized to more complex situations with a greater number of states for either aspect of the system.

In many situations there is likely to be more than two habitat states of interest (e.g., successional growth at a unit, or water levels in a pond). Extension of the above modeling to greater than two habitat states simply requires additional possible combinations of habitat and occupancy states. For example, if there were four possible habitat states, and two occupancy states, there would be eight possible habitat–occupancy combinations. Similarly, if there were multiple states of occupancy (e.g., absent, present without breeding, present with breeding; or few, some, many individuals), the number of possible habitat–occupancy combinations would increase. However, as noted in Chapter 9, there will be a practical limit to the number of possible states determined by the amount of data available for each possible transition. While one may develop a biological argument in terms of four habitat states and three occupancy states, that would define a total of 12 states with 144 possible transitions. Therefore, a very large data set would be required to reliably estimate all of the associated parameters. In addition, there is the substantial scientific burden of developing hypotheses and corresponding competing models of the important sources of variation in each of these transition probabilities. The point of these reminders is not that such complex models should not be developed, but rather that their use be accompanied by serious efforts at data collection and modeling.

A particular case of multiple occupancy states is where states are defined in terms of the presence/absence of multiple species. For example, with two species, there could be four possible states (Chapter 14). Integrated modeling of multiple species and habitat type would not only allow the colonization and extinction probabilities of each species to be different in each habitat type, but also allow for the interaction between species to be different in different habitat types (Miller et al., 2012a). This would provide a formal framework for examining, and estimating parameters associated with, models that relate community dynamics to interspecific competition, habitat heterogeneity, and disturbance (e.g., Caswell and Cohen, 1991a, 1991b).

The above modeling assumes that habitat state of a unit can be judged without error each season. However, we can also envisage cases in which the habitat assessment is subject to misclassification. Possible situations might involve the presence or absence of some critical resource (e.g., food plant, host species) or perhaps a pathogen (disease agent; see McClintock et al., 2010c) at a sample unit. In such situations, detection of the critical resource (or pathogen) would permit unambiguous classification of the unit's habitat state. However, failure to detect the resource could result from either true absence of the resource, or resource presence but nondetection. For example, consider the following situation for the habitat types:

- $A =$ food plant is absent at the unit
- $B =$ food plant is present at the unit

If the food plant is observed at a unit, that is definitive and the habitat is known to be type B, whereas if the food plant is not observed, the true habitat type may be either A or B. This situation is easily handled by modifying the detection probability matrix. When the habitat state is known without error, $\mathbf{p}_{t,j}$ was:

$$
\mathbf{p}_{t,j} =
\begin{bmatrix}
1 & 0 & 0 & 0 \\
1 - p_{t,j}^{[A]} & p_{t,j}^{[A]} & 0 & 0 \\
0 & 0 & 1 & 0 \\
0 & 0 & 1 - p_{t,j}^{[B]} & p_{t,j}^{[B]}
\end{bmatrix}.
$$

Note the 0 entries in the lower-left portion of $\mathbf{p}_{t,j}$, which indicates it is not possible to observe habitat A (first two columns) when the true state is habitat B (last two rows). When the habitat state may be misclassified as above, $\mathbf{p}_{t,j}$ could be modified to:

$$
\mathbf{p}_{t,j} =
\begin{bmatrix}
1 & 0 & 0 & 0 \\
1 - p_{t,j}^{[A]} & p_{t,j}^{[A]} & 0 & 0 \\
1 - \delta_{t,j} & 0 & \delta_{t,j} & 0 \\
\left(1 - \delta_{t,j}\right)\left(1 - p_{t,j}^{[B]}\right) & \left(1 - \delta_{t,j}\right) p_{t,j}^{[B]} & \delta_{t,j}\left(1 - p_{t,j}^{[B]}\right) & \delta_{t,j} p_{t,j}^{[B]}
\end{bmatrix},
$$

where $\delta_{t,j}$ is the probability of detecting the food plant in survey j of season t. The habitat type would have to be assessed multiple times within a season for $\delta_{t,j}$ to be estimable. It would also be possible to modify the modeling for the more general case where habitat may be misclassified in either direction (e.g., with remote sensed data where vegetation type may be misclassified) by inclusion of appropriate misclassification probability parameters, but ancillary data would likely be needed for the parameters to be estimable (e.g., from ground-truthing surveys of a subset of sample units to verify the remote sensed classification; Veran et al., 2012).

Habitat and species occurrence transitions have been expressed as first-order Markov processes in which the probability of being in a particular state at time $t + 1$ depends on state at time t. While the probability may also depend on other factors, that can be included as covariates, the probability does not depend on the state at time $t - 1, t - 2, \ldots$. Such dependence on past states would suggest a higher-order Markov process. In some cases, vegetation succession following disturbance occurs in such a way that state transition probabilities are best described as depending on the number of years that the vegetation has been in the current state or the number of years since disturbance (e.g., Beckwith, 1954; Odum, 1960). In such situations, inclusion of a higher order Markov process may be worthwhile, although such models are likely to be relatively data hungry. Another possibility is to use a first order process (our model above), but redefine habitat state to include both vegetation class and time (e.g., years) the patch has spent in this class. Such models might be especially useful for species such as the Florida scrub jay (*Aphelocoma coerulescens*) that show the best demographic performance in transient habitats that are characterized by shrub height categories that vary as a function of time since last burning (Breininger and Carter, 2003).

One special case we have chosen to highlight is where the habitat may be regarded as suitable or unsuitable for the species, where the species may be present at a unit when the habitat is *suitable*, but must be absent from the unit when the habitat is *unsuitable*. Several early theoretical investigations of joint habitat–occupancy models (e.g., Lande, 1987, 1988) defined situations of this type. In the framework detailed by MacKenzie et al. (2011), this simply reduces the number of possible states to three: (1) unsuitable habitat and species absence; (2) suitable habitat and species absence; and (3) suitable habitat and species presence. By definition, some of the rate parameters underlying occupancy dynamics must also be constrained, i.e., $\epsilon_t^{[S,U]} = 1$ where S and U denote suitable and unsuitable habitat, respectively.

13.7 EXAMPLE

To illustrate these methods, we provide an example from MacKenzie et al. (2011) investigating changes in the suitability of ponds at Patuxent Research Refuge, Maryland, USA, for breeding activity by spotted salamanders (*Ambystoma maculatum*). Files associated with this analysis are available from http://www.proteus.co.nz.

13.7.1 Patuxent Spotted Salamanders

For many amphibian species, simply monitoring breeding activity may not adequately differentiate between 'source' and 'sink' locations, as some pools may support breeding activity, but eggs or larvae may rarely survive to metamorphosis (i.e., these pools routinely experience complete reproductive failure; Taylor et al., 2006). MacKenzie et al. (2011) explored the prevalence of reproductive failure and its potential influence on the occurrence of spotted salamanders at 56 pools sampled from 2006 to 2008 at Patuxent Research Refuge (PRR) in Maryland, USA. Pools were visited twice during the breeding period (late March–mid April) by two independent observers to detect breeding activity (presence of egg masses). Occasionally, a second visit during the breeding period was not conducted if egg masses were detected during the first visit. In early June, dip-net surveys were conducted by two independent observers to detect late-stage larval salamanders prior to metamorphosis (Worthington, 1968; Petranka, 1998).

A 'season' was defined to be each visit: early spring, late spring, and early summer. The two independent observers used on each visit were considered replicate surveys within a 'season'. Therefore, habitat and occupancy dynamics could be estimated for changes within each year (i.e., early–late spring, late spring–early summer) and also between years (i.e., from early summer to early spring the following year). During each visit, a pool's habitat was considered to be 'suitable' if there was standing water in the pool, or 'unsuitable' if the pool was dry (denoted as 'S' for suitable and 'U' for unsuitable, respectively; see Fig. 13.2). The three occupancy states considered by MacKenzie et al. (2011) were therefore: (1) pool is unsuitable (pool must be unoccupied); (2) pool suitable and no breeding; and (3) pool suitable and breeding.

MacKenzie et al. (2011) fit seven models to the data and compared them by AIC. For all models they assumed that detection did not vary among years, but that detection probability of egg masses during the early and late spring surveys would be higher than detection of larvae in the early summer survey (denoted as *Stage*). One observer was highly experienced and participated in nearly all pool visits, so an observer effect was also included for detection probability (denoted *Obs*). The observer effect was assumed to be different for the surveys conducted

FIGURE 13.2   A pond at Patuxent Research Refuge in a suitable (left) and unsuitable (right) state for spotted salamanders (*Ambystoma maculatum*).

at different stages of the breeding season, i.e., an interaction between *Stage* and *Obs* was assumed for detection, denoted as *p*(*Stage* × *Obs*).

Two sub-models were considered for habitat dynamics (changes in pool suitability). The first allowed the probability of pools being suitable in period $t + 1$ to be different for each of the three possible states the pool could be in at period t (i.e., unsuitable, suitable without breeding or suitable with breeding). That is, the probability of a pool being suitable in the next period depended upon its suitability in the current period and the presence of breeding salamanders. The first sub-model also had an additive time period effect that was equal for all pools, but was independent of climatic season and year. We shall denote this sub-model as $\eta(Hab + Br + t)$ where *Hab* indicates the habitat effect, *Br* the effect of breeding salamander presence, and t for time period (which differs from the notation used by MacKenzie et al., 2011). The second sub-model allowed the probability of a pool being suitable in period $t + 1$ to be different depending on both the state of the pool in period t and the climatic season of period t (i.e., early-spring, late-spring, early-summer; denoted *CS*). It was constant across all years otherwise. We shall denote this model as $\eta((Hab + Br) \times CS)$.

Three sub-models for colonization and extinction probabilities, jointly, were considered: (1) colonization and extinction probabilities within years were zero (i.e., which pools had breeding salamanders did not change between early-spring and early-summer), but were different for each year, which we denote as $\gamma(year)\,\epsilon(year)$; (2) colonization was allowed between the early- and late-spring survey periods each year, but extinction probability was constrained to zero within a year. Colonization and extinction could also occur between years, and all effects were different in each year, denoted $\gamma(Sp \times year)\,\epsilon(year)$, where *Sp* indicates the spring interval; and (3) colonization was allowed between the early- and late-spring survey periods each year, and extinction could occur between any consecutive survey period, denoted $\gamma(Sp \times year)\,\epsilon(CS \times year)$. Note that in all sub-models it was assumed there could be no colonization of suitable pools by breeding salamanders between the late-spring and early-summer because larvae presence in a pool in early-summer indicated there must have been egg masses present in late-spring.

MacKenzie et al. (2011) fit the six models that are combinations of these sets of sub-models, plus an additional model to verify that both habitat and occupancy state in June (early-summer) influenced the probability a pool supported breeding the following year. They specified a model using the $\eta(Hab + Br + t)$ component for habitat transitions and $\gamma(Sp \times year)\epsilon(CS \times year)$ for colonization and extinction probabilities, but with an added constraint that between years the colonization probability was set equal to persistence probability, the complement of extinction probability, i.e., $\gamma = 1 - \epsilon$. This constraint means that the probability of breeding at a suitable pond in early-spring is independent of whether breeding occurred at the suitable pond in the previous summer, i.e., which suitable ponds have breeding in early-spring is a random process (with respect to previous breeding; Chapter 10). If this model was well supported by the data, it would suggest that the primary requirement for determining which ponds support breeding is that the pool have suitable habitat. We have denoted this final model as $\eta(Hab + Br + t)\gamma(Sp \times year)\epsilon(CS \times year)^b$.

Fitting the models in Program PRESENCE, MacKenzie et al. (2011) found that the model $\eta(Hab + Br + t)\gamma(Sp \times year)\epsilon(CS \times year)$ had essentially all of the support when comparing models by AIC (Table 13.2), and presented the results from only this model. Detection probabilities where estimated to be high, with $\hat{p} > 0.90$ for surveys during the spring for egg masses, and $\hat{p} > 0.75$ for surveys during the summer for larvae, regardless of observer experience.

In the early-spring of 2006, the estimated probability of a pool being suitable was $\hat{\pi}^{[S]} = 0.54$ with an associated standard error of 0.07. The probability that a suitable pool contained breeding spotted salamanders in this period was $\hat{\psi}^{[S]} = 0.28$ (SE = 0.08), therefore the estimated probability of a randomly selected pool being suitable and containing breeding salamanders would be $0.54 \times 0.28 = 0.15$. The estimated habitat transition probabilities are given in Table 13.3, which indicate that pools tend to change from being unsuitable to suitable between the June and March surveys, i.e., over fall and winter months, and some become suitable between the early- and late-spring survey periods (March and April). There tends to be a very low probability of pools becoming suitable between April and June. In March of 2007 and 2008, all pools that were suitable in the preceding June were estimated to be again suitable. In all years, most pools that were suitable in March were still suitable in April. Pools were less likely to remain suitable (i.e., transitions $[S, S]$) between the late-spring and early-summer survey periods (between April and June), particularly

TABLE 13.2 Model selection summary for models fit to spotted salamander data from 56 pools at Patuxent Research Refuge, 2006–2008. Given are the relative difference in AIC values (ΔAIC), AIC model weight (w), number of estimated parameters in the model ($Npar$), and twice the negative log-likelihood ($-2l$)

Model	ΔAIC	w	$Npar$	$-2l$
$\eta(Hab + Br + t)$ $\gamma(Sp \times year)\,\epsilon(CS \times year)$	0.00	0.99	34	907.23
$\eta(Hab + Br + t)$ $\gamma(Sp \times year)\,\epsilon(CS \times year)^b$	9.97	0.01	30	925.20
$\eta(Hab + Br + t)$ $\gamma(Sp \times year)\,\epsilon(year)$	22.30	0.00	28	939.53
$\eta((Hab + Br) \times CS)$ $\gamma(Sp \times year)\,\epsilon(CS \times year)$	32.72	0.00	33	941.95
$\eta((Hab + Br) \times CS)$ $\gamma(Sp \times year)\,\epsilon(year)$	74.10	0.00	27	995.33
$\eta(Hab + Br + t)$ $\gamma(year)\,\epsilon(year)$	1143.32	0.00	22	2074.55
$\eta((Hab + Br) \times CS)$ $\gamma(year)\,\epsilon(year)$	1192.88	0.00	21	2126.11

Source: MacKenzie et al. (2011).

in 2006 and 2007, and suitable pools without breeding salamanders were estimated to have a lower probability of remaining suitable than sites that contained breeding salamanders (i.e., $\hat{\eta}_t^{[0,S,S]} < \hat{\eta}_t^{[1,S,S]}$). Biologically, these results make some sense, and the habitat dynamics are likely a reflection of rainfall events and the system's hydroperiod. An obvious extension of this analysis would be to consider such variables as covariates for the habitat transition probabilities. Note that the yearly variation in the probability that a suitable pool in April retained water until June has important ramifications for the amount of successful breeding each year (e.g., 'boom' and 'bust' years).

The estimated probabilities of a pool changing from having no breeding spotted salamanders in period t to having breeding in period $t+1$ (i.e., colonization probabilities) are given in Table 13.4. These estimates suggest that suitable pools may be colonized by breeding spotted salamanders both before early-spring (i.e., June to March), and between early- and late-spring (i.e., March to April) in some years. Spotted salamanders appeared to delay reproduction in 2006: the probability a suitable pool contained egg masses was only $\hat{\psi}^{[S]} = 0.28$ (SE $= 0.08$) in March, but unoccupied pools that became, or remained, suitable until April had colonization probabilities of $\hat{\gamma}^{[U,S]} = 0.62$ (SE $= 0.17$) and $\hat{\gamma}^{[S,S]} = 0.47$ (SE $= 0.17$), respectively. In 2007 and 2008, most salamanders had

TABLE 13.3 Estimates of habitat dynamic transition probabilities and associated standard errors (in parentheses) for seasonal pools at Patuxent Research Refuge. Standard errors are not given for boundary estimates. Columns are notated to indicate the interval (t to $t + 1$) to which estimates correspond, in terms of month (M = March, A = April, J = June) and year (2006–2008)

Parameter	M06–A06	A06–J06	J06–M07	M07–A07	A07–J07	J07–M08	M08–A08	A08–J08
$\hat{\eta}_t^{[0,U,S]}$	0.30 (0.10)	0.00 (–)	0.83 (0.06)	0.25 (0.18)	0.00 (–)	0.81 (0.06)	0.01 (0.01)	0.03 (0.03)
$\hat{\eta}_t^{[0,S,S]}$	0.99 (0.01)	0.16 (0.08)	1.00 (–)	0.99 (0.02)	0.02 (0.02)	1.00 (–)	0.66 (0.10)	0.86 (0.15)
$\hat{\eta}_t^{[1,S,S]}$	1.00 (–)	0.70 (0.09)	1.00 (–)	1.00 (–)	0.24 (0.12)	1.00 (–)	0.96 (0.03)	0.99 (0.02)

Source: MacKenzie et al. (2011).

TABLE 13.4 Estimated colonization probabilities and associated standard errors (in parentheses) for pools at Patuxent Research Refuge. Colonization probabilities are reported for previously unoccupied pools that were either dry (unsuitable, $\hat{\gamma}_t^{[U,S]}$) or wet (suitable, $\hat{\gamma}_t^{[S,S]}$) at time t. Columns are notated to indicate the interval (t to $t+1$) to which estimates correspond, in terms of month (M = March, A = April, J = June) and year (2006–2008)

Parameter	M06–A06	J06–M07	M07–A07	J07–M08	M08–A08
$\hat{\gamma}_t^{[U,S]}$	0.62	0.47	0.00	0.40	NA[†]
	(0.17)	(0.09)	(–)	(0.07)	
$\hat{\gamma}_t^{[S,S]}$	0.47	0.91	0.05	0.00	0.19
	(0.11)	(0.08)	(0.05)	(–)	(0.10)

[†] *Estimates were not possible because very few unsuitable pools became suitable.*
Source: MacKenzie et al. (2011).

bred by the first visit in early-spring, and there was little colonization between the March and April surveys, and only at pools that were suitable in March (Table 13.4). That is, a relatively large number of suitable pools had breeding begin between the early- and late-spring survey periods during 2006, but most suitable pools had breeding begin before March in the other years. Pools that were unsuitable (dry) in June of each year had approximately a 40–47% chance of being occupied by breeding adults at the beginning of the following breeding season, while colonization probabilities for unproductive (unoccupied by larvae) but suitable pools, $\gamma^{[S,S]}$, varied widely between years (Table 13.4).

MacKenzie et al. (2011) concluded that while habitat change accounted for much of the observed reproductive failure at sampled pools (between 5 and 75% of occupied pools became unsuitable by June, Table 13.3), salamander offspring also went locally extinct at pools that remained suitable throughout the entire breeding season. Extinction probabilities were low for pools with egg masses during the spring in two of the three years (Fig. 13.3), but extinction probability estimates at pools that remained suitable between April and June visits varied considerably among years ($\hat{\epsilon}_{April}^{[S,S]}$ range: 0.00–0.61, Fig. 13.3). Pools that did produce metamorphs had high probabilities of supporting salamander reproduction the following spring: $\hat{\eta}^{[1,S,S]} \times \left(1 - \hat{\epsilon}^{[S,S]}\right) = 0.84$ and 1.0, respectively, for pools with metamorphs in 2006 and 2007.

13.8   DISCUSSION

In this chapter we have described an integrated model of habitat and occupancy dynamics, drawing primarily from MacKenzie et al. (2011). A fundamental concept of this model is the ability to account for imperfect detection of the species,

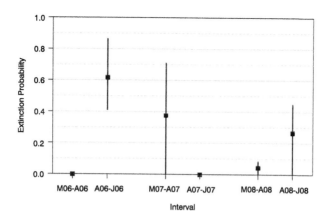

FIGURE 13.3 Estimated probability of local extinction of spotted salamander breeding from suitable habitats in Patuxent Research Refuge. Labels on the x-axis indicate the interval (t to $t + 1$) to which estimates correspond, in terms of month (M = March, A = April, J = June) and year (2006–2008). Error bars indicate approximately 95% confidence intervals; confidence intervals are not given for parameters that are near 0. *(Reproduced from: MacKenzie et al., 2011)*

which will often be a reality for many field studies. However we note that in those cases where detection is perfect, this integrated model can still be used (e.g., Martin et al., 2010). With perfect detection, the detection probability matrix **p** becomes an identity matrix (1's on the main diagonal, from top-left to bottom-right, and 0's elsewhere) which essentially removes detection from the model leaving only the dynamic parameters to be estimated. Clearly this would greatly simplify both the analysis and data requirements.

As with any maximum-likelihood based method, and particularly for the more complex models in this book, analysts need to be aware of the potential for multiple maxima on the likelihood surface. That is, the computer algorithms converge to a local maximum on the likelihood surface, but not to the *global* maximum which corresponds to our maximum likelihood estimates. Practitioners should be very aware of this point and investigate the stability of their estimates by rerunning analyses using alternative starting values for the algorithms, or consider using more general optimization routines such as simulated annealing to verify that the global maximum has been determined. In a Bayesian framework, similar behavior may translate to multi-modality of posterior distributions and may cause a lack of convergence when using Markov chain Monte Carlo procedures; chains may periodically 'jump' between different ranges of values that appear to be reasonable solutions, or different chains may initially appear to converge to different sets of values.

As noted in Chapters 11 and 12, it is important to consider habitat dynamics during the design phase of the study or monitoring program, particularly for longer-term studies. For example, when a region consists of suitable and un-suitable habitats for a species, selecting sampling units only from areas that are currently considered to be suitable will not necessarily result in a design that provides reliable information over time. Areas that are currently unsuitable may become suitable, and possibly occupied, in the future. Restricting the sampling only to areas that are suitable at the beginning of the program, will likely pro-vide a misleading view of how the system changes. In particular, based solely upon the sampled units, the proportion of suitable habitat will appear to have declined. A further aspect is that if it is recognized in the future that there have been changes in the habitat within the system, that may lead to discussions about modifying the sampling frame to accommodate those changes. Arguably a better approach is to sample from both suitable and unsuitable units from the beginning such that inferences about the complete system dynamics can be made. Another important design consideration is with respect to the definition of a sampling unit. Often habitat patches may appear to be discrete, leading to natural defini-tions of sample units, (e.g., pools within a wetland complex); however, because of system dynamics, these patches may merge or separate over time thus cre-ating confusion about unit definition. For example, three discrete pools in one year might naturally be regarded as sampling units, whereas in a wet year the pools may be merged and be defined as a single unit. In such situations it may be appropriate to instead define sampling units in terms of grid cells to ensure consistency through time, with an assessment of the habitat state in a cell each sampling season (e.g., wet or dry).

Chapter 14

Species Co-Occurrence

Chapters 4–10 (Parts II and III) have focused on the issues related to estimating and modeling occupancy patterns and dynamics for a single species. However often more than one species may be of primary interest, either in terms of studying relationships between multiple species, or investigating patterns and dynamics at the community level. As discussed in Chapter 2, there is a wide range of applications where occupancy-type thinking has been used to make inference about multiple species, with few attempts to formally incorporate detection probabilities, despite acknowledgments that they are a reality of many sampling situations. In this and the following chapter we outline how the occupancy modeling approaches, which explicitly incorporate the estimation of detection probability, can be applied to multi-species or community-level studies. This chapter deals with situations where a relatively small number of species is of interest and focuses on hypotheses about species interactions, whereas Chapter 15 focuses on entire communities with possibly large numbers of species, but does not explicitly deal with interspecific interactions. Many of the concepts and models discussed in the following chapters are the result of recent model development and did not appear in the previous edition of this book. These models are now incorporated into existing software (e.g., Programs PRESENCE and MARK, and the R packages RMark, RPresence, and unmarked), and we highlight a few applications that now appear in the literature.

In this chapter we turn our attention to the problem of studying relationships among the occurrence of multiple species from detection/nondetection data. Such approaches have a long history of use in ecology, particularly to identify nonrandom patterns in the species co-occurrence matrix (Chapter 2). A large body of literature has focused on appropriate formulations of the 'null model', associated test statistics, deriving the null distribution of the test statistics, and the resulting debate between proponents of the various concepts and ideas. However, often other factors (e.g., habitat preferences and physiological tolerances) may introduce nonrandom patterns to species incidence matrices that are unrelated to interspecific interactions. These could be incorporated into the 'null model' by identifying them *a priori*, although we prefer to think of such factors as hypotheses that one may wish to investigate (e.g., one may wish to estimate the magnitude of any such effect). A second facet of species co-occurrence data

Occupancy Estimation and Modeling. http://dx.doi.org/10.1016/B978-0-12-407197-1.00019-3
509

relevant to inferences about interspecific interactions is detectability. Very little attention had been paid to imperfect detection prior to the development of occupancy models (but see Cam et al., 2000a). By accounting for detection probability explicitly, absolute measures of co-occurrence (such as the probability that two species co-occur at a unit) can be estimated. When the intent is to compare the level of co-occurrence at multiple points in time or space, stronger inference about changes or differences can be made when detection probabilities have been incorporated into the inferential procedure. The modeling procedure detailed below can be viewed as an approach that unifies these two concepts of incidence or occurrence of different species and detection probability, within a single modeling framework for investigating species co-occurrence patterns at a single point in time. However, in tune with our arguments throughout this book, the processes underlying co-occurrence cannot be reliably inferred from observed patterns. Dynamic models that examine how co-occurrence patterns change over time when data are available from multiple sampling seasons allow stronger inference to be made about the underlying processes. Details are also given in this chapter of such models, which have been used to explore the influence of invasive species (e.g., Dugger et al., 2015), predators/competitors (e.g., Miller et al., 2012a; Robinson et al., 2014), or pathogens (e.g., Mosher et al., 2017) on species distributions while accounting for local habitat variables.

14.1 DETECTION PROBABILITY AND INFERENCES ABOUT SPECIES CO-OCCURRENCE

Imperfectly detecting species will result in misleading inferences about species co-occurrence patterns. Traditionally, the nondetection of a species at a location is interpreted as an absence in the species co-occurrence matrix. However, if some of those absences actually reflect 'false absences' (i.e., species present but undetected), both the value, and the null distribution, of the test statistic are incorrect (in comparison to the 'true' situation with respect to the presence of the species). That is, the null distribution of the test statistic for the observed data (detection/nondetection) is different than the null distribution for the true situation (presence/absence). What effect, then, would imperfect detection of the species have on our resulting inferences?

To address this, consider the simple case where the potential for interaction between two species is investigated using a simple 2×2 contingency table (e.g., Forbes, 1907; Dice, 1945; Cole, 1949; Pielou, 1977; Hayek, 1994). In Table 14.1 we give the general probability for each of the four possible outcomes for whether the two species are present at a unit where ψ^{AB} is the probability that

TABLE 14.1 Probability structure for the possible co-occurrence patterns of two species, where ψ^{AB} is the probability both species are present at a unit, ψ^A is the overall probability of species A presence at a unit, and ψ^B is the overall probability of species B presence at a unit. The structure is also expressed in terms of an alternative parameterization where ψ^{A0} is the probability of only species A present at a unit, and ψ^{0B} is the probability of only species B being present at a unit. Colors of the cells correspond to the shaded regions in the Venn diagram (Fig. 14.1)

		Species B		
		Present	*Absent*	*Overall*
Species A	*Present*	ψ^{AB}	$\psi^A - \psi^{AB}$	ψ^A
	Absent	$\psi^B - \psi^{AB}$	$1 - \psi^A - \psi^B + \psi^{AB}$	$1 - \psi^A$
	Overall	ψ^B	$1 - \psi^B$	1

		Species B		
		Present	*Absent*	*Overall*
Species A	*Present*	ψ^{AB}	ψ^{A0}	ψ^A
	Absent	ψ^{0B}	$1 - \psi^{A0} - \psi^{0B} - \psi^{AB}$	$1 - \psi^A$
	Overall	ψ^B	$1 - \psi^B$	1

both species A and B are present at a unit (i.e., the probability of co-occurrence); ψ^A is the overall probability that species A occupies a unit regardless of the presence of species B (hence $\psi^A - \psi^{AB} = \psi^{A0}$ is the probability of only species A at a unit); ψ^B is the overall probability that species B occupies a unit regardless of the presence of species A (hence $\psi^B - \psi^{AB} = \psi^{0B}$ is the probability of only species B at a unit); and $1 - \psi^A - \psi^B + \psi^{AB}$ is the probability both species are absent. These are also graphically presented in Fig. 14.1. Note that the probabilities given here are general in that they do not assume the species occur independently of one another. From this contingency table, a χ^2 test could be used to determine whether the rows and columns of the table (i.e., the presence or absence of each species) are independent. Alternatively, an odds ratio could be calculated to estimate the magnitude of any dependence between the rows and columns, which is the approach we take here (recall the odds ratio was discussed in Section 3.4.1 with respect to the logit link).

The odds that species B is present at a unit given that species A is also present is:

$$odds_{B|A} = \psi^{AB}/\left(\psi^A - \psi^{AB}\right)$$
$$= \psi^{AB}/\psi^{A0}, \tag{14.1}$$

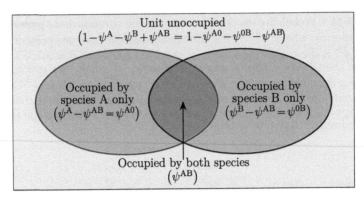

FIGURE 14.1   Venn diagram representing the probability for each occupancy state. The full red ellipse represents the probability that the unit is occupied by species A irrespective of species B, ψ^A, and the blue ellipse represents the probability that the unit is occupied by species B irrespective of species A, ψ^B. The intersection of the two ellipses is the probability that the unit is occupied by both species, ψ^{AB}. The four shaded regions (gray, red, blue, and purple) indicate the probability of the four possible occupancy states of a unit.

whereas the odds species B is present given species A is absent is (where lower case 'a' indicates absence of species A):

$$odds_{B|a} = \left(\psi^B - \psi^{AB}\right)/\left(1 - \psi^A - \psi^B + \psi^{AB}\right)$$
$$= \psi^{0B}/\left(1 - \psi^{A0} - \psi^{0B} - \psi^{AB}\right). \tag{14.2}$$

The odds ratio (*OR*) will therefore be:

$$OR = \frac{odds_{B|A}}{odds_{B|a}}$$
$$= \frac{\psi^{AB}/\left(\psi^A - \psi^{AB}\right)}{\left(\psi^B - \psi^{AB}\right)/\left(1 - \psi^A - \psi^B + \psi^{AB}\right)}$$
$$= \frac{\psi^{AB}\left(1 - \psi^A - \psi^B + \psi^{AB}\right)}{\left(\psi^A - \psi^{AB}\right)\left(\psi^B - \psi^{AB}\right)}$$
$$= \frac{\psi^{AB}\left(1 - \psi^{A0} - \psi^{0B} - \psi^{AB}\right)}{\psi^{A0}\psi^{0B}}. \tag{14.3}$$

OR will have the same form regardless of whether the odds are defined in terms of species B being present (as above) or alternatively in terms of species A.

If the two species occur independently then $OR = 1$. Under independence, the probability of both species occurring at a unit will be the product of each of the overall or marginal occurrence probabilities, i.e., $\psi^{AB} = \psi^A \times \psi^B$. By making the appropriate substitutions into Eqs. (14.1) and (14.2) and applying some algebra, under independence $odds_{B|A}$ reduces to $\psi^B/(1 - \psi^B)$ and $odds_{B|a}$ also reduces to $\psi^B/(1 - \psi^B)$. That is, if the two species occur independently, the presence or absence of species A has no effect on the odds (and therefore the probability) of species B occurring at a unit. *OR* values less than one indicate that the species have a smaller probability of co-occurring than expected under a hypothesis of independence, and values greater than one indicate a higher probability of co-occurring.

Now consider Table 14.2 which is the probability structure for a 2×2 contingency table of whether each species was *found* or not (i.e., species present and detected; detection/nondetection data, not presence/absence). Here the odds ratio is:

$$
\begin{aligned}
OR' &= \frac{\psi^{AB} p^A p^B \left(1 - \psi^A p^A - \psi^B p^B + \psi^{AB} p^A p^B\right)}{(\psi^A p^A - \psi^{AB} p^A p^B)(\psi^B p^B - \psi^{AB} p^A p^B)} \\
&= \frac{\psi^{AB} p^A p^B \left(1 - \psi^{A0} p^A - \psi^{0B} p^B - \psi^{AB} \left(p^A + p^B - p^A p^B\right)\right)}{\left(\psi^{A0} p^A + \psi^{AB} p^A (1 - p^B)\right)\left(\psi^{0B} p^B + \psi^{AB}(1 - p^A) p^B\right)}
\end{aligned}
\tag{14.4}
$$

where p^A and p^B are the probabilities of detecting each species in the survey. Clearly Eqs. (14.3) and (14.4) are not equivalent. While we do not offer any definitive proof here, when species are detected imperfectly, the odds ratio indicates the correct direction of the relationship (i.e., less than, or greater than, 1.0), but the magnitude of the relationship is underestimated (i.e., always estimated closer 1.0 than it should be). That is, if at least one of the species is detected imperfectly, when using a contingency table approach that does not account for detection probabilities, the estimated level of co-occurrence will underestimate the level of the interaction.

For example, suppose within a region the probability of species A being present at a unit (ψ^A) is 0.6 and the probability of species B being present (ψ^B) is 0.3. Further, suppose that the two species are competitors, hence do not tend to co-occur very frequently, and the probability of both species being present at a unit (ψ^{AB}) is 0.1 (therefore $\psi^{A0} = 0.5$ and $\psi^{0B} = 0.2$). Note if they occupied units independently then $\psi^{AB} = \psi^A \times \psi^B = 0.6 \times 0.3 = 0.18$. From Eq. (14.3) the true odds ratio would be:

$$
OR = \frac{0.1 \times (1 - 0.5 - 0.2 - 0.1)}{0.5 \times 0.2} = \frac{0.1 \times 0.2}{0.5 \times 0.2} = 0.2.
$$

Now suppose that in a survey for the species at a unit, both are detected imperfectly and the probability of detecting species A (p^A) is 0.4 and the probability of detecting species B (p^B) is 0.7. Then, to calculate the odds ratio for finding

TABLE 14.2 The probability of *finding* species A and/or species B given both may be detected imperfectly, where ψ^{AB} is the probability both species are present at a unit, ψ^A is the overall probability of species A presence at a unit, and ψ^B is the overall probability of species B presence at a unit. The structure is also expressed in terms of an alternative parameterization where ψ^{A0} is the probability of only species A present at a unit, and ψ^{0B} is the probability of only species B being present at a unit. The probability of detecting species A in a survey is p^A and the probability of detecting species B in a survey is p^B

		Species B		
		Found	*Not Found*	*Overall*
Species A	Found	$\psi^{AB} p^A p^B$	$\psi^A p^A - \psi^{AB} p^A p^B$	$\psi^A p^A$
	Not Found	$\psi^B p^B - \psi^{AB} p^A p^B$	$1 - \psi^A p^A - \psi^B p^B + \psi^{AB} p^A p^B$	$1 - \psi^A p^A$
	Overall	$\psi^B p^B$	$1 - \psi^B p^B$	1

		Species B		
		Found	*Not Found*	*Overall*
Species A	Found	$\psi^{AB} p^A p^B$	$\psi^{A0} p^A + \psi^{AB} p^A \left(1 - p^B\right)$	$\psi^A p^A$
	Not Found	$\psi^{0B} p^B + \psi^{AB}\left(1 - p^A\right) p^B$	$1 - \psi^{A0} p^A - \psi^{0B} p^B - \psi^{AB}\left(p^A + p^B - p^A p^B\right)$	$1 - \psi^A p^A$
	Overall	$\psi^B p^B$	$1 - \psi^B p^B$	1

the species from Eq. (14.4):

$$\psi^{AB} p^A p^B = 0.1 \times 0.4 \times 0.7$$
$$= 0.028,$$

$$1 - \psi^A p^A - \psi^B p^B + \psi^{AB} p^A p^B = 1 - 0.6 \times 0.4 - 0.3 \times 0.7$$
$$+ 0.1 \times 0.4 \times 0.7$$
$$= 1 - 0.24 - 0.21 + 0.028$$
$$= 0.578,$$

$$\psi^A p^A - \psi^{AB} p^A p^B = 0.6 \times 0.4 - 0.1 \times 0.4 \times 0.7$$
$$= 0.24 - 0.028$$
$$= 0.212,$$

$$\psi^B p^B - \psi^{AB} p^A p^B = 0.3 \times 0.7 - 0.1 \times 0.4 \times 0.7$$
$$= 0.21 - 0.028$$
$$= 0.182,$$

and

$$OR' = \frac{0.028 \times 0.578}{0.212 \times 0.182} = 0.419.$$

Hence, while we might correctly conclude some form of avoidance between the two species (as $OR' < 1.0$), the level of competition or exclusion appears to be less severe than it actually is (OR' is closer to 1.0 than the true OR).

Now consider the more general and realistic case where the probability of detecting each species depends upon whether the other species is also present at the same unit (Table 14.3). The odds ratio in this situation could be expressed as:

$$OR'' = \frac{\psi^{AB} r^A r^B \left(1 - \psi^{A0} p^A - \psi^{B0} p^B - \psi^{AB} \left(r^A + r^B - r^A r^B\right)\right)}{(\psi^{A0} p^A + \psi^{AB} r^A (1 - r^B))(\psi^{B0} p^B + \psi^{AB} (1 - r^A) r^B)}$$

where r^A and r^B are the probabilities of detecting each species, given that both species are present at a unit, and p^A and p^B are the detection probabilities given that only the one species is present. It can be shown (by considering simple examples) that the odds ratio calculated from detection/nondetection data can provide completely misleading information with regards to the actual co-occurrence of species. That is, the odds ratio may not indicate the correct direction of any relationship, or may even indicate a spurious relationship, in terms of occupancy and co-occurrence, depending on how detection probability is affected by the presence of the other species.

TABLE 14.3 The probability of *detecting* species A and/or species B given both may be detected imperfectly, where the detection of one species depends upon whether only one or both species are present at a unit. ψ^A is the overall probability of species A presence at a unit, and ψ^B is the overall probability of species B presence at a unit. The structure is also expressed in terms of an alternative parameterization where ψ^{A0} is the probability of only species A present at a unit, and ψ^{0B} is the probability of only species B being present at a unit. The probability of detecting species A in a survey, given only species A is present, is p^A, the probability of detecting species A in a survey, given both species A and B are present, is r^A, the probability of detecting species B in a survey, given only species B is present, is p^B, and the probability of detecting species B in a survey, given both species A and B are present, is r^B

		Species B		
		Found	*Not Found*	*Overall*
Species A	*Found*	$\psi^{AB} r^A r^B$	$(\psi^A - \psi^{AB}) p^A + \psi^{AB} r^A (1 - r^B)$	$\psi^A p^A - \psi^{AB}(p^A - r^A)$
	Not Found	$(\psi^B - \psi^{AB}) p^B + \psi^{AB}(1 - r^A) r^B$	$1 - (\psi^A - \psi^{AB}) p^A - (\psi^B - \psi^{AB}) p^B - \psi^{AB}(r^A + r^B - r^A r^B)$	$1 - \left(\psi^A p^A - \psi^{AB}(p^A - r^A) \right)$
	Overall	$\psi^B p^B - \psi^{AB}(p^B - r^B)$	$1 - \left(\psi^B p^B - \psi^{AB}(p^B - r^B) \right)$	1

		Species B		
		Found	*Not Found*	*Overall*
Species A	*Found*	$\psi^{AB} r^A r^B$	$\psi^{A0} p^A + \psi^{AB} r^A (1 - r^B)$	$\psi^{A0} p^A + \psi^{AB} r^A$
	Not Found	$\psi^{0B} p^B + \psi^{AB}(1 - r^A) r^B$	$1 - \psi^{A0} p^A - \psi^{0B} p^B - \psi^{AB}(r^A + r^B - r^A r^B)$	$1 - (\psi^{A0} p^A + \psi^{AB} r^A)$
	Overall	$\psi^{B0} p^B + \psi^{AB} r^B$	$1 - (\psi^{B0} p^B + \psi^{AB} r^B)$	1

14.2 A SINGLE-SEASON MODEL

MacKenzie et al. (2004a) extended the single-species model of MacKenzie et al. (2002) as a means of incorporating detection probability into inferences about co-occurrence patterns of multiple species. They noted that the model could be applied to any number of species in theory, but that due to the large number of parameters required, typically it would be most practical to model only a small number of species ($\leqslant$ three), and they presented their model for the two-species case. For a larger number of species, the majority of the parameters would be required to model high-order interactions (in the statistical sense) among a number of species (e.g., interactions among four or more species). In many real-life situations we imagine there would be insufficient data to estimate such high-order interactions reliably, and even if there were, such interactions could be very difficult to interpret. Therefore, most models and applications reduce the number of parameters and only model interactions between two or three species (e.g., Richmond et al., 2010; Waddle et al., 2010; Bailey et al., 2009). For simplicity, here we present only the two-species models initially developed by MacKenzie et al. (2004a), then reparameterized by Richmond et al. (2010) and Waddle et al. (2010), but end by highlighting recent extensions to include > three species.

14.2.1 General Sampling Situation

The general sampling situation envisaged here is very similar to the single-species case, although now data are being collected on more than one species at each unit. Multiple (not necessarily an equal number of) surveys are conducted at s units to detect the target species, with the intent of providing a snapshot of the system with respect to species co-occurrence patterns. For the duration of the surveying, units are closed to changes in the occupancy state with respect to each species (i.e., a species is either always present, or always absent from the unit over the surveying period), or changes in occupancy are completely at random for each species.

A detection history for each species at a unit can be recorded to denote the sequence of detections and nondetections at that unit. For example, the detection history $\mathbf{h}_i^A = 101$ represents that unit i was surveyed on three occasions, with species A being detected only in the first and third surveys. While the detection history $\mathbf{h}_i^B = 001$ would represent that species B was only detected in the third survey of unit i. Rather than recording separate detection histories for each species, an alternative would be to use a different system to code the outcome of each survey and use a single detection history. That is, the outcome of each survey is coded as:

- 0 if neither species detected,
- 1 if only species A detected,
- 2 if only species B detected,
- 3 if species A and B detected.

The information in the above pair of detection histories could then be recoded as $\mathbf{h}_i = 103$, i.e., only species A detected in survey 1, neither species detected in survey 2 and both species detected in survey 3.

14.2.2 Statistical Model

As already noted, there are multiple parameterizations that have been developed for the two-species co-occurrence model. Therefore, before describing the statistical model used to analyze the data, we shall provide some explanation of how the various quantities relate to one another. Conceptually, many of the ideas presented here are very similar to those already discussed with respect to the multi-state occupancy models (Chapters 5 and 9) in terms of the multinomial and conditional binomial parameterizations. Consider, again, the Venn diagram in Fig. 14.1, noting the four shaded regions (gray, red, blue, and purple), representing the probability of a unit being in each of the four possible states. Essentially, the different parameterizations are different ways of describing how the sizes of the shaded areas relate to each other.

As indicated in Fig. 14.1, one option is to parameterize the model in terms of ψ^A, ψ^B, and ψ^{AB}. ψ^A and ψ^B are the sizes of the red and blue ellipses, respectively, relative to the outer rectangle, and ψ^{AB} is the size of the overlap of the ellipses (shaded purple) relative to the rectangle. When the level of co-occurrence is greater, the size of the purple area will be larger, and under an independence assumption, the size of the overlap will be completely determined by the sizes of the two ellipses ($\psi^{AB} = \psi^A \times \psi^B$). The sizes of the remaining shaded areas are obtained by subtraction, e.g., the size of the red-shaded area is the size of the red ellipse minus the purple overlap, i.e., $\psi^A - \psi^{AB}$. Alternatively, rather than estimating the size of the shaded areas by subtraction, they could be estimated directly, which would be a multinomial parameterization. This is also indicated in Fig. 14.1 with the probabilities $1 - \psi^{A0} - \psi^{0B} - \psi^{AB}$, ψ^{A0}, ψ^{0B}, and ψ^{AB} for the gray-, red-, blue-, and purple-shaded areas, respectively.

Another option is to describe the size of the areas relative to other areas instead of relative to the size of the rectangle. For example, the size of the purple area relative to the size of the red ellipse, i.e., what fraction of the red ellipse is shaded purple. By doing so, the probabilities become conditional upon the event represented by the comparative area, e.g., the red ellipse represents species A being present, therefore the above example represents the probability of both species being present at a unit conditional upon species A being present, or the probability of species B occupying a unit given species A occupies the unit. Mathematically, this is:

$$\psi^{B|A} = \frac{\psi^{AB}}{\psi^{A}}.$$

Similarly, we could focus on the fraction of the area *outside* of the red ellipse that is blue, i.e., the probability of species B occupying a unit given species A is absent. Mathematically:

$$\psi^{B|a} = \frac{\psi^{B} - \psi^{AB}}{1 - \psi^{A}}$$

$$= \frac{\psi^{0B}}{1 - \psi^{A}},$$

where the lowercase 'a' denotes species A is absent. This suggests that the four shaded areas could be described in terms of, and the model parameterized with, ψ^{A}, $\psi^{B|A}$, and $\psi^{B|a}$. This is a conditional-binomial parameterization. Note that under independence, $\psi^{B|A} = \psi^{B|a}$, i.e., the probability of species B being present at a unit is unaffected by the presence of species A.

While the above descriptions are in terms of species occurrence, a similar situation holds for species detection when both species are present at a unit and there are more than two outcomes from a survey, e.g., detection of none, either, or both species.

MacKenzie et al. (2004a) defined the single-season, or static, co-occurrence model in terms of ψ^{A}, ψ^{B}, and ψ^{AB}. Rather than estimate ψ^{AB} directly, MacKenzie et al. (2004a) suggested estimation of a species interaction factor (SIF) they defined as:

$$\varphi = \frac{\psi^{AB}}{\psi^{A} \times \psi^{B}},$$

which may be unit-specific with the inclusion of covariates on one or more parameters. With this parameterization there are limits on the allowable values that φ may take, which depend upon the probability of occupancy for each species. This can cause some numerical difficulties in obtaining parameter estimates, and other undesirable behavior, particularly when covariates (e.g., habitat features and physiological tolerances) are incorporated into the model (see MacKenzie et al., 2004a, 2006, for details), and generally we recommend against the parameterization used by MacKenzie et al. (2004a). We provide further discussion later in the chapter. Two alternative conditional-binomial parameterizations were independently proposed (Richmond et al., 2010; Waddle et al., 2010) and the more general of the two approaches is presented here. This conditional-binomial parameterization does imply that there is a dominant and a subordinate species (Richmond et al., 2010; Waddle et al., 2010), and estimates the probability of

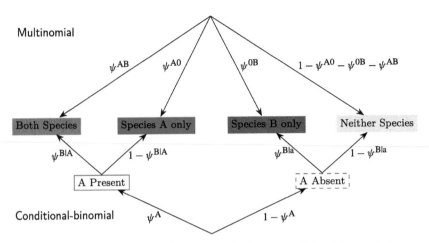

FIGURE 14.2 Diagram illustrating the multinomial and conditional-binomial parameterizations to describe the probability structure for the occupancy state of a unit, as defined by the presence or absence of two species, A and B. See text for parameter definitions. Note that $\psi^{A0} = \psi^A - \psi^{AB}$ and $\psi^{OB} = \psi^B - \psi^{AB}$.

occupancy of the subordinate species (species B) conditional on the presence, or absence, of the dominant species (species A) at a unit (Fig. 14.2). We refer to this parameterization of the two-species co-occurrence model as the RW model or RW parameterization.

Table 14.4 contains the parameters defined by Richmond et al. (2010), although our notation is slightly different, including three occupancy parameters: ψ^A, the unconditional probability that the dominant species (A) is present and two conditional probabilities, $\psi^{B|A}$, the probability that the subordinate species (B) is present, given species A is also present, and $\psi^{B|a}$, the probability that the subordinate species (B) is present, given species A is absent. Following MacKenzie et al. (2004a), Richmond et al. (2010) also defined different detection probabilities for the cases where only one of the focal species or both species are present at a unit (p and r, respectively). Note that when two species are present, a conditional-binomial parameterization was used for the detection process (i.e., probability of detecting species A, then probability of detecting species B given the detection or nondetection of species A in the same survey; Fig. 14.3). This results in an occupancy model that is very general and permits the possibility that detection probability of one species depends on whether the unit is occupied by the other species (e.g., as is hypothesized by U.S. biologists for northern spotted owls and barred owls in the Pacific northwest; see Chapter 2), and also that the detections of each species in the same may not be independent (e.g., probability of detecting species B may be lower if species A was detected in the survey, as might be expected with predator and prey species).

TABLE 14.4 Notation for the parameters used in the single-season, two-species model

Parameter	Description
ψ^A	Probability of occupancy for species A
$\psi^{B|A}$	Probability of occupancy for species B, given species A is present
$\psi^{B|a}$	Probability of occupancy for species B, given species A is absent
p_j^A	Probability of detecting species A during the jth survey, given only species A is present
p_j^B	Probability of detecting species B during the jth survey, given only species B is present
r_j^A	Probability of detecting species A during the jth survey, given both species are present
$r_j^{B|A}$	Probability of detecting species B during the jth survey, given both species are present and species A is also detected during the survey
$r_j^{B|a}$	Probability of detecting species B during the jth survey, given both species are present and species A is not detected during the survey

Source: Richmond et al. (2010).

Given the parameters defined in Table 14.4, a probability statement for the pair of species-specific detection histories observed at unit i, to be used in the observed data likelihood function, could be constructed in the same manner as for the modeling detailed in previous chapters. Namely, create a verbal description of the processes that may have resulted in a particular combination of detection histories being observed, and then translate the description into a mathematical equation involving the defined model parameters. The resultant mathematical equation is the probability of observing the combination of detection histories for the two species at that unit. For example, suppose the following pair of detection histories was observed at unit i, $\mathbf{h}_i^A = 110$ and $\mathbf{h}_i^B = 000$. A verbal description of the events giving rise to these detection histories would be:

Species A is present, was detected in surveys 1 and 2, and not detected in survey 3, and species B was also present (given A was present), but not detected in any of the surveys,

or,

only species A is present at the unit, with it being detected in surveys 1 and 2, but not in survey 3.

Translating this description into a mathematical equation using the model parameters (generically denoted as $\mathbf{\theta}$) gives the probability of observing these

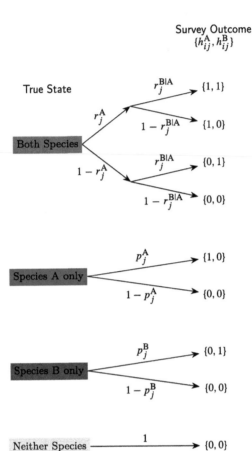

FIGURE 14.3 Diagram representing the detection probability structure of the two-species co-occurrence model of Richmond et al. (2010). The outcome of survey j can be represented as the $\{h_{ij}^A, h_{ij}^B\}$ pair, indicating detection/nondetection of species A and species B, respectively, and is conditional upon the true occupancy state. Parameters are defined in Table 14.4.

detection histories as:

$$Pr(\mathbf{h}_i^A = 110, \mathbf{h}_i^B = 000|\boldsymbol{\theta})$$
$$= \psi^A r_1^A r_2^A (1 - r_3^A) \psi^{B|A} (1 - r_1^{B|A})(1 - r_2^{B|A})(1 - r_3^{B|A})$$
$$+ \psi^A (1 - \psi^{B|A}) p_1^A p_2^A (1 - p_3^A).$$

More generally, two-species co-occurrence investigations can be cast as a special case of a multi-state problem (Chapter 5). A two-species system consists of four, mutually exclusive states (where the state is denoted as m; more generally there are 2^n possible states for n species): (1) neither species ($m = 0$); (2) occupied by species A only ($m = 1$); (3) occupied by species B only ($m = 2$);

or (4) occupied by both species A and B ($m = 3$). Define the row vector ϕ_0 that denotes the probability of a unit being in each of the four states, immediately before the unit is surveyed during the sampling season:

$$\phi_0 = \begin{bmatrix} \left(1 - \psi^A\right)\left(1 - \psi^{B|a}\right) & \psi^A\left(1 - \psi^{B|A}\right) & \left(1 - \psi^A\right)\psi^{B|a} & \psi^A\psi^{B|A} \end{bmatrix}.$$

As noted above, the outcome of survey j could be defined as having one of four possible values (l): neither species detected ($l = 0$), only species A detected ($l = 1$), only species B detected ($l = 2$), or both species detected ($l = 3$). The detection probability matrix $\mathbf{p}_j$ could be defined as:

$$\mathbf{p}_j = \begin{bmatrix} 1 & 0 & 0 & 0 \\ 1 - p_j^A & p_j^A & 0 & 0 \\ 1 - p_j^B & 0 & p_j^B & 0 \\ \left(1 - r_j^A\right)\left(1 - r_j^{B|a}\right) & r_j^A\left(1 - r_j^{B|A}\right) & \left(1 - r_j^A\right)r_j^{B|a} & r_j^A r_j^{B|A} \end{bmatrix},$$

where rows represent the true state, m, and columns the observations, l. As outlined in Chapter 5, let $\mathbf{p}_j^{[\bullet, l]}$ denote the column of $\mathbf{p}_j$ associated with observation l, and $\mathbf{p}_h$ be a column vector where each element contains the probability of observing the detection histories for the two species, given the true occupancy state of the unit. This is formed by taking the element-wise product of the respective $\mathbf{p}_j^{[\bullet, l]}$'s for each survey. For example, consider the pair of detection histories $\mathbf{h}_i^A = 110$ and $\mathbf{h}_i^B = 000$, which we shall re-express as the single history $\mathbf{h}_i = 110$ using the above 0–3 coding, representing that only species A was detected in surveys 1 and 2, and neither species in survey 3. Then:

$$\mathbf{P}_{110} = \mathbf{p}_1^{[\bullet, 1]} \odot \mathbf{p}_2^{[\bullet, 1]} \odot \mathbf{p}_3^{[\bullet, 0]}$$

$$= \begin{bmatrix} 0 \\ p_1^A \\ 0 \\ r_1^A\left(1 - r_1^{B|A}\right) \end{bmatrix} \odot \begin{bmatrix} 0 \\ p_2^A \\ 0 \\ r_2^A\left(1 - r_2^{B|A}\right) \end{bmatrix} \odot \begin{bmatrix} 1 \\ 1 - p_3^A \\ 1 - p_3^B \\ \left(1 - r_3^A\right)\left(1 - r_3^{B|a}\right) \end{bmatrix}$$

$$= \begin{bmatrix} 0 \\ p_1^A p_2^A\left(1 - p_3^A\right) \\ 0 \\ r_1^A\left(1 - r_1^{B|A}\right)r_2^A\left(1 - r_2^{B|A}\right)\left(1 - r_3^A\right)\left(1 - r_3^{B|a}\right) \end{bmatrix}.$$

That is, if only species A was present at the unit (second row), the probability of observing $\mathbf{h}_i = 110$ (or represented as the pair of histories $\mathbf{h}_i^A = 110$ and $\mathbf{h}_i^B = 000$) is $p_1^A p_2^A\left(1 - p_3^A\right)$. If both species were present at the unit (fourth row), then the probability is $r_1^A\left(1 - r_1^{B|A}\right)r_2^A\left(1 - r_2^{B|A}\right)\left(1 - r_3^A\right)\left(1 - r_3^{B|a}\right)$. Note that the first and third rows of $\mathbf{p}_{110}$ are zero. Respectively, these elements relate to the states where neither species was present, or only species B was present. As species A was detected during the surveys, which would be precluded if the unit was in either of these two states, the probability of observing these detection histories for the two species must be zero for these states.

Now consider a second example pair of histories $\mathbf{h}_i^A = 100$ and $\mathbf{h}_i^B = 001$, or $\mathbf{h}_i = 102$ using the above 0–3 coding, representing that only species A was detected in survey 1, neither species in survey 2, and only species B in survey 3. Then:

$$\mathbf{P}_{102} = \mathbf{p}_1^{[\bullet,1]} \odot \mathbf{p}_2^{[\bullet,0]} \odot \mathbf{p}_3^{[\bullet,2]}$$

$$= \begin{bmatrix} 0 \\ p_1^A \\ 0 \\ r_1^A\left(1 - r_1^{B|A}\right) \end{bmatrix} \odot \begin{bmatrix} 1 \\ 1 - p_2^A \\ 1 - p_2^B \\ \left(1 - r_2^A\right)\left(1 - r_2^{B|a}\right) \end{bmatrix} \odot \begin{bmatrix} 0 \\ 0 \\ p_3^B \\ \left(1 - r_3^A\right)r_2^{B|a} \end{bmatrix}$$

$$= \begin{bmatrix} 0 \\ 0 \\ 0 \\ r_1^A\left(1 - r_1^{B|A}\right)\left(1 - r_2^A\right)\left(1 - r_2^{B|a}\right)\left(1 - r_3^A\right)r_2^{B|a} \end{bmatrix}.$$

In this example, each species was detected at least once at the unit during the season, hence both species must be present so the final element is the only nonzero term.

Using the single-season multi-state framework, the probability statement for any pair of detection histories can then be calculated as:

$$Pr\left(\mathbf{h}_i^A, \mathbf{h}_i^B | \theta\right) = Pr(\mathbf{h}_i | \theta) = \phi_0 \mathbf{P_h},$$

for example:

$$Pr\left(\mathbf{h}_i^A = 110, \mathbf{h}_i^B = 000 | \theta\right)$$
$$= Pr(\mathbf{h}_i = 110 | \theta)$$
$$= \begin{bmatrix} \left(1 - \psi^A\right)\left(1 - \psi^{B|a}\right) & \psi^A\left(1 - \psi^{B|A}\right) & \left(1 - \psi^A\right)\psi^{B|a} & \psi^A\psi^{B|A} \end{bmatrix}$$
$$\times \begin{bmatrix} 0 \\ p_1^A p_2^A\left(1 - p_3^A\right) \\ 0 \\ r_1^A\left(1 - r_1^{B|A}\right)r_2^A\left(1 - r_2^{B|A}\right)\left(1 - r_3^A\right)\left(1 - r_3^{B|a}\right) \end{bmatrix}$$
$$= \psi^A\left(1 - \psi^{B|A}\right)p_1^A p_2^A\left(1 - p_3^A\right)$$
$$+ \psi^A\psi^{B|A}r_1^A\left(1 - r_1^{B|A}\right)r_2^A\left(1 - r_2^{B|A}\right)\left(1 - r_3^A\right)\left(1 - r_3^{B|a}\right).$$

While slightly rearranged, this is the same probability statement as that given previously for this pair of detection histories by first considering a verbal description of the data. Recall that in using a matrix formulation, the matrix multiplication is automatically summing the probabilities associated with the various possible outcomes in terms of which species may be present at a unit, given the observed data. Some further examples of detection histories and the associated probability statements are given in Table 14.5.

Assuming the detection histories collected at the s units are independent, the observed data likelihood is defined as:

$$ODL(\theta|\mathbf{h}_1, \ldots, \mathbf{h}_s) = \prod_{i=1}^{s} Pr(\mathbf{h}_i|\theta),$$

where θ denotes the set of model parameters. To obtain parameter estimates, the observed data likelihood could be maximized to give maximum likelihood estimates, or prior distributions could be assigned for each parameter and, by regarding the likelihood function as the probability of observing the data, their posterior distributions could be obtained using Bayesian estimation methods.

The single-season co-occurrence model can also be formally defined in terms of the underlying random variables. Let z_i be a categorical (latent) random variable indicating the true occupancy state of unit i, defined in terms of which species may be present at a unit, and h_{ij} be the observed outcome of survey j at unit i, the possible values of which represent different combinations of detecting each species (e.g., using the 0–3 coding described previously for two species). Then:

$$z_i \sim Categorical(\phi_0),$$

and

$$h_{ij}|z_i = m \sim Categorical(\mathbf{p}_j^{[m,\bullet]}),$$

where $\mathbf{p}_j^{[m,\bullet]}$ is the row of $\mathbf{p}_j$ corresponding to true state m. Note that defining the random variables in this manner generalizes easily to co-occurrence investigations involving a greater number of species by simply increasing the number of true and observable states. This also increases the number of parameters to estimate, although there are practical ways to limit the number of parameters (see Section 14.8). These latent and observed random variables could be used to construct the complete data likelihood as with the modeling detailed in the previous chapters.

TABLE 14.5 Example detection histories for two species (h_i^A and h_i^B), as a combined history (h_i), and resulting probability statements ($Pr(h_i | \theta)$) using the single-season, co-occurrence model. T indicates the transpose of the vector (see Appendix)

h_i^B	h_i^B	h_i	$Pr(h_i \mid \theta)$

| 011 | 010 | 031 | $\begin{bmatrix} (1-\psi^A)(1-\psi^{B\mid a}) \\ \psi^A(1-\psi^{B\mid A}) \\ (1-\psi^A)\psi^{B\mid a} \\ \psi^A\psi^{B\mid A} \end{bmatrix}^T \begin{bmatrix} 0 \\ 0 \\ 0 \\ (1-r_1^A)(1-r_1^{B\mid a})r_2^A r_2^{B\mid A}r_3^A(1-r_3^{B\mid A}) \end{bmatrix}$ |

$= \psi^A\psi^{B\mid A}(1-r_1^A)(1-r_1^{B\mid a})r_2^A r_2^{B\mid A}r_3^A(1-r_3^{B\mid A})$

| 000 | 101 | 202 | $\begin{bmatrix} (1-\psi^A)(1-\psi^{B\mid a}) \\ \psi^A(1-\psi^{B\mid A}) \\ (1-\psi^A)\psi^{B\mid a} \\ \psi^A\psi^{B\mid A} \end{bmatrix}^T \begin{bmatrix} 0 \\ 0 \\ p_1^B(1-p_2^B)p_3^B \\ (1-r_1^A)r_1^{B\mid a}(1-r_2^A)(1-r_2^{B\mid a})(1-r_3^A)r_3^{B\mid a} \end{bmatrix}$ |

$= (1-\psi^A)\psi^{B\mid a}p_1^B(1-p_2^B)p_3^B + \psi^A\psi^{B\mid A}(1-r_1^A)r_1^{B\mid a}(1-r_2^A)(1-r_2^{B\mid a})(1-r_3^A)r_3^{B\mid a}$

| 000 | 000 | 000 | $\begin{bmatrix} (1-\psi^A)(1-\psi^{B\mid a}) \\ \psi^A(1-\psi^{B\mid A}) \\ (1-\psi^A)\psi^{B\mid a} \\ \psi^A\psi^{B\mid A} \end{bmatrix}^T \begin{bmatrix} 1 \\ \prod_{j=1}^{3}(1-p_j^A) \\ \prod_{j=1}^{3}(1-p_j^B) \\ \prod_{j=1}^{3}(1-r_j^A)(1-r_j^{B\mid a}) \end{bmatrix}$ |

$= (1-\psi^A)(1-\psi^{B\mid a}) + \psi^A(1-\psi^{B\mid A})\prod_{j=1}^{3}(1-p_j^A) + (1-\psi^A)\psi^{B\mid a}\prod_{j=1}^{3}(1-p_j^B) + \psi^A\psi^{B\mid A}\prod_{j=1}^{3}(1-r_j^A)(1-r_j^{B\mid a})$

14.2.3 Derived Parameters and Alternative Parameterizations

In the parameterization we have focused on above, the occupancy probability for species A is unconditional (ψ^A) while two probabilities are associated with occupancy for species B, which are conditional on the presence or absence of species A ($\psi^{B|A}$ and $\psi^{B|a}$). As such, there is no direct estimate of the overall, or marginal, probability of species B being present at a unit. This can, however, be calculated as:

$$\psi^B = \psi^A \psi^{B|A} + \left(1 - \psi^A\right) \psi^{B|a}.$$

This also suggests that the model described above could be reparameterized such that ψ^A and ψ^B are both estimated directly along with either $\psi^{B|A}$ or $\psi^{B|a}$. Similar logic could also be applied to the detection probabilities when both species are present (i.e., the r parameters).

In the original parameterization of MacKenzie et al. (2004a), the two-species co-occurrence model was parameterized in terms of ψ^A, ψ^B, and ψ^{AB}, where ψ^{AB} is the probability of both species being at a unit. When two species co-occur at units independently, then:

$$\psi^{AB} = \psi^A \times \psi^B.$$

MacKenzie et al. (2004a) used this relationship to suggest that the level of co-occurrence between the two species could be quantified by the expression:

$$\phi = \frac{\psi^{AB}}{\psi^A \times \psi^B}.$$

That is, ϕ is the ratio of how much more or less likely the species are to co-occur at a unit compared to what would be expected if they co-occurred independently. MacKenzie et al. (2004a) termed the quantity ϕ a species interaction factor (SIF). Values of $\phi < 1$ would indicate species co-occur less frequently than if they were distributed independently (e.g., possibly exclusion or avoidance), while $\phi > 1$ would suggest a tendency for species to co-occur more frequently than expected under independence. If the species occupy units independently then $\phi = 1$. There is, however, a natural relationship among the occupancy probabilities that restricts the values that ϕ can possibly take, reflecting limits to the degree of overlap that is possible between the two species. That is:

$$\max\left(\psi^A + \psi^B - 1, 0\right) \leqslant \psi^{AB} \leqslant \min\left(\psi^A, \psi^B\right).$$

For example, if $\psi^A = \psi^B = 0.6$, then the two species must co-occur at a minimum of 20% of the locations, while if they always co-occur, then it can only be at 60% of units at most. This implies that ϕ can only take values between

$\max\left(\left(\psi^A + \psi^B - 1\right)/\left(\psi^A\psi^B\right), 0\right)$ and $\min\left(\left(1/\psi^B\right), \left(1/\psi^A\right)\right)$. In the above case, this would imply that ϕ must be between the values of 0.56 and 1.67, whereas if $\psi^A = 0.6$ and $\psi^B = 0.8$, ϕ must be between 0.83 and 1.25. This restriction must be enforced when estimating ϕ directly, which can cause numerical problems with this parameterization of the co-occurrence model, particularly when potential covariates are included for some of the model parameters. Once again, a similar parameterization could be applied to the detection probabilities r, where a detection interaction factor δ could be defined, although with similar limitations and numerical issues. It is due to these numerical issues that, generally, we do not recommend practitioners use the original parameterization of MacKenzie et al. (2004a), although there will be occasions when it is perfectly adequate.

Using the RW parameterization of the two-species co-occurrence model, which we have focused on in this chapter, one could calculate the SIF ϕ to quantify the level of dependence between the two species (Richmond et al., 2010), that is:

$$\phi = \frac{\psi^A \psi^{B|A}}{\psi^A\left(\psi^A\psi^{B|A} + \left(1 - \psi^A\right)\psi^{B|a}\right)}.$$

Alternatively, the level of dependence can be quantified in terms of $\psi^{B|A}$ and $\psi^{B|a}$, because if the two species do occur at units independently, these probabilities will be equal (i.e., the presence of species A has no effect on the probability of species B being present). While the difference in the probabilities, or their ratio, could be used to quantify the level of dependence, we suggest that it be done in terms of an odds ratio, as was done at the beginning of this chapter. The reason for this recommendation is that using an odds ratio fits naturally into use of the logit-link to model the occurrence probabilities for each species, e.g., logistic regression, which we shall now demonstrate.

In Eq. (14.3) the odds ratio is given for the presence of species B given the presence (or absence) of species A. This can be re-expressed in terms of $\psi^{B|A}$ and $\psi^{B|a}$, and will be defined as the SIF ν. That is:

$$\nu = \frac{\psi^{B|A}/\left(1 - \psi^{B|A}\right)}{\psi^{B|a}/\left(1 - \psi^{B|a}\right)},$$

where $\nu = 1$ if species occur at units independently, $\nu < 1$ indicates they occur together less often than expected under independence (e.g., some form of avoidance or different preferences), and $\nu > 1$ indicates the species co-occur more often that expected under independence. Note that ν can only take positive values. Now, ignoring detection issues for a moment, one approach to investigate the co-occurrence patterns of two species would be through logistic regression, where the observed presence/absence of one species is included as a predictor variable on the probability of occurrence of the second species. For example, an investigator might envision the relationship:

$$logit\left(\psi_i^B | z_i^A\right) = \alpha + \beta z_i^A,$$

where z_i^A is the observed presence ($z_i^A = 1$) or absence ($z_i^A = 0$) of species A at unit i. The presence of species A would have no effect on the probability of species B being present at a unit if $\beta = 0$, would lower the probability if $\beta < 0$, and increase the probability if $\beta > 0$. Hence, values of β less than, or greater than, 0 would indicate avoidance or attraction of the species, respectively. Based upon the definition of the logit-link (Section 3.4.1), then v and β are related as:

$$\beta = ln(v),$$

or conversely:

$$v = e^\beta.$$

Recall that in Chapter 3 it was suggested that covariate effect sizes on probabilities modeled with the logit-link could be interpreted in terms of an odds ratio, which is essentially what is happening with the RW parameterization. Using this parameterization with the v SIF results in equivalent inferences as when logistic regression is used with the presence of species A included as a predictor variable on the probability of presence for species B. There are further benefits when predictor variables, or covariates, are also included in the co-occurrence probabilities, as shall be explored in the next section.

The above development suggests that, instead of defining the model with respect to ψ^A, $\psi^{B|A}$, and $\psi^{B|a}$, it could be reparameterized in terms of ψ^A, v, and $\psi^{B|a}$ (or $\psi^{B|A}$), such that v is estimated directly. Note that unlike the SIF ϕ, the only limit on the allowable values v is that it must be greater than zero, which can be easily incorporated by using a log-link function (Chapter 3). Hence this parameterization is much more numerically stable than the original version of MacKenzie et al. (2004a).

This same logic can also be applied to the detection probabilities r, i.e., define the detection interaction factor ρ as:

$$\rho = \frac{r_j^{B|A}/\left(1 - r_j^{B|A}\right)}{r_j^{B|a}/\left(1 - r_j^{B|a}\right)},$$

which indicates the level of dependence in the detection of the two species (given both are present at a unit). That is, species B is less, or more, likely to be detected if species A was also detected in the same survey, when ρ is < 1, or > 1, respectively. As for v, ρ can be interpreted in terms of a logistic regression analysis on the probability of detecting species B in survey j of unit i, using the detection of species A in the same survey as a predictor variable. The detection component of the two-species co-occurrence model can therefore be reparameterized in terms of p_j^A, p_j^B, r_j^A, ρ, and $r_j^{B|a}$ (or $r_j^{B|A}$) such that ρ is estimated directly.

One disadvantage of defining the model in terms of the conditional probabilities $\psi^{B|A}$, $\psi^{B|a}$, $r_j^{B|A}$, and $r_j^{B|a}$ (or by using ν and ρ), rather than using the original parameterization of MacKenzie et al. (2004a), is that it implies a dominant and a subordinate species. That is, inference is based on how the presence of the dominant species (species A) affects the probability of presence of the subordinate species (species B); it is not a mutual effect where the presence of each species is affected by the other (and similarly for detection). This one-way effect is likely to be reasonable, and even desired, for many species co-occurrence investigations, however in other situations it may not be appropriate. The original parameterization of MacKenzie et al. (2004a) could be attempted to investigate more symmetric interactions, or alternatively Rota et al. (2016) have recently published another possible parameterization.

Above we noted that conceptually, the RW parameterization amounts to using the presence (or detection in the case of the r parameter) of species A as a covariate for the probability of species B. The parameterization of Rota et al. (2016) is similar except that the presence of each species is used as a covariate for the other, with a consistent effect size. That is:

$$logit\left(\psi_i^A | z_i^B\right) = \alpha_A + \beta^* z_i^B,$$

which gives $\psi^{A|b}$ when $z_i^B = 0$ and $\psi^{A|B}$ when $z_i^B = 1$, and:

$$logit\left(\psi_i^B | z_i^A\right) = \alpha_B + \beta^* z_i^A,$$

which gives $\psi^{B|a}$ when $z_i^A = 0$ and $\psi^{B|A}$ when $z_i^A = 1$. Note that β^* is present, and equal, in both equations. The level of co-occurrence could therefore be quantified as:

$$\nu^* = e^{\beta^*},$$

which is the odds ratio for species B being present given the presence/absence of species A, and also the odds ratio for species A being present given the presence/absence of species B. That is:

$$\nu^* = \frac{\psi^{B|A}/\left(1 - \psi^{B|A}\right)}{\psi^{B|a}/(1 - \psi^{B|a})} = \frac{\psi^{A|B}/\left(1 - \psi^{A|B}\right)}{\psi^{A|b}/(1 - \psi^{A|b})},$$

where the lowercase 'b' indicates absence of species B. This highlights the two-way nature of co-occurrence using this approach.

Importantly, note that these probabilities are conditional upon the presence or absence of the other species, whereas in order to apply the parameterization using the multi-state framework, the unconditional probabilities associated with the four regions in the Venn diagram (Fig. 14.1) need to be determined. These can be calculated as:

$$\psi^{AB} = \frac{\psi^{B|a}\psi^{A|B}\psi^{B|A}}{\psi^{B|A}(1 - \psi^{A|B}) + \psi^{A|B}\psi^{B|a}},$$

$$\psi^{A} = \frac{\psi^{B|a}\psi^{A|B}}{\psi^{B|A}(1 - \psi^{A|B}) + \psi^{A|B}\psi^{B|a}},$$

$$\psi^{B} = \frac{\psi^{B|a}\psi^{B|A}}{\psi^{B|A}(1 - \psi^{A|B}) + \psi^{A|B}\psi^{B|a}},$$

from which ψ^{A0} and ψ^{B0} can be calculated as before. There are alternative expressions for these unconditional probabilities based upon different combinations of the conditional probabilities (e.g., the above is based on $\psi^{B|a}$, $\psi^{A|B}$, and $\psi^{B|A}$), but they will all evaluate to the same numerical result. It is also possible to calculate the unconditional probabilities directly from the logistic regression coefficients (see Rota et al., 2016, for details), but the above is general and will hold if other link functions are used to model the conditional probabilities. Similar calculations can be applied to the probabilities of detection when both species are present at a unit, r, to allow for 'two-way' detection dependence between the species.

14.2.4 Covariates

Often, the probability that a species occupies a location may be affected by specific characteristics of the unit and other environmental variables. For example, some species may prefer particular habitat types over other available habitats (e.g., have a higher occupancy probability at locations near permanent water sources); require a minimum patch size for a sustainable population; or show reduced probability of occurrence in isolated patches (e.g., Verner et al., 1986; Hanski, 1999; Scott et al., 2002). Differential habitat preferences among species may be responsible for the nonrandom patterns observed in the species-occurrence matrix when using analytic methods that do not account for such factors. Similarly, the probability of detecting species at a unit may be affected by unit-specific covariates (e.g., open old growth forest vs. dense rejuvenating forest), or by factors that vary with each survey, such as air temperature, cloud cover, or time since a rain event, and the effect on species detection will likely be different for different species.

Irrespective of the parameterization used for the two-species co-occurrence model, covariates can be incorporated for any of the model parameters during an analysis through the use of appropriate link functions (Chapter 3), as demonstrated throughout this book. When using a parameterization such that probabilities relate to binary outcomes, e.g., ψ^A, ψ^B, $\psi^{B|a}$, etc., the logit link function would be one suitable option (although other options can be used, for example, the probit link). The multinomial link function can be used with parameterizations where probabilities are defined in terms of multinomial outcomes, noting that the multinomial link function involves the whole set of probabilities that relate to the multinomial outcome, e.g., ψ_{A0}, ψ_{0B}, and ψ_{AB}. Species and detection interaction factors, e.g., ϕ, ν, ρ, etc., are strictly positive values (i.e., must be > 0), hence the log link function could be used, as above. Although, recall that when using the original parameterization of MacKenzie et al. (2004a) there are limits on the values that the interaction factors ϕ and δ can take, which may vary for each unit or survey, respectively, and the limits will depend upon the values of the covariates that have been included in the respective parts of the model being fit to the data. This dependence can create numerical issues, making it difficult for optimization algorithms to converge to the maximum likelihood estimates or creating instabilities in the Markov chains when using MCMC. That is not to say the original parameterization will not work with covariates for all datasets, we are simply noting that there are practical difficulties that users may experience that can require considerable effort to overcome. As such, we generally recommend other parameterizations be used.

There are a few points that are worthy of discussion when covariates are included with the RW parameterization. Let $\mathbf{X}_i$ denote the set of covariate values measured at unit i, which form a row vector of the values [1 x_{i1} ... x_{in}], where the initial '1' denotes a constant that is required for the inclusion of an intercept term in the resultant regression equation. Further, let α and β denote sets of regression coefficients associated with each covariate in $\mathbf{X}_i$ (these are column vectors), whose values will be estimated from the data. The interpretation of α and β will depend on the structure of the specified model being fit to the data. Suppose that the following model is defined for the parameters $\psi^{B|a}$ and $\psi^{B|A}$, using a matrix based notation for convenience (see Appendix):

$$logit\left(\psi_i^{B|a}\right) = \mathbf{X}_i\alpha,$$

and

$$logit\left(\psi_i^{B|A}\right) = \mathbf{X}_i\alpha,$$

which implies that the probability of species B being present at a unit has the same covariate relationships irrespective of whether species A is absent or present at a unit. That is, the probability of occurrence for species B is unaffected

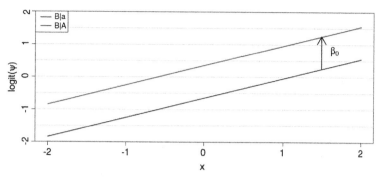

FIGURE 14.4 Illustration of the model where the parameters $\psi^{B|a}$ and $\psi^{B|A}$ are modeled with the same set of covariates with the logit link function, and an additive effect of size β_0 included for $\psi^{B|A}$.

by the presence of species A; the two species occupy units independently. This could be expressed as a single regression equation:

$$logit\left(\psi_i^B | z_i^A\right) = \mathbf{X}_i \boldsymbol{\alpha},$$

where z_i^A is the presence or absence of species A at unit i, which does not appear on the right-hand side of the equation, so has no effect on ψ^B.

Now consider the model defined by:

$$logit\left(\psi_i^{B|a}\right) = \mathbf{X}_i \boldsymbol{\alpha}, \tag{14.5}$$

and

$$logit\left(\psi_i^{B|A}\right) = \mathbf{X}_i \boldsymbol{\alpha} + \beta_0,$$

where β_0 is the first term of $\boldsymbol{\beta}$. This model implies that the probability of species B being present at a unit has the same covariate relationships regardless of the presence/absence of species A (i.e., the same effect sizes), but there is a consistent difference in the logit-probability across all units that does not depend on the covariate values (note β_0 may be estimated as a positive or a negative value). That is, the level of co-occurrence between species A and B is the same across all units, and does not depend on the covariate values. This is illustrated in Fig. 14.4, where the effect of covariate 'x' is the same in both cases (both lines have the same slope), but a consistent difference across all covariate values resulting in parallel lines. This model could be expressed as the single equation:

$$logit\left(\psi_i^B | z_i^A\right) = \mathbf{X}_i \boldsymbol{\alpha} + \beta_0 z_i^A,$$

which highlights that the effect of the presence of species A, z_i^A, on the probability of species B being present at a unit does not depend on the covariates in

$\mathbf{X}_i$, hence it is an additive effect. Note that β_0 also defines the SIF v, with:

$$v = e^{\beta_0}, \tag{14.6}$$

which is the same across all units. Therefore, this same biological model could have been fit using the ψ^A, $\psi^{B|a}$ and v parameterization, with $\psi^{B|a}$ and v defined by Eqs. (14.5) and (14.6), respectively.

Finally, consider the model defined by:

$$logit\left(\psi_i^{B|a}\right) = \mathbf{X}_i\boldsymbol{\alpha}, \tag{14.7}$$

and

$$logit\left(\psi_i^{B|A}\right) = \mathbf{X}_i(\boldsymbol{\alpha} + \boldsymbol{\beta}) = \mathbf{X}_i\boldsymbol{\alpha} + \mathbf{X}_i\boldsymbol{\beta},$$

which can also be expressed as:

$$logit\left(\psi_i^B|z_i^A\right) = \mathbf{X}_i\boldsymbol{\alpha} + \mathbf{X}_i\boldsymbol{\beta}z_i^A.$$

Here, the effect of the covariates in $\mathbf{X}_i$ on the probability of species B being present at a unit is different depending on whether species A was also present or absent at a unit, which is an interaction between the presence of species A and the other covariates, with $\boldsymbol{\beta}$ being the size of the difference. That is, the level of co-occurrence between species A and B varies between units, and depends upon the covariate values:

$$v_i = e^{\mathbf{X}_i\boldsymbol{\beta}}. \tag{14.8}$$

As above, the same biological model could have been defined using the ψ^A, $\psi^{B|a}$ and v parameterization, with $\psi^{B|a}$ and v defined by Eqs. (14.7) and (14.8), respectively, which would be a useful approach if a research question was to investigate how the co-occurrence between the species varied among units. The same biological model could also be fit to the data by defining the following structure for $\psi^{B|a}$ and $\psi^{B|A}$:

$$logit\left(\psi_i^{B|a}\right) = \mathbf{X}_i\boldsymbol{\alpha},$$

and

$$logit\left(\psi_i^{B|A}\right) = \mathbf{X}_i\boldsymbol{\beta}.$$

With this structure, the interpretation of $\boldsymbol{\beta}$ is different than above with it being the absolute effect of the covariates on the probability of occurrence for species B when species A is present, rather than the difference in the covariate effects in the presence of species A. The SIF factor would be calculated as:

$$v_i = e^{\mathbf{X}_i(\boldsymbol{\beta} - \boldsymbol{\alpha})}.$$

When using the parameterization of Rota et al. (2016) with covariates, the model could be naturally defined in terms of $\psi^{A|b}$, $\psi^{B|a}$, and ν^*. For example:

$$logit\left(\psi_i^{A|b}\right) = \mathbf{X}_i\boldsymbol{\alpha}_A,$$

$$logit\left(\psi_i^{B|a}\right) = \mathbf{X}_i\boldsymbol{\alpha}_B,$$

and

$$\nu_i^* = e^{\mathbf{X}_i\boldsymbol{\beta}},$$

where $\boldsymbol{\alpha}_A$ and $\boldsymbol{\alpha}_B$ are the effects of the covariates on the probability of occurrence for each species given the other species is absent, and $\boldsymbol{\beta}$ is the effect of the covariates on the mutual co-occurrence of the species. Note that some values in $\boldsymbol{\alpha}_A$, $\boldsymbol{\alpha}_B$, and $\boldsymbol{\beta}$ could be set as 0, i.e., some covariates in $\mathbf{X}_i$ may have no effect on certain parameters.

While the above discussion has been in terms of species occurrence, it is also relevant to the r detection parameters, with the obvious modifications, where inferences are in terms of how the detection of species during a survey is affected by covariates rather than species presence/absence at a unit.

14.2.5 Missing Observations

As in single-species studies, 'missing observations' in the detection histories may occur due to logistical constraints (it is simply not possible to survey all locations virtually simultaneously), study design, or unforeseen circumstances such as a vehicle breakdown en route. These can be easily accommodated in the multi-species modeling in the same manner as for other models. For survey occasions when no data were collected, the respective detection probabilities do not appear within the mathematical equation denoting the probability of observing that detection history. Hence, no detection probability parameters appear in the model likelihood for that unit, at that survey occasion.

Generally we imagine situations where the detection/nondetection of both species arise from the same information source (i.e., two frog species are detected/not detected during a single five-minute calling survey), hence a 'missing observation' at a particular unit at a certain survey occasion applies to both species. However, in some circumstances missing observations may occur for one species but not for the other. For example, if different field methods are required to detect each species (e.g., camera traps to detect carnivores and track plates to detect their prey), and one method functions correctly, but the other method fails, this would create a survey with a missing observation for only one (or more generally a subset) of the species. Another example might involve a host–pathogen system when information on the pathogen is only collected at a subset of units, but the host is surveyed at all study units. One view of this problem is that the second species may or may not have been detected had the data been collected. To incorporate this idea we can use the same technique we have used elsewhere when faced with the possibility of multiple explanations for the same set of data: include both possibilities within the probability statement for the detection history. Consider the detection histories $\mathbf{h}_i^A = 101$ and $\mathbf{h}_i^B = 0\text{-}0$ where there is a missing observation for species B in the second survey. A verbal description of the detection histories would be:

Both species are present at the unit with species A detected in the first survey but not species B, species A was not detected in the second survey with no survey data for species B (so it may or may not have been detected if a survey had been conducted), and in the third survey species A was detected but not species B,

or,

only species A is present at the unit, with it being detected in survey 1, not detected in survey 2 and detected in survey 3.

Translating this description into a mathematical equation using the RW parameterization:

$$Pr\left(\mathbf{h}_i^A = 101, \mathbf{h}_i^B = 0\text{-}0|\boldsymbol{\theta}\right)$$

$$= \psi^A \psi^{B|A} \left(\begin{array}{c} r_1^A\left(1 - r_1^{B|A}\right) \times \\ \left\{\left(1 - r_2^A\right) r_2^{B|a} + \left(1 - r_2^A\right)\left(1 - r_2^{B|a}\right)\right\} \times \\ r_3^A\left(1 - r_3^{B|A}\right) \end{array} \right)$$

$$+ \psi^A\left(1 - \psi^{B|A}\right) p_1^A\left(1 - p_2^A\right) p_3^A$$

$$= \psi^A \psi^{B|A} r_1^A\left(1 - r_1^{B|A}\right)\left(1 - r_2^A\right) r_3^A\left(1 - r_3^{B|A}\right)$$

$$+ \psi^A\left(1 - \psi^{B|A}\right) p_1^A\left(1 - p_2^A\right) p_3^A.$$

The term in curly brackets represents the two options for the outcomes of the second survey, which in this case, reduces to just $1 - r_2^A$ with no r probability associated with species B for the second survey. However consider a second example, where the missing observation is for species A, e.g., $\mathbf{h}_i^A = 10\text{-}$ and $\mathbf{h}_i^B = 011$. The probability statement for this pair of detection histories would be:

$$Pr\left(\mathbf{h}_i^A = 10\text{-}, \mathbf{h}_i^B = 011|\boldsymbol{\theta}\right) = \psi^A \psi^{B|A} \left(\begin{array}{c} r_1^A\left(1 - r_1^{B|A}\right) \times \\ \left(1 - r_2^A\right) r_2^{B|a} \times \\ \left\{r_3^A r_3^{B|A} + \left(1 - r_3^A\right) r_3^{B|a}\right\} \end{array} \right).$$

In this case the term in the square brackets does not simplify, hence r_3^A remains in the probability statement even though there were no survey data for species A in survey 3.

The reason for this is that with this parameterization, the detection probability for species B is conditional upon the detection of species A, given both species are present at a unit. Therefore, if the survey outcome for species A is known, as in the former example, when accounting for the two options of detection or nondetection of species B due to the missing observation, the same r parameter for species B would be used in the probability statement; $r_j^{B|a}$ if species A had not been detected in survey j, or $r_j^{B|A}$ if species A had been detected. As the same parameter is used, then some of the terms will cancel, for example:

$$\left(1 - r_2^A\right) r_2^{B|a} + \left(1 - r_2^A\right)\left(1 - r_2^{B|a}\right) = \left(1 - r_2^A\right)\left(r_2^{B|a} + \left(1 - r_2^{B|a}\right)\right)$$

$$= \left(1 - r_2^A\right).$$

When the survey outcome for species A is unknown, as in the latter example, then different detection probabilities for species B are included when accounting for the two options, which do not cancel, e.g., $r_3^A r_3^{B|A} + \left(1 - r_3^A\right) r_3^{B|a}$.

Using the parameterization of Rota et al. (2016), if there is a missing observation for only one of the two species at a survey occasion, then irrespective of which species it is, detection probabilities will be required for both species due to the mutual conditioning of the parameters (when both species are presumed to be present).

14.3 ADDRESSING BIOLOGICAL HYPOTHESES

Using the two-species co-occurrence model it is possible to address three interesting biological questions about the system by constraining various parameter values:

1. Does the occupancy of one species depend on the presence of the other species?
2. Does the detection probability of either species differ if the other species is present?
3. Does the detection probability of one species depend on the detection of the other species, when both species are present?

By fitting a set of candidate models that represent various biological hypotheses, the degree of support for each can be evaluated using formal statistical methods.

In many applications the first question will be of primary interest and is similar to searching for nonrandom patterns in the species co-occurrence matrix or investigating the potential for competitive exclusion. The degree of support for whether each species occurs at units independently can be evaluated by comparing two models. For example, with the RW parameterization: (1) a full model where each of the parameters ψ^A, $\psi^{B|A}$, and $\psi^{B|a}$ is estimated; and (2) a reduced model with $\psi^{B|A}$ and $\psi^{B|a}$ set equal. A formal comparison of these two models allows an informed decision to be made about the level of support for each hypothesis.

The second question compares the detection of a species at units where it exists alone vs. units where both species occur, i.e., does $r_j^A = p_j^A$ or $r_j^B = p_j^B$? Note that this issue is distinct from the question of whether detections of the two species occur independently during a survey given that both species are present (i.e., does $\delta = 1$?). For example, in the previously mentioned case of northern spotted owls and barred owls in the Pacific northwest, it was thought that detection probability of northern spotted owls (NSO) may be lower when barred owls are also present at a unit, i.e., $r_j^{NSO} < p_j^{NSO}$ (Bailey et al., 2009). Here there are multiple models that could be considered, as there are multiple sub-hypotheses one may be interested in. For example, it may be of interest whether the presence of each species affects the detection probability of the other species, or whether only one species affects detection of the other. That is, possible constraints that could be applied are:

1. none (i.e., r_j^A, r_j^B, p_j^A, p_j^B are all estimated separately),
2. $r_j^A = p_j^A$,
3. $r_j^B = p_j^B$,
4. $r_j^A = p_j^A$ and $r_j^B = p_j^B$,

where $r_j^B = r_j^{B|A} = r_j^{B|a}$. The level of support for each hypothesis can therefore be determined, although we advise that one should always attempt to consider *a priori* which hypotheses are of prime importance rather than considering all possible alternatives.

Finally, the third question relates to the detection of each species in a survey, at units where both species exist. The two-species co-occurrence model allows the detection probability to potentially depend upon the detection of the other species in the same survey occasion. For example, suppose the two species of interest are a predator and a prey species (e.g., a hawk and rabbits). At a unit where both species are present, if a hawk is seen during a survey, it may decrease the chances of seeing rabbits during the same survey. The level of support for this hypothesis can be assessed in the same manner as above by fitting two models, one where $r_j^{B|A}$ and $r_j^{B|a}$ are estimated separately (if using the RW parameterization), and a second where these parameters are set equal.

14.4 EXAMPLE: TERRESTRIAL SALAMANDERS IN GREAT SMOKY MOUNTAINS NATIONAL PARK

We illustrate the use of the two-species co-occurrence model, using the RW parameterization, with monitoring data collected on terrestrial salamanders in Great Smoky Mountains National Park (GSMNP), located on the Tennessee and North Carolina border, USA. Data were collected from 88 units within the Roaring Fork Watershed (GSMNP, Mt. LeConte USGS Quadrangle) that were located adjacent to trails and spaced approximately 250 m apart. Two parallel transects were sampled at each unit: a natural cover transect (50 m long × 3 m wide) and a coverboard transect consisting of five stations placed 10 m apart. We pooled the detection data from the two transects for each unit in our analysis. Because of this study design, we assumed that species were detected independently of each other in a survey due to the scale of the area being searched. If smaller areas had been searched during a survey, then it may be unreasonable to assume independent detections due to direct competition among some species for refuges.

Units were surveyed five times between 4 April 1999 and 27 June 1999, with approximately two weeks between successive survey occasions. Co-occurrence patterns for two terrestrial salamander species were considered: Jordan's salamander (*Plethodon jordani*; PJ) and members of the *Plethodon glutinosus* (PG) complex including *Plethodon glutinosus* and *Plethodon oconaluftee*. In their cursory analysis of the same data, MacKenzie et al. (2004a) explored whether there is any evidence that the two species exhibit co-occurrence patterns indicative of strong interactions, after allowing for any elevational gradient in occupancy probabilities. PG is more tolerant of dry locations found usually at lower elevations (Grover, 2000; Rissler et al., 2000), and PJ seems to have a numerical advantage in moist microhabitats common at higher elevations (Hairston, 1951; Dakin, 1978). Experimental removal studies suggest that PJ is the more dominant species (i.e., species A), with significant increases in the subordinate PG species (i.e., species B) when PJ is removed (Hairston, 1980).

Following MacKenzie et al. (2004a), we conducted the analysis in two parts to illustrate how the omission of covariates that may be related to species-specific habitat preferences or tolerances could influence one's inference about species co-occurrence. Considering a simple set of four models that assumed the occupancy and detection probabilities for both species were unaffected by elevation, only one model had substantial support (based upon AIC; Table 14.6), $\psi^A(\cdot)\,\psi^B(PJ)\,p^A(\cdot)\,p^B(\cdot)\,r^A(\cdot)\,r^B(\cdot)$. In the model notation used, inclusion of PJ as a covariate for ψ^B indicates the presence of PJ (species A) affects the occupancy of species B (PG), hence allows for possible dependence of the species occurrence at units. Where r parameters have been omitted from the model

TABLE 14.6 Summary of model fit and selection statistics for the salamander data analyzed with the RW parameterization of the two-species co-occurrence model, with no covariates. ΔAIC is the absolute difference in AIC values relative to the model with the smallest AIC, w is the AIC model weight, $Npar$ is the number of estimated parameters in the model, and $-2l$ is twice the negative log-likelihood value. 'PJ' indicates the probability depends on the presence of *Plethodon jordani* (species A)

Model	ΔAIC	w	$Npar$	$-2l$
$\psi^A(\cdot)\,\psi^B(\text{PJ})\,p^A(\cdot)\,p^B(\cdot)\,r^A(\cdot)\,r^B(\cdot)$	0.00	0.98	7	736.62
$\psi^A(\cdot)\,\psi^B(\cdot)\,p^A(\cdot)\,p^B(\cdot)\,r^A(\cdot)\,r^B(\cdot)$	8.35	0.02	6	746.97
$\psi^A(\cdot)\,\psi^B(\text{PJ})\,p^A(\cdot)\,p^B(\cdot)$	20.75	0.00	5	761.37
$\psi^A(\cdot)\,\psi^B(\cdot)\,p^A(\cdot)\,p^B(\cdot)$	33.37	0.00	4	775.99

name, that indicates that the probability of detection is the same regardless of whether only one or both species are present at a unit, i.e., the presence of the other species has no effect on the detection probability for the focal species. Hence, the top-ranked model suggests PJ and PG do not occur at units independently, and the probability of detecting each species is different when the other species is also present at a unit. It was estimated that the probability of occurrence for PG was much lower when PJ was present ($\hat{\psi}^{B|A} = 0.39$; $\hat{\text{SE}} = 0.09$) compared to units where PJ was absent ($\hat{\psi}^{B|a} = 0.75$; $\hat{\text{SE}} = 0.07$), suggesting that at least one of the two species avoided the other (SIF: $\hat{\nu} = 0.21$; $\hat{\text{SE}} = 0.11$). The estimated probability of PJ occupying a unit was $\hat{\psi}^A = 0.48$ ($\hat{\text{SE}} = 0.05$). Detection probabilities were estimated to be quite different for PJ when PG was present ($\hat{p}^A = 0.91$ and $\hat{r}^A = 0.55$) but similar for PG regardless of the presence of PJ ($\hat{p}^B = 0.54$, $\hat{r}^B = 0.48$). Note that the bottom-ranked model in Table 14.6 is equivalent to fitting a regular single-species single-season $\psi(\cdot)\,p(\cdot)$ model to the data for each species independently.

Five additional models were fit to the data representing different hypotheses about the effect of elevation (El) on the occurrence and co-occurrence of the two species (Table 14.7). For all five models elevation was included as a covariate on all detection probabilities, and detection probability for each species was assumed to be different if one species vs. two species were present at a unit, although in practice further investigation of these assumptions would be recommended. A summary of the model fitting procedure is given in Table 14.8, where the previous top-ranked model from Table 14.6 has been included for reference. Including elevation as a covariate on all of the detection probabilities resulted in models that provided much better fits to the data, with an improvement of over 80 in $-2l$ using only four additional parameters (4th- vs. 6th-ranked model). However, even after including elevation as a covariate for detection, there is still strong evidence of an apparent interaction between the species when eleva-

TABLE 14.7 Hypotheses about the occurrence (ψ^A and $\psi^{B|A}$) and co-occurrence (v) of *Plethodon jordani* (PJ; species A) and members of the *Plethodon glutinosus* complex (PG; species B), represented by five models that were fit to the data. Given are the model name and hypothesis about each parameter where '*El*' indicates an elevational relationship, 'Cons' indicates a constant effect across all units, and 'Ind' indicates independence is assumed

Model	ψ^A	$\psi^{B|A}$	v
$\psi^A(El)\,\psi^B(PJ \times El)\,p^A(El)\,p^B(El)\,r^A(El)\,r^B(El)$	El	El	El
$\psi^A(El)\,\psi^B(PJ + El)\,p^A(El)\,p^B(El)\,r^A(El)\,r^B(El)$	El	El	Cons
$\psi^A(El)\,\psi^B(El)\,p^A(El)\,p^B(El)\,r^A(El)\,r^B(El)$	El	El	Ind
$\psi^A(\cdot)\,\psi^B(PJ)\,p^A(El)\,p^B(El)\,r^A(El)\,r^B(El)$	Cons	Cons	Cons
$\psi^A(\cdot)\,\psi^B(\cdot)\,p^A(El)\,p^B(El)\,r^A(El)\,r^B(El)$	Cons	Cons	Ind

TABLE 14.8 Summary of model fit and selection statistics for the salamander data analyzed with the RW parameterization of the two-species co-occurrence model, when elevation *El* is included as a covariate for some probabilities. ΔAIC is the absolute difference in AIC values relative to the model with the smallest AIC, w is the AIC model weight, *Npar* is the number of estimated parameters in the model, and $-2l$ is twice the negative log-likelihood value. 'PJ' indicates the probability depends on the presence of *Plethodon jordani* (species A). The fitted models represent different hypotheses about occurrence and co-occurrence of the species (Table 14.7), and the top-ranked model from Table 14.6 is included for reference

Model	ΔAIC	w	*Npar*	$-2l$
$\psi^A(El)\,\psi^B(El + PJ)\,p^A(El)\,p^B(El)$ $r^A(El)\,r^B(El)$	0.00	0.38	13	613.69
$\psi^A(El)\,\psi^B(El \times PJ)\,p^A(El)\,p^B(El)$ $r^A(El)\,r^B(El)$	0.24	0.33	14	611.93
$\psi^A(El)\,\psi^B(El)\,p^A(El)\,p^B(El)$ $r^A(El)\,r^B(El)$	0.54	0.29	12	616.23
$\psi^A(\cdot)\,\psi^B(PJ)\,p^A(El)\,p^B(El)$ $r^A(El)\,r^B(El)$	35.62	0.00	11	653.31
$\psi^A(\cdot)\,\psi^B(\cdot)\,p^A(El)\,p^B(El)$ $r^A(El)\,r^B(El)$	44.30	0.00	10	663.99
$\psi^A(\cdot)\,\psi^B(PJ)\,p^A(\cdot)\,p^B(\cdot)$ $r^A(\cdot)\,r^B(\cdot)$	110.94	0.00	7	736.62

tion is not included for the occupancy probabilities (4th- vs. 5th-ranked model). When elevation is included as a covariate for the occurrence and co-occurrence parameters (top three models), there is little definitive evidence of a substantial interaction, as the AIC values are very similar for these models, and they

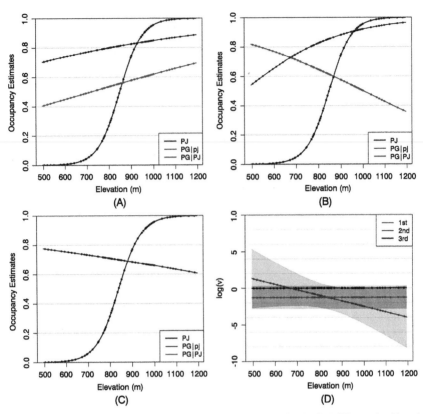

FIGURE 14.5 Estimated probabilities of occupancy for *Plethodon jordani* (PJ; species A) and members of the *Plethodon glutinosus* complex (PG; species B) as a function of elevation from the three top-ranked models in Table 14.8 (panels A–C, respectively). The lower case 'pj' denotes the absence of PJ. The estimated level of co-occurrence under each model, on the log-scale, is presented in panel D, with 95% confidence intervals. Estimated values for the survey units, at the observed elevations, are indicated on each plot.

have almost equal AIC weights. The top-ranked model assumed the level of co-occurrence was constant for all units, the second-ranked model assumed co-occurrence also varied by elevation, and the third-ranked model assumed the species co-occur independently. The estimated occupancy probabilities under each of these models are presented in panels A–C of Fig. 14.5, with the different types of co-occurrence evidenced by the different estimated probabilities of PG occurring at a unit conditional upon the presence or absence of PJ. Note that in panel C, the red and blue lines are overlapping exactly. Panel D of Fig. 14.5 presents the estimated values for v under each model, on the log-scale, with 95% confidence intervals indicated. Recall that $v = 1$ would indicate the species occur independently, which equates to a value of zero on the log-scale. Fig. 14.5D illustrates why there is no clear conclusion regarding the co-occurrence of the species from Table 14.8; zero is typically inside the 95% confidence intervals for $log(v)$ from the 1st- and 2nd-ranked models.

Using the original parameterization and allowing occupancy probabilities to vary with elevation resulted in numerical problems (details discussed in MacKenzie et al., 2004a, 2006). Because of the convergence problems, MacKenzie et al. (2004a) could not make definitive statements about the level of co-occurrence between the species once elevation was included as a covariate for occupancy. Reanalyzing the data with the RW parameterization overcame these convergence issues, not only allowing a co-occurrence SIF to be estimated, but to also consider whether it varied with elevation. However, in this case, we are still unable to make definitive statements about co-occurrence due to insufficient information in the data to discriminate between the models that included elevation as a covariate for occupancy probabilities, even though there was a consistent indication of a negative dependence in the occurrence of the species across most of the surveyed units. Most likely this is due to sample size, and increasing the number of surveyed units would reduce the uncertainty in the co-occurrence parameters. Further refinement of the model set may also yield some improvements.

We would also like to point out to readers that during the development of this example, we encountered some issues with multiple maxima in the likelihood, which is not unexpected given the underlying multi-state modeling framework. This was addressed by attempting different sets of starting values for the numerical optimization, and changing the order in which covariates were included in the model. Users need to be aware of these potential issues with their own analyses. Files associated with the analysis of this data set are available from http://www.proteus.co.nz.

14.5 EXTENSION TO MULTIPLE SEASONS

The modeling approach described in the previous sections is useful for examining patterns of how species co-occur within a general region at a specific point in time. As in the single species case, it provides little insight on the interactive processes between the species that may have taken place and produced the observed pattern. Only with the system being observed at systematic points in time can some understanding of the underlying processes be gained (Yackulic et al., 2015). In this section, we suggest how the single-season two-species models can be extended to multiple seasons. As mentioned in Chapter 2, authors have used dynamic co-occurrence models to explore the influence of invasive species (e.g., Yackulic et al., 2014; Dugger et al., 2015) and predators or competitors (e.g., Miller et al., 2012a; Robinson et al., 2014; Haynes et al., 2014) on species distributions while accounting for habitat and environmental factors that may influence dynamic and detection parameters. Host–pathogen applications have also been proposed, though care should be taken if pathogen detection is conditional on host presence and/or detection (see Mosher et al., 2017).

Here we assume that data have been collected from the same units for multiple (T) seasons, and within each season multiple surveys are conducted to detect a number of species. We develop the modeling approach for only two species, but again suggest it can be extended to a greater number of species if desired. Viewing multiple interacting species as a special case of the dynamic multi-state models (Chapter 9), it's easy to envision more complex models than we present here. In the simple, two-species case, each unit may be in one of the four mutually exclusive states described above (occupied by both species, by species A only, by species B only or occupied by neither species), the dynamic processes of change in the occupancy state of a unit between seasons can be simply represented by defining a transition probability matrix (ϕ_t), much like the multi-state dynamic model (MacKenzie et al., 2009; Chapter 9). Rows of ϕ_t denote the states of units in season t, and columns represent the states of units in season $t + 1$. Each element within ϕ_t therefore represents the probability of a unit changing from one occupancy state to another between seasons. A general representation of ϕ_t would be:

$$\phi_t = \begin{bmatrix} U \to U & U \to A & U \to B & U \to AB \\ A \to U & A \to A & A \to B & A \to AB \\ B \to U & B \to A & B \to B & B \to AB \\ AB \to U & AB \to A & AB \to B & AB \to AB \end{bmatrix}$$

where $X \to Y$ denotes the probability of transitioning from occupancy state X in season t to state Y in season $t + 1$. The state 'U' indicates that the unit is unoccupied by both species. Importantly, the elements of each row must sum to 1, as a unit must transition to one of the four states.

As in the single-species case, the dynamic processes of unit occupancy are colonization and local extinction, but in the multiple-species situation these dynamic processes can be a function of the occurrence of the opposite species. Stated differently, the vital rates (colonization and extinction) for one species may be influenced by the current (or future) occupancy status of the other species. In Table 14.9 we define parameters that could be used to capture these dynamic processes of two-species co-occurrence. We have used a conditional parameterization similar to the RW parameterization for the single-season two-species co-occurrence model, hence it is assumed there is a dominant and a subordinate species (species A and B, respectively). This parameterization is not unique, and other parameterizations could be used. Using this conditional

TABLE 14.9 Definitions of parameters to model dynamic processes of change in the occupancy state of units over time, for a multi-season two-species co-occurrence model. The defined probabilities use a conditional parameterization, and other parameterizations are possible

Parameter	Description
$\gamma_t^{A|B}$	probability species A colonizes a unit between seasons t and $t+1$, given species B is present in season t
$\gamma_t^{A|b}$	probability species A colonizes a unit between seasons t and $t+1$, given species B is absent in season t
$\gamma_t^{B|AA}$	probability species B colonizes a unit between seasons t and $t+1$, given species A is present in season t and persists to $t+1$
$\gamma_t^{B|Aa}$	probability species B colonizes a unit between seasons t and $t+1$, given species A is present in season t but goes locally extinct between t and $t+1$
$\gamma_t^{B|aa}$	probability species B colonizes a unit between seasons t and $t+1$, given species A is absent in season t and remains absent in season $t+1$
$\gamma_t^{B|aA}$	probability species B colonizes a unit between seasons t and $t+1$, given species A is absent in season t but colonizes the unit between t and $t+1$
$\epsilon_t^{A|B}$	probability species A goes locally extinct between seasons t and $t+1$, given species B is present in season t
$\epsilon_t^{A|b}$	probability species A goes locally extinct between seasons t and $t+1$, given species B is absent in season t
$\epsilon_t^{B|AA}$	probability species B goes locally extinct between seasons t and $t+1$, given species A is present in season t and persists to $t+1$
$\epsilon_t^{B|Aa}$	probability species B goes locally extinct between seasons t and $t+1$, given species A is present in season t but goes locally extinct between t and $t+1$
$\epsilon_t^{B|aa}$	probability species B goes locally extinct between seasons t and $t+1$, given species A is absent in season t and remains absent in season $t+1$
$\epsilon_t^{B|aA}$	probability species B goes locally extinct between seasons t and $t+1$, given species A is absent in season t but colonizes the unit between t and $t+1$

parameterization, the transition probability matrix would be defined as:

$$
\phi_t = \begin{bmatrix}
\left(1-\gamma_t^{A|b}\right)\left(1-\gamma_t^{B|aa}\right) & \gamma_t^{A|b}\left(1-\gamma_t^{B|aA}\right) & \left(1-\gamma_t^{A|b}\right)\gamma_t^{B|aa} & \gamma_t^{A|b}\gamma_t^{B|aA} \\
\epsilon_t^{A|b}\left(1-\gamma_t^{B|Aa}\right) & \left(1-\epsilon_t^{A|b}\right)\left(1-\gamma_t^{B|AA}\right) & \epsilon_t^{A|b}\gamma_t^{B|Aa} & \left(1-\epsilon_t^{A|b}\right)\gamma_t^{B|AA} \\
\left(1-\gamma_t^{A|B}\right)\epsilon_t^{B|aa} & \gamma_t^{A|B}\epsilon_t^{B|aA} & \left(1-\gamma_t^{A|B}\right)\left(1-\epsilon_t^{B|aa}\right) & \gamma_t^{A|B}\left(1-\epsilon_t^{B|aA}\right) \\
\epsilon_t^{A|B}\epsilon_t^{B|Aa} & \left(1-\epsilon_t^{A|B}\right)\epsilon_t^{B|AA} & \epsilon_t^{A|B}\left(1-\epsilon_t^{B|Aa}\right) & \left(1-\epsilon_t^{A|B}\right)\left(1-\epsilon_t^{B|AA}\right)
\end{bmatrix}.
$$

Note the flexible nature of this parameterization, and the interesting biology that it allows to be investigated with respect to how the dynamics of each species depend on the presence of the other species. The nature of the dependencies are slightly different for the dominant and subordinate species. For example, the probability of species A colonizing a unit between seasons t and $t + 1$ can be different depending on the presence ($\gamma_t^{A|B}$) or absence ($\gamma_t^{A|b}$) of species B in season t. However, if the presence of species B in season t has no effect on the probability of species A colonizing a unit then it would be expected that $\gamma_t^{A|B} = \gamma_t^{A|b}$. Hence, one could formally assess how much evidence there is for these two hypotheses by fitting models where these parameters are either unconstrained, or constrained to be equal, and compare the models using an appropriate method (Chapter 3). Note that with this parameterization the dynamic parameters of the dominant species may only depend on the occupancy state of the subordinate species in season t, and not season $t + 1$. However, the probabilities that model the occupancy dynamics of the subordinate species between seasons t and $t + 1$ may depend on the presence or absence of the dominant species in both seasons, e.g., $\gamma_t^{B|AA}$, $\gamma_t^{B|Aa}$, $\gamma_t^{B|aA}$, and $\gamma_t^{B|aa}$. What constraints a researcher may place upon these parameters, if any, will depend upon the hypotheses of interest. If it is suspected the presence of species A has no effect on the probability of species B colonizing a unit, for example, all of these probabilities would be set equal. Whereas, if investigators suspect the probability of species B colonizing a unit depends on the presence of species A in season $t + 1$, but not season t, the constraints $\gamma_t^{B|AA} = \gamma_t^{B|aA}$ and $\gamma_t^{B|Aa} = \gamma_t^{B|aa}$ would be made.

As with the RW parameterization for the single-season co-occurrence model, conceptually, these conditional probabilities can be expressed in terms of the occupancy latent variables for the other species. For example:

$$logit\left(\epsilon_{t,i}^A | z_{t,i}^B\right) = \alpha_{A,t} + \beta_{A,t} z_{t,i}^B, \tag{14.9}$$

which equals $\epsilon_t^{A|B}$ when $z_{t,i}^B = 1$, and $\epsilon_t^{A|b}$ when $z_{t,i}^B = 0$. Similarly, the conditional extinction probability for species B could be conceived as:

$$logit\left(\epsilon_{t,i}^B | z_{t,i}^A, z_{t+1,i}^A\right) = \alpha_{B,t} + \beta_{B1,t} z_{t,i}^A + \beta_{B2,t} z_{t+1,i}^A + \beta_{B3,t} z_{t,i}^A z_{t+1,i}^A, \tag{14.10}$$

which leads to the conditional probabilities of $\epsilon_t^{B|AA}$, $\epsilon_t^{B|Aa}$, $\epsilon_t^{B|aA}$, and $\epsilon_t^{B|aa}$ through different combinations of values for $z_{t,i}^A$ and $z_{t+1,i}^A$.

Some regression coefficients in Eqs. (14.9) and (14.10) can be fixed to zero, and this is equivalent to placing constraints on the conditional probabilities. For example, if the extinction probability between seasons t and $t + 1$ for each species was thought to only depend on the presence of the other species at season t, that model could be fit to the data by setting $\beta_{B2,t} = \beta_{B3,t} = 0$ in Eq. (14.10). This is equivalent to making the constraints $\epsilon_t^{B|AA} = \epsilon_t^{B|Aa}$ and $\epsilon_t^{B|aA} = \epsilon_t^{B|aa}$, which makes no assumption about dominant and subordinate species (e.g., Miller et al., 2012a).

As we stated previously, this conditional parameterization is not unique, and other parameterizations could be used. One alternative would be a more mutualistic version that avoids consideration of a dominant and subordinate species, similar to the approach of Rota et al. (2016) in the single-season case. That is, there is a common effect size (i.e., same regression coefficient) for the presence of one species on the colonization and/or extinction probability of the other. Furthermore, it may be the presence of the species at either the start (t) or end ($t + 1$) period, or both, associated with the dynamic parameter of interest. This allows for some fairly complex parameterizations, and, as usual, practitioners should select the approach that is most consistent with both their objectives and their views on how their specific systems operate.

Detection probabilities can be defined in the same manner as for the single-season co-occurrence model, which allows for the fact that the true state transition between seasons may not be recorded. Once the transition and detection probability matrices have been defined, the multi-season multi-state framework can be applied, and the observed data likelihood or complete data likelihood used for inference. Covariates and missing observations can also be incorporated into a multi-season two-species co-occurrence model as with the single-season modeling.

Before we present an example of the multi-season co-occurrence model, we would remind readers that these modeling approaches are not only an estimation framework for the analysis of field data, but also provide the basis for predicting future changes of the species distributions. Once the dynamic parameters have been estimated, and any interaction between the species' dynamics quantified, the transition probability matrix can be used to predict the state of each unit, i.e., which species will be present, in the next season. Researchers may also adjust the values of specific dynamic parameters to investigate what effect that may have on the long-term state of the system.

14.6 EXAMPLE: BARRED AND NORTHERN SPOTTED OWLS

Multi-season two-species models provide an opportunity to directly study the processes underlying changes in occupancy. They offer the potential to learn about species interactions even for systems near equilibrium, but learning opportunities are especially great during periods of dramatic change, for example when one species invades the range of another. This example is based on field work of Janice Reid, Eric Forsman, Raymond Davis, and their U.S. Forest Service colleagues, and is reported in Yackulic et al. (2014). The study concerns an invading species (barred owl, *Strix varia*) and a resident species of conservation

FIGURE 14.6 Male northern spotted owl (*Strix occidentalis caurina*). *Janice Reid.*

concern (northern spotted owl, *Strix occidentalis caurina*; Fig. 14.6) in western Oregon, USA.

The analysis is based on detection data from 158 contiguous survey polygons (i.e., 'patches') that covered an area of about 1000 km² on the Roseburg District of the Bureau of Land Management in western Oregon, USA (hereafter 'Tyee Study Area'). Most survey polygons were surveyed every year from 1990 to 2011, with multiple visits occurring between 1 March and 31 August to locate and band owls and record numbers of young produced (Anthony et al., 2006; Forsman et al., 2011). Surveyors targeted spotted owls by imitating their calls (Forsman, 1983; Reid et al., 1999), but barred owls were also detected during surveys because they responded aggressively to spotted owl calls (Kelly et al., 2003; Crozier et al., 2006). Survey emphasis was on spotted owl breeding pairs, so when surveyors were occasionally unable to find a mate, detections of single northern spotted owls were recorded as a nondetection (of a pair). For barred owls, detections included both paired and unpaired individuals, so occupancy was defined in terms of individual birds, rather than pairs as for spotted owls.

Survey polygons encompassed a mosaic of forest stands of various ages, ranging from recent clearcuts to old-growth forest over 200 years old. Active habitat selection can be an important factor influencing occupancy dynamics of many species, so habitat covariates were included in the modeling. Specifically, two habitat covariates performed well in a previous study of barred owl dynamics (Yackulic et al., 2012) and are similar to covariates that have been used in many previous analyses of spotted owl occupancy, as well. One covariate reflected the amount of 'old forest' within a polygon, and the other focused on proximity to streams and reflected amount of riparian forest. Both covariates were standardized to have a mean of 0.0 and a standard deviation of 0.5.

Each polygon was modeled as being in one of four mutually exclusive true states during each breeding season; both species present (state = 3), only spotted owl present (state = 2), only barred owl present (state = 1), neither species present (state = 0). Using this coding, barred owls were considered as species A, and spotted owls as species B. Observation states used the same numbering system, admitting more uncertainty with decreasing state number. For example, observation state 0 indicated no detections, and true state could have been any of the four states, 0–3. Observation state 1 indicated detection of barred owl only, with possible true states being either 1 or 3. Observation state 3 admitted no uncertainty, as both species were detected and true state was known to be 3. Uncertainty associated with the detection process was modeled using parameters $p_t^{m,l}$, indicating the probability that a patch in true state m, at time t, was observed to be in observation state l. These detection parameters were then modeled as functions of potentially relevant covariates. Based on findings of Bailey et al. (2009) and Yackulic et al. (2012), spotted owl detection probabilities were modeled as linear logistic functions of barred owl presence/absence, survey timing (night/day/dusk), and survey protocol (stopping to call at particular stations vs. continuous walks with calling). Barred owl detection probabilities were modeled as functions of survey timing and protocol, as well as survey duration and general time period (first half vs. second half of study period had very different barred owl densities).

Occupancy state dynamics are governed by state transition probability parameters, which are written as functions of species-specific probabilities of local colonization and extinction. For example, here we denote $\epsilon_t^{S|b}$ as the probability that northern spotted owls (S) are absent from a patch at time $t + 1$, given their presence in the patch at time t (local extinction), and given absence of barred owls (b) at time t. Similarly, let $\gamma_t^{B|S}$ denote the probability that barred owls are present (B) in a patch at time $t + 1$, given that they are absent from the patch at time t (colonization) and that spotted owls were present at time t. The probability of a patch transitioning from having just spotted owls (state 2) to both owl species (state 3) can be written as $\left(1 - \epsilon_t^{S|b}\right) \gamma_t^{B|S}$, indicating that between t and $t + 1$, spotted owls do not go locally extinct and barred owls colonize the patch. As introduced above, the conditioning in the superscripts indicates the dependence of these dynamic rate parameters on the presence/absence of the other species. The only difference between this notation and that of Table 14.9 is that here we condition only on the occupancy status of the other species in season t (not $t + 1$ also).

Alternative models for occupancy dynamics incorporated different ecological hypotheses about possible competitive effects (Yackulic et al., 2014). Species-specific probabilities of colonization and extinction were modeled as functions of both the habitat and occupancy state of the survey polygon. For example, under one model of competitive effects, extinction probability between times t and $t + 1$ for spotted owls could depend on whether barred owls were also present in the patch at time t, that is, on whether polygon state at time t was 2 or 3. One hypothesis of competitive effects of the dominant barred owl on spotted owl occupancy dynamics would predict higher spotted owl extinction for patches with barred owls (state 3) than without (state 2), i.e., $\epsilon_t^{S|B} > \epsilon_t^{S|b}$. In addition to competition, the modeling incorporated competing hypotheses about habitat effects and autologistic effects of neighborhood conspecific occupancy on local probabilities of extinction and colonization, i.e., incorporated spatial correlation in the co-occurrence dynamics. Neighborhood effects were found to be important to barred owl dynamics in a previous study (Yackulic et al., 2012), and some two-species models included such effects for spotted owls as well.

An example of this modeling of vital rates is provided below:

$$logit\left(\epsilon_{t,i}^{S|b}\right) = \beta_0 + \beta_1 \bar{\psi}_t^S + \beta_3 H_{t,i},$$

and

$$logit\left(\epsilon_{t,i}^{S|B}\right) = \beta_0 + \beta_1 \bar{\psi}_t^S + \beta_2 + \beta_3 H_{t,i},$$

where β_2 is the size of the competitive effect of barred owl presence in year t on spotted owl extinction. In the above expressions local extinction probability of spotted owls from polygon i between years t and $t + 1$, conditional on barred owl occupancy in year t, is written as a function of average spotted owl occupancy in the study area, $\bar{\psi}_t^S$ (the autologistic effect), barred owl occupancy for patch i, and a habitat variable associated with patch i, $H_{t,i}$. However, neither occupancy state of the hypothesized competitor species nor that of the focal species is a standard covariate, in the sense that occupancy is not directly observed, but is instead characterized by uncertainty associated with the detection process. Two-species dynamic occupancy modeling properly incorporates this uncertainty in the analysis and associated variance estimates. All modeling was conducted using Program PRESENCE (Hines, 2006).

Yackulic et al. (2014) applied a sequential approach to model selection, as a pragmatic solution to the problem of far too many potential models. The detailed process of developing and then selecting model structures is found in Yackulic et al. (2014), who also relied heavily on previous modeling of Bailey et al. (2009) and Yackulic et al. (2012). Briefly, the sequential approach of Yackulic et al. (2014) entailed first focusing on detection probability while modeling all other parameters as generally as possible, and next identifying a general baseline model for spotted owls which included effects of barred owls, but no habitat effects. The third model set included habitat effects, and the final set included full interspecific interactions, with potential reciprocal effects of spotted owls on barred owls. The models identified in this way were then modified one variable at a time, revisiting decisions made in previous steps in the analysis, mostly for the purpose of potential simplification of model structures. Perhaps the most important aspect of the model selection strategy is that all models were based on detailed *a priori* hypotheses and associated predictions about magnitudes and signs of modeled effects (Yackulic et al., 2014).

Results include substantial detail, and we simply focus on selected highlights here. An important aspect of the modeling of detection probability was the substantial reduction in spotted owl detection probability that occurred at patches also occupied by barred owls (state 3). This kind of effect is frequently predicted when one species is competitively dominant over another and becomes a very important consideration when modeling detection/nondetection data. Autologistic modeling of effects of study area occupancy on local extinction and colonization were important for barred owls as expected based on Yackulic et al. (2012). However, as predicted for species close to equilibrium, spotted owl models with autologistic effects were not highly ranked. As predicted, amount of old growth forest was positively associated with spotted owl colonization and negatively associated with extinction. Amount of riparian forest was similarly important for vital rates of barred owls (Yackulic et al., 2012).

Inferences about interspecific competition were the focus of the study. The four models receiving virtually all of the support in the final round of model selection all provided evidence that extinction probability increases for each species when the other is present, suggesting a strong role of competition in structuring occupancy dynamics. Evidence for the effects of competition on local colonization rates was equivocal, as two of the top four models suggested that spotted owls were less likely to colonize survey polygons that were already occupied by barred owls, as predicted. Contrary to predictions, there was some evidence that barred owls were more likely to colonize survey polygons already occupied by spotted owls.

Modeling provided year-specific estimates of local probabilities of extinction and colonization for spotted owls and barred owls for each polygon in the study area. Yackulic et al. (2014) used a simulation approach to compute summary statistics based on rate estimates from the top overall model, averaged over all polygons in the study area. Rates of colonization for spotted owls were highly variable over time, whereas those for barred owls steadily increased, largely because of the autologistic effect of increasing barred owl occupancy (Fig. 14.7A). Rates of local extinction increased monotonically for spotted owls, largely because of the competitive effects of increasing barred owl occupancy, whereas those for barred owls decreased, largely because of the autologistic effects of increasing conspecific occupancy (Fig. 14.7B).

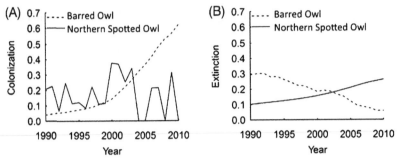

FIGURE 14.7 Mean probabilities of local (A) colonization and (B) extinction for two owl species by year. Values were calculated by averaging over all survey polygons and 1000 simulations. *(From: Yackulic et al., 2014)*

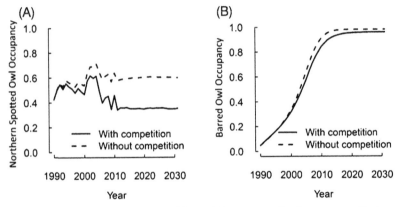

FIGURE 14.8 Projected effects of competition on (A) mean spotted owl occupancy estimates and predictions, and (B) mean barred owl occupancy estimates and predictions. *(From: Yackulic et al., 2014)*

    The two-species modeling framework also provided a nice means of projecting the population-level consequences of competitive and autologistic effects. Yackulic et al. (2014) computed average occupancy estimates and then projected occupancy dynamics for an additional 20 years. The averages and projections that included competitive effects were computed directly from estimates, whereas projections of occupancy in the absence of competition were computed by restricting spotted owl extinction and colonization probabilities to those for polygons not occupied by barred owls, $\epsilon_t^{S|b}$ and $\gamma_t^{S|b}$ (equivalent to setting β's associated with competitive effects equal to 0). Mean occupancy for spotted owls shows substantial effects of competition in recent years, with equilibrium occupancy declining from an average of 0.61 in early years to 0.35 (Fig. 14.8A). In contrast, average barred owl occupancy has increased as a consequence of autologistic effects during their invasion of the study area, with projected competitive effects of spotted owls only resulting in a drop from an equilibrium occupancy of 0.98 to 0.96 (Fig. 14.8B). Thus, although each species induces increases in local extinction probabilities of the other, competitive effects are asymmetric, with barred owls having a much larger effect on spotted owl occupancy than vice versa.

Despite the observational (non-experimental) nature of this study, the fact that this two-species system was exhibiting transient dynamics with the barred owl invasion permitted relatively strong inferences about competitive effects. In addition, resulting models led to interesting inferences not discussed here (see Yackulic et al., 2014) about competitive effects modifying apparent species–habitat relationships. Models also led to interesting inferences about the potential to manage spotted owls via efforts to remove barred owls (see Yackulic et al., 2014).

14.7 STUDY DESIGN ISSUES

We have had limited opportunity to give in-depth consideration to study design issues as they pertain to investigating relationships between species, nor do we know of any relevant work from other research groups. Many of the issues raised in Chapters 11 and 12 for single-species studies are also pertinent to the multi-species case, such as defining and selecting units, allocating effort, etc.

One aspect of study design that deserves special attention in the static multi-species case is appropriate definition of a 'season' and 'survey'. In the single-species case the main considerations for defining a season were: (1) the time period over which it is reasonable to consider the occupancy state of units as unchanged, or over which changes occur completely at random; and (2) whether 'occupancy' or 'use' provides the most relevant information with respect to the study objectives. Recall that a season represents a snapshot of the system at a particular point in time, and surveys are repeated observations of the static system; inference about the species interactions is based upon this snapshot. If the season length is too short, it may yield insufficient information about the system, but if the season is too long it may result in a distorted view of the system. These points hold true for multi-species studies, with the added consideration that interpretation of 'co-occurrence' will also be influenced by how a season and survey are defined relative to the objectives of the study and the biology of the species in the system. For example, suppose that over a shorter timeframe (e.g., a day or week) two species appear to co-occur very rarely, as they tend to be competitors for the same resources, or perhaps they may be predator and prey species, so do not tend to be at the same unit at the same time. However, over a longer timeframe (e.g., 1–3 months) they appear to co-occur more frequently, as they prefer similar habitats or are highly mobile species, hence they will both be present in the unit at some stage during the longer interval, but rarely at the same time. Inferences about how the species co-occur may be very different depending upon whether a short or long season was used to capture a snapshot of the system. In some cases, species interactions may manifest in the detection parameters because species may only avoid one another for the time-scale associated with a single survey (e.g., Lewis et al., 2015a). Using both occupancy and detection parameters, investigators can explore biological hypotheses relative to spatial and temporal avoidance (exclusion), provided an appropriate study design has been used (e.g., season length and type of repeat surveys) for the species and co-occurrence patterns of interest.

A similar argument also applies when defining a 'unit'. Here the issue is the spatial scale of co-occurrence rather than a temporal one. Too small and the resolution may be inappropriate; too large and many species will appear to co-occur. Again, as part of the study's objective, careful consideration needs to be paid to the spatial scale of the co-occurrence questions that are of interest. We emphasize that inferring process from static patterns alone can be deceiving, yielding misleading biological inferences about species interactions (Yackulic et al., 2015). When considering dynamic applications, investigators need to carefully consider the length of time between seasons. If this time period is too short, little change in occupancy states will occur, but if the time period is too long it may mask colonization and extinction processes that depend on the occurrence of other species in the system.

Finally, care should be taken to insure that all species are adequately sampled. If one of the occupancy states is unobservable, e.g., in a simple host–pathogen system, the pathogen-only state cannot be observed if detection of the pathogen is conditional on host occurrence and detection. Ignoring the unobservable state may lead to misleading conclusions regarding host–pathogen dynamics (Mosher et al., 2017).

14.8   GENERALIZING TO MORE THAN TWO SPECIES

Generalizing either the single-season, or multi-season, co-occurrence modeling to more than two species is, theoretically, very straightforward. As noted earlier in this chapter, increasing the number of species simply increases the number of possible occupancy states for a unit. Parameters associated with the occurrence, co-occurrence, transition, and detection probabilities could be duly defined. Practically, however, the problem is the potential number of parameters to estimate and that typically there will be insufficient data to reliably estimate many of those parameters. For example, suppose there were four species of interest, which would define 16 (2^4) possible states. Therefore, potentially 15 occurrence-related probabilities for the first season, 65 detection-related probabilities for each survey (if a different detection probability is allowed for each combination of species that may be present at a unit), and 240 transition probabilities per inter-season period, could be defined! Clearly, in most settings there is going to be insufficient data to estimate many of those parameters.

Progress can be made by recognizing that many of the parameters that could be defined are associated with higher-order interactions among species, and that certain parameterizations could be used so that it is not necessary to estimate all of those parameters, by applying some constraints. For example, the number of detection-related parameters could be quickly reduced by assuming one probability of detecting a focal species if it is present at a unit by itself, but allowing a different detection probability if other species within the set that are of interest are also present, without differentiating by how many, or which combination, of the other species are present.

Other types of constraints are also possible. Consider the conceptual approach used previously where the conditional occupancy probabilities were defined in terms of the presence/absence of the other species, i.e., $logit(\psi_i^A|z_i^B) = \beta_0 + \beta_1 z_i^B$. This concept can be extended for a greater number of species, for example:

$$logit(\psi_i^A|z_i^B, z_i^C, z_i^D) = \beta_0 + \beta_1 z_i^B + \beta_2 z_i^C + \beta_3 z_i^D$$
$$+ \beta_4 z_i^B z_i^C + \beta_5 z_i^B z_i^D + \beta_6 z_i^C z_i^D$$
$$+ \beta_7 z_i^B z_i^C z_i^D,$$

where β_1–β_3 are the main effects for the presence of the other three species on ψ_i^A, β_4–β_6 are the effects when two of the other three species are present (i.e., second-order interactions), and β_7 is the effect on ψ_i^A when all three of the other species are also present at the unit (i.e., a third-order interaction). The number of parameters would be greatly reduced if only the main effects of the other species were retained in the model (i.e., $\beta_4 = \beta_5 = \beta_6 = \beta_7 = 0$), so only pair-wise relationships among species would be modeled. It would also simplify interpretation. Under the full model, interpretation of what effect species B has on the probability species A occupies a unit also requires consideration of the presence of the other species because of the higher-order interactions, just as in a regular regression model that includes interactions among predictor variables.

This type of approach could be applied to both the conditional RW parameterization (by applying successive conditioning), or the more symmetric parameterization of Rota et al. (2016). It could also be applied to detection and transition probabilities.

Another, less mechanistic, approach would be to use correlated random effects to model co-occurrence patterns among a larger number of species, where the correlation would be positive if species tend to co-occur more often than expected under a hypothesis of independence, and negative if they co-occur less often than expected.

14.9   DISCUSSION

In this chapter we have presented techniques that permit investigation of relationships among multiple species, while accounting for the imperfect detection of the species. These models are very flexible and should be suitable for a wide range of applications. Not only can these models be used to determine the level of support for various biological hypotheses, but by accounting for detection probability, the levels of co-occurrence can be estimated. We believe such approaches will provide more reliable inferences about the patterns and processes of species co-occurrence than other methods that have been used to date, particularly when used in combination with a sound experimental design (Chapter 1).

The multiple-season model of Section 14.5 should be particularly relevant to researchers working with species of high conservation importance that are under threat from invasive species that prefer similar habitats or occupy a similar niche. The colonization and extinction parameters that are conditional on the occurrence of the dominant species should be of prime interest when investigating relationships between native and invasive species (e.g., Dugger et al., 2015). These models also provide the inferential basis for theoretical patch-based models that use multi-species occupancy patterns as state variables. For example, such models have been applied to investigate intraguild predation among competing species (Robinson et al., 2014) and the interaction of predation and habitat dynamics in determining structure and dynamics of multi-species communities (Miller et al., 2012a). The modeling of Section 14.5 is ideal for estimation of the various transition probabilities required for such Markov process modeling.

Finally, we acknowledge new modeling techniques that account for multiple interacting species (Rota et al., 2016), which offers one approach for applying general procedures to a large number of species in a reasonable manner. However, the suggestion of modeling only a small number of species does have its advantages. It encourages researchers to carefully consider *a priori* the species on which they wish to focus, rather than taking a 'shotgun' approach and collecting data on a large number of species and looking for 'significant' relationships that may be spurious. We would advocate the former approach as much closer to our view of science.

Chapter 15

Occupancy in Community-Level Studies

In the previous chapter we considered modeling approaches that could be used to investigate interspecific relationships among multiple species. A different kind of multi-species study focuses on the state variable species richness, defined as the number of species (either in total or within a predetermined group) within a predefined area of interest (Chapter 1). Investigations focus on changes in this state variable as a function of local rates of species colonization and extinction, without regard to how species directly interrelate. Traditionally species accumulation curves have been used to estimate species richness, where the number of species encountered is plotted as a function of search effort within the sites (see reviews in Soberón and Llorente, 1993; Colwell and Coddington, 1994; Flather, 1996). The intent is that the accumulation curve will flatten off or asymptote to the total number of species present at the site with increasing search effort. Another approach is to use closed-population capture–recapture models to estimate the number of species not encountered at a site, based upon the species that were encountered at least once (e.g., Burnham and Overton, 1979; Bunge and Fitzpatrick, 1993; Cam et al., 2000b; Williams et al., 2002).

Here we suggest that many of the methods described in Parts II and III could be applied to community-level studies as often the data collected in such studies are the presence/absence (or more correctly the detection/nondetection) of multiple species. In some instances, the basic sampling situation is different than that considered earlier and requires model parameters to be interpreted differently, but the practical application of the modeling is very similar (much like the application of capture–recapture methods to community-level studies). In other instances, we simply view the multi-species studies as the combination of a number of single-species studies, all conducted at the same set of units. We structure our thoughts by considering two different situations. The first is when sampling of the community takes place at only a single unit (or a small number of units) within a relatively small area of interest. The second situation is where the community is considered at a larger scale with sampling taking place at many units, as in the single-species situations considered earlier. We have structured the chapter in this manner because the same basic models are used, but they are applied differently in each situation.

Occupancy Estimation and Modeling. http://dx.doi.org/10.1016/B978-0-12-407197-1.00020-X

The use of occupancy models in community studies, now generally known as multi-species occupancy models (MSOMs), has greatly expanded in the last decade, with a recent review by Iknayan et al. (2014) (see also Beissinger et al., 2016, and Chapter 11 of Kéry and Royle, 2016). The MSOM framework is based on integrating an ensemble of species level occupancy models, such as described in Chapter 4, while regarding the parameters of each species-level model to be random effects, related across species with a prior distribution describing the variation among species within the community. In other words, the MSOM is a hierarchical model. Various more or less independent incarnations of the idea have appeared (Dorazio and Royle, 2005; Gelfand et al., 2005; Ovaskainen and Soininen, 2011; Warton et al., 2015). A key difference among these various incarnations has to do with whether or not the model applies to the observed community only, or whether the model permits extrapolation to the unobserved species. Only the approach by Dorazio and Royle (2005) accommodates explicit extrapolation to the unobserved portion of the community. In addition to extending inference to the full community, hierarchical models improve estimation by 'borrowing' information across species (Bayesian shrinkage), leading to improved estimates of shared parameters, and they allow modeling species level effects even for rare species. Finally, using the MSOM framework we can make explicit predictions of community structure as a function of landscape and habitat factors. The basic idea of hierarchical models constructed from elemental species-level models has also been extended in many different directions, including to multi-season studies (i.e., dynamic occupancy models for communities) and to a multi-species *abundance* modeling (MSAM) framework. We highlight some of the important methodological extensions in Sections 15.3 and 15.4 below.

15.1 INVESTIGATING THE COMMUNITY AT A SINGLE UNIT

In this section we consider a basic sampling situation where surveys are conducted for multiple species at a single sampling unit. As in the single-species case, the intent of the surveying is to establish the possible presence of the species at the unit, although the species are not detected perfectly. Hence, repeated surveys are required to estimate detection probability. Each species may be detected or not detected during a survey; therefore a detection history can be constructed denoting the sequence of detections and nondetections for each species. The period of repeated surveying is referred to as a season, and the assumption is made that the occupancy status of the unit for each species can be considered static during that period (i.e., the members of the local species pool present at the unit are constant during a season). Repeated surveys for the species may be conducted either temporally (e.g., daily) or spatially (e.g., on small plots within the unit). As for the single-species situation, the exact nature of the repeat surveys and definition of a season influence how model parameters should be interpreted. The unit may be surveyed for multiple seasons (e.g., years), where changes in the membership of the local species pool present at the unit, may occur between seasons.

When data are collected in this form, Chapter 20 of Williams et al. (2002) provides a concise summary of methods that could be used to estimate community related parameters using capture–recapture techniques. They utilize an analogy between the above sampling situation and capture–recapture studies where the detection/nondetection of species is analogous to the detection/nondetection of individual animals. The number of species present at the unit (species richness) can then be estimated using capture–recapture methods that are appropriate for estimating the number of individual animals in a study area. That is, the total number of species that may be present at a unit is estimated from data collected on the species that were detected at least once. Williams et al. (2002) give some guidance on the types of estimation methods they consider to be most appropriate depending upon whether repeated surveys have been conducted temporally or spatially, however they note that generally they would expect that heterogeneity in the detection probabilities of different species should be accounted for. If data are collected over multiple seasons, then estimates of community parameters such as rates of turnover and extinction can be obtained (Nichols et al., 1998a), and these estimates can be used to investigate hypotheses about community dynamics (e.g., Boulinier et al., 1998a; Doherty et al., 2003b, 2003a).

Species richness at a particular location is likely to be determined not only by local ecological conditions, but also by the regional species pool (e.g., see Cornell and Lawton, 1992; Cornell, 1993; Karlson and Cornell, 1998; Cam et al., 2000b). In order to focus on local ecological determinants of richness, Cam et al. (2000b) defined *relative species richness* as the ratio of richness at a unit to the number of species in the regional pool. This quantity can be viewed as a measure of community completeness and should be greater at units with favorable ecological conditions than at units with unfavorable conditions (e.g., as might be caused by urbanization). Cam et al. (2000b) estimated relative richness at a unit as the ratio of estimated richness at the unit to the known or estimated size of the local species pool.

Occupancy modeling provides a means of modeling and estimating relative richness directly, rather than via a two-step approach, as in Cam et al. (2000b). Note that the techniques used by Cam et al. (2000b) to estimate species richness at a unit (see Williams et al., 2002, and references therein) place no constraint upon the total number of species that may reside in the community. As noted above, however, in many situations it is reasonable to suggest that such a limit is known. This may arise, for example, when researchers have precompiled a list of species in which they are interested. In locations that have been studied for a long period of time, investigators will frequently have a list of the species corresponding to the species pool (e.g., Cam et al., 2000b). Similar to Williams et al. (2002), an analogy can be drawn between estimating the proportion of units occupied by a single species with estimating the proportion of species on a list of known size that occupy a single unit (i.e., each species is considered as a 'unit' in the context of Chapters 4–9). When the list is the (known) regional species pool, then the proportion of species on the list that is present at the unit is exactly the relative richness parameter of Cam et al. (2000b). The occupancy approach and methods detailed in Chapters 4–9 permit the direct estimation and modeling of this parameter, ψ, as well as of the dynamic rate parameters that cause the relative species richness and species composition to change over time. Importantly, as covariate information can be easily incorporated into the above methods, information related to the individual species (i.e., characteristics of different species such as body size or degree of specialization) can be used to model the dynamic community parameters, even for those species on the list that are never detected at the unit. This cannot be done using capture–recapture techniques, as without specification of a species list, covariate information about species that are never detected is unknown.

15.1.1 Fraction of Species Present in a Single Season

Suppose a list of s species is composed, and interest lies in what fraction of the species on the list is present at a unit. Repeated surveys are conducted for the s species, and a detection history for each of the species can be constructed (e.g., Table 15.1). Because species are detected imperfectly, some species that were not detected at the unit may have in fact been present (i.e., a false absence) while others could be genuinely absent from the unit (i.e., not part of the local community during that season). The analogy between the estimation problem here and that of Chapters 4–7 is obvious, and as such the same estimation techniques can be used. The methods of Chapter 7 may be particularly useful to account for the heterogeneity in detection probabilities among species, but the ability to incorporate covariate information for all species on the list may mitigate the usual problems caused by heterogeneity (i.e., much of the heterogeneity among species may be modeled in terms of covariates). As inference is to be made only about those species that appear on the list, the species list represents the entire population of interest. To accurately represent the uncertainty in the estimated fraction or proportion of species present, the finite population methods of Section 6.1 are particularly applicable. However, if interest is in the underlying probability of a species on the list being present at the location, then it is not

TABLE 15.1 Example of a partial species list for ten bird species with the associated detection history (h_i) for each species from four surveys conducted in the mid-Atlantic region of North America. Examples of species-specific covariates are also indicated

Species	h_i	Songbird	Body size
Northern parula, *Parula americana*	1001	yes	small
American robin, *Turdus migratorius*	0111	yes	large
Indigo bunting, *Passerina cyanea*	0100	yes	medium
Song sparrow, *Melospiza melodia*	0011	yes	medium
Northern mockingbird, *Mimus polyglottos*	0010	yes	large
Blue-winged warbler, *Vermivora pinus*	0000	yes	small
Northern cardinal, *Cardinalis cardinalis*	0000	yes	large
Mourning dove, *Zenaida macroura*	0101	no	large
Hairy woodpecker, *Picoides villosus*	0000	no	large
Ruby-throated hummingbird, *Archilochus colubris*	1000	no	small

necessary to correct for the finite population, and the methods of Section 4.4 can be used.

15.1.2 Changes in the Fraction of Species Present over Time or Space

The analogy with the single-species case can be continued to investigate changes in the members of the species pool present at a unit over time using the unconditional explicit dynamics models of MacKenzie et al. (2003) (Chapter 8). That is, changes to the species number and composition through the processes of colonization and local extinction can occur between seasons, but not within seasons. In this context, the probability of colonization (γ_t) relates to the probability that a species absent from the unit in season t, occupies the unit in season $t + 1$. Similarly, local extinction probability (ϵ_t) is the probability that a species present at the unit is season t, is absent in season $t + 1$. Reparameterized versions of this model may also be used when analogous quantities are of interest (e.g., the rate of change in the fraction of species present at the unit). This kind of dynamic model, in which members of the species pool join and leave the community at a local unit, is well known to theoretical ecology (e.g., MacArthur and Wilson, 1963, 1967; MacArthur, 1972; Ricklefs and Schluter, 1993; Boulinier et al., 2001; Hubbell, 2001). Methods such as those suggested here should permit formal inference about such topics of theoretical interest (e.g., MacArthur and Wilson, 1963, 1967; Boulinier et al., 2001; Hubbell, 2001) as the existence of dynamic equilibria (e.g., using methods of Chapters 8 and 10).

Incorporating covariate information about the species on the list, even for those species never detected, may be very useful for making reliable inference about the processes of change within the local species pool. For example, native species may have lower colonization and higher local extinction probabilities than invasive exotic species, or species with large body size may have higher local extinction probabilities than small species. Individual species characteristics cannot be incorporated into most of the community estimation and modeling methods described by Williams et al. (2002), and ability to use species covariates represents an important advantage of the occupancy estimation approach. As suggested in Chapter 7, heterogeneity in the model parameters not accounted for by covariates can be accommodated by the use of 'random effects' models.

The same modeling could also be applied to investigate how the community changes spatially along a linear feature, e.g., along a transect or a stream. The transect, for example, could be divided into segments, and within each segment repeat surveys are conducted for the species on the list of interest. Segment-specific estimates of the size of the community could be obtained assuming independence of the segments using the implicit-dynamics multi-season model (Chapter 8), with transect segments being analogous to 'seasons' in regular applications of the method. Alternatively, some spatial correlation could be accounted for by assuming that changes in the community are well approximated by a first-order Markov process, and use the explicit-dynamics multi-season model of MacKenzie et al. (2003) (Chapter 8). That is, the modeling allows the probability of a species being present in the community of segment $t + 1$ to depend on the presence of the species in the community of segment t. This implies a directional nature to the interpretation of the dynamic parameters, which would be reasonable in many applications, e.g., downstream movements, along elevational gradients, or increasing latitude.

15.2   INVESTIGATING THE COMMUNITY AT MULTIPLE UNITS

Often species richness may be of interest at more than a single unit. The methods outlined above could clearly be extended to more than one unit, where each unit may have the same or different species lists (pools). An alternative to estimating the fraction of species present for each of a large number of units, is to estimate the proportion of units occupied by each species (i.e., the initial context considered in Chapters 4–5), but model all species simultaneously. Inference about the community can then be made based upon the joint modeling of the species. *Species richness* could then be defined as the number of species present at a single unit, or as the number of species present in the community in the larger area from which the sampled units were selected. An advantage of modeling all species simultaneously is that a common parameter may be shared by different species. Therefore, it should yield gains in precision in the sense that sharing parameters may allow for improved inferences about some species relative to when each species is considered individually (e.g., MacKenzie et al., 2005), provided that the additional model assumptions hold.

In other situations, interesting biological questions involve possible similarities in species responses to environmental and habitat characteristics. For example, unit occupancy by species within the same guild (a group of species believed to exhibit similar characteristics such as foraging habits, nesting habitat, etc.) would be expected to exhibit similar relationships to habitat covariates. Guild membership can be hypothesized *a priori*, and hypotheses can be tested about similar relationships among species within, but not between, groups. Another advantage of this multi-species modeling approach is that now species-specific detection probabilities can be directly estimated, unlike using the approach suggested in the previous section for single-unit analyses.

In this section we outline our ideas on how the joint modeling of multiple species at a large number of units could be used the make inferences at the community level. First, we discuss single-season studies with two different kinds of objectives: (1) investigations directed at modeling probabilities of occupancy and/or detection as functions of species- and unit-specific covariates; and (2) the estimation of species richness, both at sampled locations and for larger areas from which these locations are selected. Then, we briefly discuss occupancy studies for multiple species over multiple seasons.

15.2.1 Single-Season Studies: Modeling Occupancy and Detection

The joint modeling of the data from each unit may be used to address interesting biological hypotheses, such as whether groups of similar species at different units have a similar occupancy probability. Discussion of covariate modeling for the single-unit investigations described in Section 15.1 emphasized species-specific covariates such as body size, specialization, and the exotic versus native dichotomy. Here we can consider analysis of multiple units with the same species pool. Such analyses would be effectively controlling for influences of the regional species pool, and covariate modeling would investigate the role of local unit characteristics (e.g., habitat, disturbance) in determining local species richness and dynamics. Because species identities are retained in these analyses, models can incorporate covariates associated with individual species as well as those associated with local units.

We consider the same basic sampling framework as that assumed in Chapters 4–7; repeated detection/nondetection surveys are conducted at s units within a single season. However, now data are collected on M species rather than just a single species. This type of data could be used to assess interspecific relationships between species (i.e., Chapter 14), but here we assume species co-occur independently. Choice of which M species to model will depend upon study objectives, but it may be: (1) a small number of indicator species that were detected at least at one of the units; (2) all species that were detected at least once; or (3) a list of species defined *a priori* that includes some species never detected at any of the units (note that in this case the detection probability for species never detected must be assumed equal to that of other species that were detected, or related by some form of detection function). Modeling of each of the M species can be conducted as in Chapters 4–7, including investigation of the effect of different covariates on different species.

Generalizing the notation used previously, let ψ_{mi} be the probability that species m is present at unit i, and p_{mij} be the probability of detecting species m in the jth survey of unit i (given presence of species m at unit i). The most general models will contain different detection and occupancy parameters for all M species. The effect of different covariates may also be different for different species. However, reduced-parameter models could be considered where different species share common parameters. For example, many frog and toad species become more or less detectable with changes in temperature due to behavioral changes. Rather than estimating a different 'temperature effect' for each species, which could be denoted as model $p(Species \times Temperature)$, it may be reasonable to consider a model where the effect of temperature is the same for all, or a subset, of species, denoted as model $p(Species + Temperature)$. That is, rather than consider a general model of the form:

$$logit(p_{mij}) = \alpha_m + \beta_m Temp_{ij},$$

where $Temp_{ij}$ is the temperature recorded during survey j of unit i, and β_m is the effect of temperature on detection probability for species m, a reduced-parameter model of the form:

$$logit(p_{mij}) = \alpha_m + \beta Temp_{ij}$$

could be considered. Under this latter model, each species may have a different average detection probability, but changes in species-specific detection with temperature occur in parallel, reflecting the similar effect of temperature. Sharing parameters among species in this manner is effectively a form of aggregating or pooling data, and in this situation model selection techniques can be used to indicate the level of aggregation among species that is best supported by the data (MacKenzie et al., 2005).

The effect of unit-specific covariates (e.g., habitat type, patch size, etc.) on occupancy probabilities can also be investigated for each species, and it may be reasonable to hypothesize that the effect of some covariates is similar across a range of species. For example, a single 'habitat effect' could be estimated that applies to the occupancy of all species in a group. Thus, we might investigate an additive model for occupancy, e.g., $\psi(Species + Habitat)$, with occupancy varying across species, but with all species (at least within a group of interest) otherwise showing the same relationship to a habitat covariate (i.e., the same slope parameter for the occupancy by habitat relationship). Note that incorporating unit-specific covariates in the analysis of multi-species data in this manner is one approach for examining factors that may affect local species richness (e.g., Boulinier et al., 1998a, 2001). Across the landscape, different species may have different habitat preferences. By modeling these preferences for multiple species within that landscape, areas that are preferred by a greater or lesser number of species could be identified.

The number of species present at unit i (unit-specific species richness; R_i) could be defined in terms of the underlying latent variable for the presence of species m at unit i, z_{mi}, as:

$$R_i = \sum_{m=1}^{M} z_{mi}.$$

This can be easily calculated when using the complete data likelihood approach with Bayesian estimation methods that impute the value of z_{mi} during the analysis. Alternatively it could be estimated as:

$$R_i = \sum_{m=1}^{M} \psi_{mi},$$

or using the results of Section 4.4.5:

$$R_i = n_i + \sum_{\{m:y_{mi}=0\}} \hat{\psi}_{condl,mi}$$

where n_i is the number of species detected at unit i, $\hat{\psi}_{condl,mi}$ is the estimated probability of occupancy for species m at unit i, conditional upon its nondetection, and $\{m : y_{mi} = 0\}$ indicates the summation is performed over those species where the number of detections at unit i (y_{mi}) was 0. An important consideration for estimation of R_i is the appropriate determination of its standard error (or other measure of uncertainty) that accounts for the finite nature of the species list. This could be approximated using similar non-Bayesian approaches to those detailed in Section 6.1, or arises naturally when using the complete data likelihood with MCMC, which is described in more detail in the next section.

15.2.2 Single-Season Studies: Multi-Species Occupancy Models (MSOMs)

Dorazio and Royle (2005) used this type of joint modeling approach when they developed a model for estimating species richness and related community parameters. Their model is based on the type of design considered previously, which is a modification of conventional designs for estimating species richness (e.g., Bunge and Fitzpatrick, 1993; Boulinier et al., 1998b; Williams et al., 2002), where s units are sampled within a region containing R distinct species. Here we describe their model assuming each unit is surveyed the same number of times (K), but this is not necessary.

A common objective of community modeling efforts is estimation of the parameter R, the total number of species in the region. However, alternative summaries of the community under study might also be of interest. For example, we might wish to estimate the number of species at a single unit or at a group of units, perhaps even the collection of sampled units. This last objective might be of interest when the samples are not randomly selected, or representative. Hence, because of the nonrandom sampling, an estimator of R may be biased. However, the number of species present at the collection of sampled units (M; which will be less than or equal to R) can be reasonably estimated regardless of the design imposed on sample unit selection. More complex summaries of community structure might also be of interest such as metrics describing the similarity of the communities among units (Dorazio et al., 2006), or a species–area curve (Yamaura et al., 2016a). Finally, there may be some interest in estimating these quantities in a manner that takes into account unit-specific habitat differences and their effects on species occurrence.

The data arising under the design described above consist of the unit- and species-specific detection histories (**h**). By assuming the probability of detecting each species was constant at each unit, the detection histories can be summarized by the detection frequencies y_{mi}, the number of times that species $m = 1, 2, \ldots, R$ was detected in K visits to unit $i = 1, 2, \ldots, s$ although note the methods of Dorazio and Royle (2005) could be extended to allow variation in detection probabilities across sampling occasions, which we demonstrate in the example below. In general, not all species in the region will be detected in the surveys. Let x (which will be less than or equal to R) be the number of distinct species detected during the visits to all units. Let $\mathbf{y}_m = (y_{m1}, y_{m2}, \ldots, y_{ms})$ denote the vector of the s unit-specific detection frequencies of species m. For our purposes, it is convenient to order the observation vectors such that $m = 1, 2, \ldots, x$ correspond to the data for the x observed species, and $m = x + 1, x + 2, \ldots, R$ correspond to the unobserved (i.e., 'all zero' encounter history) species. It is also useful to introduce an $R \times s$ matrix of the latent occupancy variables, **Z**, with elements z_{mi} indicating the presence of species m at unit i. As in the single-species case, $\mathbf{Z}$ is only partially observed in the sense that if species m is detected at unit i (i.e., $y_{mi} > 0$) then the species must be present and $z_{mi} = 1$, but if $y_{mi} = 0$ (i.e., species m is not detected at unit i), two mutually exclusive possibilities exist for the value of z_{mi}: (1) species m is present at unit i ($z_{mi} = 1$), but went undetected, or (2) species m is absent from unit i ($z_{mi} = 0$).

As in the previous section, let ψ_m denote the probability of occurrence of species m, with z_{mi}'s assumed to be independent Bernoulli random variables. Conditional on species m being present ($z_{mi} = 1$), the number of detections of species m at any unit (y_{mi}) is assumed to be a random value from a binomial distribution with index K and parameter p_m. Conversely, if species m is absent from unit i, then y_{mi} is assumed to equal zero with probability 1.0. These considerations define the joint distribution of the observations y_{mi} and the occupancy state variables z_{mi}. Because the occupancy states, z_{mi}, are only partially observed, it is convenient in some cases to remove them from the likelihood by marginalization, yielding the familiar zero-inflated binomial density used previously (Chapters 4 and 7), which we will denote by $g(y_{mi}|\psi_m, p_m)$.

Dorazio and Royle (2005) extended the single-species occupancy model to a community of species, allowing for species-specific differences in rates of occurrence and detection, by specifying normal distributions on the logits of the two parameters, detection and occupancy probability. That is:

$$logit(\psi_m) = \mu_\psi + u_m,$$

and

$$logit(p_m) = \mu_p + v_m,$$

where u_m and v_m are species-specific random effects assumed to be normally distributed with mean 0 and variances σ_u^2 and σ_v^2, respectively. Note that Dorazio and Royle (2005) provided an argument that u_m and v_m should be positively correlated (since both might be related to species-specific abundance) and their model allowed for an additional parameter $\sigma_{uv} = Cov(u_m, v_m)$. Note also that the detection model here is that used in the logistic-normal model of heterogeneous detection of species (Coull and Agresti, 1999), described in Chapter 7, except here it is applied to allow for heterogeneity among species rather than among units for a single species.

As done for the heterogeneous detection probability models, estimators of model parameters can be developed based on the *marginal probability density* of the observed data (this is an observed data likelihood approach). The marginal probability density of the observations for species m, $\mathbf{y}_m$, is obtained by calculating the integral of the probability statement for the observations, where the integration is performed over the values of the species-level random effects (u_m, v_m).

Recall that integration essentially involves calculating the product of the probability of the random variable values (u_m and v_m) and the probability statement for the observed data evaluated at those values, for all possible values of the variables and summing the result (see Appendix for a basic introduction to integration). Assuming that the $i = 1, 2, \ldots, s$ unit-level observations of each species are independent, then the marginal probability density for the encounter history of species m at unit i can be obtained by integrating the likelihood for species m:

$$\prod_{i=1}^{s} g\left(y_{mi} | \psi_m(u_m), p_m(v_m)\right),$$

over the bivariate joint distribution of the species-level random effects (u_m, v_m). The integration may be done numerically, although, as noted by Dorazio and Royle (2005), this can be computationally intensive to implement. More importantly, estimates of the u_m parameters (related to the occupancy probability of species m), and their uncertainties may be necessary for estimating summaries of community structure, e.g., Eq. (15.2), and so it is disadvantageous to remove them from the likelihood.

To resolve such estimation problems, Dorazio and Royle (2005) developed a Bayesian framework for analysis of the model based on the likelihood conditioned on the species that were detected at least once. They note that the likelihood may be factored into two components (Sanathanan, 1972), one for the detections of the observed species, conditional on x, and a second for the binomial distribution of x given the unknown community size R. Bayesian analysis based on the conditional likelihood avoids the tedious integration needed to estimate R directly by maximum likelihood, retains the species-specific parameters in the model, and yields direct estimates of them. In addition, an estimate of species richness can be computed as a function of model parameters. That is, given the parameter values (or estimates, or draws from the posterior) μ_ψ, μ_p, σ_u^2, σ_v^2, and σ_{uv}:

$$\hat{R} = \frac{x}{1 - g\left(0 | \mu_\psi, \mu_p, \sigma_u^2, \sigma_v^2, \sigma_{uv}\right)}, \tag{15.1}$$

where the term in the denominator is the marginal probability of detecting a species in the community. Thus, within the MCMC algorithm based on the conditional likelihood, $\hat{R}$ is just a function of model parameters, and MCMC samples can be obtained at each iteration of the algorithm by plugging in the current values of each model parameter into Eq. (15.1).

The motivation for introducing the latent (i.e., unobserved) occupancy state variables (the z_{mi}'s) is that estimators of many ecologically important quantities are naturally expressed as functions of them. Note that we adopted a similar construction for estimating the number of occupied units in a finite population of units (Section 6.1). If the z_{mi} variables were fully observed (i.e., if $p = 1$ for all species and units) then the presence or absence of each species would be known exactly and certain quantities of interest could be calculated directly. For example, the number of species occurring at unit i (a quantity that is conceptually similar to the number of occupied sample units considered in Section 6.1) is $R_i = \sum_{m=1}^{R} z_{mi}$. As another example, the number of species in common at two units, say i and l, is $R_{il} = \sum_{m=1}^{R} z_{mi} z_{ml}$. Estimators of such quantities may be obtained naturally from these expressions by plugging in the appropriate estimator of any unobserved quantity. For example:

$$\hat{R}_i = \sum_{m=1}^{x} \left[z_{mi} \times I(y_{mi} > 0) + \hat{z}_{mi} \times I(y_{mi} = 0) \right] + \sum_{m=x+1}^{\hat{R}} \hat{z}_{mi} \quad (15.2)$$

which is the number of species actually detected at unit i, plus the expected occurrence at unit i of species that were detected in the study, but not at unit i, plus the expected number of occurring species that were not detected at any of the s units in the sample (recall the indicator variables $I(E) = 1$ if the expression E is true, or $I(E) = 0$ if the expression E is false). A reasonable estimator of $\hat{z}_{mi}$ is its expectation conditional on the data, and this involves the species-specific occurrence probabilities, ψ_m, and detection probabilities, p_m. The Bayesian implementation yields, directly, estimates of the z_{mi}'s (or draws from the posterior distribution) and species-specific occurrence probabilities. Thus, a sample from the posterior distribution of R_i can be obtained by substituting current values of R and the unknown elements of z_{mi} into Eq. (15.2). Similar reasoning can be used to derive an estimator of the total number of species present among all s sample locations (M) and also indices of similarity in species composition among units, such as Dice's Index (Dice, 1945). See Dorazio and Royle (2005) for details.

Many generalizations of the basic MSOM of Dorazio and Royle (2005) are possible. For example, general models of species- and unit-specific parameters ψ_{mi} and p_{mi} can be developed. In many applications it will be important to model detection probabilities as a function of covariates that may change with each survey of each unit (e.g., temperature, time). One of the most useful modifications is to allow for spatial structure in species-level occupancy probability, by including explicit habitat or landscape covariates in the model for $logit(\psi_{mi})$ (Kéry and Royle, 2009; Zipkin et al., 2010; Sauer et al., 2013a). Indeed, that species should have heterogeneous effects not only in baseline occupancy but in their response to measurable landscape structure is one of the key benefits of developing community models based on species-level models of occupancy. We show an example of this below. Several applications of MSOMs have used *a priori* species groupings in developing explicit contrasts among species within a community. Such models can be developed to include the unobserved species in the community by introducing an additional species-level variable, say $g_m \sim Categorical(\pi)$ where π are the probabilities of group membership of each species. Because this group member variable is unknown for the unobserved species, and because other model parameters (occupancy, detection) may vary among the groups, this formulation allows for differential sampling of the species groups depending on such things as habitat structure and environmental conditions. See Yamaura et al. (2011), Pacifici et al. (2014), Ruiz-Gutiérrez et al. (2010), and Ruiz-Gutiérrez and Zipkin (2011) for examples.

15.2.3   Example of the Dorazio–Royle Multi-Species Occupancy Model

We provide an example here using data from Sauer et al. (2013a) on a study of bird communities on Patuxent Research Refuge, Laurel, MD, USA. The sampling protocol was based on conventional avian point counting methods. Bird point counts occurred on a subset of 316 points defined by a permanent sampling grid with vertices every 100 m. To maintain a distance of approximately 400 m, every fourth grid point was surveyed. Each observer stood at a point and recorded birds within 100 m for a duration of five minutes. A variable number of replicate surveys was made at each point (range 1–7 surveys; mean 2.54 surveys) between 31 May and 4 July 2008. A total of 17 observers was involved in the sampling, and a total of 92 species was encountered during the entire survey. We apply the MSOM approach of Dorazio and Royle (2005) implemented in the JAGS software (Plummer, 2003) using the R package jagsUI. The basic model described previously, allowing for a species-level baseline detection probability, was expanded to include observer effects in the form of an additive effect on the logit-scale of detection probability as well as a species-specific quadratic function of date:

$$logit(p_{mij}) = \mu_p + v_m + \beta_{1,m}Date_{i,j} + \beta_{2,m}Date_{i,j}^2 + \eta_{obs_{i,j}}$$

where $\beta_{1,m} \sim Normal(\mu_{\beta_1}, \sigma_{\beta_1}^2)$, $\beta_{2,m} \sim Normal(\mu_{\beta_2}, \sigma_{\beta_2}^2)$, and $\eta_{obs_{i,j}} \sim Normal(0, \sigma_{obs}^2)$. Note that $\beta_{1,m}$ and $\beta_{2,m}$ are modeled as random effects from a normal distribution rather than independent species-specific fixed effects. We

TABLE 15.2 Posterior summary statistics for the multi-species occupancy model fit to the PWRC bird survey data. Summaries are based on four chains of 24,000 iterations, burn-in = 4000 iterations, and thin rate = 4, yielding 20,000 total samples from the joint posterior

Parameter	Mean	SD	2.5%	50%	97.5%
R	109.300	9.953	96.000	107.000	135.000
μ_p	−1.196	0.207	−1.625	−1.187	−0.813
σ_v	1.174	0.171	0.897	1.157	1.577
μ_{β_1}	0.003	0.024	−0.045	0.002	0.050
σ_{β_1}	0.144	0.020	0.109	0.142	0.187
μ_{β_2}	−0.011	0.019	−0.048	−0.011	0.027
σ_{β_2}	0.115	0.015	0.090	0.114	0.146
σ_{obs}	0.446	0.090	0.307	0.434	0.656
μ_ψ	−2.186	0.590	−3.528	−2.112	−1.275
σ_u	2.949	0.457	2.216	2.894	4.014
μ_α	−0.332	0.090	−0.512	−0.329	−0.160
σ_α	0.671	0.090	0.513	0.664	0.866
deviance	23,861.490	221.380	23,439.975	23,861.130	24,301.791

also expand the model for species-level occurrence probability so that ψ_{mi} depends on local habitat variation (that is, in the vicinity of point i). For purposes of illustration we use only a single habitat covariate 'percent evergreen' which is modeled as a linear effect on the logit-scale. Note Sauer et al. (2013a) used many more covariates, but we chose just this one because we expected large differences in the responses of species to this covariate with many having very negative and many having very positive responses to evergreen percent. Thus, the model for occupancy probability has this form:

$$logit(\psi_{mi}) = \mu_\psi + u_m + \alpha_m Evergreen_i,$$

where $\alpha_m \sim Normal(\mu_\alpha, \sigma_\alpha^2)$. We standardized the variable *Evergreen* to have mean 0.0 and unit variance. The model was analyzed by MCMC methods using the jagsUI package and data augmentation (Royle and Dorazio, 2008), with four Markov chains run for 24,000 iterations each, with a burn-in of 4000 iterations and a thinning rate of four, producing posterior summaries based on 20,000 total posterior samples. Posterior summary statistics are shown in Table 15.2.

A key parameter in this model is the community size or species richness, R, which has a posterior mean of 109.3. That is, roughly 18 more species are estimated to exist at PWRC than were detected during the surveys. The posterior distribution of R is shown in the left panel of Fig. 15.1.

A key benefit of the MSOM framework is that it accommodates species level variation in detection probability. In this particular case we allowed for

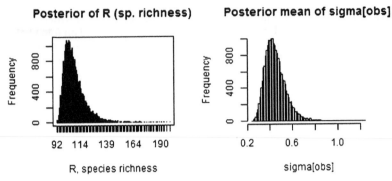

FIGURE 15.1   Posterior distributions of species richness R and observer effect standard deviation σ_{obs} in the multi-species occupancy model (MSOM) fitted to the Patuxent bird survey data.

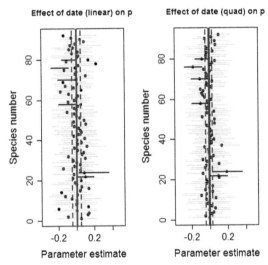

FIGURE 15.2   Summary of posterior distribution of linear and quadratic date effects for each of 92 observed species. Gray vertical lines are the posterior 95% intervals for each species, blue vertical lines are those for which the posterior interval did not include zero, black vertical line is the value $\mu_\beta = 0$, solid red line is the posterior mean of the community mean effect parameter, and the broken red line is the 95% posterior interval for the community mean effect.

a quadratic response of date of sampling, but we note that there were not generally strong effects (Fig. 15.2). Except for a handful of species, the posterior distributions of linear and quadratic date effects do not suggest a lot of variation over time which is probably sensible because the sampling happened over a relatively short four week period. The results, however, do suggest some variation among observers: the posterior mean of σ_{obs} is 0.451 (see right panel of Fig. 15.1).

Effect of % evergreen on Pr(z=1)

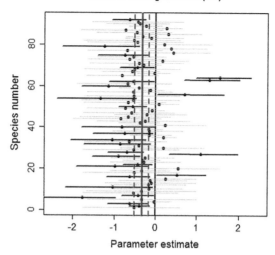

FIGURE 15.3 Posterior distribution of evergreen effect for each observed species. Gray vertical lines are the posterior 95% intervals for each species, blue vertical lines are those for which the posterior interval does not include zero, black vertical line marks $\alpha = 0$, solid red line is the posterior mean of the community mean effect, broken red line is the 95% posterior interval for the community mean effect.

The model structure of key ecological importance is the species specific effect of 'evergreen' on occupancy probability. We see that the population mean effect of evergreen (μ_α in Table 15.2) is highly negative, indicating that typical species in the community have an occupancy probability that declines with increasing percent evergreen in the vicinity of the point count location. Fig. 15.3 depicts the posterior of α_m for *each* of the observed species. It shows extreme variation with some species having a very positive effect but many also having a very negative effect. A list of all species with posterior mean at least 1.9 posterior standard deviations from zero is given in Table 15.3.

Generally these results are consistent with known habitat preferences of passerine birds in this region. Our colleague J. Sauer commented: "Pine trees at Patuxent tend to be in upland forests... pine warbler (PIWA) is always strongly associated with pine trees in forests, and ovenbird (OVEN) and Cooper's hawk (COHA) are forest bird species that use those upland habitats. American redstart (AMRE) and Louisiana waterthrush (LOWA) tend to be more common in bottomland (or floodplain) forests, and tree swallows are found over impoundments." See Sauer et al. (2013a) for additional discussion of habitat associations.

Finally we used the model to produce a prediction of the community response to evergreen habitat by computing the posterior predictive distribution of community size as a function of the % evergreen (Fig. 15.4). This shows the

TABLE 15.3 Summary of the posterior distributions for α_m for those species with the posterior mean > 1.9 standard deviations (SD) from zero

Common name	Alpha code	Mean	SD	Percentile					
				0.025	0.25	0.5	0.75	0.975
American redstart	AMRE	−1.762	0.448	−2.706	−2.047	−1.734	−1.446	−0.969
Louisiana waterthrush	LOWA	−1.338	0.491	−2.385	−1.651	−1.308	−0.989	−0.473
Tree swallow	TRES	−1.244	0.474	−2.259	−1.546	−1.214	−0.911	−0.398
Northern parula	NOPA	−1.150	0.299	−1.790	−1.333	−1.130	−0.945	−0.620
Eastern phoebe	EAPH	−1.053	0.466	−2.052	−1.344	−1.023	−0.729	−0.227
Barn swallow	BARS	−1.034	0.542	−2.184	−1.379	−0.998	−0.656	−0.072
Calliope hummingbird	CANG	−0.974	0.447	−1.927	−1.256	−0.948	−0.655	−0.177
Common grackle	COGR	−0.901	0.300	−1.532	−1.095	−0.885	−0.689	−0.366
Eastern kingbird	EAKI	−0.859	0.411	−1.735	−1.117	−0.836	−0.568	−0.128
Red-winged blackbird	RWBL	−0.742	0.311	−1.404	−0.935	−0.724	−0.524	−0.192
European starling	EUST	−0.739	0.373	−1.523	−0.977	−0.718	−0.478	−0.067
Common yellowthroat	COYE	−0.694	0.221	−1.153	−0.838	−0.685	−0.541	−0.293
Yellow-throated vireo	YTVI	−0.629	0.208	−1.058	−0.763	−0.622	−0.485	−0.246
Blue grosbeak	BLGR	−0.618	0.271	−1.188	−0.790	−0.605	−0.428	−0.130
American crow	AMCR	−0.616	0.249	−1.161	−0.765	−0.594	−0.446	−0.186
Indigo bunting	INBU	−0.546	0.175	−0.901	−0.660	−0.542	−0.427	−0.218
Orchard oriole	OROR	−0.543	0.278	−1.126	−0.722	−0.527	−0.350	−0.039
Red-bellied woodpecker	RBWO	−0.441	0.210	−0.872	−0.572	−0.435	−0.307	−0.044
Carolina wren	CARW	−0.427	0.174	−0.781	−0.541	−0.424	−0.313	−0.088
American robin	AMRO	−0.408	0.195	−0.807	−0.531	−0.403	−0.278	−0.040
Acadian flycatcher	ACFL	−0.382	0.127	−0.632	−0.466	−0.381	−0.297	−0.132
Cooper's hawk	COHA	1.100	0.419	0.328	0.817	1.080	1.361	1.973
Ovenbird	OVEN	1.299	0.338	0.696	1.060	1.275	1.513	2.026
Pine warbler	PIWA	1.550	0.339	0.983	1.309	1.515	1.752	2.301

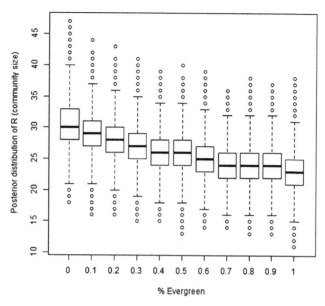

FIGURE 15.4 Posterior distribution of *R* across a gradient of % evergreen.

expected species richness (and posterior uncertainty) in the vicinity of a point with varying amounts of '% evergreen' habitat, and we see a decline in expected value from about 30 species to 22–23 species.

15.3 THE YAMAURA EXTENSIONS: MULTI-SPECIES ABUNDANCE MODELS

Hierarchical models based on elemental species-level models of occurrence provide a conceptually clear and concise framework for developing community models. The basic ideas of multi-species occupancy models have been extended to include species-level *abundance* models. These multi-species abundance models (MSAMs) include explicit models for local population size, say N_i for unit i, as well as observation models which may accommodate binary (detection/nondetection) data, simple counts, or multinomial count statistics such as encounter history frequencies or removal frequencies. The development of MSAMs initiated with Yamaura et al. (2011) who used a Royle–Nichols (Royle and Nichols, 2003) formulation of the observation model, linking detection/nondetection data to a latent abundance model. This was extended by Yamaura et al. (2012) to allow for a count observation model, Chandler et al. (2013) to allow for capture–recapture data (see also Yamaura et al., 2016a), and Sollmann et al. (2016) and Yamaura and Royle (in review) for a distance sampling protocol.

These various extensions are covered in more detail in Chapter 11 of Kéry and Royle (2016).

The MSAM proposed by Yamaura et al. (2011) is based on observed detections from K visits to a unit, i.e., standard occupancy model data. These observations are assumed to be binomial counts just as in an ordinary occupancy model:

$$y_{mi} | N_{mi} \sim Binomial(K, p_{mi})$$

where p_{mi} is the probability of detecting species m at unit i. However, the model adopts the Royle–Nichols (RN) link between detection probability and abundance:

$$p_{mi} = 1 - (1 - r_{mi})^{N_{mi}}$$

where N_{mi} is the local abundance of species m at unit i. Local abundance is assumed to vary, for example according to a Poisson distribution:

$$N_{mi} \sim Poisson(\lambda_{mi})$$

where λ_{mi} may depend on local habitat effects or other factors. At a minimum, the model allows for variation in local abundance among species, e.g., $\lambda_{mi} \equiv \lambda_m$ assumed to be a random effect as in the MSOM described in Section 15.2.2 above. Tobler et al. (2015) apply this MSAM based on the RN model. Note here that the model is formulated in terms of encounter frequencies; However, the model can be formulated in terms of site- and visit-level Bernoulli observations allowing also for the modeling of variation in p over time.

All applications of MSAMs so far have been based on Bayesian analysis using the BUGS or JAGS software. This uses a formulation of the model based on data augmentation (DA; Royle et al., 2007b; Royle and Dorazio, 2012) in which the data set is augmented with a large number of all-zero encounter histories, and then an additional latent variable w_m is added to the model where $w_m = 1$ implies that encounter history m follows the model described above and $w_m = 0$ implies that encounter history m is a deterministic 0. In effect, data augmentation casts the model for the augmented data as a zero-inflated type of model. This is accommodated in the formulation of the model in JAGS by adding the data augmentation variable to the model for p as follows:

$$p_{mi} = 1 - (1 - r_{mi})^{w_m N_{mi}}.$$

Yamaura et al. (2012) proposed an analogous MSAM but with a standard binomial counting protocol (see also Tobler et al., 2015) with:

$$y_{mi} \sim Binomial(w_m N_{mi}, p_{mi}).$$

An alternative way to formulate the model suggested by Yamaura et al. (2016a) involves a slightly different implementation of the data augmentation concept where:

$$w_m \sim Bernoulli(\omega)$$

and with the DA variable included in the latent abundance model:

$$N_{mi}|w_m \sim Poisson(w_m \lambda_{mi})$$

and then the usual binomial observation model:

$$y_{mi}|N_{mi} \sim Binomial(N_{mi}, p_{mi}).$$

15.4 MULTIPLE-SEASON MULTI-SPECIES OCCUPANCY MODELS

The approach described in Section 15.2.1 of analyzing the data for multiple species simultaneously can be used to investigate changes in the community over time for each species (e.g., an $\epsilon(Species + Time)$ model; Fig. 15.5B). This may be particularly advantageous for making inference about a species that is rarely encountered, but has a life history strategy similar to that of one or more common species. On its own, there may be insufficient data to make any reliable inferences about the former species, but by sharing parameters with one or more species that are detected with greater frequency, such inference may be possible (provided it is biologically reasonable to model those species jointly, of course). Parallel temporal variation in local extinction or colonization probabilities may also provide a means of assessing guild membership, based on the thinking that members of a guild are likely to respond to environmental variation in similar ways.

Unit- and season-specific covariates can also be used to model local rates of extinction and colonization across species. Community analyses based on aggregations of forest bird species have provided inferences about the influence of habitat covariates (e.g., fragmentation statistics) on aggregated rates of species extinction and turnover (Boulinier et al., 1998a, 2001). Analyses using additive models of species and habitat (or habitat change) would provide a more detailed assessment of this type of hypothesis. For example, the work of Boulinier et al. (1998a, 2001) relied on *a priori* groupings of forest bird species as 'area sensitive' or not. Investigation of additive species-plus-fragmentation models, e.g., $\epsilon(Species + Fragmentation)$, with common slope parameters relating species-specific extinction probabilities to fragmentation statistics would provide an empirical means of assessing species-specific group membership.

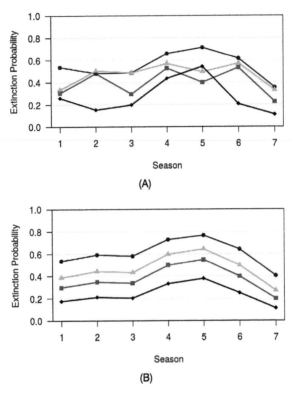

FIGURE 15.5 Example of two models that one might consider for local extinction probabilities (ϵ) when modeling multiple species simultaneously; (top panel) ϵ is species and season specific, model $\epsilon(Species \times Time)$; and (bottom panel) ϵ for the group of species varies in parallel (on the logistic scale), model $\epsilon(Species + Time)$.

The multi-species (community-based) occupancy models of Dorazio and Royle (2005) have been extended to multiple seasons as well (Dorazio et al., 2010), permitting direct estimation of changes in species richness and related parameters over time. In particular, extension of this approach will permit inference about changes in species richness and community dynamics at multiple scales ranging from the local unit, to aggregations of units, to the total area from which sampled units are selected. Such investigations should be capable of providing inferences about the relative contributions to community change of processes acting at different scales, a topic of substantial interest to community ecologists (e.g., see Cornell and Lawton, 1992; Cornell, 1993; Karlson and Cornell, 1998; Cam et al., 2000b).

The fully dynamic MSOM (Dorazio et al., 2010) follows the basic multi-season structure of Chapter 8 for a single species except it accommodates an ensemble of such single species models, with species-specific model parameters. Each species' encounter history is modeled as:

$$h_{mt,ij}|z_{mt,i}, p_{mt,ij} \sim Bernoulli(z_{mt,i}\, p_{mt,ij})$$

where $p_{mt,ij} = Pr(h_{mt,ij} = 1|z_{mt,i} = 1)$ denotes the conditional probability of detecting the mth species during the jth observation of location i and season t given that the species is present. As in other classes of models considered so far, we can model various sources of variation such as measured covariates on the logit-transformed detection probabilities $p_{mt,ij}$.

The dynamic model of occupancy state mirrors that of Chapter 8, except there is a distinct model for each species in the community (including unobserved species!). The initial occupancy state of the mth species at location i is:

$$z_{m1,i}|w_m, \psi_{m1,i} \sim Bernoulli(w_m \psi_{m1,i})$$

where $\psi_{m1,i} = Pr(z_{m1,i} = 1|w_m = 1)$ is the probability that species m is present at location i during season 1, given that the species is a member of the metacommunity being sampled. As before, w_m are the data augmentation variables that account for zero-inflation of the observed data set with a large number of all-zero encounter histories (see Dorazio et al., 2010). For subsequent states we assume:

$$z_{m,t+1,i}|w_m, z_{mt,i}, \phi_{mt,i}, \gamma_{mt,i} \sim Bernoulli\left(w_m\{\phi_{mt,i}z_{mt,i} + \gamma_{mt,i}(1 - z_{mt,i})\}\right)$$

for $t = 1, 2, \ldots, T - 1$ and where $\gamma_{mt,i} = Pr(z_{m,t+1,i} = 1|z_{mt,i} = 0, w_m = 1)$ is the conditional probability that the ith location will become occupied by species m during season $t + 1$ given that this species is a member of the metacommunity being sampled and that it was absent at that location during the previous period, i.e., local colonization. Similarly, the parameter $\phi_{mt,i} = Pr(z_{m,t+1,i} = 1|z_{mt,i} = 1, w_m = 1)$ is the local survival probability, i.e., one minus the probability of local extinction. Note that if $w_m = 0$ then $z_{mt,i} = 0$ for all units and all seasons.

There are some other versions of the multi-species multi-season occupancy models which are somewhat less general but preserve the individual species-level structure. Russell et al. (2009) develop a multi-species multi-season occupancy model which has a simplified dynamic structure, in which:

$$z_{m,t+1,i}|z_{mt,i}, \rho_{mt,i}, \gamma_{mt,i} \sim Bernoulli(\pi_{mt,i})$$

where $logit(\pi_{mt,i}) = \gamma_{mt,i} + \rho_{mt,i}z_{mt,i}$, thus accommodating persistence in the occupancy state, via the parameter $\rho_{mt,i}$, due to the net of colonization and extinction processes. Note that $\rho_{mt,i}$ is not strictly a persistence probability, so is not equivalent to ϕ above, even on the logit-scale. Russell et al. (2009) also model explicitly only the observed species, and do not use data augmentation to model the unobserved portion of the community but, instead, condition on the M observed species recorded across all surveys. Kéry et al. (2009) apply a multi-season multi-species occupancy model with a temporary emigration formulation of the dynamics. That is:

$$z_{mt,i} \sim Bernoulli\left(\psi_{mt,i}\right)$$

so that species are randomly present in the community at season t with probability $\psi_{mt,i}$. Unit- and season-varying covariates can be modeled on the temporary emigration parameter according to:

$$logit(\psi_{mt,i}) = \beta_{m1} + \beta_{m2}Covariate_{ti}.$$

A similar random temporary emigration model in the context of multi-season multi-species *abundance* models has recently been developed by Yamaura and Royle (in review).

15.5 DISCUSSION

In this chapter, we have outlined two approaches for application of the occupancy models described in Parts II and III to community level studies when interspecific relationships among species are not of interest. One approach uses a species list at a single or small number of units, whereas the other approach is based on joint modeling of multiple species at a larger number of units. Both approaches could be used to address similar questions, and the question of which approach may be more appropriate, largely depends upon the quality of the data and the objective(s) of the study. We believe the second approach will work best with data from at least a moderate number of units (≥ 20), but in general the performance of community models for estimating species richness depends on other important factors including expected local population size, detection probability and number of visits (Yamaura et al., 2016b). Depending on these various parameters, MSOMs can perform well for estimating species richness, on a cost-adjusted basis, for as few as 10–20 units (Sanderlin et al., 2014). In multi-species community studies, relatively fewer units are needed compared to the single-species case because MSOMs effectively borrow information from the ensemble of species-level data sets.

There is the potential for having a very large number of parameters in the model when using the second approach, especially when data have been collected over multiple seasons. As such there may be a seemingly infinite number of possible models that could be considered. It is very important in such cases that *a priori* reasoning, biological knowledge of the system, and hypotheses of interest are used to limit the number of models that will be fit to the data.

Provided the models are sufficiently complex to capture the main features of the data, robust inference can still be made about the system without requiring highly complex models. The fact that there may be a 'better' model outside of the model set should not necessarily be a cause of major concern when sound, rational thought has been used to construct the set of plausible models. Indeed by limiting the model set in such a manner, the likelihood is lessened of finding models that capture some random aspect of the data well, but do not portray the underlying system accurately and hence cannot be used for prediction. That said, however, one should not restrict the model set so much that the models considered only represent a single hypothesis or view of the world. In our experience we also believe that the model set should usually include models that permit variation in detection probability over survey and season.

We believe that both of the approaches presented in this chapter offer advantages over previous inferential approaches for dealing with ecological communities. We believe that the development of approaches that deal adequately with detection probabilities (e.g., Burnham and Overton, 1979; Bunge and Fitzpatrick, 1993; Williams et al., 2002) represented an important step forward. The occupancy based approach outlined in Section 15.1 can be viewed as similar to these previous methods, with a substantial advantage provided by the ability to develop species-specific covariate models. For example, these methods permit formal investigation of the relationships between species occupancy and rate of local extinction, and such species-specific characteristics as body size and degree of specialization. The other approach based on occupancy data from multiple units (Section 15.2) also offers opportunities for modeling of occupancy and related vital rates using both unit-specific and species-specific covariates. Additive models including species effects can be used to investigate such topics as guild membership and common responses to environmental and habitat changes. Studies of the relationship between species characteristics and local vital rates (extinction and colonization) have previously been based on aggregations of species (e.g., the study of avian sexual dimorphism by Doherty et al., 2003b), whereas additive models permitting species effects would represent an inferential improvement. The multi-species occupancy models (MSOMs) of Dorazio and Royle (2005) provide a useful approach to estimation of species richness at various geographic scales, ranging from local sample units to aggregations of such units to the entire area from which samples are selected. Their approach also permits inference about other community-level questions including species–area relationships and similarities in species composition across units.

While the methods described in this chapter have focused on estimation of species richness, and similar metrics, in some situations it may not be the number of species in an area of interest, but the *value* of that species community, where some species may be considered more 'valuable' than others (e.g., Yoccoz et al., 2001). A species value may be defined in a number of ways, including in economic terms, their perceived contribution to the local ecosystem, or their rarity. The overall value of the community could be easily calculated using the above methods by calculating a weighted sum of which species are present at a unit, or in the area of interest, rather than just the sum of species presence. That is: $Value = \sum_m w_m z_m$, where w_m is the weight or value assigned to species m. Calculated in this manner, the value of the community accounts for uncertainty in which species may have been present at each unit (due to detection or otherwise), sampling effort, covariate relationships, etc.

The models described in this chapter are relatively simplistic models in that they assume independence among species both in the observations and also among the latent parameters of the models. These independence assumptions are mainly for pragmatic reasons, made because the number of parameters increases rapidly with the size of the community, but, in real life, we expect species occurrence states to be dependent. In Chapter 14 we described basic models for multi-species systems which are most useful for a small number of species where parameters can be well-estimated for each species. General forms of dependence among species need to be adapted to multi-species occupancy and abundance models. To a certain extent the ideas of Chapter 14 can be adapted, but those can be highly parameterized models that are most suited for relatively few species for which there are *a priori* hypotheses about the dependence structure (in terms of which are dominant and subordinate species). How can these ideas be extended to community-level models? Some work has been done on general dependence models in the context of community abundance models (MSAMs). For example, Dorazio and Connor (2014) use morphological traits to describe dependence among species, and Dorazio et al. (2015) develop community-level models of abundance using a latent multivariate normal distribution to describe dependence among the abundance states. There is some very recent work on development of MSOMs which allow for dependence among species. Tobler et al. (unpublished) extended the latent variable (Hui et al., 2015) and multinomial probit (Pollock et al., 2014) models to a multi-species occupancy modeling framework, allowing for both imperfect detection and correlation among species. Rota et al. (2016) proposed a general formulation of multi-species occupancy models which allows for general forms of dependence among species. They exploited a newly developed distribution, the multivariate Bernoulli, which allowed them to specify interactions among species in the mean structure of an occupancy model while also modeling effects of spatially varying covariates (e.g., habitat effects). As noted in Chapter 14, one option for incorporating dependence with the multi-species co-occurrence models is to limit the dependence structure to pairwise or three-way interactions among species, and ignore higher-order interactions, which would greatly reduce the number of parameters to estimate. More development of these approaches is needed to assess whether they may be viable in the context of community-level models.

In summary, we believe that the methods presented in this chapter represent substantial improvements over previous approaches to inference in community ecology. The use of multi-species occupancy studies at many units, with repeated surveys within each season over multiple seasons, provides a rich database with which to address many of the important questions in community ecology.

Chapter 16

Final Comments

In this book we have covered a set of progressively more complicated methods, centered around an occupancy state variable that may be either dichotomous (e.g., presence/absence), or take more than two discrete values, and that is typically not observed perfectly. We began with the relatively simple case, in retrospect, of estimating the proportion of units occupied when the species is detected imperfectly, then progressed to incorporate unequal sampling effort and covariates, creating maps, multiple occupancy states, and multiple seasons. We have demonstrated that spatial correlation can be accounted for in occupancy-related variables, as can false-positive detections and state-misclassification in general. Finally we considered the joint modeling of species occurrence and habitat dynamics, species co-occurrence models for two or more species, and community-level models. For many applications we have outlined how multi-state modeling can be used as a unifying framework; conceptually, as a means for parameter estimation, and for making predictions about the state of the system at other times or places. We have also provided guidance relative to the design of studies and monitoring programs that might be based on these methods.

However, despite the breadth of the material we have covered, there is still plenty that we have not included. Primarily this was a conscious decision to avoid going off on too many tangents, as one extension will typically lead to further extensions, so at some point the decision must be made to halt.

One popular set of methods we have not discussed, in any real detail, is so-called N-mixture models (e.g., Royle, 2004b; Dail and Madsen, 2011; Kéry and Royle, 2016; Rossman et al., 2016), which can be viewed as extensions of the Royle and Nichols (2003) model, and multi-state models (e.g., Royle, 2004a; Royle and Link, 2005; Nichols et al., 2007a; MacKenzie et al., 2009). N-mixture models have a similar data structure to those models we have focused on (e.g., repeated surveys each season, possibly for multiple seasons), but where survey outcomes are counts of unique individuals rather than species-level detection/nondetection data or categories (for multi-state modeling). As with occupancy models, not all individuals at each unit will be detected, hence observed counts will be lower than the number of unique individuals that are present at a unit, and will vary among the repeat surveys. N-mixture models can

Occupancy Estimation and Modeling. http://dx.doi.org/10.1016/B978-0-12-407197-1.00021-1

provide estimates of abundance-related parameters, and individual-level detection probabilities, by assuming a spatial distribution for the number of unique individuals present at each unit. They do, however, require a more strict set of assumptions than occupancy models that practitioners must carefully evaluate. As with any modeling, assumption violations may result in misleading inferences.

We foresee that the range of modeling approaches available for occupancy-like applications will continue to grow, as researchers develop further extensions and amalgamate different sets of ideas (e.g., multi-season species co-occurrence, with joint modeling of habitat dynamics and spatial correlation), which will often reflect the types of real-world questions being asked by field ecologists. However, with the incorporation of additional complexity of the modeling comes the requirement of more information about additional model parameters from appropriately collected data sets, and increasing complexity of the analyses of those data sets. Oftentimes there are insufficient resources to collect suitable data to provide strong inference about the effect of three or four covariates on patterns of occurrence for a single-species, let alone address more complex questions. That is, while it may be easy to ask more complex and, arguably, more interesting questions about ecological systems, and formulate models that could be used to address those questions, ecologists may be disappointed by the amount of information that must be collected to provide statistically robust answers to those questions.

Another area that we believe requires further work is the provision of fair, robust assessments of the limitations of the various modeling approaches, particularly more complex ones. All statistical methods have limitations, they work well in some circumstances, but not in others. Critical assessments of the limitations provide valuable, practical insights, which can be particularly useful during the design phase of a study or monitoring program. Such assessments also provides guidance on the expectations practitioners should have about the outcomes of their studies or monitoring programs. Further analytic- and simulation-based research on these methods would be particularly useful.

We firmly believe that models and model-based inference provide the cornerstone for the conduct of science and management, but strongly advocate that such modeling must be built upon the foundations of soundly collected data. We have seen examples where, we believe, researchers have become overly reliant on the models for their inferences, overstepping what can be reliably determined based upon their data, and not critically evaluating underlying model assumptions. The conclusions drawn in such cases may be very model dependent, and therefore not as robust as those based on better quality data.

This is particularly the case where more complex models are being used, in association with Bayesian methods of inference. We think Bayesian methods are a very useful tool in the analyst's toolbox, but it must be appreciated that prior distributions for model parameters do contribute a source of information to the analysis, just as the data do. Strictly speaking, there is no such thing as an 'uninformative' prior. For example, consider the Bayesian analysis of the Mahoenui giant weta data in Chapter 4, and the sensitivity of the posterior distribution for the regression coefficients to the choice of prior distribution. This was for a relatively simple situation (single-species, single-season occupancy model, with a single covariate each for occupancy and detection probability), with a sample size that many would consider reasonable for an ecological study (72 units and 3–5 surveys per unit). There is clearly the potential for greater sensitivity for more complex situations and smaller sample sizes. Where there is not strong justification for the choice of prior distributions, sensitivity of the results to the prior distributions that have been used should be assessed. It is also important to appreciate that in using a Bayesian inferential framework, prior distributions may mask issues such as nonidentifiable parameters (parameters that are conceptualized in the model, but distinct information for them is unavailable in the data), and parameter confounding (parameters that are not separately estimable, but the value of their product, for example, is estimable). Rerunning analyses with alternative prior distributions may help identify such issues.

Further work is also required in the area of assessing model fit, both in terms of omnibus assessments and specific assumption violations. Graphical diagnostics, e.g., residual-type plots, could be particularly useful. At present there are no such methods for most of the models covered in this book, which is far from ideal, but this type of development is often viewed as a secondary, or even tertiary, issue compared to developing models that more closely represent the biological and sampling realities. That said, many of the earliest occupancy models we have discussed (e.g., MacKenzie et al., 2002, 2003; Royle and Nichols, 2003) have been around for long enough, and are sufficiently mainstream, that it is time to develop methods to assess model fit.

In short, we believe there is still plenty of scope to extend the type of modeling detailed in this book in a number of different directions, and to use these models in novel applications. However, we also believe that having good data is still vital in order to make robust inferences from model-based methods, and that study design issues, assessing the limitations of these models, and developing methods to assess model fit, should be research priorities over the next decade.

Appendix

There are some important mathematical concepts that are widely used in this book that many with stronger backgrounds in mathematics or statistics often take for granted, but we have found that they can be stumbling blocks for others. These are fundamental concepts, but have been relegated to an appendix to avoid distracting from the main focus of the book. The three topics briefly covered here are: (1) notation for summations and products, (2) vectors and matrices, and (3) differentiation and integration.

A.1 NOTATION FOR SUMMATIONS AND PRODUCTS

There is a very useful shorthand way of defining sums and products of, potentially, a large number of terms. For summations, the Σ (sigma) notation is used, which includes three main features: (1) the index variable and its first value, often placed below the Σ symbol; (2) the last value of the indexing variable, often placed above the Σ symbol; and (3) the actual term or expression that is to be summed, to the right of the Σ symbol. For example, in Eq. (A.1) '$i = 1$' indicates the indexing variable is i, starting from the value of 1, '5' indicates that terms up to $i = 5$ will be added together, and 'y_i' is the actual term to be summed:

$$\sum_{i=1}^{5} y_i = y_1 + y_2 + y_3 + y_4 + y_5. \tag{A.1}$$

Some other examples are:

$$\sum_{j=1}^{10} j = 1 + 2 + 3 + 4 + 5 + 6 + 7 + 8 + 9 + 10,$$

$$\sum_{i=1}^{n} (y_i - \bar{y})^2 = (y_1 - \bar{y})^2 + (y_2 - \bar{y})^2 + \cdots + (y_n - \bar{y})^2,$$

$$\sum_{j=1}^{3} y_{ij} = y_{i1} + y_{i2} + y_{i3}.$$

Note that in this final example, there are two potential indexing variables, i and j, but the summation is only specified (and performed) over j.

A similar notation is used for defining products or multiples of variables represented by the Π (pi) symbol. The same three main features are also present here, although now the Π symbol indicates that terms are to be multiplied together rather than summed, for example,

$$\prod_{i=1}^{5} y_i = y_1 \times y_2 \times y_3 \times y_4 \times y_5. \tag{A.2}$$

Multiple Σ or Π symbols may also be used if there is more than one indexing variable over which the summation or multiplication is to be performed. In such instances, the operation is performed for the innermost (or rightmost) symbols first, then moving outward. For example:

$$\sum_{i=1}^{3}\sum_{j=1}^{4} y_{ij} = \sum_{i=1}^{3} y_{i1} + y_{i2} + y_{i3} + y_{i4}$$
$$= (y_{11} + y_{12} + y_{13} + y_{14})$$
$$+ (y_{21} + y_{22} + y_{23} + y_{24})$$
$$+ (y_{31} + y_{32} + y_{33} + y_{34}).$$

A.2 VECTORS AND MATRICES

Vectors and matrices are simply arrays of numbers for which the numerical value and relative position within the array conveys some meaning. Each number of the array is called an *element*, which is indexed by the row and column of the array in which it appears. For example the element a_{ij} is the value in the ith row and jth column. In the context of this book, the position of the element in the array is often used to represent the occupancy state of a unit (e.g., Chapters 5, 8, and 9). Below is a very brief introduction to vectors and matrices, illustrating some key concepts that are used in this book. For a more complete introduction suitable for biologists we direct readers to Appendix B of Williams et al. (2002).

A.2.1 Vectors

A vector is simply a one-dimensional array, that either consists of a single row, or single column, e.g., a three element row vector, v, would be:

$$v = \begin{bmatrix} 5 & 1 & 3 \end{bmatrix}.$$

Note that in this book we use bold font to indicate a vector or matrix. An alternative notation is to underline the vector (or matrix) name, e.g., $\underline{v}$.

A.2.2 Matrices

Here we only consider two-dimensional matrices, but higher dimensional matrices are possible. A matrix can be considered as a series of row or column vectors. For example, a square matrix with three rows and three columns (i.e., a 3×3 matrix) would be:

$$A = \begin{bmatrix} 4 & 2 & 1 \\ 3 & 5 & 0 \\ 1 & 4 & 2 \end{bmatrix}.$$

Matrices do not have to be square. For example a matrix with three rows and two columns (i.e., a 3×2 matrix) would be:

$$B = \begin{bmatrix} -1 & 6 \\ 0 & 2 \\ 1 & -3 \end{bmatrix}.$$

A.2.3 Vector and Matrix Manipulation

To *transpose* a vector or matrix, one simply swaps the rows and columns of the array. For example, the transpose of the vector v defined above (denoted here as v^T) would be:

$$v^T = \begin{bmatrix} 5 \\ 1 \\ 3 \end{bmatrix}.$$

For the matrices defined above, the respective transposes are:

$$A^T = \begin{bmatrix} 4 & 3 & 1 \\ 2 & 5 & 4 \\ 1 & 0 & 2 \end{bmatrix} \text{ and } B^T = \begin{bmatrix} -1 & 0 & 1 \\ 6 & 2 & -3 \end{bmatrix}.$$

Note that the elements of the first, second and third rows of the original matrices, now appear as the elements in the first, second, and third columns of the transposed matrices, respectively.

Matrix (and vector) multiplication is slightly more complicated. Suppose we wish to multiply the matrices A and B to give the resulting matrix C (i.e.,

$C = AB$). The element in the ith row and jth column of C is calculated by multiplying the elements in the ith row of A with the elements in the jth column of B, and summing those product results. That is, $c_{ij} = \sum_{k=1}^{n} a_{ik}b_{kj}$. For example, in this case the element c_{11} would be:

$$c_{11} = \sum_{k=1}^{3} a_{1k}b_{k1}$$
$$= (4 \times -1) + (2 \times 0) + (1 \times 1)$$
$$= -4 + 0 + 1$$
$$= -3.$$

Clearly, a requirement for matrix multiplication is that the number of columns in the first matrix (A in this case) must be equal to the number of rows in the second matrix (here, B), otherwise the matrices are not *conformable* for multiplication. For example, A is a 3×3 matrix and B is a 3×2 matrix hence it is possible to calculate $C = AB$ as A has 3 columns and B has 3 rows. However it would not be possible to calculate $D = BA$, as B has 2 columns while A has 3 rows. This leads to another important point; that the ordering of the matrix multiplication is important, and that for two square matrices (E and F) with the same number of rows and columns, $EF \neq FE$, except where there is a special relationship between the two matrices. Below we provide some simple examples of matrix multiplication, but leave details of the working as an exercise.

$$C = AB$$
$$= \begin{bmatrix} 4 & 2 & 1 \\ 3 & 5 & 0 \\ 1 & 4 & 2 \end{bmatrix} \begin{bmatrix} -1 & 6 \\ 3 & 5 \\ 1 & -3 \end{bmatrix}$$
$$= \begin{bmatrix} 3 & 31 \\ 12 & 43 \\ 13 & 20 \end{bmatrix},$$

$$d = vA$$
$$= \begin{bmatrix} 5 & 1 & 3 \end{bmatrix} \begin{bmatrix} 4 & 2 & 1 \\ 3 & 5 & 0 \\ 1 & 4 & 2 \end{bmatrix}$$
$$= \begin{bmatrix} 26 & 27 & 11 \end{bmatrix},$$

$$e = vAv^T$$

$$= \begin{bmatrix} 5 & 1 & 3 \end{bmatrix} \begin{bmatrix} 4 & 2 & 1 \\ 3 & 5 & 0 \\ 1 & 4 & 2 \end{bmatrix} \begin{bmatrix} 5 \\ 1 \\ 3 \end{bmatrix}$$

$$= 190.$$

In practice, one would often perform matrix multiplication using computer software, either specialized mathematical software, or a spreadsheet package, as many of these include matrix functions.

It is also possible to perform element-wise multiplication (denoted with $\odot$) of two vectors or matrices that have the same dimensions. The result is a vector or matrix that has the same dimension as the original terms, with values that are the product of the corresponding elements (i.e., $c_{ij} = a_{ij}b_{ij}$). For example:

$$\mathbf{A} \odot \mathbf{A} = \begin{bmatrix} 4 & 2 & 1 \\ 3 & 5 & 0 \\ 1 & 4 & 2 \end{bmatrix} \odot \begin{bmatrix} 4 & 2 & 1 \\ 3 & 5 & 0 \\ 1 & 4 & 2 \end{bmatrix}$$

$$= \begin{bmatrix} 4 \times 4 & 2 \times 2 & 1 \times 1 \\ 3 \times 3 & 5 \times 5 & 0 \times 0 \\ 1 \times 1 & 4 \times 4 & 2 \times 2 \end{bmatrix}$$

$$= \begin{bmatrix} 16 & 4 & 1 \\ 9 & 25 & 0 \\ 1 & 16 & 4 \end{bmatrix}.$$

A.3 DIFFERENTIATION AND INTEGRATION

Differentiation and integration are very extensive topics that are presented in detail in many books on calculus (e.g., Anton, 1988). Here we do not attempt to provide details on the mechanics of these methods, but instead attempt to simply provide an interpretation of the concepts to aid the reader's understanding of why they are used in this book.

The geometric interpretation of the derivative of a function $f(x)$ is the slope of the tangent line to $f(x)$ at the point x, or the instantaneous rate of change in $f(x)$ at x (Fig. A.1). The derivative with respect to x is often denoted as $df(x)/dx$ or $f'(x)$. If the function $f(x)$ is increasing as x increases, then the slope of the tangent line, and hence the derivative, is positive, while if the function is decreasing as x increases, then the derivative will be negative. An important result in the context of this book is that at the point where $f(x)$ is maximized, the derivative of the function will be zero. This result (or the geo-

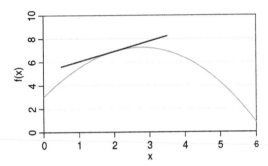

FIGURE A.1 A function $f(x)$ (gray line), and the associated tangent line at $x = 2$ (black line). The slope of the indicated tangent line is the derivative of $f(x)$ at $x = 2$.

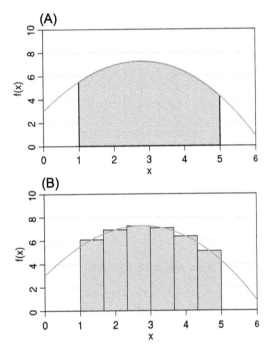

FIGURE A.2 Evaluating the area under the curve $f(x)$ between the points $x = a$ and $x = b$ using integration. Panel (A) indicates the area to be calculated and panel (B) illustrates how this may be approximated using a number of rectangular areas.

metric interpretation of it) is used to obtain estimates of the parameter values that maximize the likelihood function (i.e., obtain maximum likelihood estimates).

Integration is simply a technique for finding the area under a curve defined by the function $f(x)$ between an upper limit b and a lower limit a, of x (Fig. A.2A). This is formally notated as $Area = \int_a^b f(x)\,dx$, where 'dx' denotes that the inte-

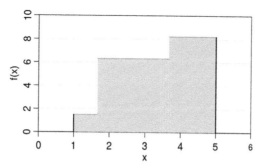

FIGURE A.3 Determining the area under the 'curve' by integration for a non-continuous function between the points $x = a$ and $x = b$.

gration is performed 'with respect to x' in case the function contains more than one variable. For many simple problems, the integration can be performed analytically using calculus, but often it must be approximated numerically. Many numerical integration algorithms are based on the view that the area under a curve can be approximated by dividing up the finite interval of x from a to b into many (say n) small subintervals that each form a small rectangular area (Fig. A.2B). The area under the curve between the limits of a and b can thus be approximated by adding together the areas of all the rectangles. Obviously the approximation improves as the rectangles become narrower by increasing the number of subintervals used (i.e., increasing n). Hence the view, as noted in Chapter 3, that integration can be considered as the sum of a very large number of small terms (as the area of each rectangle will become smaller as n is increased). Also note that while the function illustrated in Fig. A.2 is smooth and continuous between a and b, it does not necessarily have to be so for integration to be used. The function may also be stepwise as in Fig. A.3, in which case the area under the 'curve' can be well approximated by relatively few rectangular areas. Hence integration can be performed on both continuous and discrete random variables.

Finally, one common use of integration in statistics is to calculate the expected value of a random variable or a function of a random variable (see Chapter 3, Section 3.1.2). This may be done, for example, when for a given data point, the random variable itself is unobserved (i.e., latent). The probability (or the likelihood) of observing that data point is therefore dependent on the unknown value of a random variable, which may take many possible values as defined by the probability distribution for that random variable. The latent random variable can be 'removed' from the likelihood by integration, which often amounts to finding the expected value of a function of the random variable (e.g., as used in Chapter 7).

Bibliography

Abad-Franch, F., Ferraz, G., Campos, C., Palomeque, F.S., Grijalva, M.J., Aguilar, H.M., Miles, M.A., 2010. "Didn't you see that bug...?" Investigating disease vector occurrence when detection is imperfect. In: Revista da Sociedade Brasileira de Medicina Tropical, vol. 43. Sociedade Brasileira de Medicina Tropical, pp. 42–45.

Abad-Franch, F., Lima, M.M., Sarquis, O., Gurgel-Gonçalves, R., Sánchez-Martín, M., Calzada, J., Saldaña, A., Monteiro, F.A., Palomeque, F.S., Santos, W.S., Angulo, V.M., Esteban, L., Dias, F.B.S., Diotaiuti, L., Bar, M.E., Gottdenker, N.L., 2015. On palms, bugs, and Chagas disease in the Americas. Acta Tropica 151, 126–141.

Adams, M.J., Chelgren, N.D., Reinitz, D., Cole, R.A., Rachowicz, L.J., Galvan, S., McCreary, B., Pearl, C.A., Bailey, L.L., Bettaso, J., Bull, E.L., Leu, M., 2010. Using occupancy models to understand the distribution of an amphibian pathogen, *Batrachochytrium dendrobatidis*. Ecological Applications 20, 289–302.

Adams, M.J., Miller, D.A., Muths, E., Corn, P.S., Grant, E.H.C., Bailey, L.L., Fellers, G.M., Fisher, R.N., Sadinski, W.J., Waddle, H., Walls, S.C., 2013. Trends in amphibian occupancy in the United States. PLoS ONE 8, e64347.

Aing, C., Halls, S., Oken, K., Dobrow, R., Fieberg, J., 2011. A Bayesian hierarchical occupancy model for track surveys conducted in a series of linear, spatially correlated, sites. Journal of Applied Ecology 48, 1508–1517.

Akaike, H., 1973. Information theory and an extension of the maximum likelihood principle. In: Petrov, B.N., Csáaki, F. (Eds.), Second International Symposium Information Theory. Akademiai Kiado, pp. 267–281.

Al-Chokhachy, R., Ray, A.M., Roper, B.B., Archer, E., 2013. Exotic plant colonization and occupancy within riparian areas of the interior Columbia River and Upper Missouri River basins, USA. Wetlands 33, 409–420.

Alldredge, M.W., Pacifici, K., Simons, T.R., Pollock, K.H., 2008. A novel field evaluation of the effectiveness of distance and independent observer sampling to estimate aural avian detection probabilities. Journal of Applied Ecology 45, 1349–1356.

Alpizar-Jara, R., Nichols, J.D., Hines, J.E., Sauer, J.R., Pollock, K.H., Rosenberry, C.S., 2004. The relationship between species detection probability and local extinction probability. Oecologia 141, 652–660.

Alroy, J., 2010. Fair sampling of taxonomic richness and unbiased estimation of origination and extinction rates. Quantitative methods in paleobiology. The Paleontological Society Papers 16, 55–80.

Altwegg, R., Wheeler, M., Erni, B., 2008. Climate and the range dynamics of species with imperfect detection. Biology Letters 4, 581–584.

Amarasekare, P., Possingham, H., 2001. Patch dynamics and metapopulation theory: the case of successional species. Journal of Theoretical Biology 209, 333–344.

Anderson, D.R., 2008. Model Based Inference in the Life Sciences: A Primer on Evidence. Springer.

Anderson, R.M., May, R.M., 1991. Infectious Diseases of Humans: Dynamics and Control. Oxford University Press.

Anderson, R.P., 2003. Real vs artefactual absences in species distributions: tests for *Oryzomys albigularis* (Rodentia: Muridae) in Venezuela. Journal of Biogeography 30, 591–605.

Andren, H., 1994. Can one use nested subset pattern to reject the random sample hypothesis? Examples from boreal bird communities. Oikos 70, 489–491.

Andrewartha, H.G., Birch, L.C., 1954. The Distribution and Abundance of Animals. University of Chicago Press.

Anthony, R.G., Forsman, E.D., Franklin, A.B., Anderson, D.R., Burnham, K.P., White, G.C., Schwarz, C.J., Nichols, J.D., Hines, J.E., Olson, G.S., Ackers, S.H., Andrews, L.S., Biswell, B.L., Carlson, P.C., Diller, L.V., Dugger, K.M., Fehring, K.E., Fleming, T.L., Gerhardt, R.P., Gremel, S.A., Gutierrez, R.J., Happe, P.J., Herter, D.R., Higley, J.M., Horn, R.B., Irwin, L.L., Loschl, P.J., Reid, J.A., Sovern, S.G., 2006. Status and Trends in Demography of Northern Spotted Owls, 1985–2003. Wildlife Monographs, vol. 163, pp. 1–48.

Anton, H., 1988. Calculus with Analytic Geometry, 3rd edition. John Wiley and Sons.

Antón, S.C., Snodgrass, J.J., 2012. Origins and evolution of Genus *Homo*. Current Anthropology 53, S479–S496.

Araujo, M.B., Williams, P.H., 2000. Selecting areas for species persistence using occurrence data. Biological Conservation 96, 331–345.

Araujo, M.B., Williams, P.H., Fuller, R.J., 2002. Dynamics of extinction and the selection of nature reserves. Proceedings of the Royal Society, Series B 269, 1971–1980.

Armitage, D.W., Ober, H.K., 2010. A comparison of supervised learning techniques in the classification of bat echolocation calls. Ecological Informatics 5, 465–473.

Arrhenius, O., 1921. Species and area. Journal of Ecology 9, 95–99.

Azuma, D.L., Baldwin, J.A., Noon, B.R., 1990. Estimating the Occupancy of Spotted Owl Habitat Areas by Sampling and Adjusting for Bias. USDA Gen. Tech. Rep. PSW-124, Berkeley, CA.

Bailey, L.L., Hines, J.E., Nichols, J.D., MacKenzie, D.I., 2007. Sampling design trade-offs in occupancy studies with imperfect detection: examples and software. Ecological Applications 17, 281–290.

Bailey, L.L., MacKenzie, D.I., Nichols, J.D., 2014. Advances and applications of occupancy models. Methods in Ecology and Evolution 5, 1269–1279.

Bailey, L.L., Reid, J.A., Forsman, E.D., Nichols, J.D., 2009. Modeling co-occurrence of northern spotted and barred owls: accounting for detection probability differences. Biological Conservation 142, 2983–2989.

Bailey, L.L., Simons, T.R., Pollock, K.H., 2004. Estimating site occupancy and species detection probability parameters for terrestrial salamanders. Ecological Applications 14, 692–702.

Bailey, N.T.J., 1975. The Mathematical Theory of Infectious Diseases, 2nd edition. Macmillan.

Bailey, T.N., 1971. Biology of striped skunks on a southwestern Lake Erie marsh. American Midland Naturalist, 196–207.

Ball, L.C., Doherty Jr., P.F., McDonald, M.W., 2005. An occupancy modeling approach to evaluating a Palm Springs ground squirrel habitat model. Journal of Wildlife Management 69, 894–904.

Barber-Meyer, S.M., 2010. Dealing with the clandestine nature of wildlife-trade market surveys. Conservation Biology 24, 918–923.

Barbraud, C., Nichols, J.D., Hines, J.E., Hafner, H., 2003. Estimating rates of local extinction and colonization in colonial species and an extension to the metapopulation and community levels. Oikos 101, 113–126.

Barker, R.J., Sauer, J.R., 1992. Modeling population change from time series data. In: Cappucino, N., Price, P. (Eds.), Wildlife 2001: Populations. Elsevier Applied Sciences, pp. 182–194.

Barrett, R.H., 1983. Smoked aluminum track plots for determining furbearer distribution and relative abundance. California Fish and Game 69, 188–190.

Barrows, C.W., Swartz, M.B., Hodges, W.L., Allen, M.F., Rotenberry, J.T., Li, B.-L., Scott, T.A., Chen, X., 2005. A framework for monitoring multiple-species conservation plans. Journal of Wildlife Management 69, 1333–1345.

Bart, J., 1985. Causes of recording errors in singing bird surveys. The Wilson Bulletin, 161–172.

Bart, J., Klosiewski, S.P., 1989. Use of presence–absence to measure changes in avian density. Journal of Wildlife Management 53, 847–852.

Bayne, E.M., Boutin, S., Moses, R.A., 2008. Ecological factors influencing the spatial pattern of Canada lynx relative to its southern range edge in Alberta, Canada. Canadian Journal of Zoology 86, 1189–1197.

Beckwith, S.L., 1954. Ecological succession on abandoned farm lands and its relationship to wildlife management. Ecological Monographs 24, 349–376.

Beissinger, S.R., Iknayan, K.J., Guillera-Arroita, G., Zipkin, E.F., Dorazio, R.M., Royle, J.A., Kéry, M., 2016. Incorporating imperfect detection into joint models of communities: a response to Warton et al. Trends in Ecology & Evolution 31, 736–737.

Besbeas, P., Freeman, S.N., Morgan, B.J., Catchpole, E.A., 2002. Integrating mark–recapture–recovery and census data to estimate animal abundance and demographic parameters. Biometrics 58, 540–547.

Bled, F., Nichols, J.D., Altwegg, R., 2013. Dynamic occupancy models for analyzing species' range dynamics across large geographic scales. Ecology and Evolution 3, 4896–4909.

Bled, F., Royle, J.A., Cam, E., 2011. Hierarchical modeling of an invasive spread: the Eurasian Collared-Dove *Streptopelia decaocto* in the United States. Ecological Applications 21, 290–302.

Bock, C.E., 1984. Geographical correlates of rarity vs abundance in some North American winter landbirds. The Auk 101, 266–273.

Bock, C.E., Ricklefs, R.E., 1983. Range size and local abundance of some North American songbirds: a positive correlation. American Naturalist 122, 295–299.

Bodenheimer, F.S., 1938. Problems of Animal Ecology. Clarendon Press.

Bolger, D.T., Albert, A.C., Soule, M.E., 1991. Occurrence patterns of bird species in habitat fragments: sampling, extinction, and nested subsets. American Naturalist 137, 155–166.

Borchers, D.L., Buckland, S.T., Zucchini, W., 2002. Estimating Animal Abundance: Closed Populations, vol. 13. Springer Science & Business Media.

Boulinier, T., Nichols, J.D., Hines, J.E., Sauer, J.R., Flather, C.H., Pollock, K.H., 1998a. Higher temporal variability of forest breeding bird communities in fragmented landscapes. Proceedings of the National Academy of Sciences 95, 7497–7501.

Boulinier, T., Nichols, J.D., Hines, J.E., Sauer, J.R., Flather, C.H., Pollock, K.H., 2001. Forest fragmentation and bird community dynamics: inference at regional scales. Ecology 82, 1159–1169.

Boulinier, T., Nichols, J.D., Sauer, J.R., Hines, J.E., Pollock, K.H., 1998b. Estimating species richness: the importance of heterogeneity in species detectability. Ecology 79, 1018–1028.

Bradford, D.F., Neale, A.C., Nash, M.S., Sada, D.W., Jaeger, J.R., 2003. Habitat patch occupancy by toads (*Bufo punctatus*) in a naturally fragmented desert landscape. Ecology 84, 1012–1023.

Brandle, M., Stadler, J.K.J., Brandl, R., 2003. Distributional range size of weedy plant species is correlated with germination patterns. Ecology 84, 136–144.

Breininger, D.R., Carter, G.M., 2003. Territory quality transitions and source–sink dynamics in a Florida scrub-jay population. Ecological Applications 13, 516–529.

Brett, C.E., 1998. Sequence stratigraphy, paleoecology, and evolution; biotic clues and responses to sea-level fluctuations. Palaios 13, 241–262.

Britzke, E.R., Duchamp, J.E., Murray, K.L., Swihart, R.K., Robbins, L.W., 2011. Acoustic identification of bats in the eastern United States: a comparison of parametric and nonparametric methods. The Journal of Wildlife Management 75, 660–667.

Broms, K.M., Hooten, M.B., Johnson, D.S., Altwegg, R., Conquest, L.L., 2016. Dynamic occupancy models for explicit colonization processes. Ecology 97, 194–204.

Broms, K.M., Johnson, D.S., Altwegg, R., Conquest, L.L., 2014. Spatial occupancy models applied to atlas data show southern ground hornbills strongly depend on protected areas. Ecological Applications 24, 363–374.

Brown, J.H., 1984. On the relationship between abundance and distribution of species. American Naturalist 124, 255–279.

Brown, J.H., 1995. Macroecology. University of Chicago Press.

Brown, J.H., 1999. Macroecology: progress and prospect. Oikos 87, 3–14.

Brown, J.H., Kodric-Brown, A., 1977. Turnover rates in insular biogeography: effect of immigration on extinction. Ecology 58, 445–449.

Brown, J.H., Maurer, B.A., 1987. Evolution of species assemblages: effects of energetic constraints and species dynamics on the diversification of North American avifauna. American Naturalist 130, 1–17.

Brown, J.H., Maurer, B.A., 1989. Macroecology: the division of food and space among species on continents. Science 243, 1145–1150.

Brown, J.H., Stevens, G.C., Kaufman, D.M., 1996. The geographic range: size, shape, boundaries, and internal structure. Annual Review of Ecology and Systematics 27, 597–623.

Brownie, C., Anderson, D.R., Burnham, K.P., Robson, D.S., 1978. Statistical Inference from Band Recovery Data—A Handbook. U.S. Fish and Wildlife Service Resource Publications, vol. 131.

Brownie, C., Anderson, D.R., Burnham, K.P., Robson, D.S., 1985. Statistical Inference from Band Recovery Data—A Handbook, 2nd edition. U.S. Fish and Wildlife Service Resource Publications, vol. 156.

Buckland, S.T., Anderson, D.R., Burnham, K.P., Laake, J.L., 1993. Distance Sampling: Estimation of Biological Populations. Chapman and Hall.

Buckland, S.T., Anderson, D.R., Burnham, K.P., Laake, J.L., Borchers, D.L., Thomas, L., 2001. Introduction to Distance Sampling. Oxford University Press.

Buckland, S.T., Burnham, K.P., Augustin, N.H., 1997. Model selection: an integral part of inference. Biometrics, 603–618.

Buckland, S.T., Magurran, A.E., Green, R.E., Fewster, R.M., 2005. Monitoring change in biodiversity through composite indices. Philosophical Transactions of the Royal Society of London. Series B, Biological Sciences 360, 243–254.

Bunge, J., Fitzpatrick, M., 1993. Estimating the number of species – a review. Journal of the American Statistical Association 88, 364–373.

Burnham, K.P., 1993. A theory for combined analysis of ring recovery and recapture data. In: Lebreton, J.D., North, P.M. (Eds.), Marked Individuals in the Study of Bird Populations. Birkhäuser-Verlag, pp. 199–214.

Burnham, K.P., Anderson, D.R., 1998. Model Selection and Inference: A Practical Information-Theoretic Approach. Springer-Verlag.

Burnham, K.P., Anderson, D.R., 2002. Model Selection and Multi-Model Inference, 2nd edition. Springer.

Burnham, K.P., Anderson, D.R., 2004. Multimodal inference: understanding AIC and BIC in model selection. Sociological Methods & Research 33, 261–304.

Burnham, K.P., Anderson, D.R., Laake, J.L., 1980. Estimation of Density from Line-Transect Sampling of Biological Populations. Wildlife Monographs, vol. 72, pp. 3–302.

Burnham, K.P., Anderson, D.R., White, G.C., Brownie, C., Pollock, K.P., 1987. Design and Analysis of Methods for Fish Survival Experiments Based on Release–Recapture. American Fisheries Society Monographs, vol. 5, pp. 1–437.

Burnham, K.P., Overton, W.S., 1978. Estimation of the size of a closed population when capture probabilities vary among animals. Biometrika 65, 625–633.

Burnham, K.P., Overton, W.S., 1979. Robust estimation of population size when capture probabilities vary among animals. Ecology 60, 927–936.

Buzas, M.A., Koch, C.F., Culver, S.J., Sohl, N., 1982. On the distribution of species occurrence. Paleobiology 8, 143–150.

Cabeza, M., Araujo, M.B., Wilson, R.J., Thomas, C.D., Cowley, M.J.R., Moilanen, A., 2004. Combining probabilities of occurrence with spatial reserve design. Journal of Applied Ecology 41, 252–262.

Cam, E., Nichols, J.D., Hines, J.E., Sauer, J.R., 2000a. Inferences about nested subsets structure when not all species are detected. Oikos 91, 428–434.

Cam, E., Nichols, J.D., Hines, J.E., Sauer, J.R., Alpizar-Jara, R., Flather, C.H., 2002. Disentangling sampling and ecological explanations underlying species–area relationships. Ecology 83, 1118–1130.

Cam, E., Nichols, J.D., Sauer, J.R., Hines, J.E., Flather, C.H., 2000b. Relative species richness and community completeness: birds and urbanization in the mid-Atlantic states. Ecological Applications 10, 1196–1210.

Campbell, M., Francis, C.M., 2011. Using stereo-microphones to evaluate observer variation in North American Breeding Bird Survey point counts. The Auk 128, 303–312.

Carey, C., Bruzgul, J.E., Livo, L.J., Walling, M.L., Kuehl, K.A., Dixon, B.F., Pessier, A.P., Alford, R.A., Rogers, K.B., 2006. Experimental exposures of boreal toads (*Bufo boreas*) to a pathogenic chytrid fungus (*Batrachochytrium dendrobatidis*). EcoHealth 3, 5–21.

Carothers, A., 1973. Capture–recapture methods applied to a population with known parameters. Journal of Animal Ecology 42, 125–146.

Carroll, C., Zielinski, W.J., Noss, R.F., 1999. Using presence–absence data to build and test spatial habitat models for the fisher in Klamath region, U.S.A. Conservation Biology 13, 1344–1359.

Casella, G., Berger, R., 2002. Statistical Inference. Duxbury/Thompson Learning.

Casula, P., Nichols, J.D., 2003. Temporal variability of local abundance, sex ratio and activity in the Sardinian chalk hill blue butterfly. Oecologia 136, 374–382.

Caswell, H., 2001. Matrix Population Models. Wiley Online Library.

Caswell, H., Cohen, J.E., 1991a. Communities in patchy environments: a model of disturbance, competition, and heterogeneity. In: Kolasa, J., Pickett, S.T.A. (Eds.), Ecological Heterogeneity. Springer-Verlag, pp. 97–122.

Caswell, H., Cohen, J.E., 1991b. Disturbance, interspecific interaction and diversity in metapopulations. Biological Journal of the Linnean Society 42, 193–218.

Caughley, G., Grice, D., Barker, R., Brown, B., 1988. The edge of range. Journal of Animal Ecology 57, 771–785.

Ceballos, G., Ehrlich, P., 2002. Mammal population losses and the extinction crisis. Science 296, 904–907.

Chambert, T., Miller, D.A., Nichols, J.D., 2015a. Modeling false positive detections in species occurrence data under different study designs. Ecology 96, 332–339.

Chambert, T., Kendall, W.L., Hines, J.E., Nichols, J.D., Pedrini, P., Waddle, J.H., Tavecchia, G., Walls, S.C., Tenan, S., 2015b. Testing hypotheses on distribution shifts and changes in phenology of imperfectly detectable species. Methods in Ecology and Evolution 6, 638–647.

Chandler, R.B., King, D.I., Raudales, R., Trubey, R., Chandler, C., Arce Chávez, V.J., 2013. A small-scale land-sparing approach to conserving biological diversity in tropical agricultural landscapes. Conservation Biology 27, 785–795.

Chandler, R.B., Muths, E., Sigafus, B.H., Schwalbe, C.R., Jarchow, C.J., Hossack, B.R., 2015. Spatial occupancy models for predicting metapopulation dynamics and viability following reintroduction. Journal of Applied Ecology 52, 1325–1333.

Chao, A., Lee, S.-M., 1992. Estimating the number of classes via sample coverage. Journal of the American Statistical Association 87, 210–217.

Chao, A., Tsay, P., Lin, S.-H., Shau, W.-Y., Chao, D.-Y., 2001. The applications of capture–recapture models to epidemiological data. Statistics in Medicine 20, 3123–3157.

Chao, A., Yang, M.C., 1993. Stopping rules and estimation for recapture debugging with unequal failure rates. Biometrika 80, 193–201.

Chao, A., Yip, P., Lin, H.-S., 1996. Estimating the number of species via a martingale estimating function. Statistica Sinica 6, 403–418.

Chen, G., Kéry, M., Plattner, M., Ma, K., Gardner, B., 2013. Imperfect detection is the rule rather than the exception in plant distribution studies. Journal of Ecology 101, 183–191.

Chestnut, T., Anderson, C., Popa, R., Blaustein, A.R., Voytek, M., Olson, D.H., Kirshtein, J., 2014. Heterogeneous occupancy and density estimates of the pathogenic fungus *Batrachochytrium dendrobatidis* in waters of North America. PLoS ONE 9, e106790.

Chivers, D.J., 1991. Guidelines for re-introductions: procedures and problems. Symposium of the Zoological Society of London 62, 89–99.

Chown, S.L., van Rensburg, B.J., Gaston, K., Rodrigues, A.S.L., van Jaarsveld, A.S., 2003. Energy, species richness, and human population size: conservation implications at a national scale. Ecological Applications 13, 1233–1241.

Clark, C., Rosenzweig, M.L., 1994. Extinction and colonization processes: parameter estimates from sporadic surveys. American Naturalist 143, 583–596.

Clement, M.J., 2016. Designing occupancy studies when false-positive detections occur. Methods in Ecology and Evolution 7, 1538–1547.

Clement, M.J., Hines, J.E., Nichols, J.D., Pardieck, K.L., Ziolkowski, D.J., 2016. Estimating indices of range shifts in birds using dynamic models when detection is imperfect. Global Change Biology 22, 3273–3285.

Clement, M.J., Rodhouse, T.J., Ormsbee, P.C., Szewczak, J.M., Nichols, J.D., 2014. Accounting for false-positive acoustic detections of bats using occupancy models. Journal of Applied Ecology 51, 1460–1467.

Clinchy, M., Haydon, D.T., Smith, A.T., 2002. Pattern does not equal process: what does patch occupancy really tell us about metapopulation dynamics? American Naturalist 159, 351–362.

Cochran, W.G., 1977. Sampling Techniques. Wiley.

Cole, D.J., Morgan, B.J., McCrea, R.S., Pradel, R., Gimenez, O., Choquet, R., 2014. Does your species have memory? Analyzing capture–recapture data with memory models. Ecology and Evolution 4, 2124–2133.

Cole, L.C., 1949. The measurement of interspecific association. Ecology 30, 411–424.

Collier, B.A., Groce, J.E., Morrison, M.L., Newnam, J.C., Campomizzi, A.J., Farrell, S.L., Mathewson, H.A., Snelgrove, R.T., Carroll, R.J., Wilkins, R.N., 2012. Predicting patch occupancy in fragmented landscapes at the rangewide scale for an endangered species: an example of an American warbler. Diversity and Distributions 18, 158–167.

Colwell, R.K., Coddington, J.A., 1994. Estimating terrestrial biodiversity through extrapolation. Proceedings of the Royal Society, Series B 345, 101–118.

Conn, P.B., Cooch, E.G., Caley, P., 2012. Accounting for detection probability when estimating force-of-infection from animal encounter data. Journal of Ornithology 152, 511–520.

Connell, J.H., 1961. The influence of interspecific competition and other factors on the distribution of the barnacle *Chthamalus stellatus*. Ecology 42, 710–723.

Connolly, S.R., Miller, A.I., 2001a. Global Ordovician faunal transitions in the marine benthos: proximate causes. Paleobiology 27, 779–795.

Connolly, S.R., Miller, A.I., 2001b. Joint estimation of sampling and turnover rates from fossil databases: capture–mark–recapture methods revisited. Paleobiology 27, 751–767.

Connolly, S.R., Miller, A.I., 2002. Global Ordovician faunal transitions in the marine benthos: ultimate causes. Paleobiology 28, 26–40.

Connor, E.F., McCoy, E.D., 1979. The statistics and biology of the species–area relationship. American Naturalist 113, 791–833.

Connor, E.F., Simberloff, D., 1979. The assembly of species communities: chance or competition? Ecology 60, 1132–1140.

Connor, E.F., Simberloff, D., 1984. Neutral models of species co-occurrence patterns. In: Strong, D.R.J., Simberloff, D., Abele, L., Thistle, A. (Eds.), Ecological Communities: Conceptual Issues and the Evidence. Princeton University Press, pp. 316–331.

Connor, E.F., Simberloff, D., 1986. Competition, scientific method, and null models in ecology. American Scientist 74, 155–162.

Conroy, M.J., Nichols, J.D., 1984. Testing for variation in taxonomic extinction probabilities: a suggested methodology and some results. Paleobiology 10, 328–337.

Conroy, M.J., Nichols, J.D., 1996. Designing a study to assess mammalian diversity. In: Wilson, D.E., Cole, F.R., Nichols, J.D., Rudran, R., Foster, M.S. (Eds.), Measuring and Monitoring Biological Diversity: Standard Methods for Mammals. Smithsonian Institution Press, pp. 41–49.

Conroy, M.J., Runge, J.P., Barker, R.J., Schofield, M.R., Fonnesbeck, C.J., 2008. Efficient estimation of abundance for patchily distributed populations via two-phase, adaptive sampling. Ecology 89, 3362–3370.

Cooch, E.G., Conn, P.B., Ellner, S.P., Dobson, A.P., Pollock, K.H., 2012. Disease dynamics in wild populations: modeling and estimation: a review. Journal of Ornithology 152, 485–509.

Cook, R.R., Quinn, J.F., 1995. The influence of colonization in nested subsets. Oecologia 102, 413–424.

Cook, R.R., Quinn, J.F., 1998. An evaluation of randomization models for nested subset analysis. Oecologia 113, 584–592.

Cornell, H.V., 1993. Unsaturated patterns in species assemblages: the role of regional processes in setting local species richness. In: Ricklefs, R.E., Schluter, D. (Eds.), Species Diversity in Ecological Communities. Historical and Geographical Perspectives. University of Chicago Press, pp. 243–252.

Cornell, H.V., Lawton, J.H., 1992. Species interactions, local and regional processes, and limits to the richness of ecological communities: a theoretical perspective. Journal of Animal Ecology 61, 1–12.

Coull, B.A., Agresti, A., 1999. The use of mixed logit models to reflect heterogeneity in capture–recapture studies. Biometrics 55, 294–301.

Cressie, N.A., 1993. Statistics for Spatial Data. Wiley.

Crosbie, S., Manly, B., 1985. Parsimonious modelling of capture–mark–recapture studies. Biometrics, 385–398.

Crow, J.F., Kimura, M., 1970. An Introduction to Population Genetics Theory. Harper and Row.

Crowell, K.L., Pimm, S.L., 1976. Competition and niche shifts of mice introduced onto small islands. Oikos 27, 251–258.

Crozier, M.L., Seamans, M.E., Gutiérrez, R.J., Loschl, P.J., Horn, R.B., Sovern, S.G., Forsman, E.D., 2006. Does the presence of barred owls suppress the calling behavior of spotted owls? The Condor 108, 760–769.

Crum, N.J., Fuller, A.K., Sutherland, C.S., Cooch, E.G., Hurst, J., 2017. Estimating occupancy probability of moose using hunter survey data. The Journal of Wildlife Management 81, 521–534.

Curnutt, J.L., Pimm, S.L., Maurer, B.A., 1996. Population variability of sparrows in space and time. Oikos 76, 131–144.

Dail, D., Madsen, L., 2011. Models for estimating abundance from repeated counts of an open metapopulation. Biometrics 67, 577–587.

Dakin, S.F., 1978. The Influence of Interspecific Competition on the Microhabitat Distribution of Terrestrial Lungless Salamanders in the Southern Appalachian Mountains. Ph.D. thesis. Duke University.

Darwin, C., 1859. On the Origin of Species by Means of Natural Selection, or the Preservation of Favored Races in the Struggle for Life. John Murray.

Daszak, P., Cunningham, A.A., Hyatt, A.D., 2000. Emerging infectious diseases of wildlife-threats to biodiversity and human health. Science 287, 443–449.

de Valpine, P., Turek, D., Paciorek, C.J., Anderson-Bergman, C., Lang, D.T., Bodik, R., 2017. Programming with models: writing statistical algorithms for general model structures with NIMBLE. Journal of Computational and Graphical Statistics 26, 403–413.

Dempster, A.P., Laird, N.M., Rubin, D.B., 1977. Maximum likelihood from incomplete data via the EM algorithm. Journal of the Royal Statistical Society, Series B, Methodological, 1–38.

Deredec, A., Courchamp, F., 2003. Extinction thresholds in host–parasite dynamics. Annales Zoologici Fennici 40, 115–130.

Diamond, J.M., 1975. Assembly of species communities. In: Cody, M.L., Diamond, J.M. (Eds.), Ecology and Evolution of Communities. Harvard University Press, pp. 342–444.

Diamond, J.M., 1976. Island biogeography and conservation: strategy and limitations. Science 193, 1027–1029.

Diamond, J.M., 1982. Effect of species pool size on species occurrence frequencies: musical chairs on islands. Proceedings of the National Academy of Sciences 79, 2420–2424.

Diamond, J.M., Gilpin, M.E., 1982. Examination of the "null" model of Connor and Simberloff for species co-occurrences on islands. Oecologia 52, 64–74.

Diamond, J.M., May, R.M., 1977. Species turnover rates on islands: dependence on census interval. Science 197, 266–270.

DiCastri, F., Hansen, A.J., Debussche, M., 1990. Biological Invasions in Europe and the Mediterranean Basin. Kluwer Academic Publishers.

Dice, L.R., 1945. Measures of the amount of ecologic association between species. Ecology 26, 297–302.

Diller, L.V., Hamm, K.A., Early, D.A., Lamphear, D.W., Dugger, K.M., Yackulic, C.B., Schwarz, C.J., Carlson, P.C., McDonald, T.L., 2016. Demographic response of northern spotted owls to barred owl removal. The Journal of Wildlife Management 80, 691–707.

Dobson, A., Foufopoulos, J., 2001. Emerging infectious pathogens of wildlife. Philosophical Transactions of the Royal Society of London. Series B, Biological Sciences 356, 1001–1012.

Dodd, C.K., 2003. Monitoring Amphibians in Great Smoky Mountains National Park. US Geological Survey Circular, vol. 1258.

Doherty Jr., P.F., Boulinier, T., Nichols, J.D., 2003a. Extinction rates at the center and edge of species' ranges. Annales Zoologici Fennici 40, 145–153.

Doherty Jr., P.F., Sorci, G., Royle, J.A., Hines, J.E., Nichols, J.D., Boulinier, T., 2003b. Sexual selection affects local extinction and turnover in bird communities. Proceedings of the National Academy of Sciences 100, 5858–5862.

Dorazio, R.M., 2012. Predicting the geographic distribution of a species from presence-only data subject to detection errors. Biometrics 68, 1303–1312.

Dorazio, R.M., Connor, E.F., 2014. Estimating abundances of interacting species using morphological traits, foraging guilds, and habitat. PLoS ONE 9, e94323.

Dorazio, R.M., Connor, E.F., Askins, R.A., 2015. Estimating the effects of habitat and biological interactions in an avian community. PLoS ONE 10, e0135987.

Dorazio, R.M., Kéry, M., Royle, J.A., Plattner, M., 2010. Models for inference in dynamic metacommunity systems. Ecology 91, 2466–2475.

Dorazio, R.M., Royle, J.A., 2003. Mixture models for estimating the size of a closed population when capture rates vary among individuals. Biometrics 59, 351–364.

Dorazio, R.M., Royle, J.A., 2005. Estimating size and composition of biological communities by modeling the occurrence of species. Journal of the American Statistical Association 100, 389–398.

Dorazio, R.M., Royle, J.A., Söderström, B., Glimskär, A., 2006. Estimating species richness and accumulation by modeling species occurrence and detectability. Ecology 87, 842–854.

Dugger, K.M., Forsman, E.D., Franklin, A.B., Davis, R.J., White, G.C., Schwarz, C.J., Burnham, K.P., Nichols, J.D., Hines, J.E., Yackulic, C.B., Doherty Jr., P.F., Bailey, L., Clark, D.A., Ackers,

S.H., Andrews, L.S., Augustine, B., Biswell, B.L., Blakesley, J., Carlson, P.C., Clement, M.J., Diller, L.V., Glenn, E.M., Green, A., Gremel, S.A., Herter, D.R., Higley, J.M., Hobson, J., Horn, R.B., Huyvaert, K.P., McCafferty, C., McDonald, T., McDonnell, K., Olson, G.S., Reid, J.A., Rockweit, J., Ruiz, V., Saenz, J., Sovern, S.G., 2015. The effects of habitat, climate, and barred owls on long-term demography of northern spotted owls. The Condor 118, 57–116.

Dunham, J.B., Rieman, B.E., 1999. Metapopulation structure of bull trout: influence of physical, biotic, and geometrical landscape characteristics. Ecological Applications 9, 642–655.

Dunham, J.B., Rieman, B.E., Peterson, J.T., 2002. Patch-based models to predict species occurrence: lessons from salmonid fishes in streams. In: Scott, J.M., Heglund, P.J., Morrison, M.L., Haufler, J.B., Raphael, M.G., Wall, W.A., Samson, F.B. (Eds.), Predicting Species Occurrences. Island Press, pp. 327–334.

Dunn, P.K., Smyth, G.K., 1996. Randomized quantile residuals. Journal of Computational and Graphical Statistics 5, 236–244.

Dupuis, J.A., 1995. Bayesian estimation of movement and survival probabilities from capture–recapture data. Biometrika, 761–772.

Eads, D.A., Biggins, D.E., Antolin, M.F., Long, D.H., Huyvaert, K.P., Gage, K.L., 2015. Prevalence of the generalist flea *Pulex simulans* on black-tailed prairie dogs (*Cynomys ludovicianus*) in New Mexico, USA: the importance of considering imperfect detection. Journal of Wildlife Diseases 51, 498–502.

Eads, D.A., Biggins, D.E., Doherty Jr., P.F., Gage, K.L., Huyvaert, K.P., Long, D.H., Antolin, M.F., 2013. Using occupancy models to investigate the prevalence of ectoparasitic vectors on hosts: an example with fleas on prairie dogs. International Journal for Parasitology: Parasites and Wildlife 2, 246–256.

Eaton, M.J., Hughes, P.T., Hines, J.E., Nichols, J.D., 2014. Testing metapopulation concepts: effects of patch characteristics and neighborhood occupancy on the dynamics of an endangered lagomorph. Oikos 123, 662–676.

Edwards, T.C.J., Cutler, D.R., Geiser, L., Alegria, J., McKenzie, D., 2004. Assessing rarity of species with low detectability: lichens in Pacific Northwest forests. Ecological Applications 14, 414–424.

Efron, B., 1979. Computers and the theory of statistics: thinking the unthinkable. Journal of the Society for Industrial and Applied Mathematics 21, 460–480.

Elith, J., Graham, C.H., Anderson, R.P., Dudík, M., Ferrier, S., Guisan, A., Hijmans, R.J., Huettmann, F., Leathwick, J.R., Lehmann, A., Li, J., Lohmann, L.G., Loiselle, B.A., Manion, G., Moritz, C., Nakamura, M., Nakazawa, Y., Overton, J.McC., Townsend Peterson, A., Phillips, S.J., Richardson, K., Scachetti-Pereira, R., Schapire, R.E., Soberón, J., Williams, S., Wisz, M.S., Zimmermann, N.E., 2006. Novel methods improve prediction of species' distributions from occurrence data. Ecography 29, 129–151.

Elith, J., Leathwick, J.R., 2009. Species distribution models: ecological explanation and prediction across space and time. Annual Review of Ecology, Evolution, and Systematics 40, 677.

Elliott, P., Wakefield, J.C., Best, N.G., Briggs, D.J., 2001. Spatial Epidemiology: Methods and Applications. Oxford University Press.

Ellner, S.P., Fussmann, G., 2003. Effects of successional dynamics on metapopulation persistence. Ecology 84, 882–889.

Elmore, S.A., Huyvaert, K.P., Bailey, L.L., Milhous, J., Alisauskas, R.T., Gajadhar, A.A., Jenkins, E.J., 2014. *Toxoplasma gondii* exposure in arctic-nesting geese: a multi-state occupancy framework and comparison of serological assays. International Journal for Parasitology: Parasites and Wildlife 3, 147–153.

Elmore, S.A., Samelius, G., Al-Adhami, B., Huyvaert, K.P., Bailey, L.L., Alisauskas, R.T., Gajadhar, A.A., Jenkins, E.J., 2016. Estimating *Toxoplasma gondii* exposure in arctic foxes (*Vulpes lagopus*) while navigating the imperfect world of wildlife serology. Journal of Wildlife Diseases 52, 47–56.

Elton, C., 1927. Animal Ecology. Sidgwick and Jackson Limited.

Engler, R., Guisan, A., Rechsteiner, L., 2004. An improved approach for predicting the distribution of rare and endangered species from occurrence and pseudo-absence data. Journal of Applied Ecology 41, 263–274.

Enøe, C., Georgiadis, M.P., Johnson, W.O., 2000. Estimation of sensitivity and specificity of diagnostic tests and disease prevalence when the true disease state is unknown. Preventive Veterinary Medicine 45, 61–81.

Enquist, B., Jordan, M.A., Brown, J.H., 1995. Connections between ecology, biogeography, and paleobiology: relationship between local abundance and geographic distribution in fossil and recent molluscs. Evolutionary Ecology 9, 586–604.

Erwin, R.M., Nichols, J.D., Eyler, T.B., Stotts, D.B., Truitt, B.R., 1998. Modeling colony-site dynamics: a case study of gull-billed terns (*Sterna nilotica*) in coastal Virginia. The Auk 115, 970–978.

Ewen, J., Armstrong, D., Parker, K., Seddon, P. (Eds.), 2012. Reintroduction Biology: Integrating Science and Management. Wiley–Blackwell.

Fackler, P.L., Pacifici, K., Martin, J., McIntyre, C., 2014. Efficient use of information in adaptive management with an application to managing recreation near golden eagle nesting sites. PLoS ONE 9, e102434.

Fairman, C.M., Bailey, L.L., Chambers, R.M., Russell, T.M., Funk, W.C., 2013. Species-specific effects of acidity on pond occupancy in *Ambystoma* salamanders. Journal of Herpetology 47, 346–353.

Falke, J.A., Bailey, L.L., Fausch, K.D., Bestgen, K.R., 2012. Colonization and extinction in dynamic habitats: an occupancy approach for a Great Plains stream fish assemblage. Ecology 93, 858–867.

Falke, J.A., Fausch, K.D., Bestgen, K.R., Bailey, L.L., 2010. Spawning phenology and habitat use in a Great Plains, USA, stream fish assemblage: an occupancy estimation approach. Canadian Journal of Fisheries and Aquatic Sciences 67, 1942–1956.

Farmer, R.G., Leonard, M.L., Horn, A.G., 2012. Observer effects and avian-call-count survey quality: rare-species biases and overconfidence. The Auk 129, 76–86.

Fernández-Chacón, A., Stefanescu, C., Genovart, M., Nichols, J.D., Hines, J.E., Páramo, F., Turco, M., Oro, D., 2014. Determinants of extinction–colonization dynamics in Mediterranean butterflies: the role of landscape, climate and local habitat features. Journal of Animal Ecology 83, 276–285.

Ferraz, G., Nichols, J.D., Hines, J.E., Stouffer, P.C., Bierregaard, R.O., Lovejoy, T.E., 2007. A large-scale deforestation experiment: effects of patch area and isolation on Amazon birds. Science 315, 238–241.

Ferraz, G., Russell, G.J., Stouffer, P.C., Bieregaard Jr., R.O., Pimm, S.L., Lovejoy, T.E., 2003. Rates of species loss from Amazonian forest fragments. Proceedings of the National Academy of Sciences 100, 14069–14073.

Fertig, W., Reiners, W.A., 2002. Predicting presence/absence of plant species for range mapping: a case study from Wyoming. In: Scott, J.M., Heglund, P.J., Morrison, M.L., Haufler, J.B., Raphael, M.G., Wall, W.A., Samson, F.B. (Eds.), Predicting Species Occurrences. Island Press, pp. 483–489.

Field, S.A., Tyre, A.J., Possingham, H.P., 2005. Optimizing allocation of monitoring effort under economic and observational constraints. Journal of Wildlife Management 69, 473–482.

Fienberg, S.E., 1992. Bibliography on capture–recapture modelling with application to census undercount adjustment. Survey Methodology 18, 143–154.

Fienberg, S.E., Johnson, M.S., Junker, B.W., 1999. Classical multilevel and Bayesian approaches to population size estimation using multiple lists. Journal of the Royal Statistical Society, Series A 163, 383–405.

Fischer, J., Lindenmayer, D.B., Cowling, A., 2004. The challenge of managing multiple species at multiple scales: reptiles in an Australian grazing landscape. Journal of Applied Ecology 41, 32–44.

Fisher, J.T., Wheatley, M., MacKenzie, D.I., 2014. Spatial patterns of breeding success of grizzly bears derived from hierarchical multistate models. Conservation Biology 28, 1249–1259.

Fisher, R.A., 1947. The Design of Experiments, 4th edition. Hafner.

Fisher, R.A., Corbet, A.S., Williams, C.B., 1943. The relation between the number of species and the number of individuals in a random sample of an animal population. Journal of Animal Ecology 12, 42–58.

Fiske, I., Chandler, R., 2011. unmarked: an R package for fitting hierarchical models of wildlife occurrence and abundance. Journal of Statistical Software 43, 1–23.

Fitzpatrick, M.C., Preisser, E.L., Ellison, A.M., Elkinton, J.S., 2009. Observer bias and the detection of low-density populations. Ecological Applications 19, 1673–1679.

Flanders, N.P., Gardner, B.A., Winiarski, K.J., Paton, P.W., Allison Jr., T., O'Connell, A.F., 2015. Key seabird areas in southern New England identified using a community occupancy model. Marine Ecology. Progress Series 533, 277–290.

Flather, C., 1996. Fitting species-accumulation functions and assessing regional land use impacts on avian diversity. Journal of Biogeography 23, 155–168.

Fleishman, E., MacNally, R.F.J.P., Murphy, D.D., 2001. Modeling and predicting species occurrence using broad-scale environmental variables: an example with butterflies of the Great Basin. Conservation Biology 15, 1674–1685.

Fleishman, E., Murphy, D.D., 1999. Patterns and processes of nestedness in a Great Basin butterfly community. Oecologia 119, 133–139.

Fletcher, D., 2012. Estimating overdispersion when fitting a generalized linear model to sparse data. Biometrika 99, 230.

Foote, M., 2001. Inferring temporal patterns of preservation, origination, and extinction from taxonomic survivorship analysis. Paleobiology 27, 602–630.

Foote, M., Raup, D.M., 1996. Fossil preservation and the stratigraphic ranges of taxa. Paleobiology 22, 121–140.

Forbes, S.A., 1907. On the local distribution of certain Illinois fishes: an essay in statistical ecology. Bulletin of the Illinois State Laboratory of Natural History 7, 273–303.

Forman, R.T.T., Galli, A.E., Leck, C.F., 1976. Forest size and avian diversity in New Jersey woodlots with some land-use applications. Oecologia 26, 1–8.

Forsman, E.D., 1983. Methods and Materials for Locating and Studying Spotted Owls. USDA Forest Service, Pacific Northwest Forest and Range Experiment Station.

Forsman, E.D., Anthony, R.G., Dugger, K.M., Glenn, E.M., Franklin, A.B., White, G.C., Schwarz, C.J., Burnham, K.P., Anderson, D.R., Nichols, J.D., Hines, J.E., Lint, J.B., Davis, R.J., Ackers, S.H., Andrews, L.S., Biswell, B.L., Carlson, P.C., Diller, L.V., Gremel, S.A., Herter, D.R., Higley, J.M., Horn, R.B., Reid, J.A., Rockweit, J., Schaberl, J.P., Snetsinger, T.J., Sovern, S.G., 2011. Population Demography of Northern Spotted Owls. University of California Press.

Franklin, A.B., Anderson, D.R., Forsman, E.D., Burnham, K.P., Wagner, F.W., 1996. Methods for collecting and analyzing demographic data on the northern spotted owl. Studies in Avian Biology 17, 12–20.

Franklin, A.B., Anderson, D.R., Gutierrez, R.J., Burnham, K.P., 2000. Climate, habitat quality, and fitness in northern spotted owl populations in northwestern California. Ecological Monographs 70, 539–590.

Franklin, A.B., Guttierez, R.J., Nichols, J.D., Seamans, M.E., White, G.C., Zimmerman, G.S., Hines, J.E., Munton, T.E., LaHaye, W.S., Blakesley, J.A., Steger, G.N., Noon, B.R., Shaw, D.W.H., Keane, J.J., McDonald, T.L., Britting, S., 2004. Population Dynamics of the California Spotted Owl: A Meta Analysis. American Ornithologists' Union Monographs, vol. 54, pp. 1–54.

Fretwell, S.D., 1972. Populations in a Seasonal Environment. Princeton University Press.

Fretwell, S.D., Lucas, H.R., 1969. On territorial behavior and other factors influencing habitat distribution in birds. I. Theoretical development. Acta Biotheoretica 19, 16–36.

Fritts, T.H., Rodda, G.H., 1998. The role of introduced species in the degradation of island ecosystems: a case study of Guam. Annual Review of Ecology and Systematics 29, 113–140.

Fujiwara, M., Caswell, H., 2002. Estimating population projection matrices from multi-stage mark–recapture data. Ecology 83, 3257–3265.

Fuller, A.K., Linden, D.W., Royle, J.A., 2016. Management decision making for fisher populations informed by occupancy modeling. The Journal of Wildlife Management 80, 794–802.

Gardner, C.L., Lawler, J.P., Ver Hoef, J.M., Magoun, A.J., Kellie, K.A., 2010. Coarse-scale distribution surveys and occurrence probability modeling for wolverine in interior Alaska. Journal of Wildlife Management 74, 1894–1903.

Gaston, K.J., 1990. Patterns in the geographical ranges of species. Biological Reviews 65, 105–129.

Gaston, K.J., 1991. How large is a species' geographic range? Oikos 61, 434–438.

Gaston, K.J., 1994. Rarity. Chapman and Hall.

Gaston, K.J., 1996. The multiple forms of the interspecific abundance–distribution relationship. Oikos 75, 211–220.

Gaston, K.J., 1998. Species–range size distributions: products of speciation, extinction and transformation. Proceedings of the Royal Society, Series B 353, 219–230.

Gaston, K.J., Blackburn, T.M., 1996. Global scale macroecology: interactions between population size, geographic rangesize and body size in the Anseriformes. Journal of Animal Ecology 65, 701–714.

Gaston, K.J., Blackburn, T.M., 1999. A critique for macroecology. Oikos 84, 353–368.

Gaston, K.J., Blackburn, T.M., Gregory, R., 1997a. Interspecific abundance–range size relationships: range position and phylogeny. Ecography 20, 390–399.

Gaston, K.J., Blackburn, T.M., Lawton, J.H., 1997b. Interspecific abundance–range size relationships: an appraisal of mechanisms. Journal of Animal Ecology 66, 579–601.

Geissler, P.H., Fuller, M.R., 1987. Estimation of the proportion of area occupied by an animal species. In: Proceedings of the Section on Survey Research Methods of the American Statistical Association, 1986, pp. 533–538.

Gelfand, A.E., Schmidt, A.M., Wu, S., Silander, J.A., Latimer, A., Rebelo, A.G., 2005. Modelling species diversity through species level hierarchical modelling. Journal of the Royal Statistical Society. Series C. Applied Statistics 54, 1–20.

Gelman, A., Carlin, J.B., Stern, H.S., Rubin, D.B., 2004. Bayesian Data Analysis, 2nd edition. Chapman and Hall.

Gelman, A., Jakulin, A., Pittau, M.G., Su, Y.-S., 2008. A weakly informative default prior distribution for logistic and other regression models. Annals of Applied Statistics 2, 1360–1383.

Gelman, A., Meng, X.-L., Stern, H., 1996. Posterior predictive assessment of model fitness via realized discrepancies. Statistica Sinica, 733–760.

Gelman, A., Rubin, D.B., 1992. Inference from iterative simulation using multiple sequences. Statistical Science, 457–472.

Genet, K.S., Sargent, L.G., 2003. Evaluation of methods and data quality from a volunteer-based amphibian call survey. Wildlife Society Bulletin, 703–714.

Gibson, L.A., Wilson, B.A., Cahill, D.M., Hill, J., 2004. Spatial prediction of rufous bristlebird habitat in a coastal heathland: a GIS-based approach. Journal of Applied Ecology 41, 213–223.

Gilbert, A.T., O'Connell Jr., A.F., Annand, E.M., Talancy, N.W., Sauer, J.R., Nichols, J.D., 2008. An Inventory of Terrestrial Mammals at National Parks in the Northeast Temperate Network and Sagamore Hill National Historic Site. Tech. Rep. US Geological Survey.

Gilpin, M.E., Diamond, J.M., 1982. Factors contributing to nonrandomness in species co-occurrences on islands. Oecologia 52, 75–84.

Gilpin, M.E., Diamond, J.M., 1984. Are species co-occurrences on islands non-random, and are null hypotheses useful in community ecology. In: Strong, D.R.J., Simberloff, D., Abele, L.G., Thistle, A.B. (Eds.), Ecological Communities: Conceptual Issues and the Evidence. Princeton University Press, pp. 297–315.

Glazier, D.S., 1986. Temporal variability of abundance and the distribution of species. Oikos 47, 309–314.

Goijman, A.P., Conroy, M.J., Bernardos, J.N., Zaccagnini, M.E., 2015. Multi-season regional analysis of multi-species occupancy: implications for bird conservation in agricultural lands in east-central Argentina. PLoS ONE 10, e0130874.

Gómez-Díaz, E., Doherty Jr., P.F., Duneau, D., McCoy, K.D., 2010. Cryptic vector divergence masks vector-specific patterns of infection: an example from the marine cycle of Lyme borreliosis. Evolutionary Applications 3, 391–401.

Goswami, V.R., Medhi, K., Nichols, J.D., Oli, M.K., 2015. Mechanistic understanding of human–wildlife conflict through a novel application of dynamic occupancy models. Conservation Biology 29, 1100–1110.

Gotelli, N.J., 2000. Null model analysis of species co-occurrence patterns. Ecology 81, 2606–2621.

Gotelli, N.J., Buckley, N.J., Wiens, J.A., 1997. Co-occurrence of Australian birds: Diamond's assembly rules revisited. Oikos 80, 311–324.

Gotelli, N.J., Graves, G.R., 1996. Null Models in Ecology. Smithsonian Institution Press.

Gotelli, N.J., McCabe, D.J., 2002. Species co-occurrence: a meta-analysis of J.M. Diamond's assembly rules model. Ecology 83, 2091–2096.

Gotelli, N.J., Simberloff, D., 1987. The distribution and abundance of tallgrass prairie plants: a test of the core-satellite hypothesis. American Naturalist 130, 18–35.

Gould, W.R., Patla, D.A., Daley, R., Corn, P.S., Hossack, B.R., Bennetts, R., Peterson, C.R., 2012. Estimating occupancy in large landscapes: evaluation of amphibian monitoring in the Greater Yellowstone Ecosystem. Wetlands 32, 379–389.

Govindan, B.N., Swihart, R.K., 2015. Community structure of acorn weevils (*Curculio*): inferences from multispecies occupancy models. Canadian Journal of Zoology 93, 31–39.

Grant, E.H.C., Miller, D.A., Schmidt, B.R., Adams, M.J., Amburgey, S.M., Chambert, T., Cruickshank, S.S., Fisher, R.N., Green, D.M., Hossack, B.R., Johnson, P.T.J., Joseph, M.B., Rittenhouse, T.A.G., Ryan, M.E., Waddle, J.H., Walls, S.C., Bailey, L.L., Fellers, G.M., Gorman, T.A., Ray, A.M., Pilliod, D.S., Price, S.J., Saenz, D., Sadinski, W., Muths, E., 2016. Quantitative evidence for the effects of multiple drivers on continental-scale amphibian declines. Scientific Reports 6, 25625.

Gray, J.S., Elliott, M., 2009. Ecology of Marine Sediments: From Science to Management. Oxford University Press on Demand.

Grayson, D.K., Livingston, S.D., 1993. Missing mammals on Great Basin Mountains: Holocene extinctions and inadequate knowledge. Conservation Biology 7, 527–532.

Green, A.W., Bailey, L.L., Nichols, J.D., 2011. Exploring sensitivity of a multistate occupancy model to inform management decisions. Journal of Applied Ecology 48, 1007–1016.

Green, R.H., 1979. Sampling Design and Statistical Methods for Environmental Biologists. John Wiley and Sons.

Greenwood, P.J., Harvey, P.H., 1982. The natal and breeding dispersal of birds. Annual Review of Ecology and Systematics 13, 1–21.

Griffith, B., Scott, J.M., Carpenter, J.W., Reed, C., 1989. Translocation as a species conservation tool: status and strategy. Science 245, 477–480.

Groce, M.C., Bailey, L.L., Fausch, K.D., 2012. Evaluating the success of Arkansas darter translocations in Colorado: an occupancy sampling approach. Transactions of the American Fisheries Society 141, 825–840.

Grover, M.C., 2000. Determinants of salamander distributions along moisture gradients. Copeia 2000, 156–168.

Gu, W., Swihart, R.K., 2004. Absent or undetected? Effects of non-detection of species occurrence on wildlife-habitat models. Biological Conservation 116, 195–203.

Guillera-Arroita, G., 2011. Impact of sampling with replacement in occupancy studies with spatial replication. Methods in Ecology and Evolution 2, 401–406.

Guillera-Arroita, G., Morgan, B.J.T., Ridout, M.S., Linkie, M., 2011. Species occupancy modeling for detection data collected along a transect. Journal of Agricultural, Biological, and Environmental Statistics 16, 301–317.

Guillera-Arroita, G., Ridout, M.S., Morgan, B.J.T., 2010. Design of occupancy studies with imperfect detection. Methods in Ecology and Evolution 1, 131–139.

Hackett, H.M., Lesmeister, D.B., Desanty-Combes, J., Montague, W.G., Millspaugh, J.J., Gompper, M.E., 2007. Detection rates of eastern spotted skunks (*Spilogale putorius*) in Missouri and Arkansas using live-capture and non-invasive techniques. The American Midland Naturalist 158, 123–131.

Haila, Y., Hanski, I.K., Raivio, S., 1993. Turnover of breeding birds in small forest fragments: the "sampling" colonization hypothesis corroborated. Ecology 74, 714–725.

Haines, L.M., 2016. A note on the Royle–Nichols model for repeated detection–nondetection data. Journal of Agricultural, Biological, and Environmental Statistics 21, 588–598.

Hairston, N.G., 1951. Interspecies competition and its probable influence upon the vertical distribution of Appalachian salamanders of the genus *Plethodon*. Ecology 32, 266–274.

Hairston, N.G., 1980. The experimental test of an analysis of field distributions: competition in terrestrial salamanders. Ecology 61, 817–826.

Hall, R.J., Langtimm, C.A., 2001. The U.S. national amphibian research and monitoring initiative and the role of protected area. George Wright Forum 18, 14–25.

Hames, R.S., Rosenberg, K.V., Lowe, J.D., Barker, S.E., Dhondt, A.A., 2002. Adverse effects of acid rain on the distribution of the wood thrush *Hylocichla mustelina* in North America. Proceedings of the National Academy of Sciences 99, 11235–11240.

Hames, R.S., Rosenberg, K.V., Lowe, J.D., Dhondt, A.A., 2001. Site reoccupation in fragmented landscapes: testing predictions of metapopulation theory. Journal of Animal Ecology 70, 182–190.

Hansen, T.A., 1980. Influence of larval dispersal and geographic distribution on species longevity in neo-gastropods. Paleobiology 6, 193–207.

Hanski, I., 1982. Dynamics of regional distribution: the core and satellite species hypothesis. Oikos, 210–221.

Hanski, I., 1991. Single-species metapopulation dynamics: concepts, models and observations. Biological Journal of the Linnean Society 42, 17–38.

Hanski, I., 1992. Inferences from ecological incidence functions. American Naturalist 139, 657–662.

Hanski, I., 1994a. A practical model of metapopulation dynamics. Journal of Animal Ecology 63, 151–162.

Hanski, I., 1994b. Patch-occupancy dynamics in fragmented landscapes. Trends in Ecology & Evolution 9, 131–135.

Hanski, I., 1997. Metapopulation dynamics: from concepts and observations to predictive models. In: Hanski, I.A., Gilpin, M.E. (Eds.), Metapopulation Biology: Ecology, Genetics, and Evolution. Academic Press, pp. 69–91.

Hanski, I., 1998. Metapopulation dynamics. Nature 396, 41–49.

Hanski, I., 1999. Metapopulation Ecology. Oxford University Press.

Hanski, I., Gaggiotti, O.E., 2004. Ecology, Genetics and Evolution. Elsevier Academic Press.

Hanski, I., Gilpin, M.E. (Eds.), 1997. Metapopulation Biology. Academic Press.

Hanski, I., Kouki, J., Halkka, A., 1993. Three explanations of the positive relationship between distribution and abundance of species. In: Ricklefs, R.E., Schluter, D. (Eds.), Species Diversity in Ecological Communities. Historical and Geographical Perspectives. University of Chicago Press, pp. 108–116.

Hanski, I., Moilanen, A., Pakkala, T., Kuussaari, M., 1996. The quantitative incidence function model and persistence of an endangered butterfly population. Conservation Biology 10, 578–590.

Hanski, I., Ovaskainen, O., 2000. The metapopulation capacity of a fragmented landscape. Nature 404, 755–758.

Hanski, I., Pakkala, T., Kuussaari, M., Lei, G., 1995. Metapopulation persistence of an endangered butterfly in a fragmented landscape. Oikos 72, 21–28.

Harris, L.D., 1984. The Fragmented Forest. University of Chicago Press.

Harrison, S., Maron, J., Huxel, G., 2000. Regional turnover and fluctuation in populations of five plants confined to serpentine seeps. Conservation Biology 14, 769–779.

Havel, J.E., Shurin, J.B., Jones, J.R., 2002. Estimating dispersal from patterns of spread: spatial and local control of lake invasions. Ecology 83, 3306–3318.

Hayek, L.-A.C., 1994. Analysis of amphibian biodiversity data. In: Heyer, W., Donnelly, M., McDiarmid, R., Hayek, L.-A.C., Foster, M. (Eds.), Measuring and Monitoring Biological Diversity: Standard Methods for Amphibians. Smithsonian Institution Press, pp. 207–273.

Haynes, T.B., Schmutz, J.A., Lindberg, M.S., Wright, K.G., Uher-Koch, B.D., Rosenberger, A.E., 2014. Occupancy of yellow-billed and Pacific loons: evidence for interspecific competition and habitat mediated co-occurrence. Journal of Avian Biology 45, 296–304.

He, F., Gaston, K.J., 2000. Estimating species abundance from occurrence. American Naturalist 156, 553–559.

He, F., Gaston, K.J., 2003. Occupancy, spatial variance, and the abundance of species. American Naturalist 162, 366–375.

Hecnar, S.J., M'Closkey, R.T., 1996. Regional dynamics and the status of amphibians. Ecology 77, 2091–2097.

Heer, P., Pellet, J., Sierro, A., Arlettaz, R., 2013. Evidence-based assessment of butterfly habitat restoration to enhance management practices. Biodiversity and Conservation 22, 239–252.

Hengeveld, R., 1990. Dynamic Biogeography. Cambridge University Press.

Henneman, C., Andersen, D.E., 2009. Occupancy models of nesting-season habitat associations of red-shouldered hawks in central Minnesota. The Journal of Wildlife Management 73, 1316–1324.

Hestbeck, J.B., Nichols, J.D., Malecki, R.A., 1991. Estimates of movement and site fidelity using mark resight data of wintering Canada geese. Ecology 72, 523–533.

Hewitt, O.H., 1967. A road-count index to breeding populations of red-winged blackbirds. Journal of Wildlife Management 31, 39–47.

Hilborn, R., Mangel, M., 1997. The Ecological Detective. Confronting Models with Data. Princeton University Press.

Hilborn, R., Walters, C.J., 1992. Quantitative Fisheries Stock Assessment: Choice, Dynamics, and Uncertainty. Chapman and Hall.

Hines, J.E., 2006. Program PRESENCE – software to estimate patch occupancy and related parameters. http://www.mbr-pwrc.usgs.gov/software/presence.html.

Hines, J.E., Nichols, J.D., Collazo, J.A., 2014. Multiseason occupancy models for correlated replicate surveys. Methods in Ecology and Evolution 5, 583–591.

Hines, J.E., Nichols, J.D., Royle, J.A., MacKenzie, D.I., Gopalaswamy, A., Kumar, N.S., Karanth, K., 2010. Tigers on trails: occupancy modeling for cluster sampling. Ecological Applications 20, 1456–1466.

Hirzel, A.H., Hausser, J., Chessel, D., Perrin, N., 2002. Ecological-niche factor analysis: how to compute habitat-suitability maps without absence data. Ecology 83, 2027–2036.

Holling, C.S., 1978. Adaptive Environmental Assessment and Management. Wiley.

Holt, A.R., Gaston, K.J., He, F.H., 2002. Occupancy–abundance relationships and spatial distribution. Basic and Applied Ecology 3, 1–13.

Holt, B.G., Rioja-Nieto, R., Aaron MacNeil, M., Lupton, J., Rahbek, C., 2013. Comparing diversity data collected using a protocol designed for volunteers with results from a professional alternative. Methods in Ecology and Evolution 4, 383–392.

Hosmer, D.W., Lemeshow, S., 1989. Applied Logistic Regression. John Wiley and Sons.

Hubbell, S.P., 2001. The Unified Neutral Theory of Biodiversity and Biogeography. Princeton University Press.

Hudson, P.J., Rizzoli, A., Grenfell, B.T., Heesterbeek, H., Dobson, A.P., 2002. The Ecology of Wildlife Diseases. Oxford University Press, Oxford.

Hughes, J., Haran, M., 2013. Dimension reduction and alleviation of confounding for spatial generalized linear mixed models. Journal of the Royal Statistical Society, Series B, Statistical Methodology 75, 139–159.

Hui, F.K., Taskinen, S., Pledger, S., Foster, S.D., Warton, D.I., 2015. Model-based approaches to unconstrained ordination. Methods in Ecology and Evolution 6, 399–411.

Hunt, S.D., Guzy, J.C., Price, S.J., Halstead, B.J., Eskew, E.A., Dorcas, M.E., 2013. Responses of riparian reptile communities to damming and urbanization. Biological Conservation 157, 277–284.

Hurlbert, S.H., 1984. Pseudoreplication and the design of ecological field experiments. Ecology 54, 187–211.

Hutchinson, G.E., 1957. Concluding remarks. Cold Spring Harbor Symposia on Quantitative Biology 22, 415–427.

Ihaka, R., Gentleman, R., 1996. R: a language for data analysis and graphics. Journal of Computational and Graphical Statistics 5, 299–314.

Iknayan, K.J., Tingley, M.W., Furnas, B.J., Beissinger, S.R., 2014. Detecting diversity: emerging methods to estimate species diversity. Trends in Ecology & Evolution 29, 97–106.

Jathanna, D., Karanth, K.U., Kumar, N.S., Karanth, K.K., Goswami, V.R., 2015. Patterns and determinants of habitat occupancy by the Asian elephant in the Western Ghats of Karnataka, India. PLoS ONE 10, e0133233.

Jeffress, M.R., Paukert, C.P., Sandercock, B.K., Gipson, P.S., 2011. Factors affecting detectability of river otters during sign surveys. The Journal of Wildlife Management 75, 144–150.

Jennelle, C.S., Cooch, E.G., Conroy, M.J., Senar, J.C., 2007. State-specific detection probabilities and disease prevalence. Ecological Applications 17, 154–167.

Johnson, C.M., Johnson, L.B.R.C., Beasley, V., 2002. Predicting the occurrence of amphibians: an assessment of multiple-scale models. In: Scott, J.M., Heglund, P.J., Morrison, M.L., Haufler, J.B., Raphael, M.G., Wall, W.A., Samson, F.B. (Eds.), Predicting Species Occurrences. Island Press, pp. 157–170.

Johnson, D.S., Conn, P.B., Hooten, M.B., Ray, J.C., Pond, B.A., 2013. Spatial occupancy models for large data sets. Ecology 94, 801–808.

Johnson, D.S., Nichols, J.D., Schwarz, M., 1992. Population dynamics of breeding waterfowl. In: Batt, B., Afton, A., Anderson, M., Ankney, C., Johnson, D., Kadlec, J., Krapu, G. (Eds.), Ecology and Management of Breeding Waterfowl. University of Minnesota Press, pp. 446–485.

Johnson, F.A., Boomer, G.S., Williams, B.K., Nichols, J.D., Case, D.J., 2015. Multilevel learning in the adaptive management of waterfowl harvests: 20 years and counting. Wildlife Society Bulletin 39, 9–19.

Johnson, F.A., Breininger, D.R., Duncan, B.W., Nichols, J.D., Runge, M.C., Williams, B.K., 2011. A Markov decision process for managing habitat for Florida scrub-jays. Journal of Fish and Wildlife Management 2, 234–246.

Johnson, F.A., Fackler, P.L., Boomer, G.S., Zimmerman, G.S., Williams, B.K., Nichols, J.D., Dorazio, R.M., 2016. State-dependent resource harvesting with lagged information about system states. PLoS ONE 11, e0157373.

Johnson, F.A., Moore, C.T., Kendall, W.L., Dubosky, J.A., Caithamer, D.F., Kelley Jr., J.R., Williams, B.K., 1997. Uncertainty and the management of mallard harvests. Journal of Wildlife Management 61, 202–216.

Jones, K.E., Patel, N.G., Levy, M.A., Storeygard, A., Balk, D., Gittleman, J.L., Daszak, P., 2008. Global trends in emerging infectious diseases. Nature 451, 990–993.

Karanth, K.K., Nichols, J.D., Karanth, K.U., Hines, J.E., Christensen, N.L., 2010. The shrinking ark: patterns of large mammal extinctions in India. Proceedings of the Royal Society of London. Series B, Biological Sciences, rspb.2010.0171.

Karanth, K.K., Nichols, J.D., Kumar, N., Link, W.A., Hines, J.E., 2004. Tigers and their prey: predicting carnivore densities from prey abundance. Proceedings of the National Academy of Sciences of the United States of America 101, 4854–4858.

Karanth, K.K., Nichols, J.D., Sauer, J.R., Hines, J.E., 2006. Comparative dynamics of avian communities across edges and interiors of North American ecoregions. Journal of Biogeography 33, 674–682.

Karanth, K.U., Gopalaswamy, A.M., Kumar, N.S., Vaidyanathan, S., Nichols, J.D., MacKenzie, D.I., 2011. Monitoring carnivore populations at the landscape scale: occupancy modelling of tigers from sign surveys. Journal of Applied Ecology 48, 1048–1056.

Karanth, K.U., Nichols, J.D., 2010. Non-invasive survey methods for assessing tiger populations. In: Tigers of the World: The Science, Politics and Conservation of *Panthera tigris*. University of Minnesota Press, pp. 241–261.

Karanth, K.U., Sunquist, M.E., 2000. Behavioural correlates of predation by tiger, leopard and dhole in Nagarahole, India. Journal of Zoology 250, 255–265.

Karlson, R.H., Cornell, H.V., 1998. Scale-dependent variation in local vs. regional effects on coral species richness. Ecological Monographs 68, 259–274.

Kawanishi, K., Sunquist, M.E., 2004. Conservation status of tigers in a primary rainforest of Peninsular Malaysia. Biological Conservation 120, 329–344.

Keddy, P.A., Drummond, C.G., 1996. Ecological properties for the evaluation, management, and restoration of temperate deciduous forest ecosystems. Ecological Applications 6, 748–762.

Keesing, F., Belden, L.K., Daszak, P., Dobson, A., Harvell, C.D., Holt, R.D., Hudson, P., Jolles, A., Jones, K.E., Mitchell, C.E., Myers, S.S., Bogich, T., Ostfeld, R.S., 2010. Impacts of biodiversity on the emergence and transmission of infectious diseases. Nature 468, 647–652.

Kelly, E.G., Forsman, E.D., Anthony, R.G., 2003. Are barred owls displacing spotted owls? The Condor 105, 45–53.

Kelt, D.A., Taper, M.L., Mesevre, P.L., 1995. Assessing the impact of competition on community assembly: a case study using small mammals. Ecology 76, 1283–1296.

Kendall, W.L., 1999. Robustness of closed capture–recapture methods to violations of the closure assumption. Ecology 80, 2517–2525.

Kendall, W.L., 2001. Using models to facilitate complex decisions. In: Shenk, T.M., Franklin, A.B. (Eds.), Modeling in Natural Resource Management. Island Press, pp. 147–170.

Kendall, W.L., 2009. One size does not fit all: adapting mark–recapture and occupancy models for state uncertainty. In: Thomson, D.L., Cooch, E.G., Conroy, M.J. (Eds.), Modeling Demographic Processes in Marked Populations. Springer, pp. 765–780.

Kendall, W.L., Bjorkland, R., 2001. Using open robust design models to estimate temporary emigration from capture–recapture data. Biometrics 57, 1113–1122.

Kendall, W.L., Hines, J.E., 1999. Program RDSURVIV: an estimation tool for capture–recapture data collected under Pollock's robust design. Bird Study 46, 32–38.

Kendall, W.L., Hines, J.E., Nichols, J.D., 2003. Adjusting multistate capture–recapture models for misclassification bias: manatee breeding proportions. Ecology 84, 1058–1066.

Kendall, W.L., Hines, J.E., Nichols, J.D., Grant, E.H.C., 2013. Relaxing the closure assumption in occupancy models: staggered arrival and departure times. Ecology 94, 610–617.

Kendall, W.L., Nichols, J.D., 1995. On the use of secondary capture–recapture samples to estimate temporary emigration and breeding proportions. Journal of Applied Statistics 22, 751–762.

Kendall, W.L., Nichols, J.D., 2004. On the estimation of dispersal and movement of birds. Condor 106, 720–731.

Kendall, W.L., Nichols, J.D., Hines, J.E., 1997. Estimating temporary emigration using capture–recapture data with Pollock's robust design. Ecology 78, 563–578.

Kendall, W.L., White, G.C., 2009. A cautionary note on substituting spatial subunits for repeated temporal sampling in studies of site occupancy. Journal of Applied Ecology 46, 1182–1188.

Kennedy, C.M., Grant, E.H.C., Neel, M.C., Fagan, W.F., Marra, P.P., 2011. Landscape matrix mediates occupancy dynamics of Neotropical avian insectivores. Ecological Applications 21, 1837–1850.

Kéry, M., 2004. Extinction rate estimates for plant populations in revisitation studies: importance of detectability. Conservation Biology 18, 570–574.

Kéry, M., 2010. Introduction to WinBUGS for Ecologists. Academic, Burlington.

Kéry, M., Royle, J.A., 2008. Hierarchical Bayes estimation of species richness and occupancy in spatially replicated surveys. Journal of Applied Ecology 45, 589–598.

Kéry, M., Royle, J.A., 2009. Inference about species richness and community structure using species-specific occupancy models in the national Swiss breeding bird survey MHB. In: Thomson, D.L., Cooch, E.G., Conroy, M.J. (Eds.), Modeling Demographic Processes in Marked Populations. Springer, pp. 639–656.

Kéry, M., Royle, J.A., 2016. Applied Hierarchical Modeling in Ecology: Analysis of Distribution, Abundance and Species Richness in R and BUGS: Volume 1. Academic Press.

Kéry, M., Royle, J.A., Plattner, M., Dorazio, R.M., 2009. Species richness and occupancy estimation in communities subject to temporary emigration. Ecology 90, 1279–1290.

Kéry, M., Royle, J.A., Schmid, H., Schaub, M., Volet, B., Haefliger, G., Zbinden, N., 2010. Site-occupancy distribution modeling to correct population-trend estimates derived from opportunistic observations. Conservation Biology 24, 1388–1397.

Kéry, M., Schaub, M., 2012. Bayesian Population Analysis Using WinBUGS. A Hierarchical Approach. Academic Press, San Diego, CA.

Kimura, M., Weiss, G.H., 1964. The stepping stone model of population structure and the decrease of genetic correlations with distance. Genetics 49, 561–576.

Klute, D., Lovallo, M.J., Tzilkowski, W.M., 2002. Autologistic regression modeling of American woodcock habitat use with spatially dependent data. In: Scott, J.M., Heglund, P.J., Morrison, M.L., Haufler, J.B., Raphael, M.G., Wall, W.A., Samson, F.B. (Eds.), Predicting Species Occurrences. Island Press, pp. 335–343.

Kodric-Brown, A., Brown, J.H., 1993. Incomplete data sets in community ecology and biogeography: a cautionary tale. Ecological Applications 3, 736–742.

Krebs, C.J., 1972. Ecology. Harper and Row.

Krebs, C.J., 1991. The experimental paradigm and long-term population studies. Ibis 133, 3–8.

Krebs, C.J., 2001. Ecology: the Experimental Analysis of Distribution and Abundance: Hands-on Field Package, 5th edition. Benjamin Cummings.

Kriger, K.M., Hero, J.-M., Ashton, K.J., 2006. Cost efficiency in the detection of chytridiomycosis using PCR assay. Diseases of Aquatic Organisms 71, 149–154.

Kunin, W.E., 1998. Extrapolating species abundance across spatial scales. Science 281, 1513–1515.

Kunin, W.E., Hartley, S., Lennon, J.J., 2000. Scaling down: on the challenge of estimating abundance from occurrence patterns. American Naturalist 156, 560–566.

Kuo, L., Mallick, B., 1998. Variable selection for regression models. Sankhyā: The Indian Journal of Statistics, Series B, 65–81.

Lachish, S., Gopalaswamy, A.M., Knowles, S.C., Sheldon, B.C., 2012. Site-occupancy modelling as a novel framework for assessing test sensitivity and estimating wildlife disease prevalence from imperfect diagnostic tests. Methods in Ecology and Evolution 3, 339–348.

Lancaster, T., Imbens, G., 1996. Case-control studies with contaminated controls. Journal of Econometrics 71, 145–160.

Lancia, R.A., Kendall, W.L., Pollock, K.H., Nichols, J.D., 2005. Estimating the number of animals in wildlife populations. In: Braun, C.E. (Ed.), Research and Management Techniques for Wildlife and Habitats. The Wildlife Society, pp. 105–153.

Lancia, R.A., Nichols, J.D., Pollock, K.H., 1994. Estimating the number of animals in wildlife populations. In: Bookhout, T. (Ed.), Research and Management Techniques for Wildlife and Habitats. The Wildlife Society, pp. 215–253.

Lande, R., 1987. Extinction thresholds in demographic models of territorial populations. American Naturalist 130, 624–635.

Lande, R., 1988. Demographic models of the northern spotted owl (*Strix occidentalis caurina*). Oecologia 75, 601–607.

LaRoe, E., Farris, G., Puckett, C., Doran, P., Mac, M., 1995. Our Living Resources: A Report to the Nation on the Distribution, Abundance, and Health for U.S. Plants, Animals, and Ecosystems. U.S. Department of the Interior, National Biological Service.

Lebreton, J.D., Burnham, K.P., Clobert, J., Anderson, D.R., 1992. Modeling survival and testing biological hypotheses using marked animals – a unified approach with case-studies. Ecological Monographs 62, 67–118.

Lebreton, J.D., Pradel, R., 2002. Multistate recapture models: modelling incomplete individual histories. Journal of Applied Statistics 29, 353–369.

Lele, S.R., Keim, J.L., 2006. Weighted distributions and estimation of resource selection probability functions. Ecology 87, 3021–3028.

Lele, S.R., Moreno, M., Bayne, E., 2012. Dealing with detection error in site occupancy surveys: what can we do with a single survey? Journal of Plant Ecology 5, 22–31.

Leopold, A., 1933. Game Management. Charles Scribner's Sons.

Levins, R., 1969. Some demographic and genetic consequences of environmental heterogeneity for biological control. Bulletin of the Entomological Society of America 15, 237–240.

Levins, R., 1970. Extinction. In: Gustenhaver, M. (Ed.), Some Mathematical Questions in Biology, vol. II. American Mathematical Society, pp. 77–107.

Lewis, J.S., Bailey, L.L., VandeWoude, S., Crooks, K.R., 2015a. Interspecific interactions between wild felids vary across scales and levels of urbanization. Ecology and Evolution 5, 5946–5961.

Lewis, T.L., Lindberg, M.S., Schmutz, J.A., Bertram, M.R., Dubour, A.J., 2015b. Species richness and distributions of boreal waterbird broods in relation to nesting and brood-rearing habitats. The Journal of Wildlife Management 79, 296–310.

Linden, D.W., Fuller, A.K., Royle, J.A., Hare, M.P., 2017. Examining the occupancy–density relationship for a low-density carnivore. Journal of Applied Ecology.

Link, W.A., Sauer, J.R., 1997. New approaches to the analysis of population trends in land birds: comment. Ecology 78, 2632–2634.

Link, W.A., 2003. Nonidentifiability of population size from capture–recapture data with heterogeneous detection probabilities. Biometrics 59, 1123–1130.

Link, W.A., Barker, R.J., 2006. Model weights and the foundations of multimodel inference. Ecology 87, 2626–2635.

Link, W.A., Barker, R.J., 2009. Bayesian Inference: With Ecological Applications. Academic Press.

Link, W.A., Barker, R.J., Sauer, J.R., Droege, S., 1994. Within-site variability in surveys of wildlife populations. Ecology 74, 1097–1108.

Link, W.A., Doherty Jr., P.F., 2002. Scaling in sensitivity analysis. Ecology 83, 3299–3305.

Link, W.A., Nichols, J.D., 1994. On the importance of sampling variance to investigations of temporal variation in animal population size. Oikos 69, 539–544.

Link, W.A., Sauer, J.R., 2002. A hierarchical analysis of population change with application to cerulean warbler. Ecology 83, 2832–2840.

Liow, L.H., 2013. Simultaneous estimation of occupancy and detection probabilities: an illustration using Cincinnatian brachiopods. Paleobiology 39, 193–213.

Liow, L.H., Nichols, J.D., 2010. Estimating rates and probabilities of origination and extinction using taxonomic occurrence data: capture–mark–recapture (CMR) approaches. In: Alroy, J., Hunt, G. (Eds.), Quantitative Methods in Paleobiology. The Paleontological Society, pp. 81–94.

Lobo, J.M., Jiménez-Valverde, A., Real, R., 2008. AUC: a misleading measure of the performance of predictive distribution models. Global Ecology and Biogeography 17, 145–151.

Lomolino, M.V., 1996. Investigating causality of nestedness of insular communities: selective immigrations or extinctions? Journal of Biogeography 23, 699–703.

Lomolino, M.V., Brown, J.H., Davis, R., 1989. Island biogeography of montane forest mammals in the American southwest. Ecology 70, 180–194.

Lotz, A., Allen, C.R., 2007. Observer bias in anuran call surveys. The Journal of Wildlife Management 71, 675–679.

Lunn, D.J., Jackson, C., Best, N., Thomas, A., Spiegelhalter, D., 2012. The BUGS Book: A Practical Introduction to Bayesian Analysis. CRC Press.

Lunn, D.J., Thomas, A., Best, N., Spiegelhalter, D., 2000. WinBUGS—a Bayesian modelling framework: concepts, structure, and extensibility. Statistics and Computing 10, 325–337.

Lynch, J.F., Whigham, D.F., 1984. Effects of forest fragmentation on breeding bird communities in Maryland, USA. Biological Conservation 28, 287–324.

MacArthur, R.H., 1972. Geographical Ecology. Harper and Row.

MacArthur, R.H., Wilson, E.O., 1963. An equilibrium theory of insular zoogeography. Evolution 17, 373–387.

MacArthur, R.H., Wilson, E.O., 1967. The Theory of Island Biogeography. Princeton University Press.

MacKenzie, D.I., 2002. Assessing the Fit of Mark–Recapture Models. Ph.D. thesis. University of Otago.

MacKenzie, D.I., 2005a. Was it there? Dealing with imperfect detection for species presence/absence data. Australian & New Zealand Journal of Statistics 47, 65–74.

MacKenzie, D.I., 2005b. What are the issues with presence–absence data for wildlife managers? Journal of Wildlife Management 69, 849–860.

MacKenzie, D.I., 2006. Modeling the probability of resource use: the effect of, and dealing with, detecting a species imperfectly. Journal of Wildlife Management 70, 367–374.

MacKenzie, D.I., Bailey, L.L., 2004. Assessing the fit of site occupancy models. Journal of Agricultural, Biological, and Environmental Statistics 9, 300–318.

MacKenzie, D.I., Bailey, L.L., Hines, J.E., Nichols, J.D., 2011. An integrated model of habitat and species occurrence dynamics. Methods in Ecology and Evolution 2, 612–622.

MacKenzie, D.I., Bailey, L.L., Nichols, J.D., 2004a. Investigating species co-occurrence patterns when species are detected imperfectly. Journal of Animal Ecology 73, 546–555.

MacKenzie, D.I., Kendall, W.K., 2002. How should detection probability be incorporated into estimates of relative abundance? Ecology 83, 3532.

MacKenzie, D.I., Nichols, J.D., 2004. Occupancy as a surrogate for abundance estimation. Animal Biodiversity and Conservation 27, 461–467.

MacKenzie, D.I., Nichols, J.D., Hines, J.E., Knutson, M.G., Franklin, A.B., 2003. Estimating site occupancy, colonization, and local extinction when a species is detected imperfectly. Ecology 84, 2200–2207.

MacKenzie, D.I., Nichols, J.D., Lachman, G.B., Droege, S., Royle, J.A., Langtimm, C.A., 2002. Estimating site occupancy rates when detection probabilities are less than one. Ecology 83, 2248–2255.

MacKenzie, D.I., Nichols, J.D., Royle, J.A., Pollock, K.H., Bailey, L.L., Hines, J.E., 2006. Occupancy Estimation and Modeling: Inferring Patterns and Dynamics of Species Occurrence. Elsevier, Amsterdam, Boston.

MacKenzie, D.I., Nichols, J.D., Seamans, M.E., Gutiérrez, R.J., 2009. Modeling species occurrence dynamics with multiple states and imperfect detection. Ecology 90, 823–835.

MacKenzie, D.I., Nichols, J.D., Sutton, N., Kawanishi, K., Bailey, L.L., 2005. Improving inferences in population studies of rare species that are detected imperfectly. Ecology 86, 1101–1113.

MacKenzie, D.I., Royle, J.A., 2005. Designing occupancy studies: general advice and allocating survey effort. Journal of Applied Ecology 42, 1105–1114.

MacKenzie, D.I., Royle, J.A., Brown, J.A., Nichols, J.D., 2004b. Occupancy estimation and modeling for rare and elusive populations. In: Thompson, W.L. (Ed.), Sampling Rare or Elusive Species: Concepts, Designs, and Techniques for Estimating Population Parameters. Island Press, pp. 149–172.

MacKenzie, D.I., Seamans, M.E., Gutiérrez, R.J., Nichols, J.D., 2012. Investigating the population dynamics of California spotted owls without marked individuals. Journal of Ornithology 152, 597–604.

Magoun, A.J., Ray, J.C., Johnson, D.S., Valkenburg, P., Dawson, F.N., Bowman, J., 2007. Modeling wolverine occurrence using aerial surveys of tracks in snow. The Journal of Wildlife Management 71, 2221–2229.

Manley, P.N., Schlesinger, M.D., Roth, J.K., Van Horne, B., 2005. A field-based evaluation of a presence–absence protocol for monitoring ecoregional-scale biodiversity. Journal of Wildlife Management 69, 950–966.

Manley, P.N., Zielinski, W.J., Schlesinger, M.D., Mori, S.R., 2004. Evaluation of a multiple-species approach to monitoring species at the ecoregional scale. Ecological Applications 14, 296–310.

Manly, B.F.J., 1992. The Design and Analysis of Research Studies. Cambridge University Press.

Manly, B.F.J., 1995. A note on the analysis of species co-occurrences. Ecology 76, 1109–1115.

Manly, B.F.J., 1997. Randomization, Bootstrap and Monte Carlo Methods in Biology, 2nd edition. Chapman and Hall.

Manly, B.F.J., McDonald, L., Thomas, D., McDonald, T.L., Erickson, W., 2002. Resource Selection by Animals: Statistical Design and Analysis for Field Studies. Kluwer Academic Publishers.

Marcelli, M., Poledník, L., Poledníková, K., Fusillo, R., 2012. Land use drivers of species re-expansion: inferring colonization dynamics in Eurasian otters. Diversity and Distributions 18, 1001–1012.

Marra, P.P., Griffing, S., Caffrey, C., Kilpatrick, A.M., McLean, R., Brand, C., Saito, E., Dupuis, A.P., Kramer, L., Novak, R., 2004. West Nile virus and wildlife. Bioscience 54, 393–402.

Martin, J., Chamaillé-Jammes, S., Nichols, J.D., Fritz, H., Hines, J.E., Fonnesbeck, C.J., MacKenzie, D.I., Bailey, L.L., 2010. Simultaneous modeling of habitat suitability, occupancy, and relative abundance: African elephants in Zimbabwe. Ecological Applications 20, 1173–1182.

Martin, J., Fackler, P.L., Nichols, J.D., Runge, M.C., McIntyre, C.L., Lubow, B.L., McCluskie, M.C., Schmutz, J.A., 2011. An adaptive-management framework for optimal control of hiking near golden eagle nests in Denali National Park. Conservation Biology 25, 316–323.

Martin, J., McIntyre, C.L., Hines, J.E., Nichols, J.D., Schmutz, J.A., MacCluskie, M.C., 2009a. Dynamic multistate site occupancy models to evaluate hypotheses relevant to conservation of golden eagles in Denali National Park, Alaska. Biological Conservation 142, 2726–2731.

Martin, J., Nichols, J.D., McIntyre, C.L., Ferraz, G., Hines, J.E., 2009b. Perturbation analysis for patch occupancy dynamics. Ecology 90, 10–16.

Martinez-Solano, I., Bosch, J., Garcia-Paris, M., 2003. Demographic trends and community stability in a montane amphibian assemblage. Conservation Biology 17, 238–244.

Mattfeldt, S.D., Bailey, L.L., Grant, E.H.C., 2009. Monitoring multiple species: estimating state variables and exploring the efficacy of a monitoring program. Biological Conservation 142, 720–737.

Mattfeldt, S.D., Grant, E.H.C., 2007. Are two methods better than one? Area constrained transects and leaf litterbags for sampling stream salamanders. Herpetological Review 38, 43–45.

Mattsson, B.J., Zipkin, E.F., Gardner, B., Blank, P.J., Sauer, J.R., Royle, J.A., 2013. Explaining local-scale species distributions: relative contributions of spatial autocorrelation and landscape heterogeneity for an avian assemblage. PLoS ONE 8, e55097.

Maurer, B.A., 1994. Geographical Population Analysis. Blackwell Scientific.

Mazerolle, M.J., Bailey, L.L., Kendall, W.L., Andrew Royle, J., Converse, S.J., Nichols, J.D., 2007. Making great leaps forward: accounting for detectability in herpetological field studies. Journal of Herpetology 41, 672–689.

Mazerolle, M.J., Desrochers, A., Rochefort, L., 2005. Landscape characteristics influence pond occupancy by frogs after accounting for detectability. Ecological Applications 15, 824–834.

McArdle, B., 1990. When are rare species not there? Oikos 57, 276–277.

McClintock, B.T., Bailey, L.L., Pollock, K.H., Simons, T.R., 2010a. Experimental investigation of observation error in anuran call surveys. The Journal of Wildlife Management 74, 1882–1893.

McClintock, B.T., Bailey, L.L., Pollock, K.H., Simons, T.R., 2010b. Unmodeled observation error induces bias when inferring patterns and dynamics of species occurrence via aural detections. Ecology 91, 2446–2454.

McClintock, B.T., Nichols, J.D., Bailey, L.L., MacKenzie, D.I., Kendall, W., Franklin, A.B., 2010c. Seeking a second opinion: uncertainty in disease ecology. Ecology Letters 13, 659–674.

McCullagh, P., Nelder, J.A., 1989. Generalized Linear Models. Chapman and Hall.

McCullough, D.R., 1996. Metapopulations and Wildlife Conservation. Island Press.

McGowan, C.P., Lyons, J.E., Smith, D.R., 2015a. Developing objectives with multiple stakeholders: adaptive management of horseshoe crabs and red knots in the Delaware Bay. Environmental Management 55, 972–982.

McGowan, C.P., Smith, D.R., Nichols, J.D., Lyons, J.E., Sweka, J., Kalasz, K., Niles, L.J., Wong, R., Brust, J., Davis, M., Spear, B., 2015b. Implementation of a framework for multi-species, multi-objective adaptive management in Delaware Bay. Biological Conservation 191, 759–769.

McNaughton, S., 1983. Compensatory plant growth as a response to herbivory. Oikos 40, 329–336.

McNaughton, S., Banyikwa, F., McNaughton, M., 1997. Promotion of the cycling of diet-enhancing nutrients by African grazers. Science 278, 1798–1800.

McNaughton, S., Wolf, L.L., 1970. Dominance and the niche in ecological systems. Science 167, 131–139.

Mehlman, D.W., 1997. Change in avian abundance across the geographical range in response to environmental change. Ecological Applications 7, 614–624.

Merila, J., Kotze, J.D. (Eds.), 2003. Extinction Thresholds. Annales Zoologici Fennici 40, 69–245.

Mihaljevic, J.R., Joseph, M.B., Johnson, P.T., 2015. Using multispecies occupancy models to improve the characterization and understanding of metacommunity structure. Ecology 96, 1783–1792.

Millar, R.B., 2009. Comparison of hierarchical Bayesian models for overdispersed count data using DIC and Bayes' factors. Biometrics 65, 962–969.

Miller, D.A., 2012. General methods for sensitivity analysis of equilibrium dynamics in patch occupancy models. Ecology 93, 1204–1213.

Miller, D.A., Bailey, L.L., Grant, E.H.C., McClintock, B.T., Weir, L.A., Simons, T.R., 2015. Performance of species occurrence estimators when basic assumptions are not met: a test using field data where true occupancy status is known. Methods in Ecology and Evolution 6, 557–565.

Miller, D.A., Brehme, C.S., Hines, J.E., Nichols, J.D., Fisher, R.N., 2012a. Joint estimation of habitat dynamics and species interactions: disturbance reduces co-occurrence of non-native predators with an endangered toad. Journal of Animal Ecology 81, 1288–1297.

Miller, D.A., Grant, E.H.C., 2015. Estimating occupancy dynamics for large-scale monitoring networks: amphibian breeding occupancy across protected areas in the northeast United States. Ecology and Evolution 5, 4735–4746.

Miller, D.A., Nichols, J.D., Gude, J.A., Rich, L.N., Podruzny, K.M., Hines, J.E., Mitchell, M.S., 2013. Determining occurrence dynamics when false positives occur: estimating the range dynamics of wolves from public survey data. PLoS ONE 8, e65808.

Miller, D.A., Nichols, J.D., McClintock, B.T., Grant, E.H.C., Bailey, L.L., Weir, L.A., 2011. Improving occupancy estimation when two types of observational error occur: non-detection and species misidentification. Ecology 92, 1422–1428.

Miller, D.A., Talley, B.L., Lips, K.R., Campbell Grant, E.H., 2012b. Estimating patterns and drivers of infection prevalence and intensity when detection is imperfect and sampling error occurs. Methods in Ecology and Evolution 3, 850–859.

Miller, D.A., Weir, L.A., McClintock, B.T., Grant, E.H.C., Bailey, L.L., Simons, T.R., 2012c. Experimental investigation of false positive errors in auditory species occurrence surveys. Ecological Applications 22, 1665–1674.

Moilanen, A., 1999. Patch occupancy models of metapopulation dynamics: efficient parameter estimation using implicit statistical inference. Ecology 80, 1031–1043.

Moilanen, A., 2002. Implications of empirical data quality to metapopulation model parameter estimation and application. Oikos 96, 516–530.

Moilanen, A., Cabeza, M., 2002. Single-species dynamic site selection. Ecological Applications 12, 913–926.

Moilanen, A., Hanski, I., 1995. Habitat destruction and coexistence of competitors in a spatially realistic metapopulation model. Journal of Animal Ecology 64, 141–144.

Molinari-Jobin, A., Kéry, M., Marboutin, E., Molinari, P., Koren, I., Fuxjäger, C., Breitenmoser-Würsten, C., Wölfl, S., Fasel, M., Kos, I., Wölfl, M., Breitenmoser, U., 2012. Monitoring in the presence of species misidentification: the case of the Eurasian lynx in the Alps. Animal Conservation 15, 266–273.

Mondol, S., Kumar, N.S., Gopalaswamy, A., Sunagar, K., Karanth, K.U., Ramakrishnan, U., 2015. Identifying species, sex and individual tigers and leopards in the Malenad–Mysore Tiger Landscape, Western Ghats, India. Conservation Genetics Resources 7, 353–361.

Mooney, H.A., Drake, J.A., 1986. Ecology of Biological Invasions in North America and Hawaii. Springer-Verlag.

Moore, N.W., Hooper, M.D., 1975. On the number of bird species in British woods. Biological Conservation 8, 239–250.

Moore, J.F., Mulindahabi, F., Masozera, M.K., Nichols, J.D., Hines, J.E., Turikunkiko, E., Oli, M.K., 2017. Are ranger patrols effective in reducing poaching-related threats in protected areas? Journal of Applied Ecology. http://dx.doi.org/10.1111/1365-2664.12965.

Mordecai, R.S., Mattsson, B.J., Tzilkowski, C.J., Cooper, R.J., 2011. Addressing challenges when studying mobile or episodic species: hierarchical Bayes estimation of occupancy and use. Journal of Applied Ecology 48, 56–66.

Moreno, M., Lele, S.R., 2010. Improved estimation of site occupancy using penalized likelihood. Ecology 91, 341–346.

Moritz, C., Patton, J.L., Conroy, C.J., Parra, J.L., White, G.C., Beissinger, S.R., 2008. Impact of a century of climate change on small-mammal communities in Yosemite National Park, USA. Science 322, 261–264.

Morris, W., Doak, D.F., 2002. Quantitative Conservation Biology. Sinauer.

Mosher, B.A., Bailey, L.L., Hubbard, B.A., Huyvaert, K.P., 2017. Inferential biases linked to unobservable states in complex occupancy models. Ecography. http://dx.doi.org/10.1111/ecog.02849.

Mowat, G., Paetkau, D., 2002. Estimating marten *Martes americana* population size using hair capture and genetic tagging. Wildlife Biology 8, 201–209.

Muths, E., Jung, R.E., Bailey, L.L., Adams, M.J., Corn, P.S., Dodd, C.K., Fellers, G.M., Sadinski, W.J., Schwalbe, C.R., Walls, S.C., Fisher, R.N., Gallant, A.L., Battaglin, W.A., Green, D.E.,

2005. Amphibian Research and Monitoring Initiative (ARMI): a successful start to a national program in the United States. Applied Herpetology 2, 355–371.

Nee, S., 1994. How populations persist. Nature 367, 123–124.

Newman, K., Buckland, S., Morgan, B., King, R., Borchers, D., Cole, D., Besbeas, P., Gimenez, O., Thomas, L., 2014. Modelling Population Dynamics. Springer.

Nichols, J.D., 1991. Extensive monitoring programs viewed as long-term population studies: the case of North American waterfowl. Ibis 133, 89–98.

Nichols, J.D., 1996. Sources of variation in migratory movements of animal populations: statistical inference and a selective review of empirical results for birds. In: Rhodes, O.E.J., Chesser, R.K., Smith, M.H. (Eds.), Population Dynamics in Ecological Space and Time. University of Chicago Press, pp. 147–197.

Nichols, J.D., 2001. Using models in the conduct of science and management of natural resources. In: Shenk, T.M., Franklin, A.B. (Eds.), Modeling in Natural Resource Management: Development, Interpretation and Application. Island Press, pp. 11–34.

Nichols, J.D., Bailey, L.L., O'Connell Jr., A.F., Talancy, N.W., Grant, E.H.C., Gilbert, A.T., Annand, E.M., Husband, T.P., Hines, J.E., 2008. Multi-scale occupancy estimation and modelling using multiple detection methods. Journal of Applied Ecology 45, 1321–1329.

Nichols, J.D., Boulinier, T., Hines, J.E., Pollock, K.H., Sauer, J.R., 1998a. Estimating rates of local species extinction, colonization, and turnover in animal communities. Ecological Applications 8, 1213–1225.

Nichols, J.D., Boulinier, T., Hines, J.E., Pollock, K.H., Sauer, J.R., 1998b. Inference methods for spatial variation in species richness and community composition when not all species are detected. Conservation Biology 12, 1390–1398.

Nichols, J.D., Conroy, M.J., 1996. Estimation of species richness. In: Wilson, D.W., Coles, F.R., Nichols, J.D., Rudran, R., Foster, M.S. (Eds.), Measuring and Monitoring Biological Diversity: Standard Methods for Mammals.. Smithsonian Institution Press, pp. 226–234.

Nichols, J.D., Hines, J.E., Lebreton, J.-D., Pradel, R., 2000a. The relative contributions of demographic components to population growth: a direct estimation approach based on reverse-time capture–recapture. Ecology 81, 3362–3376.

Nichols, J.D., Hines, J.E., MacKenzie, D.I., Seamans, M.E., Gutiérrez, R.J., 2007a. Occupancy estimation and modeling with multiple states and state uncertainty. Ecology 88, 1395–1400.

Nichols, J.D., Hines, J.E., Sauer, J.R., Fallon, F.W., Fallon, J.E., Heglund, P.J., 2000b. A double-observer approach for estimating detection probability and abundance from point counts. The Auk 117, 393–408.

Nichols, J.D., Johnson, F.A., 1989. Evaluation and experimentation with duck harvest management strategies. Transactions of the North American Wildlife and Natural Resources Conference 54, 566–593.

Nichols, J.D., Johnson, F.A., Williams, B.K., 1995. Managing North American waterfowl in the face of uncertainty. Annual Review of Ecology and Systematics 26, 177–199.

Nichols, J.D., Johnson, F.A., Williams, B.K., Boomer, G.S., 2015. On formally integrating science and policy: walking the walk. Journal of Applied Ecology 52, 539–543.

Nichols, J.D., Karanth, K.U., 2002. Statistical concepts: assessing spatial distributions. In: Karanth, K.U., Nichols, J.D. (Eds.), Monitoring Tigers and Their Prey: A Manual for Wildlife Managers, Researchers, and Conservationists. Centre for Wildlife Studies, pp. 29–38.

Nichols, J.D., Kendall, W.L., Hines, J.E., Spendelow, J.A., 2004. Estimation of sex-specific survival from capture–recapture data when sex is not always known. Ecology 85, 3192–3201.

Nichols, J.D., Morris, R.W., Brownie, C., Pollock, K.H., 1986. Sources of variation in extinction rates, turnover and diversity of marine invertebrate families during the Paleozoic. Paleobiology 12, 421–432.

Nichols, J.D., Pollock, K.H., 1983. Estimating taxonomic diversity, extinction rates and speciation rates from fossil data using capture–recapture models. Paleobiology 9, 150–163.

Nichols, J.D., Runge, M.C., Johnson, F.A., Williams, B.K., 2007b. Adaptive harvest management of North American waterfowl populations: a brief history and future prospects. Journal of Ornithology 148, 343–349.

Nichols, J.D., Stokes, S.L., Hines, J.E., Conroy, M.J., 1982. Additional comments on the assumption of homogeneous survival rates in modern band recovery estimation models. Journal of Wildlife Management 46, 953–962.

Nichols, J.D., Thomas, L., Conn, P.B., 2009. Inferences about landbird abundance from count data: recent advances and future directions. In: Thomson, D.L., Cooch, E.G., Conroy, M.J. (Eds.), Modeling Demographic Processes in Marked Populations. Springer, pp. 201–235.

Nichols, J.M., Murphy, K.D., 2016. Modeling and Estimation of Structural Damage. John Wiley & Sons.

Noon, B.R., Bailey, L.L., Sisk, T.D., McKelvey, K.S., 2012. Efficient species-level monitoring at the landscape scale. Conservation Biology 26, 432–441.

Noon, B.R., McKelvey, K.S., 1997. A common framework for conservation planning: linking individual and metapopulation models. In: McCullough, D. (Ed.), Metapopulations and Wildlife Conservation. Island Press, pp. 138–165.

Norris, J.L., Pollock, K.H., 1996. Nonparametric MLE under two closed-capture models with heterogeneity. Biometrics 52, 639–649.

Northrup, J.M., Gerber, B.D., in review. A comment on priors for Bayesian occupancy models. PLoS ONE. http://biorxiv.org/content/early/2017/05/17/138735.

O'Connell Jr., A.F., Bailey, L.L., 2011. Inference for occupancy and occupancy dynamics. In: O'Connell Jr., A.F., Nichols, J.D., Karanth, K.U. (Eds.), Camera Traps in Animal Ecology. Springer, pp. 191–204.

O'Connell Jr., A.F., Talancy, N.W., Bailey, L.L., Sauer, J.R., Cook, R., Gilbert, A.T., 2006. Estimating site occupancy and detection probability parameters for meso- and large mammals in a coastal ecosystem. Journal of Wildlife Management 70, 1625–1633.

Odum, E.P., 1960. Organic production and turnover in old field succession. Ecology 41, 34–49.

Olson, G.S., Anthony, R.G., Forsman, E.D., Ackers, S.H., Loschl, P.J., Reid, J.A., Dugger, K.M., Glenn, E.M., Ripple, W.J., 2005. Modeling of site occupancy dynamics for northern spotted owls, with emphasis on the effects of barred owls. Journal of Wildlife Management 69, 918–932.

Otis, D.L., Burnham, K.P., White, G.C., Anderson, D.R., 1978. Statistical Inference from Capture Data on Closed Animal Populations. Wildlife Monographs, vol. 62, pp. 3–135.

Otto, C.R., Bailey, L.L., Roloff, G.J., 2013. Improving species occupancy estimation when sampling violates the closure assumption. Ecography 36, 1299–1309.

Ovaskainen, O., Sato, K., Bascompte, J., Hanski, I., 2002. Metapopulation models for extinction threshold in spatially correlated landscapes. Journal of Theoretical Biology 215, 95–108.

Ovaskainen, O., Soininen, J., 2011. Making more out of sparse data: hierarchical modeling of species communities. Ecology 92, 289–295.

Pacifici, K., Reich, B.J., Dorazio, R.M., Conroy, M.J., 2016. Occupancy estimation for rare species using a spatially-adaptive sampling design. Methods in Ecology and Evolution 7, 285–293.

Pacifici, K., Zipkin, E.F., Collazo, J.A., Irizarry, J.I., DeWan, A., 2014. Guidelines for *a priori* grouping of species in hierarchical community models. Ecology and Evolution 4, 877–888.

Padilla-Torres, S.D., Ferraz, G., Luz, S.L., Zamora-Perea, E., Abad-Franch, F., 2013. Modeling dengue vector dynamics under imperfect detection: three years of site-occupancy by Aedes aegypti and Aedes albopictus in urban Amazonia. PLoS ONE 8, e58420.

Patil, G., Taillie, C., 1979. An overview of diversity. In: Grassle, J., Patil, G., Smith, W., Taillie, C. (Eds.), Ecological Diversity in Theory and Practice. In: Statistical Ecology, vol. 6. International Co-operative Publishing House, pp. 3–27.

Patterson, B.D., 1987. The principle of nested subsets and its implications for biological conservation. Conservation Biology 1, 323–334.

Patterson, B.D., 1990. On the temporal development of nested subset patterns of species composition. Oikos 59, 330–342.

Patterson, B.D., Atmar, W., 1986. Nested subsets and the structure of insular mammalian faunas and archipelagos. Biological Journal of the Linnean Society 28, 65–82.

Pavlacky, D.C., Blakesley, J.A., White, G.C., Hanni, D.J., Lukacs, P.M., 2012. Hierarchical multi-scale occupancy estimation for monitoring wildlife populations. The Journal of Wildlife Management 76, 154–162.

Pellet, J., Fleishman, E., Dobkin, D.S., Gander, A., Murphy, D.D., 2007. An empirical evaluation of the area and isolation paradigm of metapopulation dynamics. Biological Conservation 136, 483–495.

Peltonen, A., Hanski, I., 1991. Patterns of island occupancy explained by colonization and extinction rates in shrews. Ecology 72, 1698–1708.

Peoples, B.K., Frimpong, E.A., 2016. Biotic interactions and habitat drive positive co-occurrence between facilitating and beneficiary stream fishes. Journal of Biogeography 43, 923–931.

Pepper, M.A., Herrmann, V., Hines, J.E., Nichols, J.D., Kendrot, S.R., 2017. Evaluation of nutria (*Myocastor coypus*) detection methods in Maryland, USA. Biological Invasions 19, 831–841.

Peres-Neto, P., Olden, J.D., Jackson, D.A., 2001. Environmentally constrained null models: site suitability as occupancy criterion. Oikos 93, 110–120.

Peterjohn, B.G., Sauer, J.R., 1993. North American Breeding Bird Survey annual summary 1990–1991. Bird Populations 1, 52–67.

Peters, S.E., Ausich, W.I., 2008. A sampling-adjusted macroevolutionary history for Ordovician–Early Silurian crinoids. Paleobiology 34, 104–116.

Peterson, C.R., Dorcas, M.E., 1994. Automated data acquisition. In: Heyer, W.R., Donnelly, M.A., McDiarmid, R.W., Hayek, L.-A., Foster, S.M. (Eds.), Measuring and Monitoring Biological Diversity: Standard Methods for Amphibians. Smithsonian Institution Press, pp. 47–57.

Petranka, J.W., 1998. Salamanders of the United States and Canada. Smithsonian Institution Press.

Phillips, S.J., Anderson, R.P., Schapire, R.E., 2006. Maximum entropy modeling of species geographic distributions. Ecological Modelling 190, 231–259.

Phillips, S.J., Dudík, M., Elith, J., Graham, C.H., Lehmann, A., Leathwick, J., Ferrier, S., 2009. Sample selection bias and presence-only distribution models: implications for background and pseudo-absence data. Ecological Applications 19, 181–197.

Pielou, E.C., 1975. Ecological Diversity. Wiley.

Pielou, E.C., 1977. Mathematical Ecology, 2nd edition. Wiley.

Pillay, R., Miller, D.A., Hines, J.E., Joshi, A.A., Madhusudan, M., 2014. Accounting for false positives improves estimates of occupancy from key informant interviews. Diversity and Distributions 20, 223–235.

Pilliod, D.S., Muths, E., Scherer, R.D., Bartelt, P.E., Corn, P.S., Hossack, B.R., Lambert, B.A., Mccaffery, R., Gaughan, C., 2010. Effects of amphibian chytrid fungus on individual survival probability in wild boreal toads. Conservation Biology 24, 1259–1267.

Pirsig, R.M., 1974. Zen and the Art of Motorcycle Maintenance. Bantam Books.

Platt, J.R., 1964. Strong inference. Science 146, 347–353.

Pledger, S., 2000. Unified maximum likelihood estimates for closed capture–recapture models using mixtures. Biometrics 56, 434–442.

Pledger, S., Pollock, K.H., Norris, J.L., 2003. Open capture–recapture models with heterogeneity: I. Cormack–Jolly–Seber model. Biometrics 59, 786–794.

Plummer, M., 2003. JAGS: a program for analysis of Bayesian graphical models using Gibbs sampling. In: Proceedings of the 3rd International Workshop on Distributed Statistical Computing, vol. 124. Technische Universität Wien, Vienna, Austria, p. 125.

Pollock, K.H., 1974. The Assumption of Equal Catchability of Animals in Tag–Recapture Experiments. Ph.D. thesis. Cornell University.

Pollock, K.H., 1982. A capture–recapture design robust to unequal probability of capture. Journal of Wildlife Management 46, 752–757.

Pollock, K.H., Nichols, J.D., Brownie, C., Hines, J.E., 1990. Statistical Inference for Capture–Recapture Experiments. Wildlife Society Monographs, vol. 107, pp. 1–97.

Pollock, K.H., Nichols, J.D., Simons, T.R., Farnsworth, G.L., Bailey, L.L., Sauer, J.R., 2002. Large scale wildlife monitoring studies: statistical methods for design and analysis. Environ-Metrics 13, 105–119.

Pollock, K.H., Raveling, D.G., 1982. Assumptions of modern band recovery models with emphasis on heterogeneous survival rates. Journal of Wildlife Management 46, 88–98.

Pollock, L.J., Tingley, R., Morris, W.K., Golding, N., O'Hara, R.B., Parris, K.M., Vesk, P.A., Mc-Carthy, M.A., 2014. Understanding co-occurrence by modelling species simultaneously with a Joint Species Distribution Model (JSDM). Methods in Ecology and Evolution 5, 397–406.

Pradel, R., 1996. Utilization of capture–mark–recapture for the study of recruitment and population growth rate. Biometrics 52, 703–709.

Pradel, R., 2005. Multievent: an extension of multistate capture–recapture models to uncertain states. Biometrics 61, 442–447.

Preatoni, D.G., Nodari, M., Chirichella, R., Tosi, G., Wauters, L.A., Martinoli, A., 2005. Identifying bats from time-expanded recordings of search calls: comparing classification methods. Journal of Wildlife Management 69, 1601–1614.

Preston, F.W., 1948. The commonness and rarity of species. Ecology 29, 254–283.

Pulliam, H.R., 1988. Sources, sinks, and population regulation. American Naturalist 132, 652–661.

Rapoport, E.H., 1982. Areography. Pergamon Press.

Redgwell, R.D., Szewczak, J.M., Jones, G., Parsons, S., 2009. Classification of echolocation calls from 14 species of bat by support vector machines and ensembles of neural networks. Algorithms 2, 907–924.

Reid, J.A., Horn, R.B., Forsman, E.D., 1999. Detection rates of spotted owls based on acoustic-lure and live-lure surveys. Wildlife Society Bulletin 27, 986–990.

Repasky, R.R., 1991. Temperature and the northern distributions of wintering birds. Ecology 72, 2274–2285.

Reunanen, P., Nikula, A., Monkkonen, M., Hurme, E., Nivala, V., 2002. Predicting occupancy for the Siberian flying squirrel in old-growth forest patches. Ecological Applications 12, 1188–1198.

Rich, L.N., Russell, R.E., Glenn, E.M., Mitchell, M.S., Gude, J.A., Podruzny, K.M., Sime, C.A., Laudon, K., Ausband, D.E., Nichols, J.D., 2013. Estimating occupancy and predicting numbers of gray wolf packs in Montana using hunter surveys. The Journal of Wildlife Management 77, 1280–1289.

Richmond, O.M., Hines, J.E., Beissinger, S.R., 2010. Two-species occupancy models: a new parameterization applied to co-occurrence of secretive rails. Ecological Applications 20, 2036–2046.

Ricklefs, R.E., Schluter, D., 1993. Species Diversity in Ecological Communities: Historical and Geographical Perspectives. The University of Chicago Press.

Risk, B.B., de Valpine, P., Beissinger, S.R., 2011. A robust-design formulation of the incidence function model of metapopulation dynamics applied to two species of rails. Ecology 92, 462–474.

Rissler, L.J., Barber, A.M., Wilbur, H.M., Baker, A., 2000. Spatial and behavioral interactions between a native and introduced salamander species. Behavioral Ecology and Sociobiology 48, 61–68.

Robbins, C.S., Bystrak, D., Geissler, P.H., 1986. The Breeding Bird Survey: Its First Fifteen Years, 1965–1979. U.S. Fish and Wildlife Service. Resource Publication 157.

Robbins, C.S., Dawson, D.K., Dowell, B.A., 1989. Habitat Area Requirements of Breeding Forest Birds in the Middle Atlantic States. Wildlife Monographs, vol. 103, pp. 3–34.

Robinson, Q.H., Bustos, D., Roemer, G.W., 2014. The application of occupancy modeling to evaluate intraguild predation in a model carnivore system. Ecology 95, 3112–3123.

Rohde, K., 1999. Latitudinal gradients in species diversity and Rapoport's rule revisited: a review of recent work and what can parasites teach us about the causes of the gradients? Ecography 22, 593–613.

Romesburg, H.C., 1981. Wildlife science: gaining reliable knowledge. Journal of Wildlife Management 45, 293–313.

Root, T., 1988a. Energy constraints on avian distributions and abundance. Ecology 69, 330–339.

Root, T., 1988b. Environmental factors associated with avian distributional boundaries. Journal of Biogeography 15, 489–505.

Rosenzweig, M.L., 1995. Species Diversity in Space and Time. Cambridge University Press.

Rosenzweig, M.L., Clark, C.W., 1994. Island extinction rates from regular censuses. Conservation Biology 8, 491–494.

Rossman, S., Yackulic, C.B., Saunders, S.P., Reid, J., Davis, R., Zipkin, E.F., 2016. Dynamic N-occupancy models: estimating demographic rates and local abundance from detection–nondetection data. Ecology 97, 3300–3307.

Rota, C.T., Ferreira, M.A., Kays, R.W., Forrester, T.D., Kalies, E.L., McShea, W.J., Parsons, A.W., Millspaugh, J.J., 2016. A multispecies occupancy model for two or more interacting species. Methods in Ecology and Evolution 7, 1164–1173.

Rota, C.T., Fletcher Jr., R.J., Dorazio, R.M., Betts, M.G., 2009. Occupancy estimation and the closure assumption. Journal of Applied Ecology 46, 1173–1181.

Royle, J.A., 2004a. Modeling abundance index data from anuran calling surveys. Conservation Biology 18, 1378–1385.

Royle, J.A., 2004b. N-mixture models for estimating population size from spatially replicated counts. Biometrics 60, 108–115.

Royle, J.A., 2006. Site occupancy models with heterogeneous detection probabilities. Biometrics 62, 97–102.

Royle, J.A., Chandler, R.B., Sollmann, R., Gardner, B., 2014. Spatial Capture–Recapture. Academic Press.

Royle, J.A., Chandler, R.B., Yackulic, C., Nichols, J.D., 2012. Likelihood analysis of species occurrence probability from presence-only data for modelling species distributions. Methods in Ecology and Evolution 3, 545–554.

Royle, J.A., Dorazio, R.M., 2008. Hierarchical Modeling and Inference in Ecology: The Analysis of Data from Populations, Metapopulations and Communities. Academic Press.

Royle, J.A., Dorazio, R.M., 2012. Parameter-expanded data augmentation for Bayesian analysis of capture–recapture models. Journal of Ornithology 152, 521–537.

Royle, J.A., Dorazio, R.M., Link, W.A., 2007a. Analysis of multinomial models with unknown index using data augmentation. Journal of Computational and Graphical Statistics 16, 67–85.

Royle, J.A., Kéry, M., 2007. A Bayesian state-space formulation of dynamic occupancy models. Ecology 88, 1813–1823.

Royle, J.A., Kéry, M., Gautier, R., Schmid, H., 2007b. Hierarchical spatial models of abundance and occurrence from imperfect survey data. Ecological Monographs 77, 465–481.

Royle, J.A., Link, W.A., 2005. A general class of multinomial mixture models for anuran calling survey data. Ecology 86, 2505–2512.

Royle, J.A., Link, W.A., 2006. Generalized site occupancy models allowing for false positive and false negative errors. Ecology 87, 835–841.

Royle, J.A., Nichols, J.D., 2003. Estimating abundance from repeated presence–absence data or point counts. Ecology 84, 777–790.

Rue, H., Held, L., 2005. Gaussian Markov Random Fields: Theory and Applications. CRC Press.

Ruiz-Gutiérrez, V., Zipkin, E.F., 2011. Detection biases yield misleading patterns of species persistence and colonization in fragmented landscapes. Ecosphere 2, 1–14.

Ruiz-Gutiérrez, V., Zipkin, E.F., Dhondt, A.A., 2010. Occupancy dynamics in a tropical bird community: unexpectedly high forest use by birds classified as non-forest species. Journal of Applied Ecology 47, 621–630.

Russell, J.C., Stjernman, M., Lindström, Å., Smith, H.G., 2015. Community occupancy before-after-control-impact (CO-BACI) analysis of Hurricane Gudrun on Swedish forest birds. Ecological Applications 25, 685–694.

Russell, R.E., Royle, J.A., Saab, V.A., Lehmkuhl, J.F., Block, W.M., Sauer, J.R., 2009. Modeling the effects of environmental disturbance on wildlife communities: avian responses to prescribed fire. Ecological Applications 19, 1253–1263.

Sanathanan, L., 1972. Estimating the size of a multinomial population. The Annals of Mathematical Statistics 43, 142–152.

Sanderlin, J.S., Block, W.M., Ganey, J.L., 2014. Optimizing study design for multi-species avian monitoring programmes. Journal of Applied Ecology 51, 860–870.

Sanders, N.J., Gotelli, N.J., Heller, N.E., Gordon, D.M., 2003. Community disassembly by an invasive species. Proceedings of the National Academy of Sciences 100, 2474–2477.

Sargeant, G.A., Sovada, M.A., Slivinski, C.C., Johnson, D.H., 2005. Markov chain Monte Carlo estimation of species distributions: a case study of the swift fox in western Kansas. Journal of Wildlife Management 69, 483–497.

Sauer, J.R., Blank, P.J., Zipkin, E.F., Fallon, J.E., Fallon, F.W., 2013a. Using multi-species occupancy models in structured decision making on managed lands. The Journal of Wildlife Management 77, 117–127.

Sauer, J.R., Link, W.A., Fallon, J.E., Pardieck, K.L., Ziolkowski Jr., D.J., 2013b. The North American breeding bird survey 1966–2011: summary analysis and species accounts. North American Fauna 79, 1–32.

Schaub, M., Abadi, F., 2011. Integrated population models: a novel analysis framework for deeper insights into population dynamics. Journal of Ornithology 152, 227–237.

Schmid, H., Zbinden, N., Keller, V., 2004. Überwachung der Bestandsentwicklung häufiger Brutvögel in der Schweiz. Swiss Ornithological Institute.

Schmidt, B.R., Kéry, M., Ursenbacher, S., Hyman, O.J., Collins, J.P., 2013. Site occupancy models in the analysis of environmental DNA presence/absence surveys: a case study of an emerging amphibian pathogen. Methods in Ecology and Evolution 4, 646–653.

Schoener, T., 1974. Competition and the form of habitat shift. Theoretical Population Biology 6, 265–307.

Schwarz, C.J., Arnason, A.N., 1996. A general methodology for the analysis of capture–recapture experiments in open populations. Biometrics, 860–873.

Scott, J.M., Davis, F., Csuti, B., Noss, R., Butterfield, B., Groves, C., Anderson, H., Caicco, S., D'Erchia, F., Edwards, T.C.J., Ulliman, J., Wright, R.G., 1993. Gap Analysis: A Geographic Approach to Protection of Biological Diversity. Wildlife Monographs, vol. 123, pp. 3–41.

Scott, J.M., Heglund, P.J., Morrison, M.L., Haufler, J.B., Raphael, M.G., Wall, W.A., Samson, F.B. (Eds.), 2002. Predicting Species Occurrences. Island Press.

Seamans, M.E., 2005. Population Biology of the California Spotted Owl in the Central Sierra Nevada. Ph.D. thesis. University of Minnesota.

Seber, G.A.F., 1973. The Estimation of Animal Abundance and Related Parameters. Griffen.

Seber, G.A.F., 1982. The Estimation of Animal Abundance and Related Parameters, 2nd edition. MacMillan.

Self, S.G., Liang, K.Y., 1987. Asymptotic properties of maximum likelihood estimators and likelihood ratio tests under non-standard conditions. Journal of the American Statistical Association 82, 605–610.

Senar, J.C., Conroy, M.J., 2004. Multi-state analysis of the impacts of avian pox on a population of serins (Serinus serinus): the importance of estimating recapture rates. Animal Biodiversity and Conservation 27, 133–146.

Sepkoski Jr., J.J., 1975. Stratigraphic biases in the analysis of taxonomic survivorship. Paleobiology, 343–355.

Simberloff, D.S., 1969. Experimental zoogeography of islands: a model for insular colonization. Ecology 50, 296–314.

Simberloff, D.S., Connor, E.F., 1981. Missing species combinations. American Naturalist 118, 215–239.

Simons, T.R., Alldredge, M.W., Pollock, K.H., Wettroth, J.M., Dufty Jr., A., 2007. Experimental analysis of the auditory detection process on avian point counts. The Auk 124, 986–999.

Sjogren-Gulve, P., Ray, C., 1996. Using logistic regression to model metapopulation dynamics: large-scale forestry extirpates the pool frog. In: McCullough, D.R. (Ed.), Metapopulations and Wildlife Conservation. Island Press, pp. 111–137.

Skalski, J.R., 1994. Estimating wildlife populations based on incomplete area surveys. Wildlife Society Bulletin 22, 192–203.

Skalski, J.R., Robson, D.S., 1992. Techniques for Wildlife Investigations: Design and Analysis of Capture Data. Academic Press.

Skelly, D.K., Werner, E.E., Cortwright, S.A., 1999. Long-term distributional dynamics of a Michigan amphibian assemblage. Ecology 80, 2326–2337.

Smith, A.T., Gilpin, M.E., 1997. Spatially correlated dynamics in a pika metapopulation. In: Hanski, I.A., Gilpin, M.E. (Eds.), Metapopulation Biology: Ecology, Genetics, and Evolution. Academic Press, pp. 401–428.

Smith, L.L., Barichivich, W.J., Staiger, J.S., Smith, K.G., Dodd Jr., C.K., 2006. Detection probabilities and site occupancy estimates for amphibians at Okefenokee National Wildlife Refuge. The American Midland Naturalist 155, 149–161.

Soberón, J., Llorente, J., 1993. The use of species accumulation functions for the prediction of species richness. Conservation Biology 7, 480–488.

Sollmann, R., Gardner, B., Williams, K.A., Gilbert, A.T., Veit, R.R., 2016. A hierarchical distance sampling model to estimate abundance and covariate associations of species and communities. Methods in Ecology and Evolution 7, 529–537.

Sørensen, L.L., Coddington, J.A., Scharff, N., 2002. Inventorying and estimating subcanopy spider diversity using semiquantitative sampling methods in an Afromontane forest. Environmental Entomology 31, 319–330.

Spiegelhalter, D.J., Best, N.G., Carlin, B.P., Van Der Linde, A., 2002. Bayesian measures of model complexity and fit. Journal of the Royal Statistical Society, Series B, Statistical Methodology 64, 583–639.

Stauffer, H.B., Ralph, C.J., Miller, S.L., 2002. Incorporating detection uncertainty into presence–absence surveys for marbled murrelet. In: Scott, J.M., Heglund, P.J., Morrison, M.L., Haufler, J.B., Raphael, M.G., Wall, W.A., Samson, F.B. (Eds.), Predicting Species Occurrences. Island Press, pp. 357–366.

Stauffer, H.B., Ralph, C.J., Miller, S.L., 2004. Ranking habitat for marbled murrelets: new conservation approach for species with uncertain detection. Ecological Applications 14, 1374–1383.

Stevens, G.C., 1989. The latitudinal gradient in geographic range: how so many species coexist in the tropics. American Naturalist 133, 240–256.

Stevens, G.C., 1992. The elevational gradient in altitudinal range: an extension of Rapoport's latitudinal rule to altitude. American Naturalist 140, 893–911.

Stith, B., Kumar, N., 2002. Spatial distributions of tigers and prey: mapping and the use of GIS. In: Karanth, K.U., Nichols, J.D. (Eds.), Monitoring Tigers and Their Prey: A Manual for Wildlife Managers, Researchers, and Conservationists. Centre for Wildlife Studies, pp. 51–59.

Stolen, E.D., Oddy, D.M., Legare, M.L., Breininger, D.R., Gann, S.L., Legare, S.A., Weiss, S.K., Holloway-Adkins, K.G., Schaub, R., 2014. Preventing tracking-tube false detections in occupancy modeling of southeastern beach mouse. Journal of Fish and Wildlife Management 5, 270–281.

Stone, L., Roberts, A., 1990. The checkerboard score and species distributions. Oecologia 85, 74–79.

Stone, L., Roberts, A., 1992. Competitive exclusion, or species aggregation: an aid in deciding. Oecologia 91, 419–424.

Strong, D.R.J., Simberloff, D., Abele, L.G., Thistle, A.B. (Eds.), 1984. Ecological Communities: Conceptual Issues and the Evidence. Princeton University Press.

Sutherland, C.S., Elston, D.A., Lambin, X., 2014. A demographic, spatially explicit patch occupancy model of metapopulation dynamics and persistence. Ecology 95, 3149–3160.

Sweitzer, R., Furnas, B., Barrett, R., Purcell, K., Thompson, C., 2016. Landscape fuel reduction, forest fire, and biophysical linkages to local habitat use and local persistence of fishers (*Pekania pennanti*) in Sierra Nevada mixed-conifer forests. Forest Ecology and Management 361, 208–225.

Syms, C., Jones, G.P., 2000. Disturbance, habitat structure, and the dynamics of a coral-reef fish community. Ecology 81, 2714–2729.

Taylor, B.E., Scott, D.E., Gibbons, J.W., 2006. Catastrophic reproductive failure, terrestrial survival, and persistence of the marbled salamander. Conservation Biology 20, 792–801.

Tenan, S., O'Hara, R.B., Hendriks, I., Tavecchia, G., 2014. Bayesian model selection: the steepest mountain to climb. Ecological Modelling 283, 62–69.

ter Braak, C.J., Etienne, R.S., 2003. Improved Bayesian analysis of metapopulation data with an application to a tree frog metapopulation. Ecology 84, 231–241.

Thomas, C.D., 1994. Extinction, colonization, and metapopulations: environmental tracking by rare species. Conservation Biology 8, 373–378.

Thomas, C.D., Hanski, I., 1997. Butterfly populations. In: Hanski, I.A., Gilpin, M.E. (Eds.), Metapopulation Biology: Ecology, Genetics, and Evolution. Academic Press, pp. 359–386.

Thompson, K.G., 2007. Use of site occupancy models to estimate prevalence of *Myxobolus cerebralis* infection in trout. Journal of Aquatic Animal Health 19, 8–13.

Thompson, R.L., Gidden, C., 1972. Territorial basking counts to estimate alligator populations. Journal of Wildlife Management 36, 1081–1088.

Thompson, S.K., 1992. Sampling. John Wiley and Sons.

Thompson, S.K., 2002. Sampling. Wiley.

Thompson, S.K., Seber, G.A.F., 1996. Adaptive Sampling. Wiley.

Thompson, W.L., White, G.C., Gowan, C., 1998. Monitoring Vertebrate Populations. Academic Press.

Thomson, G.M., 1922. The Naturalization of Animals and Plants in New Zealand. Cambridge University Press.

Tingley, M.W., Beissinger, S.R., 2013. Cryptic loss of montane avian richness and high community turnover over 100 years. Ecology 94, 598–609.

Tobalske, C., 2002. Effects of spatial scale on the predictive ability of habitat models for the green woodpecker in Switzerland. In: Scott, J.M., Heglund, P.J., Morrison, M.L., Haufler, J.B., Raphael, M.G., Wall, W.A., Samson, F.B. (Eds.), Predicting Species Occurrences. Island Press, pp. 197–204.

Tobler, M.W., Kéry, M., Sattler, T., Knaus, P., unpublished. Joint species distribution models with interactions and imperfect detection.

Tobler, M.W., Zúñiga Hartley, A., Carrillo-Percastegui, S.E., Powell, G.V., 2015. Spatiotemporal hierarchical modelling of species richness and occupancy using camera trap data. Journal of Applied Ecology 52, 413–421.

Tosh, C.A., Reyers, B., van Jaarsveld, A.S., 2004. Estimating the abundance of large herbivores in the Kruger National Park using presence–absence data. Animal Conservation 7, 55–61.

Trenham, P., Koenig, W.D., Mossman, M.J., Stark, S.L., Jagger, L.A., 2003. Regional dynamics of wetland-breeding frogs and toads: turnover and synchrony. Ecological Applications 13, 1522–1532.

Tuljapurkar, S., 1990. Population Dynamics in Variable Environments. Springer-Verlag, Berlin, Heidelberg.

Tyre, A.J., Possingham, H.P., Lindenmayer, D.B., 2001. Inferring process from pattern: can territory occupancy provide information about life history parameters? Ecological Applications 11, 1722–1737.

Tyre, A.J., Tenhumberg, B., Field, S.A., Niejalke, D., Parris, K., Possingham, H.P., 2003. Improving precision and reducing bias in biological surveys: estimating false-negative error rates. Ecological Applications 13, 1790–1801.

Udvardy, M.D.F., 1969. Dynamic Zoogeography: With Special Reference to Land Animals. Van Nostrand–Reinhold.

Urquhart, N.S., Kincaid, T.M., 1999. Designs for detecting trend from repeated surveys of ecological resources. Journal of Agricultural, Biological, and Environmental Statistics 4, 404–414.

U.S. Fish and Wildlife Service, 1990. Endangered and threatened wildlife and plants: 16 determination of threatened status for the northern spotted owl. Federal Register 55, 26114–26194.

Van Horne, B., 1983. Density as a misleading indicator of habitat quality. Journal of Wildlife Management 47, 893–901.

Veran, S., Kleiner, K.J., Choquet, R., Collazo, J.A., Nichols, J.D., 2012. Modeling habitat dynamics accounting for possible misclassification. Landscape Ecology 27, 943–956.

Verboom, J., Schotman, A., Opdam, P., Metz, A.J., 1991. European nuthatch metapopulations in a fragmented agricultural landscape. Oikos 61, 149–156.

Verner, J., Morrison, M.L., Ralph, C.J. (Eds.), 1986. Wildlife 2000: Modeling Habitat Relationships of Terrestrial Vertebrates. University of Wisconsin Press.

Vetaas, O.R., 2002. Realized and potential climate niches: a comparison of four *Rhododendron* tree species. Journal of Biogeography 29, 545–554.

Waddle, J.H., Dorazio, R.M., Walls, S.C., Rice, K.G., Beauchamp, J., Schuman, M.J., Mazzotti, F.J., 2010. A new parameterization for estimating co-occurrence of interacting species. Ecological Applications 20, 1467–1475.

Wahlberg, N., Klemetti, T., Hanski, I., 2002. Dynamic populations in a dynamic landscape: the metapopulation structure of the marsh fritillary butterfly. Ecography 25, 224–232.

Wahlberg, N., Moilanen, A., Hanski, I., 1996. Predicting the occurrence of endangered species in fragmented landscapes. Science 273, 1536–1538.

Walters, C.J., 1986. Adaptive Management of Renewable Resources. MacMillan.

Walters, C.L., Freeman, R., Collen, A., Dietz, C., Brock Fenton, M., Jones, G., Obrist, M.K., Puechmaille, S.J., Sattler, T., Siemers, B.M., Parsons, S., Jones, K.E., 2012. A continental-scale tool for acoustic identification of European bats. Journal of Applied Ecology 49, 1064–1074.

Walther, B.A., Cotgreave, P., Price, R.D., Gregory, R.D., Clayton, D.H., 1995. Sampling effort and parasite species richness. Parasitology Today 11, 306–310.

Warren, M., McGeoch, M.A., Chown, S.L., 2003. Predicting abundance from occupancy: a test for an aggregated insect assemblage. Journal of Animal Ecology 72, 468–477.

Warton, D.I., Blanchet, F.G., O'Hara, R.B., Ovaskainen, O., Taskinen, S., Walker, S.C., Hui, F.K., 2015. So many variables: joint modeling in community ecology. Trends in Ecology & Evolution 30, 766–779.

Warton, D.I., Stoklosa, J., Guillera-Arroita, G., MacKenzie, D.I., Welsh, A.H., 2017. Graphical diagnostics for occupancy models with imperfect detection. Methods in Ecology and Evolution 8, 408–419.

Webb, M.H., Wotherspoon, S., Stojanovic, D., Heinsohn, R., Cunningham, R., Bell, P., Terauds, A., 2014. Location matters: using spatially explicit occupancy models to predict the distribution of the highly mobile, endangered swift parrot. Biological Conservation 176, 99–108.

Weber, D., Hinterman, U., Zangger, A., 2004. Scale and trends in species richness: considerations for monitoring biological diversity for political purposes. Global Ecology and Biogeography 13, 97–104.

Weir, L.A., Fiske, I.J., Royle, J.A., 2009. Trends in anuran occupancy from northeastern states of the North American Amphibian Monitoring Program. Herpetological Conservation and Biology 4, 389–402.

Weir, L.A., Mossman, M.J., 2005. North American Amphibian Monitoring Program (NAAMP). In: Amphibian Declines: The Conservation Status of United States Species, 1st edition. University of California Press, pp. 307–313.

Weir, L.A., Royle, J.A., Gazenski, K.D., Villena, O., 2014. Northeast regional and state trends in anuran occupancy from calling survey data (2001–2011) from the North American Amphibian Monitoring Program. Herpetological Conservation and Biology 9, 223–245.

Weir, L.A., Royle, J.A., Nanjappa, P., Jung, R.E., 2005. Modeling anuran detection and site occupancy on North American Amphibian Monitoring Program (NAAMP) routes in Maryland. Journal of Herpetology 39, 627–639.

Wenger, S.J., Freeman, M.C., 2008. Estimating species occurrence, abundance, and detection probability using zero-inflated distributions. Ecology 89, 2953–2959.

Whitcomb, S.D., Servello, F.A., O'Connell Jr., A.F., 1996. Patch occupancy and dispersal of spruce grouse on the edge of its range in Maine. Canadian Journal of Zoology 74, 1951–1955.

White, G.C., Anderson, D.R., Burnham, K., Otis, D.L., 1982. Capture–Recapture and Removal Methods for Sampling Closed Populations. Los Alamos National Laboratory.

White, G.C., Burnham, K.P., 1999. Program MARK: survival estimation from populations of marked animals. Bird Study 46, 120–139.

White, G.C., Burnham, K.P., Anderson, D.R., 2002. Advanced features of Program MARK. In: Warren, R.J., Okarma, H., Sievert, P.R. (Eds.), Wildlife, Land, and People: Priorities for the 21st Century. Proceedings of the Second International Wildlife Management Congress. The Wildlife Society, pp. 368–377.

White, G.C., Cooch, E.G., 2017. Population abundance estimation with heterogeneous encounter probabilities using numerical integration. The Journal of Wildlife Management 81, 368–377.

Whittaker, R.H., 1956. Vegetation of the Great Smoky Mountains. Ecological Monographs 22, 1–44.

Wiens, J.A., Crist, T.O., Day, R.H., Murphy, S.M., Hayward, G.D., 1996. Effects of the Exxon Valdez oil spill on marine bird communities in Prince William Sound, Alaska. Ecological Applications 6, 828–841.

Wikle, C.K., 2003. Hierarchical Bayesian models for predicting the spread of ecological processes. Ecology 84, 1382–1394.

Williams, B.K., 1997. Logic and science in wildlife biology. Journal of Wildlife Management 61, 1007–1015.

Williams, B.K., Nichols, J.D., Conroy, M.J., 2002. Analysis and Management of Animal Populations. Academic Press, San Diego, CA.

Williams, B.K., Szaro, R.C., Shapiro, C.D., 2007. Adaptive Management: The US Department of the Interior Technical Guide. Adaptive Management Working Group, US Department of the Interior, Washington, DC.

Williams, C.B., 1964. Patterns in the Balance of Nature. Academic Press.

Williams, M., 1981. The Duckshooter's Bag. The Wetland Press.

Williams, P.H., Araujo, M.B., 2000. Using probabilities of persistence to identify important areas for biodiversity conservation. Proceedings of the Royal Society, Series B 267, 1959–1966.

Williamson, M., 1996. Biological Invasions. Chapman and Hall.

Willis, J.C., 1922. Age and Area. Cambridge University Press.

Wilson, E.O., Willis, E.O., 1975. Applied biogeography. In: Cody, M.L., Diamond, J.M. (Eds.), Ecology and Evolution of Communities. Harvard University Press, pp. 522–534.

Wintle, B.A., McCarthy, M.A., Parris, K.M., Burgman, M.A., 2004. Precision and bias of methods for estimating point survey detection probabilities. Ecological Applications 14, 703–712.

Worthington, R.D., 1968. Observations on the relative sizes of three species of salamander larvae in a Maryland pond. Herpetologica 24, 242–246.

Wright, D.H., Patterson, B.D., Mikkelson, G.M., Cutler, A., Atmar, W., 1998. A comparative analysis of nested subset patterns of species composition. Oecologia 113, 1–20.

Wright, D.H., Reeves, J.H., 1992. On the meaning and measurement of nestedness of species assemblages. Oecologia 92, 416–428.

Wright, S., 1931. Evolution in Mendelian populations. Genetics 16, 97–159.

Wright, S., 1951. The genetical structure of populations. Annals of Eugenics 15, 323–354.

Yackulic, C.B., Chandler, R., Zipkin, E.F., Royle, J.A., Nichols, J.D., Grant, E.H.C., Veran, S., 2013. Presence-only modelling using MAXENT: when can we trust the inferences? Methods in Ecology and Evolution 4, 236–243.

Yackulic, C.B., Nichols, J.D., Reid, J., Der, R., 2015. To predict the niche, model colonization and extinction. Ecology 96, 16–23.

Yackulic, C.B., Reid, J., Davis, R., Hines, J.E., Nichols, J.D., Forsman, E., 2012. Neighborhood and habitat effects on vital rates: expansion of the barred owl in the Oregon Coast Ranges. Ecology 93, 1953–1966.

Yackulic, C.B., Reid, J., Nichols, J.D., Hines, J.E., Davis, R., Forsman, E., 2014. The roles of competition and habitat in the dynamics of populations and species distributions. Ecology 95, 265–279.

Yamaura, Y., Connor, E.F., Royle, J.A., Itoh, K., Sato, K., Taki, H., Mishima, Y., 2016a. Estimating species–area relationships by modeling abundance and frequency subject to incomplete sampling. Ecology and Evolution 6, 4836–4848.

Yamaura, Y., Kery, M., Royle, J.A., 2016b. Study of biological communities subject to imperfect detection: bias and precision of community N-mixture abundance models in small-sample situations. Ecological Research 31, 289–305.

Yamaura, Y., Royle, J.A., in review. Community distance sampling models allowing for imperfect detection and temporary emigration. Ecosphere.

Yamaura, Y., Royle, J.A., Kuboi, K., Tada, T., Ikeno, S., Makino, S., 2011. Modelling community dynamics based on species-level abundance models from detection/nondetection data. Journal of Applied Ecology 48, 67–75.

Yamaura, Y., Royle, J.A., Shimada, N., Asanuma, S., Sato, T., Taki, H., Makino, S., 2012. Biodiversity of man-made open habitats in an underused country: a class of multispecies abundance models for count data. Biodiversity and Conservation 21, 1365–1380.

Yoccoz, N.G., Nichols, J.D., Boulinier, T., 2001. Monitoring of biological diversity in space and time. Trends in Ecology & Evolution 16, 446–453.

Yule, G.U., 1926. Why do we sometimes get nonsense-correlations between time-series? A study in sampling and the nature of time-series. Journal of the Royal Statistical Society 89, 1–69.

Zielinski, W.J., Kucera, T.E., 1995. American Marten, Fisher, Lynx, and Wolverine: Survey Methods for Their Detection. Tech. Rep. PSW-GTR-157. Pacific Southwest Research Station, US Forest Station, Albany, CA, USA.

Zielinski, W.J., Stauffer, H.B., 1996. Monitoring *Martes* populations in California: survey design and power analysis. Ecological Applications 6, 1254–1267.

Zipkin, E.F., DeWan, A., Andrew Royle, J., 2009. Impacts of forest fragmentation on species richness: a hierarchical approach to community modelling. Journal of Applied Ecology 46, 815–822.

Zipkin, E.F., Grant, E.H.C., Fagan, W.F., 2012. Evaluating the predictive abilities of community occupancy models using AUC while accounting for imperfect detection. Ecological Applications 22, 1962–1972.

Zipkin, E.F., Royle, J.A., Dawson, D.K., Bates, S., 2010. Multi-species occurrence models to evaluate the effects of conservation and management actions. Biological Conservation 143, 479–484.

Zonneveld, C., Longcore, T., Mulder, C., 2003. Optimal schemes to detect the presence of insect species. Conservation Biology 17, 476–487.

Zylstra, E.R., Steidl, R.J., 2009. Habitat use by Sonoran desert tortoises. The Journal of Wildlife Management 73, 747–754.

Index